Springer Texts in Statistics

Advisors:
Stephen Fienberg   Ingram Olkin

Springer Science+Business Media, LLC

Springer Texts in Statistics

Alfred: Elements of Statistics for the Life and Social Sciences

Berger: An Introduction to Probability and Stochastic Processes

Blom: Probability and Statistics: Theory and Applications

Brockwell and Davis: An Introduction to Times Series and Forecasting

Chow and Teicher: Probability Theory: Independence, Interchangeability, Martingales, Second Edition

Christensen: Plane Answers to Complex Questions: The Theory of Linear Models, Second Edition

Christensen: Linear Models for Multivariate, Time Series, and Spatial Data

Christensen: Log-Linear Models

Creighton: A First Course in Probability Models and Statistical Inference

du Toit, Steyn and Stumpf: Graphical Exploratory Data Analysis

Edwards: Introduction to Graphical Modelling

Finkelstein and Levin: Statistics for Lawyers

Jobson: Applied Multivariate Data Analysis, Volume I: Regression and Experimental Design

Jobson: Applied Multivariate Data Analysis, Volume II: Categorical and Multivariate Methods

Kalbfleisch: Probability and Statistical Inference, Volume I: Probability, Second Edition

Kalbfleisch: Probability and Statistical Inference, Volume II: Statistical Inference, Second Edition

Karr: Probability

Keyfitz: Applied Mathematical Demography, Second Edition

Kiefer: Introduction to Statistical Inference

Kokoska and Nevison: Statistical Tables and Formulae

Lehmann: Testing Statistical Hypotheses, Second Edition

Lindman: Analysis of Variance in Experimental Design

Madansky: Prescriptions for Working Statisticians

McPherson: Statistics in Scientific Investigation: Its Basis, Application, and Interpretation

Mueller: Basic Principles of Structural Equation Modeling

Nguyen and Rogers: Fundamentals of Mathematical Statistics: Volume I: Probability for Statistics

Nguyen and Rogers: Fundamentals of Mathematical Statistics: Volume II: Statistical Inference

Noether: Introduction to Statistics: The Nonparametric Way

Peters: Counting for Something: Statistical Principles and Personalities

Pfeiffer: Probability for Applications

Pitman: Probability

Robert: The Bayesian Choice: A Decision-Theoretic Motivation

Continued at end of book

David J. Saville   Graham R. Wood

Statistical Methods:
The Geometric Approach

With 143 Figures

Springer

David J. Saville
Biometrics Unit
AgResearch, P.O. Box 60
Lincoln, New Zealand

Graham R. Wood
Department of Mathematics
University of Canterbury
Christchurch, New Zealand

Table 7.5 was reproduced from *Statistics for Biology*, 3rd ed., by O.N. Bishop, 1980, with permission from Longman Group UK, Ltd.

Mathematics Subject Classification (1991): 62-01, 62J05, 62J10, 62P10, 62-07, 62A20, 62E15, 62F03, 62F10, 62F25, 62J05, 62K10, 62K15

Library of Congress Cataloging-in-Publication Data
Saville, David J.
    Statistical methods: the geometric approach/David J. Saville,
Graham R. Wood.
        p.   cm. — (Springer texts in statistics)
    Includes bibliographical references and index.

    1. Analysis of variance.   2. Regression analysis.   3. Geometry.
I. Wood, Graham R.   II. Title   III. Series.
QA279.S25   1991
519.5′38 — dc20   90-28347

Printed on acid-free paper.

© 1991 Springer Science+Business Media New York
Originally published by Springer-Verlag Berlin Heidelberg New York in 1991
Softcover reprint of the hardcover 1st edition 1991

Production managed by Francine McNeill; manufacturing supervised by Vincent Scelta.
Photocomposed copy prepared using LaTeX.

9 8 7 6 5 4 3 (Corrected third printing, 1997)

ISBN 978-1-4612-6965-6   ISBN 978-1-4612-0971-3 (eBook)
DOI 10.1007/978-1-4612-0971-3

To all who have weathered the preparation of this book: our wives,
Sandra and Stephanie; our children, Jordan, Zoë, Verity, and Jessica
and
to our parents.

Preface

This book is a novel exposition of the traditional workhorses of statistics: analysis of variance and regression. The key feature is that these tools are viewed in their natural mathematical setting, the geometry of finite dimensions.

The Authors

To introduce ourselves, Dave Saville is a practicing statistician working in agricultural research; Graham Wood is a university lecturer involved in the teaching of statistical methods. Each of us has worked for sixteen years in our current field.

Features of the Book

People like pictures. One picture can present a set of ideas at a glance, while a series of pictures, each building on the last, can unify a wealth of ideas. Such a series we present in this text by means of a systematic geometric approach to the presentation of the theory of basic statistical methods. This approach fills the void between the traditional extremes of the "cookbook" approach and the "matrix algebra" approach, providing an elementary but at the same time rigorous view of the subject. It combines the virtues of the traditional methods, while avoiding their vices.

At the same time we present much of the practical folklore that is passed verbally from one generation of statisticians to the next. Problem-solving methods used by practicing statisticians, such as the use of transformations or techniques for handling missing values, are introduced throughout the text. Real-life datasets are used where possible, and any assumptions made are carefully checked. Suggestions are given, generally in the form of a table or figure, on how to present the final results of an analysis. Chapters conclude with a summary of the key results.

Class exercises have been included in certain key chapters. The authors find these particularly valuable for teaching "frequentist" statistical methods: Students can experience at first hand the variation from one sample to another. In addition to the class exercises, a wealth of general exercises is provided. These are mainly agricultural or biological in origin, though a few medical and industrial examples are also included.

Computing is made simple by the geometric approach, with any multiple regression program sufficing to handle all our calculations. We have chosen to use the widely available teaching package Minitab for this purpose.

Reading the Book

We are careful in the text to use the minimal set of geometric and statistical tools. Ideally the reader should have a background in elementary vector geometry and have taken a first statistics course. Chapters 2 and 3 do, however, provide a convenient summary of this core material.

We recommend that you begin by reading the introductory Chapter 1. Chapter 4 provides an overview of our approach and should be tackled next, dipping back to the reference Chapters 2 and 3 as necessary. Chapter 5 contains the rudimentary analysis of variance material, while Chapter 15 forms the basis for regression.

Experience has taught us that new ideas are made more palatable when presented in the context of an example. For this reason, following Chapter 4, the first section of each chapter is generally devoted to analyzing an example in detail, whereas later sections deal with the material in full generality.

Lecture Courses

The book has been based on lecture course material developed by the authors over the past six years. Two main lecture courses have been taught over this period. The first is a course for second-year undergraduate statistics students at the University of Canterbury in Christchurch, New Zealand. This has been taught using the geometric approach since 1984. The second was a course for graduate-level agriculture students, taught in 1985, at the University of California, Davis. Both courses have been very well received.

Our courses have consisted of about 30 one-hour lectures, plus 10 two-hour practical periods. We typically spend about three lecture hours on the introductory material in Part I, fourteen hours on the analysis of variance material of Parts II and III, three hours on the blocking ideas of Part IV, seven hours on the regression material in Part V, and three hours on revision and testing.

Acknowledgments

We have prepared the book ourselves using LATEX, and are most grateful for typing assistance from Mrs. Beverley Haberfield and Mrs. Ann Tindall. We are also particularly grateful to the many students who have contributed to this evolution of the geometric approach. Mrs. Val Elley is thanked for her comments on Chapters 1 to 5, and our employers, the Mathematics Department of the University of Canterbury, Christchurch, and the Ministry of Agriculture and Fisheries, Lincoln, are thanked for their support.

Christchurch, New Zealand

David J. Saville
Graham R. Wood

Contents

Preface   vii

I   Basic Ideas   1

1   Introduction   3
   1.1   Why Use Geometry? 3
   1.2   A Simple Illustration 4
   1.3   Tradition and Practice 7
   1.4   How to Read This Book 8
   Exercise 9

2   The Geometric Tool Kit   10
   2.1   Introducing Vectors 11
   2.2   Putting Vectors Together 15
   2.3   Angles Between Vectors 20
   2.4   Projections 26
   2.5   Sums of Squares 30
   Exercises 32
   Solutions to the Reader Exercises 35

3   The Statistical Tool Kit   39
   3.1   Basic Ideas 39
   3.2   Combining Variables 45
   3.3   Estimation 48
   3.4   Reference Distributions 51
   Solutions to the Reader Exercises 54

4   Tool Kits At Work   **55**
   4.1   The Scientific Method   55
   4.2   Statistical Analysis   58
   Exercises   63

II   Introduction to Analysis of Variance   **65**

5   Single Population Questions   **67**
   5.1   An Illustrative Example   67
   5.2   General Case   77
   5.3   Virtues of Our Estimates   83
   5.4   Summary   86
   Class Exercise   87
   Exercises   90
   Solutions to the Reader Exercises   94

6   Questions About Two Populations   **97**
   6.1   A Case Study   98
   6.2   General Case   112
   6.3   Computing   119
   6.4   Summary   123
   Class Exercise   124
   Exercises   127
   Solution to the Reader Exercise   131

7   Questions About Several Populations   **133**
   7.1   A Simple Example   133
   7.2   Types of Contrast   136
   7.3   The Overview   144
   7.4   Summary   148
   Solutions to the Reader Exercises   150

III   Orthogonal Contrasts   **153**

8   Class Comparisons   **155**
   8.1   Analyzing Example A   155
   8.2   General Case   169
   8.3   Summary   175
   Class Exercise   176
   Exercises   179

9 Factorial Contrasts **187**

   9.1 Introduction 187

   9.2 Analyzing Example B 188

   9.3 Analyzing Example C 203

   9.4 Generating Factorial Contrasts 213

   9.5 Summary 218

   Exercises 218

10 Polynomial Contrasts **224**

   10.1 Analyzing Example D 224

   10.2 Consolidating the Ideas 248

   10.3 A Case Study 258

   10.4 Summary 262

   Exercises 263

   Solutions to the Reader Exercises 267

11 Pairwise Comparisons **271**

   11.1 Analyzing Example E 271

   11.2 Least Significant Difference 283

   11.3 Multiple Comparison Procedures 284

   11.4 Summary 290

   Class Exercise 291

   Exercises 292

IV Introducing Blocking **297**

12 Randomized Block Design **299**

   12.1 Illustrative Example 300

   12.2 General Discussion 310

   12.3 A Realistic Case Study 312

   12.4 Why and How to Block 321

   12.5 Summary 324

   Class Exercise 325

   Exercises 327

13 Latin Square Design **340**

   13.1 Illustrative Example 340

   13.2 General Discussion 349

   13.3 Summary 353

   Exercise 353

14 Split Plot Design **354**

14.1 Introduction 354

14.2 Analysis 356

14.3 Discussion 369

14.4 Summary 372

Exercises 373

Solutions to the Reader Exercises 378

V   Fundamentals of Regression **381**

15 Simple Regression **383**

15.1 Illustrative Example 383

15.2 General Case 399

15.3 Confidence Intervals 402

15.4 Correlation Coefficient 408

15.5 Pitfalls for the Unwary 414

15.6 Summary 419

Class Exercise 421

Exercises 425

Solutions to the Reader Exercises 430

16 Polynomial Regression **431**

16.1 No Pure Error Term 431

16.2 Pure Error Term 448

16.3 Summary 456

Exercises 457

17 Analysis of Covariance **461**

17.1 Illustrative Example 461

17.2 Independent Lines 473

17.3 Use of ANCOVA 479

17.4 Summary 485

Exercises 488

Solutions to the Reader Exercises 492

18 General Summary **493**

18.1 Review 493

18.2 Where to from Here? 499

18.3 Summary 503

Appendices 504

A Unequal Replications: Two Populations   **504**
  A.1 Illustrative Example 504
  A.2 General Case 510
  Exercises 513

B Unequal Replications: Several Populations   **515**
  B.1 Class Comparisons 515
  B.2 Factorial Contrasts 522
  B.3 Other Cases 526
  B.4 Summary 527
  Exercises 528

C Alternative Factorial Notation   **530**
  Solution to the Reader Exercise 534

D Regression Through the Origin   **535**

E Confidence Intervals   **538**
  E.1 General Theory 538

T Statistical Tables   **542**

References   **550**

Index   **552**

Part I
Basic Ideas

This first part of the book comprises four chapters which lay the groundwork for the more detailed developments of later parts. The brief introductory Chapter 1 offers motivation, and includes an informal example of the power of the geometric approach. Chapters 2 and 3 are reference chapters which present the geometric and statistical tools we shall continually need. Lastly, in Chapter 4, we provide an overview of the basic geometric method which will be employed throughout.

Chapter 1

Introduction

In §1 of this chapter we explain why geometry is important for an understanding of statistics. We investigate a simple example in §2 to convince you of the virtues of thinking with pictures. How does this geometric method link with the more traditional approaches? The answer is presented in §3. We wrap up in §4 with a few hints on how to read the book efficiently.

1.1 Why Use Geometry?

Why use geometry in statistics? Won't this make a difficult subject even more difficult? The answer is to the contrary: use of geometry clarifies and unifies our understanding of the basic statistical techniques of analysis of variance and regression.

How does geometry *clarify* basic statistics? Geometry provides pictures which serve as powerful and easily grasped summaries of a problem. In turn the pictures suggest, after a little reflection, the solution to the problem. The example in the next section is an illustration of this process. If you like it, then this book is for you. Our experience has shown that students do like a picture early in the learning process: it means more, and it is more easily remembered, than a set of algebraic expressions. A picture forms a natural bridge between problem and solution.

Geometry also serves to *unify* the many aspects of analysis of variance and regression. In the geometric approach such aspects are seen as variations upon a common theme, rather than as an unrelated collection of ad hoc techniques. This also means that the computing of solutions can be carried out in a single unified manner.

Historically, analysis of variance and regression methods were developed in the early decades of this century, using geometry. Their founder, R.A. Fisher, was an astute geometer and derived his inventive inspiration from thinking in pictures. This is evident from his writings and is clearly spelled out in the biography of Fisher written by his daughter, Joan Fisher Box (1978). Unfortunately Fisher found it difficult to explain his geometric proofs to his colleagues, and in the main they decided the geometry was too difficult. For this reason, and because easy computing formulae were needed in the early days, the tradition arose of expressing the geometric results in algebraic form.

The 19th century was the golden age of geometry, whereas the 20th century has been the golden age of algebra. Our approach derives its inspiration from geometry, but relies on simple linear algebra for formal proofs. The result is a unified framework which enables basic statistical methods to be taught rigorously at an elementary level. The book aims to retrieve Fisher's lost insight, and to present it in a dish palatable to all and even delicious to many.

1.2   A Simple Illustration

Here we use a simple example to convey the flavor of the geometric approach.

We begin, as always, with a question: "Does a strong cup of tea really increase your pulse rate?" In order to investigate this question you could take, say, ten people, settle them for ten minutes, and take their pulse rates. Then offer them a strong cup of tea and take their pulse rates five minutes later, after giving the tea time to act. You'll end up with ten pulse rate changes, $y_1, \ldots, y_{10}$, the responses of the ten individuals to their tea. How do you decide whether the tea really does change pulse rate?

With small samples like this statisticians use the t-test. You calculate " $\bar{y}/(s/\sqrt{10})$ ", where $\bar{y}$ and s are the sample mean and standard deviation, and compare the value with that in standard tables of t-distributions. But what does all this mystery amount to?

Imagine that you carry out the above study many times, using only two people for each study, to simplify our explanation. Also imagine that all of the people cheat by not drinking their cup of tea, so that the changes in pulse rate average to $\mu = 0$ in the long run. Then each pair of observations y_1, y_2 , such as the pair shown in Figure 1.1(a), can be plotted as a single point ($\times$) on a 2-dimensional graph, as shown in Figure 1.1(b). Over many repetitions of the study, these points will be dotted around the origin, as also shown in Figure 1.1(b).

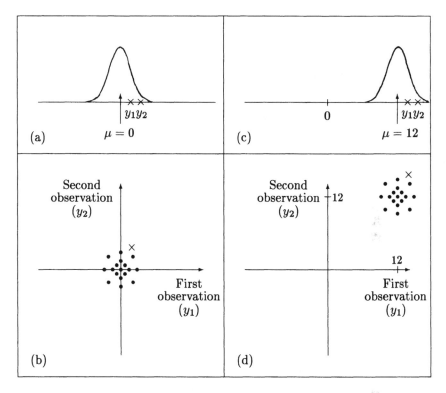

Figure 1.1: The correspondence between samples of size two and points (y_1, y_2) in 2-space. The scatter of points resulting from many repetitions of the tea-drinking study will be centered on the origin if $\mu = 0$, and will be centered on the point $(\mu, \mu) = (12, 12)$ if $\mu = 12$.

Now imagine that nobody cheats, and that the changes in pulse rate average to $\mu = 12$ beats per minute in the long run. Then the corresponding pictures are Figures 1.1(c) and (d). A typical sample point ($\times$) is now one in a scatter of points which is centered on the point $(\mu, \mu) = (12, 12)$.

With this picture at our disposal, we are now able to translate the original question "Does a strong cup of tea really alter your pulse rate" into a geometric question. In a real situation you will have only one set of experimental results, say $(y_1, y_2) = (13, 15)$. The equivalent geometric question is, "Does the point belong to a cloud centered on the origin, or does it belong to one centered away from the origin?". More tersely, is $\mu = 0$ or is $\mu \neq 0$? What we need is a measure, termed a *test statistic*, which will distinguish the two cases. It must be small if $\mu = 0$ and large if $\mu \neq 0$.

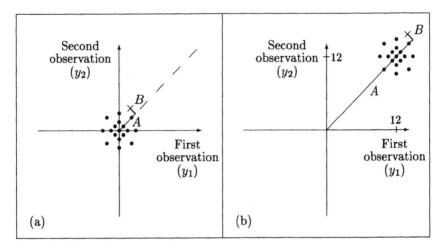

Figure 1.2: Distinguishing between the cases (a) $\mu = 0$ and (b) $\mu \neq 0$ using the test statistic A/B.

Figure 1.2 provides the clues we require. When $\mu \neq 0$ the scatter of points is moved up the equiangular line, the line at 45^0 to both axes. For a typical sample point, such as "×", this has the effect of making the ratio of the distance "up the line" (A), to the distance "from the line" (B) greater when $\mu \neq 0$ than when $\mu = 0$. This suggests that we use the ratio A/B as our test statistic. Note that this statistic works no matter what the spread of the points, and that it does not depend on our units of measurement, two very important considerations.

For our sample point $(y_1, y_2) = (13, 15)$ the test statistic takes the value $14\sqrt{2}/\sqrt{2} = 14$, as illustrated in Figure 1.3. In Chapter 5 we shall see that if the mean μ really is zero, then our test statistic will follow the well-known "Student's t_1" distribution, so-called after the pen-name of its discoverer, W.S. Gosset. Gosset discovered the distribution while employed as an experimental scientist at Guinness Breweries in Dublin in 1908. (To be strictly correct here, we must regard distances A down-left of the origin as negative.)

Knowledge of the true distribution enables us to be precise about a test statistic being "large". When $\mu = 0$ the absolute value of the t_1 statistic is less than 12.7 in 95% of cases. Statisticians declare these values to be "small", with values greater than 12.7 being declared "large". This means that our value of 14 would be declared large, so that we would have evidence to support the notion that a strong cup of tea increases your pulse rate.

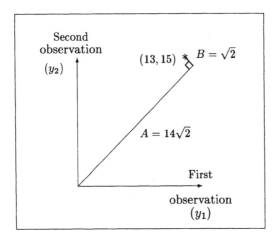

Figure 1.3: The sample point $(y_1, y_2) = (13, 15)$, the perpendicular onto the equiangular line, and the associated distances A and B.

1.3 Tradition and Practice

Traditional approaches to the teaching of statistics fall into two broad categories. The first is the elementary but mysterious *cookbook* approach. Here mathematical recipes are provided in the form of algebraic formulae for the most commonly occurring cases. These formulae can be derived from the geometry presented in this book. An able pure mathematician, after two years spent teaching elementary statistics cookbook style, approached the authors and with a note of despair in his voice asked "What on earth are degrees of freedom?". This illustrates how the cookbook approach at best provides only semiplausible explanations of concepts such as degrees of freedom, orthogonality in experimental design, orthogonal contrasts and orthogonal polynomials.

The second is the rigorous, but inaccessible, *matrix algebra* approach. Here matrices are used to give a very streamlined presentation of the N-dimensional geometry. Unfortunately, however, the mathematics is so sophisticated that the non-mathematician cannot gain a feel for the subject.

There appears to be a real need for an approach which bridges the gap between the two traditional approaches: a method that conveys an understanding of the underlying mathematical principles at an elementary level. Our experience shows that the geometric method provides such an approach, with vectors serving as a halfway house between the numbers of the cookbook method and the matrices of the matrix method. Thus a major goal of this book is to illuminate the bond between vector geometry and statistical theory, as illustrated in Figure 1.4.

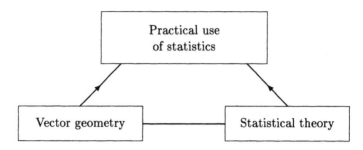

Figure 1.4: Links between geometry, statistical theory, and statistical practice.

Practice and theory go hand in hand. An equally important goal of this book is to illuminate the link between the theory and practice of statistics, as also illustrated in Figure 1.4.

1.4 How to Read This Book

Home base for this text is Chapter 4. It is the place where you should start, and the place to which you should periodically return. In its few pages you are introduced to the general method of working with geometry to solve statistical problems, so it is worthy of thorough study. You'll need, however, some background to handle it. If you are lucky you may have the prerequisites already. These are laid out in Chapters 2 and 3, the geometric and statistical toolkits. We suggest that you glance through the framed summary boxes in these chapters, and study only those ideas which are unfamiliar.

Chapter 5 contains the fundamental material on analysis of variance. In particular, §5.1 contains a single population case study which encapsulates the essential features. This should be studied extensively, since a good understanding of this example will ease your passage through every other chapter.

Chapter 15 contains the fundamental material on regression. Again, §15.1 contains a straight line regression example which is core reading.

The remainder of the book expands these basic ideas. Chapters 6 to 11 are concerned with comparisons among two or more populations, using analysis of variance techniques. Chapters 12 to 14 are concerned with the use of blocking to increase efficiency of design. Your own interests will determine how much time to spend on these chapters. Chapter 16 deals with polynomial regression, and Chapter 17 with analysis of covariance, the fitting of parallel straight lines. Our final Chapter 18 offers an overall perspective, and could be tackled early on by some readers.

Experience has taught us that statistical ideas are best learnt in the context of a real problem. For this reason we use a "question – theory – answer" format as much as possible. Our discussion will include assumption checks, and a reworking of analyses where necessary. We also offer interpretations of our results, and discuss practical aspects of the problem. A brief summary is provided at the end of each of Chapters 5 to 17.

Highly recommended are the class exercises. These allow the students to gain first-hand experience of the variation involved in sampling, and have been found particularly valuable by the authors. General exercises are also provided, mostly using real data.

Computing asides are periodically included. We have settled on the widely used statistical package Minitab, having found that it blends in very well with our approach.

Testing Versus Estimation

In this book our presentation leads initially into hypothesis testing, then into estimation of confidence intervals for parameters. The reason for choosing this ordering is that hypothesis testing ideas lead more naturally to such traditional items as ANOVA tables. We emphasise, however, that the geometric approach yields the material for lecture courses which stress either hypothesis testing or estimation. Confidence interval estimation is covered throughout the text, and the underlying theory is unified in Appendix E.

Exercise

(1.1) Follow the approach given in §1.2 to test the hypothesis that $\mu = 0$, where μ is the mean of the population of pulse rate changes. Assume that the data from the two strong tea drinkers were pulse rate changes of 10 and 16 beats per minute.

Chapter 2

The Geometric
Tool Kit

This book is about *visualizing* the classical methods of statistical analysis. The critical link between the numbers of a data set and the picture we analyze is the notion of a vector, a visual representation of that set of numbers. This chapter introduces you to the basic ideas of the geometry of vectors. Vectors are defined in §1, then in §2 we demonstrate how to combine them. Angles between vectors are defined in §3, projections of vectors are discussed in §4, and we conclude in §5 with a discussion of Pythagoras' Theorem. Rest assured that we discuss only those ideas which are necessary for the statistics we do later!

Each new geometric tool is introduced in a two-sided table. On the left will be the formal definition, and on the right will be a picture to help you visualize this definition. The left side summarizes the formal mathematics, while the right side provides insight and inspiration. Incidentally, there is growing evidence that the hemispheres of our brain operate in this way: the left may be largely responsible for symbolic processing while the right may be responsible for visual processing. We suggest that initially you glance through the framed tables, then return to those that are unfamiliar and study them using the surrounding explanation.

2.1 Introducing Vectors

Numbers in Space

A data set $\{v_1, v_2, \cdots, v_N\}$ can be represented in N-space by a point which is associated with a vector v.

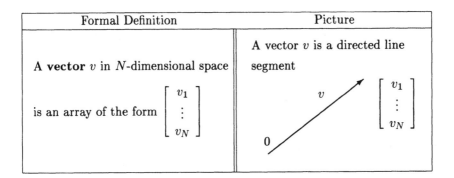

Formal Definition	Picture
A vector v in N-dimensional space is an array of the form $\begin{bmatrix} v_1 \\ \vdots \\ v_N \end{bmatrix}$	A vector v is a directed line segment

Any pair of numbers can be thought of as a point in the plane. For example, the pair of numbers (2,1) can be represented as the point "2 along" and "1 up". In the form $\begin{bmatrix} 2 \\ 1 \end{bmatrix}$ this is called a *vector*. Pictorially we represent this vector by the line segment joining the origin to the point $x = 2$, $y = 1$ as illustrated in Figure 2.1.

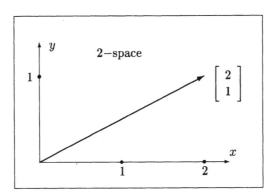

Figure 2.1: A vector in 2-space.

Note that in order to establish this association we have fixed a point in the plane, the *origin*, and a pair of lines at right angles through the point, the x and y *coordinate axes*.

In our familiar three dimensional world we can also draw a realistic pic-
ture of a vector. For example, the vector $v = \begin{bmatrix} 8 \\ 4.5 \\ 6 \end{bmatrix}$ can be drawn as a line
segment joining the origin to the point with coordinates $x = 8$, $y = 4.5$ and
$z = 6$, as shown in Figure 2.2.

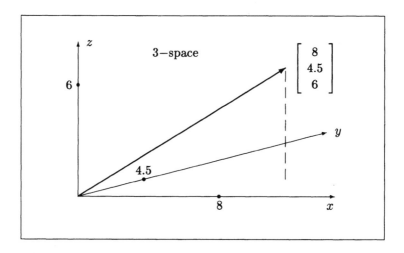

Figure 2.2: The 3-space vector $[8, 4.5, 6]^T$.

A word on a convenient notation. When vectors appear in a paragraph
we shall write them horizontally, so that the vector illustrated above will
appear as $[8, 4.5, 6]^T$. The "T" denotes *transpose*, indicating that the vector
has been turned on its side.

In higher dimensions vectors cannot be shown pictorially in a strictly
correct manner. Examples of such vectors are

$$p = \begin{bmatrix} 4 \\ 1 \\ 5 \\ -7 \end{bmatrix} \quad \text{and} \quad q = \begin{bmatrix} 0.79 \\ 8.57 \\ 1.21 \\ 0.15 \\ -0.19 \end{bmatrix}$$

in 4-dimensional and 5-dimensional space respectively. Nevertheless it is still
very helpful to draw pictures treating vectors as directed line segments.

The Length of a Vector

Squared lengths of vectors are the building blocks of statistical tests.

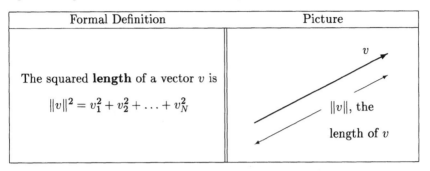

Formal Definition	Picture
The squared **length** of a vector v is $$\|v\|^2 = v_1^2 + v_2^2 + \ldots + v_N^2$$	v $\|v\|$, the length of v

In two dimensions the length of the vector $v = [2,1]^T$ can be deduced from Figure 2.3 using Pythagoras' Theorem: in a right-angled triangle the square on the hypotenuse is equal to the sum of the squares on the other two sides, or $a^2 = b^2 + c^2$ in Figure 2.3. That is, the length of v is $\|v\| = \sqrt{2^2 + 1^2} = \sqrt{5}$. Notice that the symbol $\|\ \ \|$ is a shorthand for "length".

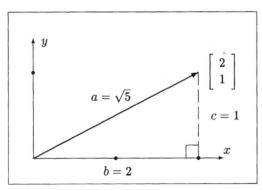

Figure 2.3: The length of the vector $[2,1]^T$.

Similarly, in three dimensions the vector $v = [8, 4.5, 6]^T$ can be shown by Pythagoras' Theorem to have length

$$\|v\| = \sqrt{8^2 + 4.5^2 + 6^2} = \sqrt{64 + 20.25 + 36} = 10.97 \text{(to two decimal places)}$$

This motivates the formal definition for higher dimensions given in the framed table. Thus the vectors

$$p = \begin{bmatrix} 4 \\ 1 \\ 5 \\ -7 \end{bmatrix} \quad \text{and} \quad q = \begin{bmatrix} 0.79 \\ 8.57 \\ 1.21 \\ 0.15 \\ -0.19 \end{bmatrix} \quad \text{have lengths}$$

$$\|p\| = \sqrt{16 + 1 + 25 + 49} = \sqrt{91} = 9.54 \quad \text{and}$$
$$\|q\| = \sqrt{(0.79)^2 + \cdots + (-.19)^2} = 8.69 \quad \text{respectively.}$$

Exercise for the reader

(2.1) Find the lengths of the following vectors:

$$\begin{bmatrix} 4 \\ -1 \end{bmatrix} \qquad \begin{bmatrix} 6 \\ 1 \\ 0 \\ -2 \\ 4 \end{bmatrix} \qquad \begin{bmatrix} 1 \\ 1 \\ -2 \end{bmatrix} \qquad \begin{bmatrix} 1 \\ 1 \\ 1 \\ 1 \\ 1 \end{bmatrix}$$

Answers appear at the end of the chapter.

Directions in Space

In general, vectors have both magnitude and direction. Thus vectors standardized to a length of one can be used to specify directions in space.

Formal Definition	Picture
A unit vector, U, is a vector of length 1; that is, $\|U\| = 1$	U is a unit vector $\|U\|=1$

The unit vector in the direction of any particular vector can be obtained by dividing each component in the vector by its length. Thus the unit vector in the direction of $v = [2, 1]^T$ is

$$U = \begin{bmatrix} 2/\sqrt{5} \\ 1/\sqrt{5} \end{bmatrix} = \frac{1}{\sqrt{5}} \begin{bmatrix} 2 \\ 1 \end{bmatrix} = \frac{v}{\|v\|}$$

Similarly, the unit vector in the direction of $p = [4, 1, 5, -7]^T$ is

$$U = \frac{p}{\|p\|} = \frac{1}{\sqrt{91}} [4, 1, 5, -7]^T$$

Here are further important examples in two dimensions:

$[1, 0]^T$ is the unit vector in the direction of the x axis

$[0, 1]^T$ is the unit vector in the direction of the y axis

$[1, 1]^T/\sqrt{2}$ is the unit vector at 45^0 to these axes.

In N-dimensional space

$$\begin{bmatrix} 1 \\ 0 \\ 0 \\ \vdots \\ 0 \end{bmatrix} \quad \begin{bmatrix} 0 \\ 1 \\ 0 \\ \vdots \\ 0 \end{bmatrix} \quad \cdots \quad \begin{bmatrix} 0 \\ 0 \\ 0 \\ \vdots \\ 1 \end{bmatrix}$$

are unit vectors. These form a natural coordinate system.

Exercise for the reader

(2.2) Make the following vectors into unit vectors:

$$\begin{bmatrix} 1 \\ 1 \\ 1 \\ 1 \\ 1 \end{bmatrix} \quad \begin{bmatrix} 1 \\ -1 \\ 0 \\ 0 \end{bmatrix} \quad \begin{bmatrix} 3 \\ 3 \\ -2 \\ -2 \\ -2 \end{bmatrix} \quad \begin{bmatrix} 1 \\ -1 \\ 1 \\ -1 \end{bmatrix}$$

2.2   Putting Vectors Together

Adding Vectors

In statistical work, the observations are viewed as a single vector, whereas the model is specified as a combination of vectors. This section introduces you to methods for combining vectors. We begin with addition.

Formal Definition	Picture
Vector addition is defined componentwise by $$\begin{bmatrix} v_1 \\ \vdots \\ v_n \end{bmatrix} + \begin{bmatrix} w_1 \\ \vdots \\ w_N \end{bmatrix} = \begin{bmatrix} v_1 + w_1 \\ \vdots \\ v_N + w_N \end{bmatrix}$$	Vectors are added by placing them "head to tail" 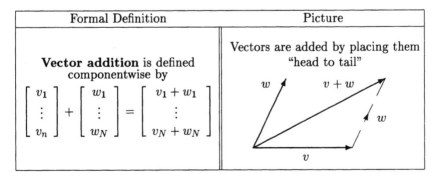

For example, the sum of the vectors $\begin{bmatrix} 3 \\ 1 \end{bmatrix}$ and $\begin{bmatrix} 1 \\ 2 \end{bmatrix}$ is the vector $\begin{bmatrix} 4 \\ 3 \end{bmatrix}$.

Pictorially, we move the vector $[1, 2]^T$ to the tip of the vector $[3, 1]^T$, and the sum of the vectors is the resultant vector $[4, 3]^T$, as shown in Figure 2.4.

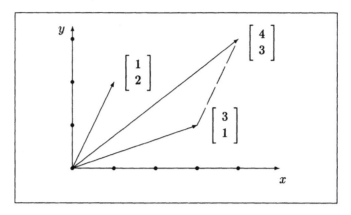

Figure 2.4: Adding the vectors $[3,1]^T$ and $[1,2]^T$.

As a second example, in 4-dimensional space the sum of the vectors $[3,9,7,0]^T$ and $[9,1,5,8]^T$ is the vector

$$
\begin{bmatrix} 3 \\ 9 \\ 7 \\ 0 \end{bmatrix} + \begin{bmatrix} 9 \\ 1 \\ 5 \\ 8 \end{bmatrix} = \begin{bmatrix} 12 \\ 10 \\ 12 \\ 8 \end{bmatrix}
$$

Note that vectors in spaces of differing dimensions cannot be added. For example, $[4,7]^T$ cannot be added to $[1,5,2]^T$.

Altering the Length of a Vector

In producing unit vectors, we altered our vector to give it a length of one. We now define this length changing process more generally.

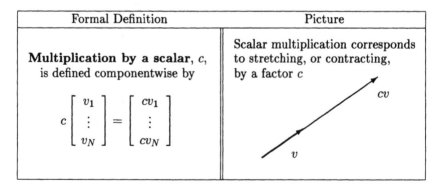

For a positive scalar, multiplication contracts the vector when the scalar is less than one, and stretches the vector when the scalar is greater than one.

If the scalar is negative, then the direction is reversed as well. For example:

$$\text{Direction unchanged:} \qquad 4 \begin{bmatrix} 1 \\ 2 \end{bmatrix} = \begin{bmatrix} 4 \times 1 \\ 4 \times 2 \end{bmatrix} = \begin{bmatrix} 4 \\ 8 \end{bmatrix}$$

$$\text{Direction reversed:} \qquad -\frac{1}{2} \begin{bmatrix} 1 \\ 2 \end{bmatrix} = \begin{bmatrix} -\frac{1}{2} \\ -1 \end{bmatrix}$$

Exercise for the reader

(2.3) Simplify the following expressions:

$$\text{(a)} \quad 4 \begin{bmatrix} 1 \\ 1 \\ 0 \\ 0 \\ 0 \end{bmatrix} + 5 \begin{bmatrix} 0 \\ 0 \\ 1 \\ 1 \\ 1 \end{bmatrix} \qquad\qquad \text{(b)} \quad 3 \begin{bmatrix} 1 \\ -1 \\ 0 \\ 0 \end{bmatrix} - 2 \begin{bmatrix} 0 \\ 0 \\ 1 \\ -1 \end{bmatrix}$$

Spaces Inside Other Spaces

Informally, a set of vectors forms a *subspace* if you can't escape from the set by adding or subtracting two vectors in the set, or by contracting or stretching a vector in the set.

Formal Definition	Picture
A set of vectors S is called a **subspace** if (1) for all vectors v and w in S their sum $v + w$ is also in S, and (2) for all vectors v in S all scalar multiples cv are also in S	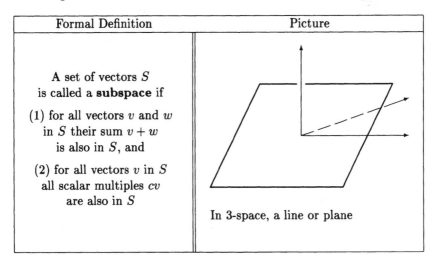 In 3-space, a line or plane

In the plane, the only interesting subspaces are lines through the origin; in 3-space the interesting subspaces are either lines or planes passing through the origin. Here are a couple of examples.

$$S_1 = \{\text{all vectors } c \begin{bmatrix} 4 \\ -1 \\ 2 \end{bmatrix}, \text{ where } c \text{ is a real number}\}$$

Here S_1 comprises all multiples of the vector $[4, -1, 2]^T$, including the zero vector (the origin) and all negative multiples. The subspace S_1 is a line in 3-dimensional space.

$$S_2 = \{\text{all vectors } \begin{bmatrix} c_1 \\ c_2 \\ c_2 \end{bmatrix}, \text{ where } c_1 \text{ and } c_2 \text{ are real numbers}\}$$

The simplest vectors in this subspace are the vectors

$$\begin{bmatrix} 1 \\ 0 \\ 0 \end{bmatrix} \text{ and } \begin{bmatrix} 0 \\ 1 \\ 1 \end{bmatrix}$$

All others can be written in terms of these two vectors; for example,

$$\begin{bmatrix} 3 \\ -2 \\ -2 \end{bmatrix} = 3 \begin{bmatrix} 1 \\ 0 \\ 0 \end{bmatrix} - 2 \begin{bmatrix} 0 \\ 1 \\ 1 \end{bmatrix}$$

The subspace S_2 is a two-dimensional subspace lying in 3-space.

Vector Spans

An important fact has just been illustrated: for any subspace there is always a finite number of basic vectors which generate all other vectors in the subspace. These basic vectors are said to *span* the subspace.

Formal Definition	Picture
The **span** of a set of vectors, $v_1, \ldots, v_k$ is the set of all vectors of the form $c_1 v_1 + \ldots + c_k v_k$ that is, all **linear combinations** of $v_1, \ldots, v_k$	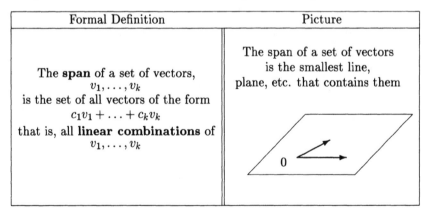 The span of a set of vectors is the smallest line, plane, etc. that contains them

The span of $[4, -1, 2]^T$ is the subspace S_1, while the span of $[1, 0, 0]^T$ and $[0, 1, 1]^T$ is the subspace S_2. This suggests an alternative description of a subspace: a subspace is always the span of a finite number of vectors.

In 2-space, the span of any two vectors is the space itself, provided that one vector is not a scalar multiple of the other. For example, the vectors $[5,1]^T$ and $[3,2]^T$ span the entire plane, since any vector $[v_1,v_2]^T$ can be written in terms of these two vectors. To express $[2,6]^T$ in these terms, for example, we need to find constants a and b for which

$$a\begin{bmatrix}5\\1\end{bmatrix}+b\begin{bmatrix}3\\2\end{bmatrix}=\begin{bmatrix}2\\6\end{bmatrix}$$

This necessitates equating the components and solving the simultaneous equations:

$$5a+3b=2 \qquad a+2b=6$$

The solution is $a=-2$ and $b=4$. That is,

$$-2\begin{bmatrix}5\\1\end{bmatrix}+4\begin{bmatrix}3\\2\end{bmatrix}=\begin{bmatrix}2\\6\end{bmatrix}$$

In other words, to reach the point $[2,6]^T$ we must go four times the length of $[3,2]^T$ then drop back twice the length of $[5,1]^T$.

Taking another example, in 5-space, the span of any three vectors is a 3-dimensional subspace provided one of the vectors cannot be written in terms of the other two. Thus

$$\begin{bmatrix}1\\0\\0\\0\\0\end{bmatrix}\quad\begin{bmatrix}0\\1\\0\\0\\0\end{bmatrix}\quad\text{and}\quad\begin{bmatrix}0\\0\\0\\0\\1\end{bmatrix}$$

span a 3-dimensional subspace of 5-space. However,

$$\begin{bmatrix}1\\1\\0\\0\\0\end{bmatrix}\quad\begin{bmatrix}0\\0\\1\\0\\1\end{bmatrix}\quad\text{and}\quad\begin{bmatrix}-2\\-2\\3\\0\\3\end{bmatrix}$$

do not span a 3-dimensional subspace, since the third vector can be written as three times the second vector minus twice the first vector. In this case the span is just a 2-dimensional subspace.

Exercises for the reader

(2.4) Write the vector $v=[4,4,0,0]^T$ in terms of the vectors spanning the subspace $S=\text{span}\{[0,1,0,1]^T,[1,0,0,-1]^T\}$.

(2.5) Show that the vector $w=[-1,1,0,0,0]^T$ does not belong to the subspace $T=\text{span}\{[1,1,0,0,0]^T,[0,0,1,1,1]^T\}$.

2.3   Angles Between Vectors

Introducing Angles

We now introduce the idea of the angle between two vectors in N-space. This leads to the definition of orthogonality, a word frequently used in statistical work.

Formal Definition	Picture
The **angle**, θ, between two vectors v and w is given by $$\cos\theta = \frac{v_1 w_1 + \cdots + v_N w_N}{\|v\| \, \|w\|}$$	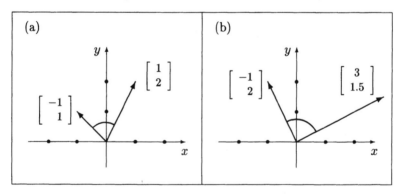 θ is the angle between v and w

Is the definition reasonable in two dimensions? Consider Figure 2.5(a) where the definition gives the angle between the two vectors as

$$\cos\theta = \frac{1\times(-1) + 2\times 1}{\sqrt{1^2 + 2^2} \times \sqrt{(-1)^2 + 1^2}} = \frac{1}{\sqrt{10}}$$

Hence $\theta = \cos^{-1}(1/\sqrt{10}) = 72^0$, which is visually reasonable.

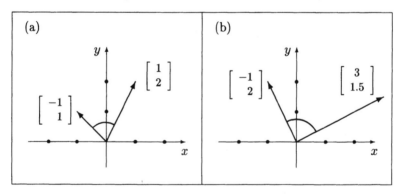

Figure 2.5: The angle θ between the two vectors (a) $[1, 2]^T$, $[-1, 1]^T$ and (b) $[3, 1.5]^T$ and $[-1, 2]^T$.

Consider now Figure 2.5(b), where

$$\cos\theta = \frac{3\times(-1) + 1.5\times 2}{\sqrt{3^2 + 1.5^2} \times \sqrt{(-1)^2 + 2^2}} = 0$$

Hence $\theta = \cos^{-1} 0 = 90^0$, which is again visually reasonable.

Here's a higher dimensional example: calculate the angle θ between the vectors

$$v = \begin{bmatrix} -2 \\ 1 \\ 5 \\ 4 \\ -3 \end{bmatrix} \quad \text{and} \quad w = \begin{bmatrix} 4 \\ -3 \\ -7 \\ -2 \\ 6 \end{bmatrix}$$

Notice that all of the negative components of v are positive in w, and vice versa, so we expect the vectors to be in roughly opposite directions. In fact $\|v\| = \sqrt{55}$ and $\|w\| = \sqrt{114}$, so

$$\cos \theta = \frac{-72}{\sqrt{55}\sqrt{114}} = -.909,$$

whence $\theta = \cos^{-1}(-.909) = 155^0$, as expected.

The mathematically inclined reader may be wondering how you would demonstrate once and for all that the formula for the angle is correct. The key tool is the cosine rule: apply it to the triangle formed by the vectors v, w and $v - w$ and the formula drops out.

A formal notation

The expression

$$v.w = v_1 w_1 + \cdots + v_N w_N$$

appearing in the angle formula, is termed the *dot product* of the vectors $v = [v_1, \ldots, v_N]^T$ and $w = [w_1, \ldots, w_N]^T$. This is just shorthand. For example, we can now write $\cos \theta = \frac{v.w}{\|v\|\|w\|}$.

Exercises for the reader

Find the angles between the following pairs of vectors:
(2.6) $[2, 1]^T$ and $[1, 3]^T$ in the plane.
(2.7) $[1, 1]^T$ and $[1, -1]^T$ in the plane.
(2.8) $[1, 1, 1, 1, 1, 1]^T$ and $[0, 4, 0, 0, -7, 0]^T$ in 6-space.
(2.9) $[5, 2, 1, 3, 7, 0, 2]^T$ and $[1, 3, 5, 0, 2, 1, 1]^T$ in 7-space.

Right Angles

We now formalize the notion of a right angle between two vectors.

Formal Definition	Picture
Vectors v and w are **orthogonal** if $\theta = 90^0$ This occurs when $v.w = 0$	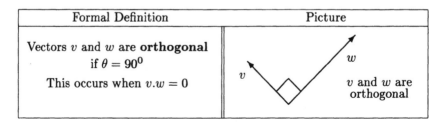 w v v and w are orthogonal

For example, $\begin{bmatrix} 2 \\ 0 \\ 1 \\ 0 \end{bmatrix}$ and $\begin{bmatrix} -2 \\ 0 \\ 4 \\ 0 \end{bmatrix}$ are an orthogonal pair of vectors, since

$$\cos\theta = \frac{2 \times (-2) + 0 \times 0 + 1 \times 4 + 0 \times 0}{\sqrt{2^2 + 1^2}\sqrt{(-2)^2 + 4^2}} = 0$$

To test two vectors for orthogonality, simply check whether their dot product is zero. Other orthogonal pairs are $v = [1,0,0,0]^T$ and $w = [0,1,0,0]^T$, and $v = [-2,0,1,4,0]^T$ and $w = [2,5,0,1,7]^T$.

Exercises for the reader

Are the following sets of vectors orthogonal?

(2.10) $\begin{bmatrix} 1 \\ 1 \\ 1 \\ 1 \\ 1 \end{bmatrix}$ $\begin{bmatrix} 2 \\ 2 \\ 2 \\ -3 \\ -3 \end{bmatrix}$ $\begin{bmatrix} 1 \\ -1 \\ 0 \\ 0 \\ 0 \end{bmatrix}$ $\begin{bmatrix} 1 \\ 1 \\ -2 \\ 0 \\ 0 \end{bmatrix}$ $\begin{bmatrix} 0 \\ 0 \\ 0 \\ 1 \\ -1 \end{bmatrix}$

(2.11) $\begin{bmatrix} -1 \\ 1 \\ -1 \\ 1 \end{bmatrix}$ $\begin{bmatrix} -1 \\ -1 \\ 1 \\ 1 \end{bmatrix}$ $\begin{bmatrix} -1 \\ 1 \\ 0 \\ 0 \end{bmatrix}$ $\begin{bmatrix} 0 \\ 0 \\ -1 \\ 1 \end{bmatrix}$

(2.12) $\begin{bmatrix} 1 \\ 1 \\ 1 \\ 1 \end{bmatrix}$ $\begin{bmatrix} -1 \\ 1 \\ -1 \\ 1 \end{bmatrix}$ $\begin{bmatrix} -1 \\ -1 \\ 1 \\ 1 \end{bmatrix}$ $\begin{bmatrix} 1 \\ -1 \\ -1 \\ 1 \end{bmatrix}$

(2.13) Find a vector orthogonal to $[3,4]^T$.

Orthogonal Coordinate Systems

In later chapters we frequently set up a complete system of vectors at right angles, different from the natural coordinate system. These new directions will be suggested by the hypotheses of interest, and will lead naturally into the statistical analysis.

Formal Definition	Picture
An **orthogonal coordinate system** for N-space is a set of N orthogonal unit vectors $U_1, \ldots, U_N$	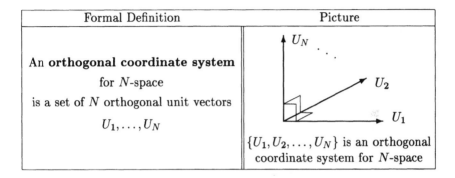 $\{U_1, U_2, \ldots, U_N\}$ is an orthogonal coordinate system for N-space

The natural coordinate system in 2-space comprises the unit vectors $U_1 = [1, 0]^T$ and $U_2 = [0, 1]^T$ as shown in Figure 2.6(a). An alternative system, consisting of the unit vectors $U_1 = [3, 2]^T / \sqrt{13}$ and $[-2, 3]^T / \sqrt{13}$, is shown in Figure 2.6(b). In fact there is an infinite number of such systems, corresponding to all possible angles that U_1 can make with the x-axis.

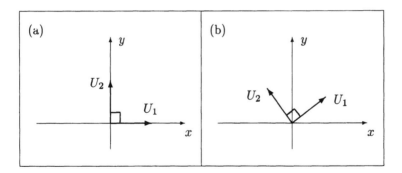

Figure 2.6: Orthogonal coordinate systems for the plane.

In 3-space the natural coordinate system consists of the vectors $U_1 = [1, 0, 0]^T$, $U_2 = [0, 1, 0]^T$ and $U_3 = [0, 0, 1]^T$. In higher dimensions, the natural coordinate system is $[1, 0, \ldots, 0]^T, \ldots, [0, 0, \ldots, 1]^T$.

In all dimensions, the systems most useful in statistics are those which include $U_1 = [1, \ldots, 1]^T / \sqrt{N}$, the unit vector in the *equiangular* direction, so-called since U_1 is equally angled to each coordinate axis.

In 4-space one such system is the set:

$$U_1 = \frac{1}{\sqrt{4}}\begin{bmatrix} 1 \\ 1 \\ 1 \\ 1 \end{bmatrix}, \quad U_2 = \frac{1}{\sqrt{4}}\begin{bmatrix} -1 \\ -1 \\ 1 \\ 1 \end{bmatrix}, \quad U_3 = \frac{1}{\sqrt{4}}\begin{bmatrix} -1 \\ 1 \\ -1 \\ 1 \end{bmatrix}, \quad U_4 = \frac{1}{\sqrt{4}}\begin{bmatrix} 1 \\ -1 \\ -1 \\ 1 \end{bmatrix}$$

Another useful system is the set:

$$U_1 = \frac{1}{\sqrt{4}}\begin{bmatrix} 1 \\ 1 \\ 1 \\ 1 \end{bmatrix}, \quad U_2 = \frac{1}{\sqrt{2}}\begin{bmatrix} 1 \\ -1 \\ 0 \\ 0 \end{bmatrix}, \quad U_3 = \frac{1}{\sqrt{6}}\begin{bmatrix} 1 \\ 1 \\ -2 \\ 0 \end{bmatrix}, \quad U_4 = \frac{1}{\sqrt{12}}\begin{bmatrix} 1 \\ 1 \\ 1 \\ -3 \end{bmatrix}$$

Yet another, in 5-space, is the set $U_1, \ldots, U_5$:

$$\frac{1}{\sqrt{5}}\begin{bmatrix} 1 \\ 1 \\ 1 \\ 1 \\ 1 \end{bmatrix}, \quad \frac{1}{\sqrt{30}}\begin{bmatrix} 2 \\ 2 \\ 2 \\ -3 \\ -3 \end{bmatrix}, \quad \frac{1}{\sqrt{2}}\begin{bmatrix} 1 \\ -1 \\ 0 \\ 0 \\ 0 \end{bmatrix}, \quad \frac{1}{\sqrt{6}}\begin{bmatrix} 1 \\ 1 \\ -2 \\ 0 \\ 0 \end{bmatrix}, \quad \frac{1}{\sqrt{2}}\begin{bmatrix} 0 \\ 0 \\ 0 \\ 1 \\ -1 \end{bmatrix}$$

Exercise for the reader

(2.14) Which of the following sets of vectors make up an orthogonal coordinate system? If not, why?

(a) $\frac{1}{\sqrt{3}}[1, 1, 1]^T, \quad \frac{1}{\sqrt{6}}[-2, 1, 1]^T, \quad \frac{1}{\sqrt{14}}[2, 1, 3]^T$

(b) $\frac{1}{\sqrt{4}}[1, 1, 1, 1]^T, \quad \frac{1}{\sqrt{20}}[-3, -1, 1, 3]^T, \quad \frac{1}{\sqrt{4}}[1, -1, -1, 1]^T$

(c) $\frac{1}{\sqrt{3}}[1, 1, 1]^T, \quad \frac{1}{\sqrt{74}}[3, -7, 4]^T, \quad \frac{1}{\sqrt{222}}[11, -1, -10]^T.$

The Size of a Subspace

At this stage we pause to clarify the meaning of the word "dimension".

Formal Definition	Picture
The **dimension** of a subspace S is the smallest number k of vectors $U_1, \ldots, U_k$, for which $\text{span}\{U_1, \ldots, U_k\} = S$	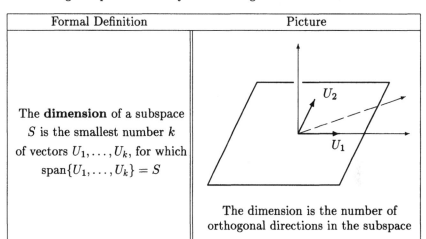 The dimension is the number of orthogonal directions in the subspace

By way of example, suppose we wish to find the dimension of the subspace

$$S = \{\text{all vectors of the form } \begin{bmatrix} a \\ a \\ b \\ c \\ b \end{bmatrix} \text{ where } a, b \text{ and } c \text{ are real numbers}\}$$

Notice that a general vector in S can be written as

$$\begin{bmatrix} a \\ a \\ b \\ c \\ b \end{bmatrix} = a \begin{bmatrix} 1 \\ 1 \\ 0 \\ 0 \\ 0 \end{bmatrix} + b \begin{bmatrix} 0 \\ 0 \\ 1 \\ 0 \\ 1 \end{bmatrix} + c \begin{bmatrix} 0 \\ 0 \\ 0 \\ 1 \\ 0 \end{bmatrix}$$

Here $U_1 = [1, 1, 0, 0, 0]^T / \sqrt{2}$, $U_2 = [0, 0, 1, 0, 1]^T / \sqrt{2}$ and $U_3 = [0, 0, 0, 1, 0]^T$ are the perpendicular directions in S. Hence the subspace S is of dimension three.

Exercise for the reader

(2.15) Find the dimension of the subspace

$$S = \text{span} \left\{ \begin{bmatrix} 1 \\ 0 \\ 0 \\ 0 \end{bmatrix}, \begin{bmatrix} 1 \\ 3 \\ 2 \\ 2 \end{bmatrix}, \begin{bmatrix} 0 \\ 1 \\ 0 \\ 0 \end{bmatrix}, \begin{bmatrix} 0 \\ 0 \\ 1 \\ 1 \end{bmatrix} \right\}$$

2.4   Projections

Projecting onto Directions

The ability to project a vector onto a specified direction is essential in the fitting of statistical models. In later chapters this vector will be the observation vector, y, so we begin now with this notation.

Formal Definition	Picture
The **projection** of an arbitrary vector y onto a unit vector U is the vector $(y.U)U$	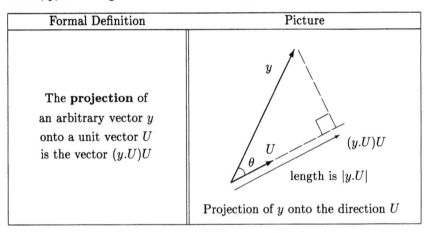 Projection of y onto the direction U

In the diagram the length of the projection of y onto U is

$$\|y\| \cos \theta = \|y\| \left(\frac{y.U}{\|y\| \, \|U\|} \right) = \frac{y.U}{\|U\|} = y.U$$

since U is a unit vector. Hence the projection of y onto U is the vector $(y.U)U$, which has length $y.U$ and is in the direction of U. Note that $y.U$ can be negative, so to be strictly correct we should say that the length of $(y.U)U$ is $|y.U|$.

The result is conveniently simple. To calculate the length of a projection we simply take a dot product. For example, in Figure 1.3 we wanted to find the lengths of the projections of

$$y = \begin{bmatrix} 13 \\ 15 \end{bmatrix} \text{ onto the directions } U_1 = \frac{1}{\sqrt{2}} \begin{bmatrix} 1 \\ 1 \end{bmatrix} \text{ and } U_2 = \frac{1}{\sqrt{2}} \begin{bmatrix} -1 \\ 1 \end{bmatrix}$$

From our dot product rule, these are simply

$$y.U_1 = \frac{13 + 15}{\sqrt{2}} = \frac{28}{\sqrt{2}} = 14\sqrt{2} \quad \text{and} \quad y.U_2 = \frac{-13 + 15}{\sqrt{2}} = \frac{2}{\sqrt{2}} = \sqrt{2},$$

as redisplayed in Figure 2.7.

The first projection vector in Figure 2.7 is therefore

$$(y.U_1)U_1 = (14\sqrt{2}) \frac{1}{\sqrt{2}} \begin{bmatrix} 1 \\ 1 \end{bmatrix} = 14 \begin{bmatrix} 1 \\ 1 \end{bmatrix} = \begin{bmatrix} 14 \\ 14 \end{bmatrix}$$

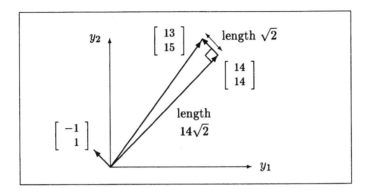

Figure 2.7: Projection vectors and their lengths for the heartbeat data as displayed in Figure 1.3.

The second projection vector is

$$(y.U_2)U_2 = (\sqrt{2})\frac{1}{\sqrt{2}}\begin{bmatrix} -1 \\ 1 \end{bmatrix}$$

Notice that the vector y is the sum of these two projection vectors, that is,

$$\begin{bmatrix} 13 \\ 15 \end{bmatrix} = \begin{bmatrix} 14 \\ 14 \end{bmatrix} + \begin{bmatrix} -1 \\ 1 \end{bmatrix}$$

To emphasize this point in a diagram, we shall often move the second projection vector $[-1, 1]^T$ from the origin to the tip of the first projection vector, as shown in Figure 2.7.

As a second example, suppose that in 4-space we wish to project the vector

$$y = \begin{bmatrix} 7 \\ 8 \\ 3 \\ 1 \end{bmatrix} \text{ onto the directions } U_1 = \frac{1}{\sqrt{2}}\begin{bmatrix} 1 \\ 1 \\ 0 \\ 0 \end{bmatrix} \text{ and } U_2 = \frac{1}{\sqrt{2}}\begin{bmatrix} 0 \\ 0 \\ 1 \\ 1 \end{bmatrix}$$

First we calculate $y.U_1 = (7+8)/\sqrt{2} = 15/\sqrt{2}$ and $y.U_2 = (3+1)/\sqrt{2} = 4/\sqrt{2}$. The projection vectors are then

$$(y.U_1)U_1 = \frac{15}{\sqrt{2}}\frac{1}{\sqrt{2}}\begin{bmatrix} 1 \\ 1 \\ 0 \\ 0 \end{bmatrix} = \frac{15}{2}\begin{bmatrix} 1 \\ 1 \\ 0 \\ 0 \end{bmatrix} = \begin{bmatrix} 7.5 \\ 7.5 \\ 0 \\ 0 \end{bmatrix}$$

$$\text{and } (y.U_2)U_2 = \frac{4}{\sqrt{2}}\frac{1}{\sqrt{2}}\begin{bmatrix} 0 \\ 0 \\ 1 \\ 1 \end{bmatrix} = \frac{4}{2}\begin{bmatrix} 0 \\ 0 \\ 1 \\ 1 \end{bmatrix} = \begin{bmatrix} 0 \\ 0 \\ 2 \\ 2 \end{bmatrix}$$

A word on terminology. In mathematics, *projection* carries the meaning *projection vector*, so the word *vector* is strictly unnecessary. For clarity, however, we shall slip in the word *vector* whenever it is helpful to do so.

Exercise for the reader

(2.16) In the last example, the directions $U_3 = [1, -1, 0, 0]^T/\sqrt{2}$ and $U_4 = [0, 0, 1, -1]^T/\sqrt{2}$ complete an orthogonal system for 4-space. Find the projections of y onto the directions U_3 and U_4. Remember that the answers are vectors.

Projecting onto Subspaces

We shall often wish to project a vector onto a subspace, the model space in our statistical work. In order to do this we shall project the vector onto each orthogonal coordinate axis in the subspace, then sum the resulting individual projection vectors. In doing this, we are finding the nearest vector in the subspace to our vector of observations.

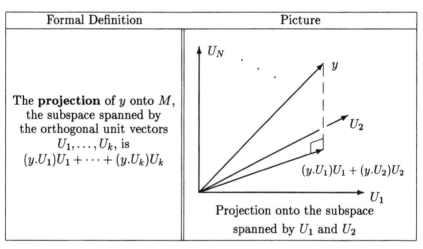

Formal Definition	Picture
The **projection** of y onto M, the subspace spanned by the orthogonal unit vectors $U_1, \ldots, U_k$, is $(y.U_1)U_1 + \cdots + (y.U_k)U_k$	Projection onto the subspace spanned by U_1 and U_2

As an example, suppose we wish to project

$$y = \begin{bmatrix} 7 \\ 8 \\ 3 \\ 1 \end{bmatrix} \text{ onto the subspace } M = \text{vectors of the form } \left\{ \begin{bmatrix} a \\ a \\ b \\ b \end{bmatrix} \right\}$$

First we need to find an orthogonal coordinate system for M. The obvious choice is

$$U_1 = \frac{1}{\sqrt{2}} \begin{bmatrix} 1 \\ 1 \\ 0 \\ 0 \end{bmatrix}, \quad U_2 = \frac{1}{\sqrt{2}} \begin{bmatrix} 0 \\ 0 \\ 1 \\ 1 \end{bmatrix}$$

We then calculate

$$(y.U_1)U_1 = \begin{bmatrix} 7.5 \\ 7.5 \\ 0 \\ 0 \end{bmatrix} \text{ and } (y.U_2)U_2 = \begin{bmatrix} 0 \\ 0 \\ 2 \\ 2 \end{bmatrix}$$

The required projection is the sum

$$(y.U_1)U_1 + (y.U_2)U_2 = \begin{bmatrix} 7.5 \\ 7.5 \\ 0 \\ 0 \end{bmatrix} + \begin{bmatrix} 0 \\ 0 \\ 2 \\ 2 \end{bmatrix} = \begin{bmatrix} 7.5 \\ 7.5 \\ 2 \\ 2 \end{bmatrix}$$

Exercise for the reader

(2.17) In the last example, a less obvious choice of coordinate system for the subspace M is $U_1 = [1, 1, 1, 1]^T / \sqrt{4}$, $U_2 = [1, 1, -1, -1]^T / \sqrt{4}$. Use this alternative coordinate system to calculate the projection of y onto M. Is your answer the same?

Breaking a Vector into Perpendicular Components

Any vector can be expressed as the sum of its projections onto the complete set of coordinate axes $U_1, U_2, \ldots, U_N$. This result can be viewed as an application of the previous framed table to the case where the subspace is the whole space.

Formal Definition	Picture
The **orthogonal decomposition** of an arbitrary vector y, using such a coordinate system is $$y = (y.U_1)U_1 + \cdots + (y.U_N)U_N.$$	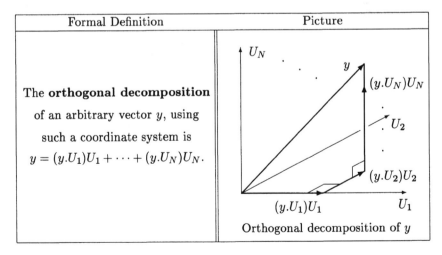 Orthogonal decomposition of y

In 2-space, the result has already been illustrated in Figure 2.7, where we saw that

$$\begin{bmatrix} 13 \\ 15 \end{bmatrix} = \begin{bmatrix} 14 \\ 14 \end{bmatrix} + \begin{bmatrix} -1 \\ 1 \end{bmatrix}$$

In 4-space, we now illustrate the result using $y = [7, 8, 3, 1]^T$ and

$$U_1 = \frac{1}{\sqrt{4}}\begin{bmatrix} 1 \\ 1 \\ 1 \\ 1 \end{bmatrix}, U_2 = \frac{1}{\sqrt{4}}\begin{bmatrix} 1 \\ 1 \\ -1 \\ -1 \end{bmatrix}, U_3 = \frac{1}{\sqrt{2}}\begin{bmatrix} 1 \\ -1 \\ 0 \\ 0 \end{bmatrix}, U_4 = \frac{1}{\sqrt{2}}\begin{bmatrix} 0 \\ 0 \\ 1 \\ -1 \end{bmatrix}$$

Here $(y.U_1)U_1 + (y.U_2)U_2 + (y.U_3)U_3 + (y.U_4)U_4$

$$= \begin{bmatrix} 4.75 \\ 4.75 \\ 4.75 \\ 4.75 \end{bmatrix} + \begin{bmatrix} 2.75 \\ 2.75 \\ -2.75 \\ -2.75 \end{bmatrix} + \begin{bmatrix} -.5 \\ .5 \\ 0 \\ 0 \end{bmatrix} + \begin{bmatrix} 0 \\ 0 \\ 1 \\ -1 \end{bmatrix} = \begin{bmatrix} 7 \\ 8 \\ 3 \\ 1 \end{bmatrix} = y$$

That is, the decomposition holds true, as expected.

2.5   Sums of Squares

A key tool in our work is the ability to split the squared length of the observation vector into a sum of squares using Pythagoras' Theorem. This will serve as the cornerstone for hypothesis testing.

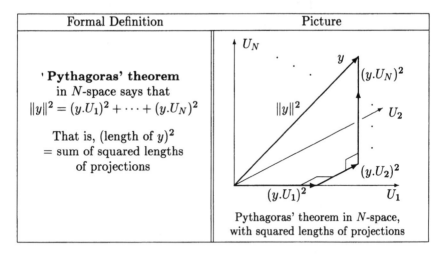

Formal Definition	Picture
' Pythagoras' theorem in N-space says that $\|y\|^2 = (y.U_1)^2 + \cdots + (y.U_N)^2$ That is, (length of y)^2 = sum of squared lengths of projections	Pythagoras' theorem in N-space, with squared lengths of projections

This extends Pythagoras' Theorem from two and three dimensions to the general case of N dimensions. As an example of its use in the plane, consider again Figure 2.7 and the orthogonal decomposition

$$y = (y.U_1)U_1 + (y.U_2)U_2$$

$$\text{or} \quad \begin{bmatrix} 13 \\ 15 \end{bmatrix} = \begin{bmatrix} 14 \\ 14 \end{bmatrix} + \begin{bmatrix} -1 \\ 1 \end{bmatrix}$$

Pythagoras' Theorem tells us that

$$\left\|\begin{bmatrix} 13 \\ 15 \end{bmatrix}\right\|^2 = \left\|\begin{bmatrix} 14 \\ 14 \end{bmatrix}\right\|^2 + \left\|\begin{bmatrix} -1 \\ 1 \end{bmatrix}\right\|^2$$

that is, $(13^2 + 15^2) = (14^2 + 14^2) + ((-1)^2 + 1^2)$

or, $\qquad\quad 394 \quad = \quad\ 392 \quad + \quad\ 2$

which is clearly true. Notice that in this example it was easier to directly calculate the squared lengths of the projection vectors, instead of using the formula $(y.U_i)^2$.

As a second example, we illustrate Pythagoras' Theorem for the orthogonal decomposition of the example of the last subsection, that is, for

$$\begin{bmatrix} 7 \\ 8 \\ 3 \\ 1 \end{bmatrix} = \begin{bmatrix} 4.75 \\ 4.75 \\ 4.75 \\ 4.75 \end{bmatrix} + \begin{bmatrix} 2.75 \\ 2.75 \\ -2.75 \\ -2.75 \end{bmatrix} + \begin{bmatrix} -.5 \\ .5 \\ 0 \\ 0 \end{bmatrix} + \begin{bmatrix} 0 \\ 0 \\ 1 \\ -1 \end{bmatrix}$$

Pythagoras' Theorem states that the squared length of y is the sum of the squares of the lengths of the projections, or

$$7^2 + 8^2 + 3^2 + 1^2 = 4(4.75)^2 + 4(2.75)^2 + 2(.5)^2 + 2(1)^2$$

or $\qquad\quad 123 \qquad = \ \ 90.25 \ + \ \ 30.25 \ + \ \ .5 \ + \ \ 2$

which again is clearly true!

Exercise for the reader

(2.18) In 4-space consider

$$y = \begin{bmatrix} 7 \\ 8 \\ 3 \\ 1 \end{bmatrix}, U_1 = \frac{1}{\sqrt{2}}\begin{bmatrix} 1 \\ 1 \\ 0 \\ 0 \end{bmatrix}, U_2 = \frac{1}{\sqrt{2}}\begin{bmatrix} 0 \\ 0 \\ 1 \\ 1 \end{bmatrix}, U_3 = \frac{1}{\sqrt{2}}\begin{bmatrix} 1 \\ -1 \\ 0 \\ 0 \end{bmatrix}, U_4 = \frac{1}{\sqrt{2}}\begin{bmatrix} 0 \\ 0 \\ 1 \\ -1 \end{bmatrix}$$

For this alternative orthogonal coordinate system, write out in full:

(a) the orthogonal decomposition of y with respect to $U_1, \ldots, U_4$, and

(b) the breakup of $\|y\|^2$ in terms of squared lengths of projections onto $U_1, \ldots, U_4$.

Exercises

(2.1) Find the length of each of the following vectors, and the unit vector in the same direction as each vector:

(a) $\begin{bmatrix} 3 \\ 4 \end{bmatrix}$ (b) $\begin{bmatrix} 0 \\ 5 \\ 0 \\ -12 \end{bmatrix}$ (c) $\begin{bmatrix} 6 \\ -3 \\ 2 \\ 5 \\ 1 \end{bmatrix}$ (d) $\begin{bmatrix} 1 \\ -1 \\ -1 \\ 1 \end{bmatrix}$

(2.2) Simplify the following expressions:

(a) $\begin{bmatrix} 4 \\ 2 \\ 3 \end{bmatrix} + \begin{bmatrix} 1 \\ 5 \\ 9 \end{bmatrix}$ (b) $\begin{bmatrix} 6.4 \\ 7.6 \\ 5.9 \end{bmatrix} + 3 \begin{bmatrix} -6.2 \\ 1.1 \\ 4.1 \end{bmatrix}$

(c) $3 \begin{bmatrix} 0 \\ 5 \\ 2 \\ 4 \end{bmatrix} - 2 \begin{bmatrix} 1 \\ 3 \\ -1 \\ 4 \end{bmatrix}$ (d) $5 \begin{bmatrix} 1 \\ 4 \\ 2 \end{bmatrix} - 3 \begin{bmatrix} 5 \\ 4 \end{bmatrix}$

(2.3) Which of the following vectors belong to the subspace S, where

$$S = \text{span} \left\{ \begin{bmatrix} 1 \\ 1 \\ 0 \\ 0 \\ 0 \end{bmatrix}, \begin{bmatrix} 0 \\ 0 \\ 1 \\ 1 \\ 1 \end{bmatrix} \right\}? \quad \text{Justify your answers.}$$

(a) $\begin{bmatrix} 2 \\ 2 \\ -8 \\ -8 \\ -8 \end{bmatrix}$ (b) $\begin{bmatrix} -5 \\ -5 \\ 0 \\ 1 \\ 0 \end{bmatrix}$ (c) $\begin{bmatrix} 0 \\ 0 \\ 0 \\ -2 \\ 1 \end{bmatrix}$ (d) $\begin{bmatrix} -3 \\ -3 \\ 1 \\ 1 \\ 1 \end{bmatrix}$

(2.4) Work out the dimension of the following subspaces:

(a) $S = \text{span} \left\{ \begin{bmatrix} 1 \\ 1 \\ 1 \\ 1 \end{bmatrix}, \begin{bmatrix} -1 \\ -1 \\ 2 \\ 0 \end{bmatrix}, \begin{bmatrix} 4 \\ 4 \\ -8 \\ 0 \end{bmatrix}, \begin{bmatrix} 0 \\ 0 \\ 3 \\ 1 \end{bmatrix} \right\}$

(b) $S = \text{span} \left\{ \text{all vectors} \begin{bmatrix} a \\ a - b \\ b \end{bmatrix} \text{ where } a \text{ and } b \text{ are real numbers} \right\}$

(c) $S = \text{span}\left\{ \begin{bmatrix} -1 \\ 1 \\ 0 \\ 0 \\ 0 \end{bmatrix}, \begin{bmatrix} 0 \\ 0 \\ 2 \\ -1 \\ -1 \end{bmatrix}, \begin{bmatrix} 0 \\ 0 \\ 0 \\ 1 \\ -1 \end{bmatrix}, \begin{bmatrix} 5 \\ -5 \\ 2 \\ 0 \\ -2 \end{bmatrix} \right\}$

(2.5) Find the lengths of the following vectors and the angle between each pair of vectors:

(a) $\begin{bmatrix} 3 \\ 4 \end{bmatrix} \begin{bmatrix} 6 \\ 7 \end{bmatrix}$
(b) $\begin{bmatrix} 12 \\ -5 \end{bmatrix} \begin{bmatrix} 5 \\ 12 \end{bmatrix}$
(c) $\begin{bmatrix} -2 \\ -1 \\ 0 \\ 1 \\ 2 \end{bmatrix} \begin{bmatrix} 2 \\ -1 \\ -2 \\ -1 \\ 2 \end{bmatrix}$

(d) $\begin{bmatrix} 5 \\ 7 \\ 3 \\ 2 \end{bmatrix} \begin{bmatrix} 6 \\ 6 \\ 2 \\ 3 \end{bmatrix}$

(2.6) Which of the following sets of vectors are mutually orthogonal?

(a) $\begin{bmatrix} 1 \\ 1 \end{bmatrix} \quad \begin{bmatrix} 1.7 \\ -1.7 \end{bmatrix}$

(b) $\begin{bmatrix} 1 \\ 2 \\ 3 \\ 4 \end{bmatrix} \quad \begin{bmatrix} -2 \\ 1 \\ 4 \\ -3 \end{bmatrix} \quad \begin{bmatrix} -1 \\ 1 \\ 1 \\ -1 \end{bmatrix}$

(c) $\begin{bmatrix} -1 \\ 2 \\ -1 \end{bmatrix} \quad \begin{bmatrix} 1 \\ 1 \\ 1 \end{bmatrix} \quad \begin{bmatrix} 1 \\ 0 \\ -1 \end{bmatrix}$

(2.7) Write down four alternative orthogonal coordinate systems U_1, U_2, U_3 for 3-space, all of which include a coordinate axis in the direction of the vector $[1, 1, 1]^T$.

(2.8) Calculate the length of the projection of $y = [4, 5, 7, 8]^T$ onto the direction of the following vectors:

(a) $\frac{1}{2}\begin{bmatrix} 1 \\ 1 \\ 1 \\ 1 \end{bmatrix}$
(b) $\begin{bmatrix} 1 \\ 1 \\ 0 \\ 0 \end{bmatrix}$
(c) $\begin{bmatrix} 10 \\ -10 \\ 0 \\ 0 \end{bmatrix}$
(d) $\begin{bmatrix} 1 \\ -1 \\ -1 \\ 1 \end{bmatrix}$

(2.9) Project the vector

$$y = \begin{bmatrix} 7.1 \\ 8.7 \\ 10.2 \\ 11.4 \\ 9.9 \end{bmatrix} \text{ onto the subspace } S = \text{span} \left\{ \begin{bmatrix} 1 \\ 1 \\ 0 \\ 0 \\ 0 \end{bmatrix}, \begin{bmatrix} 0 \\ 0 \\ 1 \\ 1 \\ 1 \end{bmatrix} \right\}$$

Express your answer as a single vector.

(2.10) Given the vector y and orthogonal coordinate system

$$y = \begin{bmatrix} 3 \\ 5 \\ 7 \\ 13 \end{bmatrix} \quad U_1 = \frac{1}{2}\begin{bmatrix} 1 \\ 1 \\ 1 \\ 1 \end{bmatrix} \quad U_2 = \frac{1}{2}\begin{bmatrix} 1 \\ 1 \\ -1 \\ -1 \end{bmatrix} \quad U_3 = \frac{1}{2}\begin{bmatrix} 1 \\ -1 \\ 1 \\ -1 \end{bmatrix} \quad U_4 = \frac{1}{2}\begin{bmatrix} 1 \\ -1 \\ -1 \\ 1 \end{bmatrix}$$

write y in the form

$$y = \text{constant} \times U_1 + \text{constant} \times U_2 + \text{constant} \times U_3 + \text{constant} \times U_4$$

(2.11) Find the constants c_1 and c_2 for which:

$$\text{(a)} \quad c_1 \begin{bmatrix} 1 \\ 1 \end{bmatrix} + c_2 \begin{bmatrix} 1 \\ -1 \end{bmatrix} = \begin{bmatrix} 8 \\ 2 \end{bmatrix}$$

$$\text{(b)} \quad c_1 \begin{bmatrix} 1 \\ 2 \end{bmatrix} + c_2 \begin{bmatrix} 3 \\ 4 \end{bmatrix} = \begin{bmatrix} 3 \\ 2 \end{bmatrix}$$

$$\text{(c)} \quad c_1 \begin{bmatrix} 1 \\ 1 \\ 1 \\ 1 \end{bmatrix} + c_2 \begin{bmatrix} -1 \\ -1 \\ 1 \\ 1 \end{bmatrix} = \begin{bmatrix} 1 \\ 1 \\ 7 \\ 7 \end{bmatrix}$$

(2.12) (a) Find the vectors P_1y, P_2y, P_3y, P_4y corresponding to the projection of $y = [6, 5, 9, 12]^T$ onto the orthogonal unit vectors

$$U_1 = \frac{1}{\sqrt{4}}\begin{bmatrix} 1 \\ 1 \\ 1 \\ 1 \end{bmatrix}, \quad U_2 = \frac{1}{\sqrt{20}}\begin{bmatrix} -3 \\ -1 \\ 1 \\ 3 \end{bmatrix}, \quad U_3 = \frac{1}{\sqrt{4}}\begin{bmatrix} 1 \\ -1 \\ -1 \\ 1 \end{bmatrix}, \quad U_4 = \frac{1}{\sqrt{20}}\begin{bmatrix} -1 \\ 3 \\ -3 \\ 1 \end{bmatrix}$$

(b) Write y as the sum of these projection vectors.

(c) Calculate the squared length of y and of each projection vector.

(d) Hence confirm Pythagoras' Theorem,

$$\|y\|^2 = \|P_1y\|^2 + \|P_2y\|^2 + \|P_3y\|^2 + \|P_4y\|^2$$

(e) Also confirm that

$$\|P_1 y\|^2 = (y.U_1)^2, \quad \|P_2 y\|^2 = (y.U_2)^2, \quad \|P_3 y\|^2 = (y.U_3)^2$$

$$\text{and } \|P_4 y\|^2 = (y.U_4)^2$$

(2.13) (a) Find the projection vectors $P_1 y$, $P_2 y$, $P_3 y$, and $P_4 y$ corresponding to the projection of $y = [5, 9, 7, 11]^T$ onto the coordinate axes

$$U_1 = \frac{1}{\sqrt{4}} \begin{bmatrix} 1 \\ 1 \\ 1 \\ 1 \end{bmatrix} \quad U_2 = \frac{1}{\sqrt{4}} \begin{bmatrix} -1 \\ -1 \\ 1 \\ 1 \end{bmatrix} \quad U_3 = \frac{1}{\sqrt{2}} \begin{bmatrix} -1 \\ 1 \\ 0 \\ 0 \end{bmatrix} \quad U_4 = \frac{1}{\sqrt{2}} \begin{bmatrix} 0 \\ 0 \\ -1 \\ 1 \end{bmatrix}$$

(b) Check that $y - P_1 y = P_2 y + P_3 y + P_4 y$

(c) What is $\|P_1 y\|^2$?

(d) Calculate $\|y - P_1 y\|^2$ in two ways: (i) directly from $y - P_1 y$

(ii) by Pythagoras' Theorem.

Solutions to the Reader Exercises

(2.1) The lengths are
$$\sqrt{4^2 + 1^2} = \sqrt{16 + 1} = \sqrt{17},$$
$$\sqrt{6^2 + 1^2 + 0^2 + (-2)^2 + 4^2} = \sqrt{36 + 1 + 4 + 16} = \sqrt{57},$$
$$\sqrt{1^2 + 1^2 + (-2)^2} = \sqrt{1 + 1 + 4} = \sqrt{6}, \text{ and}$$
$$\sqrt{1^2 + 1^2 + 1^2 + 1^2 + 1^2} = \sqrt{5}.$$

(2.2) The unit vectors are

$$\frac{1}{\sqrt{5}} \begin{bmatrix} 1 \\ 1 \\ 1 \\ 1 \\ 1 \end{bmatrix}, \quad \frac{1}{\sqrt{2}} \begin{bmatrix} 1 \\ -1 \\ 0 \\ 0 \end{bmatrix}, \quad \frac{1}{\sqrt{30}} \begin{bmatrix} 3 \\ 3 \\ -2 \\ -2 \\ -2 \end{bmatrix} \quad \text{and } \frac{1}{2} \begin{bmatrix} 1 \\ -1 \\ 1 \\ -1 \end{bmatrix}$$

(2.3)

(a) $4 \begin{bmatrix} 1 \\ 1 \\ 0 \\ 0 \\ 0 \end{bmatrix} + 5 \begin{bmatrix} 0 \\ 0 \\ 1 \\ 1 \\ 1 \end{bmatrix} = \begin{bmatrix} 4 \\ 4 \\ 0 \\ 0 \\ 0 \end{bmatrix} + \begin{bmatrix} 0 \\ 0 \\ 5 \\ 5 \\ 5 \end{bmatrix} = \begin{bmatrix} 4 \\ 4 \\ 5 \\ 5 \\ 5 \end{bmatrix}$

(b) $3 \begin{bmatrix} 1 \\ -1 \\ 0 \\ 0 \end{bmatrix} - 2 \begin{bmatrix} 0 \\ 0 \\ 1 \\ -1 \end{bmatrix} = \begin{bmatrix} 3 \\ -3 \\ 0 \\ 0 \end{bmatrix} + \begin{bmatrix} 0 \\ 0 \\ -2 \\ 2 \end{bmatrix} = \begin{bmatrix} 3 \\ -3 \\ -2 \\ 2 \end{bmatrix}$

$$(2.4) \quad \begin{bmatrix} 4 \\ 4 \\ 0 \\ 0 \end{bmatrix} = 4 \begin{bmatrix} 0 \\ 1 \\ 0 \\ 1 \end{bmatrix} + 4 \begin{bmatrix} 1 \\ 0 \\ 0 \\ -1 \end{bmatrix}$$

(2.5) Suppose there are constants, a and b, such that

$$\begin{bmatrix} -1 \\ 1 \\ 0 \\ 0 \\ 0 \end{bmatrix} = a \begin{bmatrix} 1 \\ 1 \\ 0 \\ 0 \\ 0 \end{bmatrix} + b \begin{bmatrix} 0 \\ 0 \\ 1 \\ 1 \\ 1 \end{bmatrix}$$

Then equating each component in turn we get the five simultaneous equations

$$\begin{aligned} -1 &= a \\ 1 &= a \\ 0 &= b \\ 0 &= b \\ 0 &= b \end{aligned}$$

The equations $a = -1$ and $a = 1$ are incompatible, so we conclude that the vector w is not in the subspace.

(2.6) $\quad \cos\theta = \dfrac{2+3}{\sqrt{5}\sqrt{10}} = \dfrac{5}{\sqrt{50}} = \dfrac{1}{\sqrt{2}}$. Thus $\theta = 45^0$.

(2.7) $\quad \begin{bmatrix} 1 \\ 1 \end{bmatrix} \cdot \begin{bmatrix} 1 \\ -1 \end{bmatrix} = 0$, so $\cos\theta = 0$ and $\theta = 90^0$.

(2.8) $\quad \cos\theta = \dfrac{4-7}{\sqrt{6}\sqrt{65}} = -0.15$, so $\theta = 99^0$.

(2.9) $\quad \cos\theta = \dfrac{32}{\sqrt{92}\sqrt{41}} = 0.52$, so $\theta = 59^0$.

(2.10) $\qquad$ Yes. There are $_5C_2 = 10$ pairs to check.

(2.11) $\qquad$ No. If labelled U_1, U_2, U_3 and U_4, U_1 is not orthogonal to U_3 or U_4. All other pairs are orthogonal.

(2.12) $\qquad$ Yes. There are $_4C_2 = 6$ pairs to check.

(2.13) We want a vector $\begin{bmatrix} a \\ b \end{bmatrix}$ such that $\begin{bmatrix} a \\ b \end{bmatrix} \cdot \begin{bmatrix} 3 \\ 4 \end{bmatrix} = 0$. That is, we require $3a + 4b = 0$, or $b = -3a/4$. Any vector of the form $\begin{bmatrix} a \\ -3a/4 \end{bmatrix}$ will suffice. This is equivalent to the form $\begin{bmatrix} 4a \\ -3a \end{bmatrix}$. Sample answers for $a = 1/4$, $a = 1$ and $a = -1$ are $\begin{bmatrix} 1 \\ -3/4 \end{bmatrix}, \begin{bmatrix} 4 \\ -3 \end{bmatrix}, \begin{bmatrix} -4 \\ 3 \end{bmatrix}$.

(2.14) (a) This is not an orthogonal coordinate system since

$$\frac{1}{\sqrt{3}}\begin{bmatrix} 1 \\ 1 \\ 1 \end{bmatrix} \cdot \frac{1}{\sqrt{14}}\begin{bmatrix} 2 \\ 1 \\ 3 \end{bmatrix} = \frac{1}{\sqrt{42}}(2 + 1 + 3) \neq 0$$

That is, the first and third vectors are not orthogonal.

(b) This is not an orthogonal coordinate system since there are only three coordinate axes. Four are the minimum needed for 4-space.

(c) This is an orthogonal coordinate system.

(2.15) The vector $\begin{bmatrix} 1 \\ 3 \\ 2 \\ 2 \end{bmatrix}$ equals $\begin{bmatrix} 1 \\ 0 \\ 0 \\ 0 \end{bmatrix} + 3\begin{bmatrix} 1 \\ 1 \\ 0 \\ 0 \end{bmatrix} + 2\begin{bmatrix} 0 \\ 0 \\ 1 \\ 0 \end{bmatrix}$, so that the sub-

space can be written more concisely as

$$U = \text{span} \left\{ \begin{bmatrix} 1 \\ 0 \\ 0 \\ 0 \end{bmatrix}, \begin{bmatrix} 0 \\ 1 \\ 0 \\ 0 \end{bmatrix}, \begin{bmatrix} 0 \\ 0 \\ 1 \\ 1 \end{bmatrix} \right\}$$

None of these three vectors can be written in terms of the other two vectors, so the dimension of the subspace is three.

(2.16)

$$y.U_3 = \begin{bmatrix} 7 \\ 8 \\ 3 \\ 1 \end{bmatrix} \cdot \frac{1}{\sqrt{2}}\begin{bmatrix} 1 \\ -1 \\ 0 \\ 0 \end{bmatrix} = \frac{1}{\sqrt{2}}(7 - 8 + 0 + 0) = \frac{-1}{\sqrt{2}}$$

$$y.U_4 = \begin{bmatrix} 7 \\ 8 \\ 3 \\ 1 \end{bmatrix} \cdot \frac{1}{\sqrt{2}}\begin{bmatrix} 0 \\ 0 \\ 1 \\ -1 \end{bmatrix} = \frac{1}{\sqrt{2}}(0 + 0 + 3 - 1) = \frac{2}{\sqrt{2}}$$

Hence the projection of y onto U_3 is

$$(y.U_3)U_3 = \left(\frac{-1}{\sqrt{2}}\right)\frac{1}{\sqrt{2}}\begin{bmatrix} 1 \\ -1 \\ 0 \\ 0 \end{bmatrix} = -\frac{1}{2}\begin{bmatrix} 1 \\ -1 \\ 0 \\ 0 \end{bmatrix} = \begin{bmatrix} -1/2 \\ 1/2 \\ 0 \\ 0 \end{bmatrix}$$

and the projection of y onto U_4 is

$$(y.U_4)U_4 = \left(\frac{2}{\sqrt{2}}\right)\frac{1}{\sqrt{2}}\begin{bmatrix} 0 \\ 0 \\ 1 \\ -1 \end{bmatrix} = \frac{2}{2}\begin{bmatrix} 0 \\ 0 \\ 1 \\ -1 \end{bmatrix} = \begin{bmatrix} 0 \\ 0 \\ 1 \\ -1 \end{bmatrix}$$

(2.17) The projection of y onto M is

$$(y.U_1)U_1 + (y.U_2)U_2 = \left(\frac{19}{\sqrt{4}}\right)\frac{1}{\sqrt{4}}\begin{bmatrix} 1 \\ 1 \\ 1 \\ 1 \end{bmatrix} + \left(\frac{11}{\sqrt{4}}\right)\frac{1}{\sqrt{4}}\begin{bmatrix} 1 \\ 1 \\ -1 \\ -1 \end{bmatrix}$$

$$= \frac{19}{4}\begin{bmatrix} 1 \\ 1 \\ 1 \\ 1 \end{bmatrix} + \frac{11}{4}\begin{bmatrix} 1 \\ 1 \\ -1 \\ -1 \end{bmatrix}$$

$$= \begin{bmatrix} 4.75 \\ 4.75 \\ 4.75 \\ 4.75 \end{bmatrix} + \begin{bmatrix} 2.75 \\ 2.75 \\ -2.75 \\ -2.75 \end{bmatrix} = \begin{bmatrix} 7.5 \\ 7.5 \\ 2 \\ 2 \end{bmatrix}$$

Yes, the answer is the same.

(2.18) (a) With the alternative coordinate system the orthogonal decomposition can be written as

$$y = \frac{15}{\sqrt{2}}U_1 + \frac{4}{\sqrt{2}}U_2 + \frac{-1}{\sqrt{2}}U_3 + \frac{2}{\sqrt{2}}U_4$$

or more fully in the form

$$\begin{bmatrix} 7 \\ 8 \\ 3 \\ 1 \end{bmatrix} = \begin{bmatrix} 7.5 \\ 7.5 \\ 0 \\ 0 \end{bmatrix} + \begin{bmatrix} 0 \\ 0 \\ 2 \\ 2 \end{bmatrix} + \begin{bmatrix} -0.5 \\ 0.5 \\ 0 \\ 0 \end{bmatrix} + \begin{bmatrix} 0 \\ 0 \\ 1 \\ -1 \end{bmatrix}$$

(b) Pythagoras' Theorem applied to these decompositions yields

$$\|y\|^2 = \left(\frac{15}{\sqrt{2}}\right)^2 + \left(\frac{4}{\sqrt{2}}\right)^2 + \left(\frac{-1}{\sqrt{2}}\right)^2 + \left(\frac{2}{\sqrt{2}}\right)^2, \text{ or}$$

$$\|y\|^2 = 2 \times 7.5^2 + 2 \times 2^2 + 2 \times .5^2 + 2 \times 1^2$$

Both these expressions lead to

$$\|y\|^2 = 112.5 + 8 + 0.5 + 2$$

This is the required breakup in terms of squared lengths of projections onto the unit vectors U_1, U_2, U_3 and U_4. As a check we can calculate $\|y\|^2$ directly as $7^2 + 8^2 + 3^2 + 1^2 = 123$, the value on the right hand side of our equation.

Chapter 3

The Statistical
Tool Kit

We complement the previous chapter by now summarizing the *statistical* concepts and techniques which will be used in succeeding chapters. We begin in §1 with the notion of a population, a sample, a random variable and a probability distribution. Properties of combinations of random variables are dealt with in §2. We discuss estimation in §3, and finish in §4 with the definitions of the F and t distributions.

The central ideas are again placed in framed boxes. We suggest that initially you glance through these frames to absorb the flavor of the material, returning to study unfamiliar concepts in more detail. If you find these statistical ideas hard to grasp skip across to Chapter 4, and the applications in Chapters 5 to 18, since statistical ideas are more easily understood in the context of a real problem. Refer back to this chapter as necessary during your reading of the book.

3.1   Basic Ideas

Populations

In daily life we use the word *population* to refer to a group of living things. For example we speak of a population of people, or a population of insects. In statistical work the word has a special connotation.

> **A population** is a class of measurements or
> counts on a set of individuals of interest

For example, the population may be the set of heights of all people residing in the U.S. on January 1 this year. Or the population may be the set of weight losses, over a specified two month period, of the ten people enrolled in a particular weight-watching class. Whatever the population, it is important that it be specified as clearly as possible.

Samples

Usually the population is too large for us to be able to obtain every individual measurement. In this case we have to be content with a subset of the measurements, this subset being known as a sample. For this sample to be useful for testing ideas about the population, we need it to be as representative as possible. To achieve this we take a random sample.

> A **random sample** of a given size from a population
> is one chosen in such a way that each set of that size
> is as likely to be chosen as any other set

Random Variables

When we take a sample we are frequently more interested in some combination of the sample values than just the raw values themselves. We formalize this process in the notion of a random variable.

> A **random variable** is a rule for obtaining a
> real number via a random sampling scheme

For example, the rule could be: take a random sample of size six from a population, and average the sample values. Or it might be: take a sample of size four from each of two populations, then calculate the difference between the two sample means.

Each repetition of the rule will, in general, produce a different real number. These numbers are termed *realized values.* To distinguish these realized values from the random variable we now introduce a useful convention.

Notation

> To distinguish a random variable from one of its realized values, we use upper case characters for the random variable, and lower case characters for the realized value.

In the first example given above, we use $\overline{Y}$ to denote the random variable, and $\bar{y}$ to denote the sample mean. In the second example, if we denote a difference between two sample means by $\bar{y}_1 - \bar{y}_2$, then we denote the associated random variable by $\overline{Y}_1 - \overline{Y}_2$.

Here is another frequent use of the convention. If the N observations in a random sample from a population are denoted $y_1, y_2, \ldots, y_N$, then we shall need to think of y_1 as just one realization of the random variable Y_1 for which the rule is: take a sample of size N from a population, and pick out the first observation. Similarly, $y_2, y_3, \ldots, y_N$ will each be thought of as realized values of the corresponding random variables $Y_2, Y_3, \ldots, Y_N$.

In our geometric approach we shall always view our sample as a vector $[y_1, y_2, \ldots, y_N]^T$. If $U = [a_1, a_2, \ldots, a_N]^T$ is a unit vector, then $y.U = a_1 y_1 + \cdots + a_N y_N$, the signed length of the projection of y onto the direction U, should be thought of as just one realized value of the random variable $Y.U = a_1 Y_1 + \cdots + a_N Y_N$.

Exercise for the reader

(3.1) Suppose that $[y_1, y_2, y_3, y_4]^T$ is a vector of observations in which y_1 and y_2 are drawn from one population, and y_3 and y_4 are drawn from a second population. Also, suppose that $U = [1, 1, -1, -1]^T/2$. Write out in words the rule for obtaining a real number for the random variable $Y.U$.

Probability Distributions

In any measuring process, values occur with different frequencies. The relative frequencies of these different values are conveniently summarized in the probability distribution. For discrete random variables the distribution can be summarized in a finite list of relative frequencies. For example, if the random variable is the number of heads in two tosses of an unbiased coin, the relative frequencies of 0, 1 and 2 heads are 1/4, 1/2 and 1/4 respectively. For a continuous random variable, Y, the probability distribution cannot be summarized in such a finite list, since there is an infinite number of possible values, with a zero probability of observing a particular value. Instead the distribution is best summarized using the density of probability around each value. This takes the form of the probability distribution, or probability density function, $f(y)$.

> The **probability distribution** of a random variable Y, $f(y)$, assigns probabilities to the realized values, y

Normal Distributions

In this book we always deal with a particular type of probability distribution called the normal, or Gaussian, distribution. This is the distribution which occurs most commonly in biological work. As an example, Figure 3.1 shows the relative frequency distribution of refractometer readings of 10,000 onion bulbs, together with the fitted normal distribution. The distribution is bell-shaped, with medium values occurring more often than low or high values. The probability density function is

$$f(y) = \frac{1}{\sqrt{2\pi}\sigma} e^{-\frac{1}{2}\left(\frac{y-\mu}{\sigma}\right)^2}$$

where μ and σ^2 are key parameters.

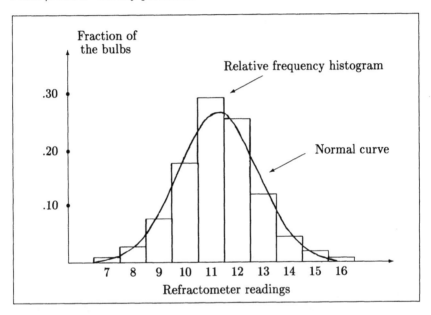

Figure 3.1: The relative frequency distribution of refractometer readings of 10,000 onion bulbs with the theoretical normal distribution curve. Data from "Agricultural Experimentation: Design and Analysis", by T.M. Little and F.J. Hills (1978), reproduced with kind permission from J. Wiley and Sons.

Mean and Variance

Normally distributed random variables are easy to describe since their distributions depend on just two parameters, each of which has a simple interpretation. The parameter μ measures the center, or balance point, of the distribution, whereas σ^2 measures the spread, or variability. More formally, μ is called the *mean* and σ^2 the *variance*. We now define these parameters for a general random variable Y.

The **mean** of a random variable is

$$\mu = \frac{\text{sum of all measurements}}{\text{number of measurements}}$$
$$= \sum \text{measurement} \times \text{relative frequency of measurement}$$
$$= \int y f(y) dy$$
$$= E(Y)$$

Here $E(Y)$ is an alternative expression for μ, which we read as the *expected value of Y*.

The **variance** of a random variable Y is

$$\sigma^2 = \frac{\text{sum of (measurement} - \text{mean)}^2}{\text{number of measurements}}$$
$$= \sum (\text{measurement} - \text{mean})^2 \times \text{relative frequency of measurement}$$
$$= \int (y - \mu)^2 f(y) dy$$
$$= Var Y$$

Note that $Var Y = E[(Y - \mu)^2]$. That is, the measure of spread is the average squared distance of a measurement from the expected value.

For our population of 10,000 bulbs displayed in Figure 3.1, μ and σ^2 have been calculated using these formulae. The normal curve has then been added to the figure by calculating $f(y)$ values using the formula given on the previous page.

Notation

We interpret the statement

$$Y \sim N[\mu, \sigma^2]$$

as "the random variable Y has a normal distribution with mean μ and variance σ^2".

Derived Populations

In some textbooks a distinction is made between the original, or parent, population, and derived populations. To illustrate this, consider a large flock of sheep depicted at the bottom left of Figure 3.2.

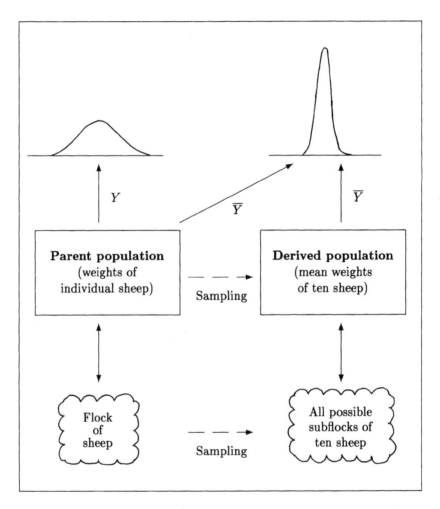

Figure 3.2: The relationship between the tangible population, the population of measurements, and the distribution of the measurements, for individual sheep and subflocks of ten sheep.

The set of weights of these sheep is the parent population. Consider also the mean random variable, $\overline{Y}$, for samples of size ten. This random variable can be viewed as either a direct map from the parent population to the set of real numbers, or as a composite map via the population of mean weights of ten sheep, as shown in Figure 3.2. This latter population is known as a derived population.

Note that the distinction between parent and derived populations is artificial, since the definitions depend on our starting point. For example, if we had started with the set of all possible subflocks of ten sheep, we would have referred to the population of mean weights as the parent population.

3.2 Combining Variables

Our main aim in this section is to determine the distribution of the projection length random variable, $Y.U = a_1Y_1 + \cdots + a_NY_N$. Towards this end we firstly define statistical independence, and write down the rules concerning linear combinations of random variables.

Independence

Broadly speaking, two random variables are statistically independent if the outcome of one has no influence on the outcome of the other. Mathematical aptitude, and time taken to sprint 100m are probably independent variables. On the other hand, human height and weight are dependent variables. Statistical independence is formally defined in terms of probabilities of events in a sample space, an area we have no wish to delve into here. Fortunately, for normally distributed random variables, independence is equivalent to zero *covariance*, a concept we can more easily express.

The covariance of W and X is defined as the expected value of the product of their deviations from the mean, that is,

$$Covariance(W, X) = Cov(W, X) = E[(W - E(W))(X - E(X))]$$

> A pair of normally distributed random variables W and X are **statistically independent** precisely when their covariance is zero

Suppose that we draw a random sample of size N, $y_1, \ldots, y_N$, from a $N[\mu, \sigma^2]$ population. As described in §3.1, we use $Y_1, \ldots, Y_N$ to denote the sample random variables. These have two properties

(i) each $Y_i \sim N[\mu, \sigma^2]$

(ii) any pair Y_i, Y_j are independent

We shall use these properties of $Y_1, \ldots, Y_N$ often in future.

Distributions of Sums and Multiples of Random Variables

Typically we shall calculate other quantities using these observations, $y_1, \ldots, y_N$. One such quantity is the sample mean, $\bar{y} = (y_1 + \cdots + y_N)/N$, a realized value of the random variable $\bar{Y} = (Y_1 + \cdots + Y_N)/N$. Such a combination random variable itself has a mean and variance. How do we compute such a mean and variance, knowing only the means and variances of the component random variables, $Y_1, \ldots, Y_N$? The rules we require are shown in the next frame.

Let $W = a_1 Y_1 + \cdots + a_N Y_N$, a combination of the random variables $Y_1, \ldots, Y_N$. Then

Rule 1: If $Y_1, \ldots, Y_N$ are normally distributed, then the random variable W is also normally distributed

Rule 2: The mean, or expected value, of W is given by

$$E(W) = a_1 E(Y_1) + \cdots + a_N E(Y_N)$$

Rule 3: The variance of W is given by

$$Var(W) = a_1^2\, Var(Y_1) + \cdots + a_N^2\, Var(Y_N)$$

provided $Y_1, \ldots, Y_N$ are independent random variables

These rules will not be proved in this text. The interested reader can consult a textbook on probability theory. To illustrate their use we now work out the distribution of the sample mean random variable, $\bar{Y} = (Y_1 + \cdots + Y_N)/N$. In this instance each a_i is just $1/N$, so Rule 1 assures us that $\bar{Y}$ follows a normal distribution. For the mean of $\bar{Y}$ we have

$$
\begin{aligned}
E(\bar{Y}) &= \frac{E(Y_1) + \cdots + E(Y_N)}{N} \quad \text{using Rule 2} \\
&= \frac{\mu + \cdots + \mu}{N} = \frac{N\mu}{N} = \mu
\end{aligned}
$$

while the variance of $\bar{Y}$ is:

$$
\begin{aligned}
Var(\bar{Y}) &= \frac{Var(Y_1) + \cdots + Var(Y_N)}{N^2} \quad \text{using Rule 3} \\
&= \frac{\sigma^2 + \cdots + \sigma^2}{N^2} = \frac{N\sigma^2}{N^2} = \frac{\sigma^2}{N}
\end{aligned}
$$

To summarize, we write

$$\bar{Y} \sim N[\mu, \sigma^2/N]$$

Distributions of Projection Coefficients

In our geometric approach we find ourselves primarily interested in the distributions of projection coefficients. These have the form

$$y.U = a_1 y_1 + \cdots + a_N y_N, \quad \text{where } U = \begin{bmatrix} a_1 \\ \vdots \\ a_N \end{bmatrix} \text{ is a unit vector.}$$

All such distributions arise as applications of the following general result.

If $Y_1, \ldots, Y_N$ are independent normal random variables with a common variance of σ^2 then $Y.U = a_1 Y_1 + \cdots + a_N Y_N$, where $a_1^2 + \cdots + a_N^2 = 1$, has a normal distribution with

$$E(Y.U) = a_1 E(Y_1) + \cdots + a_N E(Y_N), \quad Var(Y.U) = \sigma^2$$

The only surprising aspect of this result is that no matter which unit vector we choose, the variance of the projection coefficient is σ^2. Why? Here is the answer:

$$
\begin{aligned}
Var(Y.U) &= Var(a_1 Y_1 + \cdots + a_N Y_N) \\
&= a_1^2 Var(Y_1) + \cdots + a_N^2 Var(Y_N) \quad \text{using Rule 3} \\
&= a_1^2 \sigma^2 + \cdots + a_N^2 \sigma^2 \\
&= (a_1^2 + \cdots + a_N^2)\sigma^2 \\
&= \sigma^2 \quad \text{since } a_1^2 + \cdots + a_N^2 = 1
\end{aligned}
$$

This is a crucial result. In succeeding chapters we rely heavily on the fact that for $U_1, \ldots, U_N$ an orthogonal coordinate sytem for N-space, the random variables $Y.U_1, \ldots, Y.U_N$ have a common variance σ^2. On the other hand, the expected values of the random variables $Y.U_i$ will vary.

Example

For the example in §1.2 the appropriate orthogonal coordinate system is

$$U_1 = \frac{1}{\sqrt{2}} \begin{bmatrix} 1 \\ 1 \end{bmatrix} \quad \text{and} \quad U_2 = \frac{1}{\sqrt{2}} \begin{bmatrix} -1 \\ 1 \end{bmatrix}$$

Hence $Y.U_1 = (Y_1 + Y_2)/\sqrt{2}$ and $Y.U_2 = (Y_2 - Y_1)/\sqrt{2}$

Thus $E(Y.U_1) = (\mu + \mu)/\sqrt{2} = \sqrt{2}\mu$ while $E(Y.U_2) = (\mu - \mu)/\sqrt{2} = 0$

Also $Var(Y.U_1) = (Var(Y_1) + Var(Y_2))/2 = (\sigma^2 + \sigma^2)/2 = \sigma^2$, and similarly $Var(Y.U_2) = \sigma^2$. In summary,

$$Y.U_1 \sim N[\sqrt{2}\mu, \sigma^2] \quad \text{and} \quad Y.U_2 \sim N[0, \sigma^2]$$

Exercise for the reader

(3.2) To continue with reader exercise (3.1), suppose that the means of the two populations are μ_1 and μ_2. Also suppose that both populations are normally distributed, and have common variance σ^2. From first principles, find the mean and variance of the random variable $Y.U$.

Independence and Orthogonality

When are the random variables associated with two directions in space independent? Formally, under what conditions are the random variables $Y.U_c$ and $Y.U_d$ independent, where $U_c = [c_1, \ldots, c_N]^T$ and $U_d = [d_1, \ldots, d_N]^T$? We provide the answer as follows:

$$
\begin{aligned}
Cov(Y.U_c, Y.U_d) &= Cov(c_1 Y_1 + \ldots + c_N Y_N, d_1 Y_1 + \ldots + d_N Y_N) \\
&= \sum_{i=1}^{N}\sum_{j=1}^{N} c_i d_j \, Cov(Y_i, Y_j) \\
&= \sum_{i=1}^{N} c_i d_i \, Var(Y_i) \quad \text{since the } Y_i's \text{ are independent} \\
&= \left(\sum_{i=1}^{N} c_i d_i\right)\sigma^2 \quad \text{since for each } i, \ Var Y_i = \sigma^2 \\
&= (U_c.U_d)\sigma^2 \\
&= 0 \qquad \text{if and only if } U_c \text{ is orthogonal to } U_d
\end{aligned}
$$

> Random variables $Y.U_c$ and $Y.U_d$ are statistically independent if and only if the unit vectors U_c and U_d are orthogonal

This fact will be used later to establish a one-to-one correspondence between a set of independent hypothesis tests and the vectors in an orthogonal coordinate system.

3.3 Estimation

Estimators and Estimates

Our test statistics and confidence intervals will invariably be built from estimates of population parameters.

> An **unbiased estimator** of a population parameter is a random variable whose expected value is the parameter. For example, W is an unbiased estimator of μ if $E(W) = \mu$.
>
> A realized value of such a random variable is called an **unbiased estimate**.

In the case of an independently drawn sample $y_1, \ldots, y_N$ from a $N[\mu, \sigma^2]$ distribution, the usual estimate of the mean μ is the sample mean $\bar{y}$, with corresponding estimator $\overline{Y}$. In Section 3.2 we saw that $E(\overline{Y}) = \mu$, so that $\overline{Y}$ is unbiased. Note that there is a plentiful supply of alternative unbiased estimators of μ. For example, Y_1, the rule which picks out the first observation in the sample, is an estimator with $E(Y_1) = \mu$. Its variance, however, is σ^2, so that it is a more variable estimator than $\overline{Y}$, whose variance is σ^2/N. Similarly, $W = (Y_1 + 3Y_2)/4$ is an unbiased estimator of μ. Its variance is $((\frac{1}{4})^2 + (\frac{3}{4})^2)\sigma^2 = \frac{10}{16}\sigma^2$, so it is also more variable than $\overline{Y}$. In fact, if we examine all linear combinations

$$W = a_1 Y_1 + \cdots + a_N Y_N, \quad \text{where } a_1 + \cdots + a_N = 1$$

so ensuring that $E(W) = \mu$, we find that the weightings which minimize the variance of W are $a_1 = \ldots = a_N = 1/N$. That is,

$$\overline{Y} = \frac{Y_1 + \cdots + Y_N}{N}$$

is the *best linear unbiased estimator*, in that it has a lower variance than any of its competitors.

Exercise for the reader

(3.3) For the case of a sample of size $N = 5$, show that the random variable $W = (Y_1 + 2Y_2 + Y_3 + 3Y_4 + 4Y_5)/11$ is

(a) an unbiased estimator of μ,

(b) of higher variance than the estimator $\overline{Y}$.

Estimation of σ^2

Of critical importance in our work is the estimation of the common variance σ^2 of our N sample random variables. We describe now the method used to estimate σ^2. This method will underly every situation we treat.

In succeeding chapters the random variables $Y.U_1, \ldots, Y.U_N$ fall neatly into two categories: those associated with unit vectors $U_1, \ldots, U_p$ lying in what we shall call the **model space**, and those associated with unit vectors $U_{p+1}, \ldots, U_N$ in what we shall call the **error space**. Variables in the first category will, in general, have non-zero means, whereas variables in the second category will always have zero mean. These latter $N - p$ variables are used in the estimation of σ^2.

How are they used to estimate σ^2? For U_i in the error space,

$$
\begin{aligned}
Var(Y.U_i) &= E\left\{[Y.U_i - E(Y.U_i)]^2\right\} \quad \text{by the variance definition} \\
&= E\left[(Y.U_i)^2\right] \qquad\qquad \text{since } E(Y.U_i) = 0
\end{aligned}
$$

But $Var(Y.U_i)$ is also always equal to σ^2, from Section 3.2. Hence

$$
E\left[(Y.U_i)^2\right] = \sigma^2
$$

for each of the unit vectors in the error space. Therefore each random variable $(Y.U_i)^2$ is an unbiased estimator of σ^2.

Now for each random variable $Y.U_i$ we have only one realized value, $y.U_i$, so if we want to estimate σ^2 we can substitute this value, giving $(y.U_i)^2$. That is, $(y.U_{p+1})^2, \ldots, (y.U_N)^2$ all estimate σ^2. We pool these to produce the best estimate, namely

$$
s^2 = \frac{(y.U_{p+1})^2 + \cdots + (y.U_N)^2}{N - p}
$$

That is,

$$
S^2 = \frac{(Y.U_{p+1})^2 + \cdots + (Y.U_N)^2}{N - p}
$$

is the best unbiased estimator of σ^2.

Example

In the example of §1.2 both the model space and error space are one-dimensional. The model space is the equiangular line spanned by $U_1 = [1,1]^T/\sqrt{2}$ while the error space is the line perpendicular to this, spanned by $U_2 = [-1,1]^T/\sqrt{2}$.

The best estimate of σ^2 is therefore simply $s^2 = (y.U_2)^2 = (y_2 - y_1)^2/2$, the realized value of $(Y.U_2)^2$.

If the population mean μ is non-zero then $(Y.U_1)^2$ is biased as an estimator of σ^2, as the following calculation shows:

$$
\begin{aligned}
Var(Y.U_1) &= E\left\{[Y.U_1 - E(Y.U_1)]^2\right\} \\
&= E\left[(Y.U_1)^2\right] - [E(Y.U_1)]^2 \\
&= E\left[(Y.U_1)^2\right] - (\sqrt{2}\mu)^2 \\
&= \sigma^2 \qquad \text{since } Var(Y.U_1) = \sigma^2 \text{ from §3.2}
\end{aligned}
$$

Hence $\qquad\qquad E\left[(Y.U_1)^2\right] = \sigma^2 + 2\mu^2,$

so $(Y.U_1)^2$ is positively biased if used as an estimator of σ^2. Of course, if the population mean μ really is zero, then $(Y.U_1)^2$ is an unbiased estimator of σ^2 also.

3.4 Reference Distributions

The F Distributions

Our standard method for testing hypotheses will be to calculate ratios of averages of squared projection lengths, $(Y.U_i)^2$. In all cases these ratios will follow the distributions we now describe.

The random variable

$$F_{p,q} = \cfrac{\left[\dfrac{W_1^2 + \cdots + W_p^2}{p}\right]}{\left[\dfrac{W_{p+1}^2 + \cdots + W_{p+q}^2}{q}\right]}$$

where $W_1, \ldots, W_{p+q}$ are independent $N[0, \sigma^2]$ random variables, is called the F **statistic with p and q degrees of freedom.**

In the example at the end of §3.3 the ratio $(Y.U_1)^2/(Y.U_2)^2$ is an $F_{1,1}$ statistic if $\mu = 0$, since then $Y.U_1$ and $Y.U_2$ are both $N[0, \sigma^2]$ random variables.

The $F_{p,q}$ statistic can alternatively be defined in terms of $Z_1, \ldots, Z_{p+q}$, independently distributed $N[0, 1]$ variables. This follows because the σ^2 can be cancelled from the numerator and denominator of the ratio.

The statistic follows what is called the $F_{p,q}$ distribution: each pair (p, q) yields a distinct distribution. As we have introduced it, both the numerator and denominator can be thought of as unbiased estimators of σ^2. For low values of p and q, these estimators are relatively variable, causing the $F_{p,q}$ statistic to have relatively high variability. In the extreme case where $p = q = 1$, the $F_{1,1}$ statistic is so variable that its 95 percentile is 161 and its 99 percentile is 4052! As p and q grow larger, the numerator and denominator become less variable, with their realized values clustering closer and closer around σ^2; this causes the $F_{p,q}$ statistic to concentrate around the value of 1. This is illustrated in Figure 3.3.

Because of their key role in hypothesis testing, the two critical (95 and 99) percentiles of the $F_{p,q}$ distributions are tabulated in many statistical textbooks, Table T.2 in this volume. Realized values greater than these percentiles are considered *unusual* and *very unusual* respectively, and are grounds for rejecting a test hypothesis.

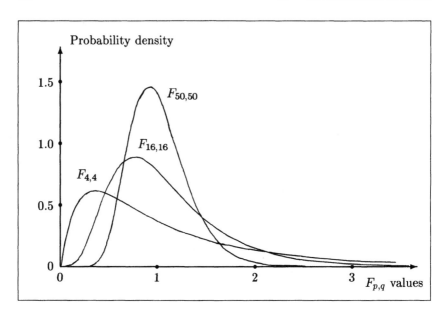

Figure 3.3: A selection of $F_{p,q}$ distributions.

The t Distributions

A special case of the $F_{p,q}$ distribution, when $p = 1$, occurs so frequently that a particularly convenient alternative is often used. It is especially convenient for the production of confidence intervals.

The random variable (or statistic)

$$t_q = \frac{W_1}{\sqrt{\dfrac{W_2^2 + \cdots + W_{q+1}^2}{q}}}$$

where $W_1, \ldots, W_{q+1}$ are independent $N[0, \sigma^2]$ random variables, is called the t **statistic with** q **degrees of freedom.**

Note that the t random variable breaks with convention, and is written in lower case, for the reason of common usage.

In our example, the ratio $(Y.U_1)/ \mid Y.U_2 \mid$ is a t_1 statistic if $\mu = 0$. Note that the ratio can be rewritten as $\sqrt{2}\bar{y}/s = \bar{y}/(s/\sqrt{2})$, the traditional expression.

Alternatively, the t_q statistic can be defined in terms of $Z_1, \ldots, Z_{q+1}$ which follow independent $N[0, 1]$ distributions. The distribution of the t_q statistic is called the t_q distribution. Note that $t_q^2 = F_{1,q}$.

Unlike the $F_{p,q}$ statistic, the t_q statistic can take on positive and negative values. The numerator follows a normal distribution, so t_q is symmetric about 0. The denominator is the square root of an estimator of σ^2. As the number of degrees of freedom (q) increases, the variability of this estimator decreases to zero, so the denominator more and more closely approximates the constant σ^2. The upshot is that as q increases, the t_q statistic increasingly approximates the standard normal random variable, as illustrated in Figure 3.4.

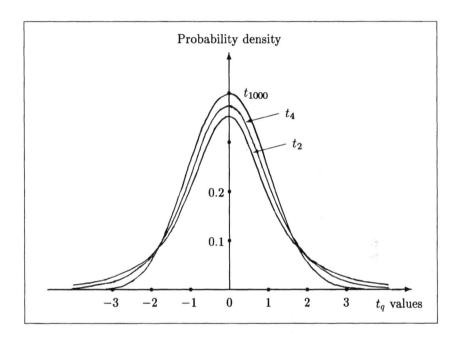

Figure 3.4: A selection of t_q distributions.

The middle 95% of values from a t_q distribution lie between the 2.5 and 97.5 percentiles: values outside this range are unusual. In the framework of hypothesis testing, unusual values of t_q lead to rejection of the test hypothesis. You will find that tables of percentiles give only the 97.5 value. In fact this suffices, since the 2.5 percentile is just minus this value. For example, the 97.5 percentile of the t_4 distribution is 2.776, so the 2.5 percentile is -2.776. Table T.2 tabulates the 97.5 and 99.5 percentiles for the t_q distributions, also known as the 5% and 1% *critical values*.

In later applications, where we typically set up a hypothesis in which a parameter is assumed to be zero, the t statistic can be written in the form

$$t = \frac{\text{estimate of the parameter}}{\text{standard error of the estimator}}$$

This simple format makes the t statistic very useful.

Solutions to the Reader Exercises

(3.1) A realized value of the random variable $Y.U$ is

$$y.U = \frac{1}{2}(y_1 + y_2 - y_3 - y_4) = \bar{y}_{1.} - \bar{y}_{2.}$$

where $\bar{y}_{1.}$ and $\bar{y}_{2.}$ denote the means of the observations from the first and second populations.

Hence the rule defining the random variable $Y.U$ is "take a random sample of size two from each of the two populations, and subtract the second sample mean from the first sample mean".

(3.2) We know $Y.U = \frac{1}{2}(Y_1 + Y_2 - Y_3 - Y_4)$. Hence the mean and variance of $Y.U$ are

$$
\begin{aligned}
E(Y.U) &= \frac{1}{2}\left[E(Y_1) + E(Y_2) - E(Y_3) - E(Y_4)\right] \\
&= \frac{1}{2}\left[\mu_1 + \mu_1 - \mu_2 - \mu_2\right] \\
&= (\mu_1 - \mu_2) \quad \text{and} \\
Var(Y.U) &= \frac{1}{4}\left[Var(Y_1) + Var(Y_2) + Var(Y_3) + Var(Y_4)\right] \\
&= \frac{1}{4}(\sigma^2 + \sigma^2 + \sigma^2 + \sigma^2) \\
&= \sigma^2
\end{aligned}
$$

(3.3) (a)

$$
\begin{aligned}
E(W) &= \frac{E(Y_1) + 2E(Y_2) + E(Y_3) + 3E(Y_4) + 4E(Y_5)}{11} \\
&= \frac{\mu + 2\mu + \mu + 3\mu + 4\mu}{11} \\
&= \mu
\end{aligned}
$$

Hence W is an unbiased estimator of μ.

(b) $Var(W)$ is given by

$$\left(\frac{1}{11}\right)^2 \left[Var(Y_1) + 4\,Var(Y_2) + Var(Y_3) + 9\,Var(Y_4) + 16\,Var(Y_5)\right]$$

$$= \left(\frac{1}{11}\right)^2 \left[\sigma^2 + 4\sigma^2 + \sigma^2 + 9\sigma^2 + 16\sigma^2\right] \quad = \quad 0.26\sigma^2$$

By comparison, $Var(\bar{Y}) = \sigma^2/5 = 0.20\sigma^2$. Hence the random variable W has a higher variance than the random variable $\bar{Y}$.

Chapter 4

Tool Kits At Work

Here we set the scene for the remainder of the book. We begin in §1 with an overview of the scientific method, or question – research – answer process, in which statistical data analysis plays an integral part. In §2 we home in upon the novel contribution of this text, namely a consistent geometric approach to data analysis. While this is a feature which sets this book apart from other texts, we emphasize that throughout we shall view data analysis as just a part of the design and analysis of each research study.

4.1 The Scientific Method

In each study that we carry out, we follow the *scientific method*. This is the following three stage process:

1. Having an idea, or formulating a hypothesis

2. Checking out the idea, or testing the hypothesis

3. Revising the idea, or modifying the hypothesis

Figure 4.1 illustrates the cyclic nature of this process, in which new answers generate further questions.

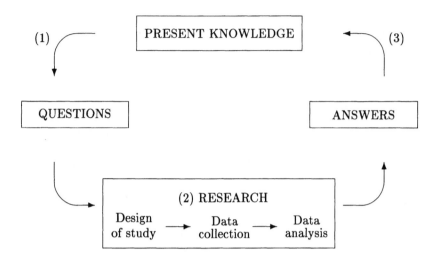

Figure 4.1: The research cycle, illustrating the way in which knowledge is continually updated by ongoing research.

We now fill out in more detail the steps of the scientific method, in the context of a well conducted research study:

1. **Having the idea**   An educational research officer poses the question "Do first year male statistics students spend the same number of hours per week working on statistics as do female students?"

2. **Checking out the idea**

   a) The first step in designing the study is to establish the *study populations*. In this example we settle on all first year male statistics students, and all first year female statistics students, at Canterbury University in Christchurch, New Zealand.

   b) The *method of data collection* is decided. For example, male and female students may be drawn randomly from a list of all Canterbury University first year statistics students. Alternatively, male and female twins may be sought out for the study.

   c) A *model* is assumed, specifying the distributions of the study populations. Firstly, the expected values of these populations are specified. These are in part suggested by the method of data collection. For example, if the students were chosen completely at random, the expected value for males would be just μ_M and for females μ_F. On the other hand, if the students sampled were twin pairs, the expected values would be $\mu_M + \beta_j$ and $\mu_F + \beta_j$, where μ_M and μ_F are as before and β_j represents the difference in work hours between the jth twin pair and an average twin pair. Secondly, populations are assumed to be normal, with a common variance of σ^2.

d) *Hypotheses of interest* are formulated in terms of the parameters of the model. These hypotheses follow from the questions of interest. In our example, the only hypothesis of interest is $H_0 : \mu_M = \mu_F$.

e) The *sample size* required to meet the objectives as efficiently as possible can now be calculated. This calculation requires knowledge of the minimum size of difference in which the researcher is interested, as well as an estimate of the underlying variability in the study.

f) *Data is collected*, with notes taken of any problems encountered.

g) The *analysis of the data* is performed as described in §4.2.

h) The analysis is subjected to *quality control*. This includes checking of data entry, examination of residuals and checking that the model is reasonable. The latter includes checking the appropriateness of the model of expected values, and the assumptions of independence, normality and constant variance.

i) Analyses may require *revision* if assumptions are violated. Certain populations may need to be excluded from the analysis, such as highly variable "control" groups, or a transformation to a more appropriate model may be required.

j) Last, but not least, answers or *conclusions* should be written out in everyday language.

3. **Revising the idea**   The study revealed that the females spend longer studying than do the males. The "present knowledge" of the education research officer has been increased.

4.2   Statistical Analysis

The Key Idea

Data analysis in this text will be performed using the geometric method. The
crux of this method is the following idea. In analysis of variance, regression
and analysis of covariance, *hypothesis testing is simply a comparison* of

the squared length of the projection of the observation vector
$y = [y_1, \ldots, y_N]^T$ onto the direction of the hypothesis

with

the average of the squared lengths of the projections of y onto
the orthogonal directions in the error space.

This is illustrated in Figure 4.2. The error space is the space of directions
in which the projection coefficient of y has zero mean.

In our tea drinking example of §1.2 the hypothesis test involves the comparison $A^2/B^2 = (A/B)^2$, and the error space is just the line perpendicular
to the equiangular line.

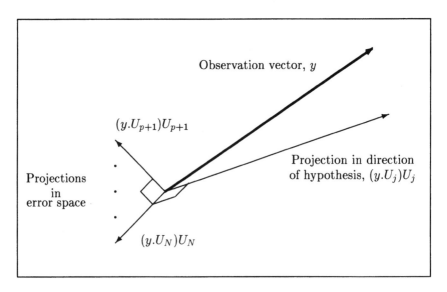

Figure 4.2: Illustration of hypothesis testing. A hypothesis is tested by dividing the squared length of the projection in the direction of the hypothesis,
by the average of the squared lengths of the projections in the error space.

If the hypothesis is false the squared length of the projection in the direction of the hypothesis will in general be considerably greater than the
average of the squared lengths of the projections onto the error space coordinate axes. If the hypothesis is true then the former will in general be

similar to the latter. This provides us with a method (not foolproof) of distinguishing truth from falsehood; if the ratio of the former to the latter is "large" we decide the hypothesis is false, and if the ratio is "small" we decide the hypothesis is true.

The Geometric Method

Routine application of the method involves isolating three objects and then using two processes. We present these objects and processes in summary form in Figure 4.3.

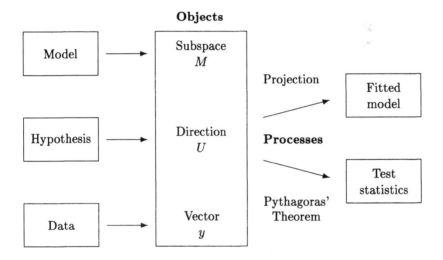

Figure 4.3: The three objects and two processes of the geometric method.

Let's explain this. Firstly, the model determines a subspace of N-space, $M = span\{U_1, \ldots, U_p\}$. This is the subspace consisting of all the possible *model vectors*, where a model vector is a vector whose components are the expected values of our observations. For example, suppose that in our study of male and female work hours we decided to sample two males and two females. We would record observations of y_1 and y_2 study hours per week for the males and y_3 and y_4 for the females. The expected values of the observations are μ_M, μ_M, μ_F and μ_F, provided that our samples are drawn in a purely random manner. So any vector of the form $[\mu_M, \mu_M, \mu_F, \mu_F]^T$ is a possible model vector. Such a model vector can always be written as

$$\begin{bmatrix} \mu_M \\ \mu_M \\ \mu_F \\ \mu_F \end{bmatrix} = \mu_M \begin{bmatrix} 1 \\ 1 \\ 0 \\ 0 \end{bmatrix} + \mu_F \begin{bmatrix} 0 \\ 0 \\ 1 \\ 1 \end{bmatrix}$$

which is a linear combination of two fixed vectors.

Hence the model space is $M = \mathrm{span} \left\{ \begin{bmatrix} 1 \\ 1 \\ 0 \\ 0 \end{bmatrix}, \begin{bmatrix} 0 \\ 0 \\ 1 \\ 1 \end{bmatrix} \right\}.$

This is a 2-dimensional subspace of 4-space, spanned by the unit vectors $U_1 = [1, 1, 0, 0]^T / \sqrt{2}$ and $U_2 = [0, 0, 1, 1]^T / \sqrt{2}$.

Second, associated with every hypothesis of interest is a *direction* lying within the model space M. Each hypothesis takes the form that some measurable quantity equals zero. The corresponding direction is then determined by the requirement that the projection of y onto this direction must average to a scalar multiple of the true value of the quantity. For example, in our study the only hypothesis of interest is $H_0 : \mu_M - \mu_F = 0$.

Hence the hypothesis direction is $U = \dfrac{1}{\sqrt{4}} \begin{bmatrix} 1 \\ 1 \\ -1 \\ -1 \end{bmatrix},$

since $y.U = \frac{1}{2}(y_1 + y_2 - y_3 - y_4) = \bar{y}_M - \bar{y}_F$, where $\bar{y}_M$ and $\bar{y}_F$ are the mean work hours of the males and females respectively, so the projection coefficient $Y.U$ averages to $\mu_M - \mu_F$. Note that the unit vector U is uniquely determined by the requirements that it must lie in the model space, and that $E(Y.U) = k(\mu_M - \mu_F)$, where k is a constant. In our example $k = 1$.

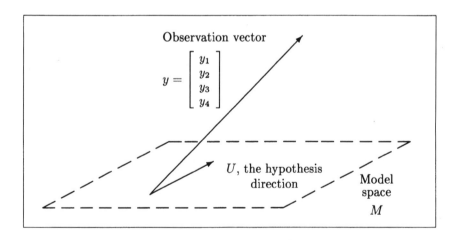

Figure 4.4: The three objects: the model space M, the direction associated with the hypothesis U, and the observation vector y.

The third and final object is our data, viewed in space as an *observation vector* $y = [y_1, \ldots, y_N]^T$. For our example,

the observation vector is $y = \begin{bmatrix} y_1 \\ y_2 \\ y_3 \\ y_4 \end{bmatrix}$.

Here y_1, y_2 and y_3, y_4 would be the work hours per week of the two males and two females respectively.

This concludes our description of the three essential objects, summarized in a picture in Figure 4.4.

We turn now to the two essential processes involved in data analysis. The first process is the fitting of the model by *projecting* y onto M, forming $(y.U_1)U_1 + \ldots + (y.U_p)U_p$, the fitted model. In our example the observation vector $y = [y_1, y_2, y_3, y_4]^T$ comes from a distribution centered on an unknown vector $[\mu_M, \mu_M, \mu_F, \mu_F]^T$ in the model space. A reasonable estimate of that model vector is the nearest vector to y in M, the projection of y on M. This is

$$(y.U_1)U_1 + (y.U_2)U_2 = \begin{bmatrix} \bar{y}_M \\ \bar{y}_M \\ 0 \\ 0 \end{bmatrix} + \begin{bmatrix} 0 \\ 0 \\ \bar{y}_F \\ \bar{y}_F \end{bmatrix} = \begin{bmatrix} \bar{y}_M \\ \bar{y}_M \\ \bar{y}_F \\ \bar{y}_F \end{bmatrix}$$

yielding estimates $\bar{y}_M$ and $\bar{y}_F$ of μ_M and μ_F respectively. This is illustrated in Figure 4.5.

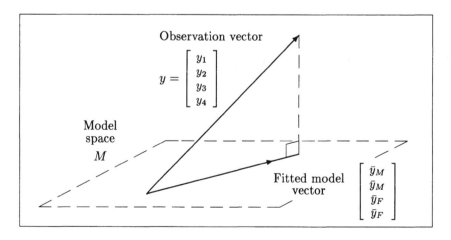

Figure 4.5: The first process: fitting the model by projecting the observation vector y onto the model space M.

The second process involves applying Pythagoras' Theorem and calculating appropriate test statistics. To see how this works in our example, we
write down an appropriate coordinate system for 4-space

$$U_1 = \frac{1}{\sqrt{4}}\begin{bmatrix} 1 \\ 1 \\ 1 \\ 1 \end{bmatrix}, \quad U_2 = \frac{1}{\sqrt{4}}\begin{bmatrix} 1 \\ 1 \\ -1 \\ -1 \end{bmatrix}, \quad U_3 = \frac{1}{\sqrt{2}}\begin{bmatrix} 1 \\ -1 \\ 0 \\ 0 \end{bmatrix}, \quad U_4 = \frac{1}{\sqrt{2}}\begin{bmatrix} 0 \\ 0 \\ 1 \\ -1 \end{bmatrix}$$

Here U_1 and U_2 are new unit vectors spanning the model space M, chosen
to include the direction associated with the hypothesis, $U_2 = U$. The unit
vectors U_3 and U_4 are any pair which complete an orthogonal coordinate
system for 4-space. The space they span is called the *error space*. The
corresponding Pythagorean breakup is

$$\|y\|^2 = (y.U_1)^2 + (y.U_2)^2 + (y.U_3)^2 + (y.U_4)^2$$

or, $y_1^2 + y_2^2 + y_3^2 + y_4^2$

$$= 4\bar{y}^2 + (\bar{y}_M - \bar{y}_F)^2 + (y_1 - y_2)^2/2 + (y_3 - y_4)^2/2$$

as shown in Figure 4.6.

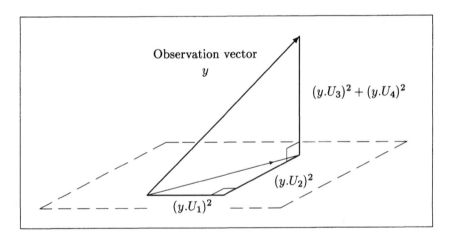

Figure 4.6: The second process: testing a hypothesis by comparing the
squared length of the projection of y onto the direction of the hypothesis,
$(y.U_2)^2$, with the average of the squared lengths of the projections onto the
error space directions, $[(y.U_3)^2 + (y.U_4)^2]/2$.

To test our hypothesis that $\mu_M = \mu_F$ we divide the squared length of
the projection in the direction of the hypothesis, $(y.U_2)^2$, by the average of
the squared lengths of the projections onto the error space coordinate axes.

The resulting test statistic is

$$F = \frac{(y.U_2)^2}{\left[(y.U_3)^2 + (y.U_4)^2\right]/2} = \frac{(\bar{y}_M - \bar{y}_F)^2}{\left[(y_1 - y_2)^2 + (y_3 - y_4)^2\right]/4}$$

If our hypothesis that $\mu_M = \mu_F$ is true, then the top and bottom lines will in general be roughly comparable, whereas if our hypothesis is false the top line is enlarged. This provides us with a test to distinguish the two situations. Specifically, if $\mu_M = \mu_F$, the ratio is a realized value of the $F_{1,2}$ random variable; we therefore reject our hypothesis H_0 if the observed value is outside the usual range of values of the $F_{1,2}$ distribution.

In summary, *three objects*, M from the assumed model, directions U from the hypotheses, and y from the data, together with *two processes*, projection and application of Pythagoras' Theorem, are sufficient for the statistical analysis of a wide variety of data sets. This pattern of analysis will be followed throughout, so should be firmly embedded in the reader's mind by the end of the book. For further simple examples, however, see Saville and Wood (1986).

Exercises

(4.1) Suppose y in Exercise 2.13 is the vector of observations from an experiment in which four randomly selected sheep were fed alfalfa for a month. The observations are the changes in the level of a certain hormone.

We wish to test the hypothesis of no change in the hormone levels in the population of sheep, $H_0 : \mu = 0$. Our model space M is the span of $U_1 = [1, 1, 1, 1]^T/\sqrt{4}$, where U_1 is also the direction associated with the hypothesis $H_0 : \mu = 0$.

(a) Calculate the average of the last three squared projection lengths, $(\|P_2y\|^2 + \|P_3y\|^2 + \|P_4y\|^2)/3$, using your answer from 2.13(d). This is an unbiased estimate of σ^2.

(b) Calculate the F ratio $= \dfrac{\|P_1y\|^2}{(\|P_2y\|^2 + \|P_3y\|^2 + \|P_4y\|^2)/3}$

(c) If H_0 is true the calculated F ratio follows the $F_{1,3}$ distribution. Look up the critical values for this distribution and decide whether to accept or reject H_0. If H_0 is rejected, how strong is the evidence?

(4.2) Using the same observation vector y from Exercise 2.13, suppose that the first two sheep were fed Moapa cultivar of alfalfa, and the last two sheep Caliverde cultivar of alfalfa. There are now two observations from each of two populations. Our model space M is now the span of U_1 and U_2, where U_1 is the direction associated with the hypothesis $H_0 : \mu = 0$, and U_2 is the direction associated with the hypothesis $H_0 : \mu_M = \mu_C$.

(a) Recalculate the estimate of σ^2 as $(\|P_3 y\|^2 + \|P_4 y\|^2)/2$.

(b) Test the hypothesis $H_0 : \mu = 0$, where $\mu = (\mu_M + \mu_C)/2$, using

$$F = \frac{\|P_1 y\|^2}{(\|P_3 y\|^2 + \|P_4 y\|^2)/2}$$

Note that if H_0 is true, the calculated F ratio follows the $F_{1,2}$ distribution.

(c) Test the hypothesis $H_0 : \mu_M = \mu_C$, using

$$F = \frac{\|P_2 y\|^2}{(\|P_3 y\|^2 + \|P_4 y\|^2)/2}$$

If this hypothesis is true the calculated F ratio follows the $F_{1,2}$ distribution.

Part II
Introduction to
Analysis of Variance

This part of the book addresses questions concerning the means of populations. In Chapter 5 we answer questions about the mean of a single population. Problems about two populations are dealt with in Chapter 6, while in Chapter 7 we deal with the general case of several populations. In the process we begin the development of traditional *analysis of variance* techniques.

Chapter 5

Single Population Questions

What is the mean of the single population which we are studying? Is it reasonable that the mean is zero? These are the types of question we deal with in this chapter.

We introduce the main ideas carefully in §1, using a study which has been reduced in size to simplify our exposition. In §2 we describe how these questions are answered in general, and relate the approach of this chapter to the intuitive result we discovered in §1.2. In §3 we point to the desirable properties of our answers, and in §4 we offer an easy-reading summary.

A class exercise, included at the end of the chapter, is designed to serve as an introduction to the chapter. In the authors' experience class time is as well spent on such a class exercise as on a careful exposition of theory.

5.1 An Illustrative Example

In this section we introduce the fundamental methods for dealing with single populations. We do so using a study which looks at the effect of a fungicide seed treatment on the yield of wheat.

Objectives

The study objectives are as follows:

1. To estimate the average response of wheat to fungicide seed treatment, in Methven County, Canterbury, New Zealand, in the spring of 1986.

2. To test whether the average response could be zero.

Population of Interest

The population we study is a population of differences. Here we look at the differences in wheat yield between fungicide treated wheat and untreated wheat, called *responses* in the world of agriculture. Of interest is the population of all possible responses which could be obtained by treating seed on farms throughout Methven County.

Design of Study

Within the Methven wheat growing district three farms are chosen at random for the study. On each farm two similar adjacent areas are selected and the two treatments

1. Seed not treated (control)

2. Seed treated with fungicide

allocated at random to the two plots as shown in Figure 5.1. The plan at harvest time is to use a combine harvester to clear the guard area and to harvest the harvest area in the centre of each plot, hence avoiding the possibility of edge effects. On each farm the resulting grain yields are then differenced to give a response to fungicide treatment. Our observation vector will be provided by a sample of three such responses, $y = [y_1, y_2, y_3]^T$.

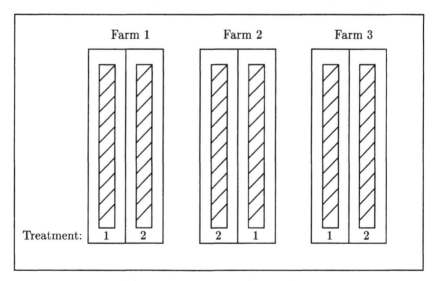

Figure 5.1: Layout of the six experimental plots. In each plot the harvest area is shaded and the guard area is unshaded.

Model

We assume our population of responses is normally distributed with unknown mean μ and unknown variance σ^2. Thus the average value of each observation, over many repetitions of the study, is μ. Model vectors therefore have the form $[\mu, \mu, \mu]^T$, so that the model space M is the one-dimensional subspace

$$M = \text{span}\left\{ \begin{bmatrix} 1 \\ 1 \\ 1 \end{bmatrix} \right\}$$

We can then write our observation vector y as

$$\begin{bmatrix} y_1 \\ y_2 \\ y_3 \end{bmatrix} = \mu \begin{bmatrix} 1 \\ 1 \\ 1 \end{bmatrix} + \begin{bmatrix} e_1 \\ e_2 \\ e_3 \end{bmatrix}$$

where $e_i = y_i - \mu$ for each $i = 1, 2, 3$. That is, the model demands that we express y as a scalar multiple of $[1, 1, 1]^T$, together with an error vector.

Hypothesis

One of our study objectives is to test whether the true response to fungicide is zero. Formally, we wish to test the null hypothesis $H_0 : \mu = 0$ against the alternative hypothesis $H_1 : \mu \neq 0$. For this, we need to find the direction within the model space which is associated with the hypothesis $H_0 : \mu = 0$. Here the general procedure, as outlined in §4.2, is to look for a direction U lying in the model space, such that the length of the projection of y onto U, given by $y.U$, has an expected value of $k\mu$, where k is a constant.

In this simple case there is only one direction in the model space M. We take $U_1 = [1, 1, 1]^T/\sqrt{3}$ as the direction associated with the hypothesis. To check that this is correct we calculate

$$y.U_1 = \begin{bmatrix} y_1 \\ y_2 \\ y_3 \end{bmatrix} \cdot \frac{1}{\sqrt{3}} \begin{bmatrix} 1 \\ 1 \\ 1 \end{bmatrix} = \frac{1}{\sqrt{3}}(y_1 + y_2 + y_3) = \frac{3\bar{y}}{\sqrt{3}} = \sqrt{3}\,\bar{y}$$

where $\bar{y}$ denotes the sample mean. It has the property that we require, since its expected value is $\sqrt{3}\mu$. If $\mu = 0$ the projection coefficient $y.U_1$ will be relatively small, averaging to zero over many repetitions of the study, whereas if $\mu \neq 0$ the projection coefficient $y.U_1$ will be relatively large, averaging to the nonzero quantity, $\sqrt{3}\mu$ (see Figure 5.2). This knowledge provides us with a method of distinguishing between the cases $\mu = 0$ and $\mu \neq 0$.

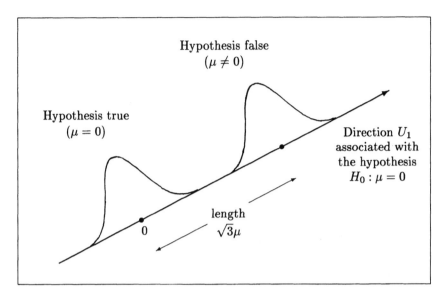

Figure 5.2: Distribution of projection coefficient $Y.U_1$ when hypothesis is true ($\mu = 0$) and when the hypothesis is false ($\mu \neq 0$). If $\mu = 0$ the realized value $y.U_1$ will be relatively small, whereas if $\mu \neq 0$ the realized value $y.U_1$ will be relatively large.

Data

The harvest area was measured in each case and the weight of harvested grain scaled so that the data was in tonnes/ha. The resulting data is shown in Table 5.1.   Our observation vector is thus $y = [1.1, 1.4, 1.1]^T$.

	Farm 1	Farm 2	Farm 3
Control	5.0	4.3	5.9
Treated	6.1	5.7	7.0
Response	1.1	1.4	1.1

Table 5.1: Grain yields in tonnes per hectare.

The collection of our three objects is now complete. The model space is $M = \text{span}\{[1, 1, 1]^T\}$, the direction associated with the hypothesis $H_0 : \mu = 0$ is $U_1 = [1, 1, 1]^T/\sqrt{3}$ and the observation vector is $y = [1.1, 1.4, 1.1]^T$.

Fitting the Model

The basic model assumed here tells us that each observation has the form $y_i = \mu + e_i$, where e_1, e_2 and e_3 are independent $N[0, \sigma^2]$ values. So as mentioned in our discussion of the model, we must express y as a point in the model space, together with an error vector. How can we best approximate this desired decomposition of y, or in the usual phrasing, how do we fit the model? The answer is that we choose $\hat{\mu}$, our estimate of μ, so that $\hat{\mu}[1,1,1]^T$ is the model vector closest to our observation vector y.

How do we find this nearest point? It is obtained by projecting the observation vector y onto the model space $M = \text{span}\left\{[1,1,1]^T\right\}$, yielding

$$(y.U_1)U_1 = \left(\begin{bmatrix} 1.1 \\ 1.4 \\ 1.1 \end{bmatrix} \cdot \frac{1}{\sqrt{3}}\begin{bmatrix} 1 \\ 1 \\ 1 \end{bmatrix}\right)\frac{1}{\sqrt{3}}\begin{bmatrix} 1 \\ 1 \\ 1 \end{bmatrix} = 1.2\begin{bmatrix} 1 \\ 1 \\ 1 \end{bmatrix}$$

Hence $\hat{\mu} = \bar{y} = 1.2$, the sample mean, is our estimate of μ. We have met our first study objective: the average response to fungicide seed treatment in Methven County in the spring of 1986 is estimated to be 1.2 tonnes per hectare.

We have also arrived at the fitted model

$$\begin{bmatrix} 1.1 \\ 1.4 \\ 1.1 \end{bmatrix} = 1.2\begin{bmatrix} 1 \\ 1 \\ 1 \end{bmatrix} + \begin{bmatrix} -.1 \\ .2 \\ -.1 \end{bmatrix}$$

or more generally

$$\underset{\substack{\text{observation} \\ \text{vector}}}{\begin{bmatrix} y_1 \\ y_2 \\ y_3 \end{bmatrix}} = \underset{\substack{\text{fitted model} \\ \text{vector}}}{\bar{y}\begin{bmatrix} 1 \\ 1 \\ 1 \end{bmatrix}} + \underset{\substack{\text{fitted error} \\ \text{vector}}}{\begin{bmatrix} y_1 - \bar{y} \\ y_2 - \bar{y} \\ y_3 - \bar{y} \end{bmatrix}}$$

which we abbreviate to $y = \bar{y} + (y - \bar{y})$. Here $\bar{y}$, the fitted model vector, is also called the *mean vector*.

To summarize, we have broken y into two orthogonal components, a fitted model vector and an error vector, as shown in Figure 5.3. Note that our fitted error vector is the shortest possible such vector: it is the vector $y - \hat{\mu}$ for which $\|y - \hat{\mu}\|^2 = (y_1 - \hat{\mu})^2 + (y_2 - \hat{\mu})^2 + (y_3 - \hat{\mu})^2$ is minimized. For this reason the process is termed the method of *least squares*.

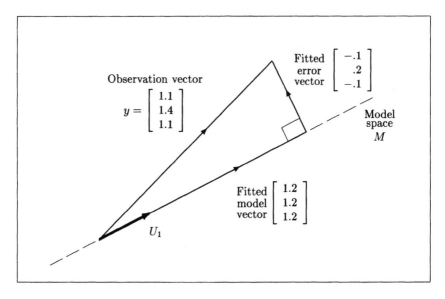

Figure 5.3: The least squares fit of the model, achieved by projecting the observation vector y onto the model space M.

Testing the Hypothesis

We now investigate the second study objective: Is the true mean response to fungicide zero? We have formalized this to testing the null hypothesis $H_0 : \mu = 0$ against the alternative hypothesis, $H_1 : \mu \neq 0$. This involves checking whether the squared projection coefficient $(y.U_1)^2$ is similar to, or considerably greater than, the average of the squares of the error space projection coefficients. To do this we need to write down axes U_2 and U_3 for the error space. A simple choice of such axes yields the orthogonal coordinate system

$$U_1 = \frac{1}{\sqrt{3}}\begin{bmatrix} 1 \\ 1 \\ 1 \end{bmatrix}, \quad U_2 = \frac{1}{\sqrt{2}}\begin{bmatrix} 1 \\ -1 \\ 0 \end{bmatrix}, \quad U_3 = \frac{1}{\sqrt{6}}\begin{bmatrix} 1 \\ 1 \\ -2 \end{bmatrix}$$

The corresponding orthogonal decomposition of the observation vector is

$$\begin{aligned} y &= (y.U_1)U_1 & + & (y.U_2)U_2 & + & (y.U_3)U_3 \\ &= (3.6/\sqrt{3})U_1 & - & (0.3/\sqrt{2})U_2 & + & (0.3/\sqrt{6})U_3 \end{aligned}$$

as illustrated in Figure 5.4.

The associated Pythagorean breakup is

$$\|y\|^2 = (y.U_1)^2 + (y.U_2)^2 + (y.U_3)^2$$

or, $4.38 = 4.32 + .045 + .015$

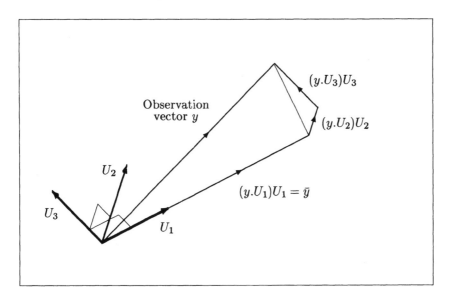

Figure 5.4: Orthogonal decomposition of the observation vector, y, in terms of the projection vectors $(y.U_i)U_i$.

To test the hypothesis $H_0 : \mu = 0$ we simply compare the squared projection length $(y.U_1)^2$ with the average of the squared lengths in the error space. The test statistic is therefore

$$F = \frac{(y.U_1)^2}{[(y.U_2)^2 + (y.U_3)^2]/2} = \frac{4.32}{(.045 + .015)/2} = 144$$

Here $Y.U_1$ is a $N(\sqrt{3}\mu, \sigma^2)$ random variable, while $Y.U_2$ and $Y.U_3$ are independent $N(0, \sigma^2)$ random variables, from §3.2. If $\mu = 0$, the test statistic follows the $F_{1,2}$ distribution described in §3.4; if $\mu \neq 0$ the test statistic will be inflated. Our test procedure is therefore as follows:

1. If our test statistic is within the usual range of values for the $F_{1,2}$ statistic, accept that the hypothesis $H_0 : \mu = 0$ may be true.

2. If the test statistic is outside the usual range of values for the $F_{1,2}$ statistic, reject the hypothesis $H_0 : \mu = 0$.

For a 5% level test the acceptable range is zero up to the 95 percentile of $F_{1,2}$; that is, the interval from 0 to 18.5. For a 1% level test the acceptable range is zero up to the 99 percentile; that is, the interval from 0 to 98.5. These percentiles are obtained from Table T.2, and are displayed in Figure 5.5.

Our test value of 144 is outside the acceptable range for both the 5% and 1% level tests. We therefore reject the hypothesis $H_0 : \mu = 0$ at the 1% level

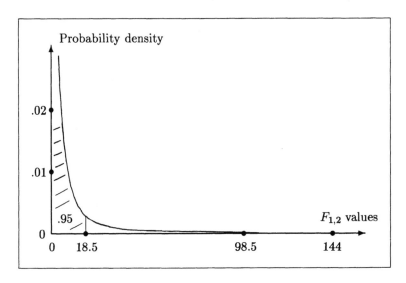

Figure 5.5: The $F_{1,2}$ distribution, showing our test value of 144 in relation to the 95 and 99 percentiles (18.5 and 98.5, respectively).

of significance. We conclude that on average over the entire population, seed treatment with fungicide increased the yield of wheat.

Exercises for the reader

(5.1) Calculate the three projection vectors $(y.U_1)U_1$, $(y.U_2)U_2$ and $(y.U_3)U_3$. Check your calculations by substituting into the orthogonal decomposition

$$y = (y.U_1)U_1 + (y.U_2)U_2 + (y.U_3)U_3$$

(5.2) Confirm that $E(Y.U_1) = \sqrt{3}\mu$, $E(Y.U_2) = 0$, $E(Y.U_3) = 0$ and that $Var(Y.U_1) = Var(Y.U_2) = Var(Y.U_3) = \sigma^2$, by expanding each random variable as a linear combination of Y_1, Y_2 and Y_3 and using the rules given in §3.2.

(5.3) Suppose that U_2 and U_3 had been chosen differently, to provide the new coordinate system

$$U_1 = \frac{1}{\sqrt{3}} \begin{bmatrix} 1 \\ 1 \\ 1 \end{bmatrix}, \quad U_2 = \frac{1}{\sqrt{14}} \begin{bmatrix} -2 \\ -1 \\ 3 \end{bmatrix}, \quad U_3 = \frac{1}{\sqrt{42}} \begin{bmatrix} 4 \\ -5 \\ 1 \end{bmatrix}$$

Confirm that the sum $(y.U_2)^2 + (y.U_3)^2$ remains 0.06.

ANOVA Table

The orthogonal breakup of $\|y\|^2$ using Pythagoras' theorem, and the subsequent calculation of our test statistic, are traditionally summarized in an analysis of variance (ANOVA) table as shown in Table 5.2.

Source of Variation	Degrees of freedom (df)	Sum of Squares (SS)	Mean Square (MS)	F
Subspace	Dimension of subspace	Sum of squared projection lengths	Mean squared length	Ratio of mean squares
Mean	1	4.32	4.32	144(**)
Error	2	0.06	0.03	
Total	3	4.38		

Table 5.2: Traditional ANOVA table for our single population example, together with the geometric meanings of the traditional terms.

Study this in conjunction with Figure 5.4, and you realize that it is a tabular summary of that picture. In this first appearance of an ANOVA table we have given the traditional expressions together with their geometric interpretations (for example, Source of Variation, and Subspace). In later tables we shall just use the traditional expressions or their abbreviations.

Test Statistic in Simpler Form

In Figures 5.3 and 5.4 we can see that $(y.U_1)^2 = \|\bar{y}\|^2$ and that $(y.U_2)^2 + (y.U_3)^2 = \|y - \bar{y}\|^2$, so we can rewrite our Pythagorean breakup as $\|y\|^2 = \|\bar{y}\|^2 + \|y - \bar{y}\|^2$ and our test statistic as

$$F = \frac{\|\bar{y}\|^2}{\|y - \bar{y}\|^2 / 2} = \frac{4.32}{.06/2} = 144$$

In fact, the usual method of analysis is to use these simplified forms. Thus for routine analysis we do not need to write down coordinate axes for the error space; in fact, all that we need to know is the number of such axes, two in this example.

Estimation of σ^2

In our example we have transformed our original set of independent, identically distributed random variables Y_1, Y_2, $Y_3 \sim N(\mu, \sigma^2)$, into a new set of independent random variables,

$$Y.U_1 \sim N(\sqrt{3}\mu, \sigma^2), \quad Y.U_2 \sim N(0, \sigma^2), \quad Y.U_3 \sim N(0, \sigma^2)$$

The first of these was used to estimate μ, while the last two were used to estimate σ^2 via

$$s^2 = \frac{(y.U_2)^2 + (y.U_3)^2}{2} = \frac{\|y - \bar{y}\|^2}{2} = \frac{0.06}{2} = 0.03$$

Here $y.U_2$ and $y.U_3$ are both observations from $N(0, \sigma^2)$ distributions, so their squares will both average to σ^2 over many repeats of the experiment, as described in §3.3. This variance estimate, s^2, is given the label, Error MS, in the ANOVA table, Table 5.2.

Checking of Assumptions

Do the assumptions of normality and independence seem reasonable, in retrospect? In the last section we saw that the deviations of the observations from the fitted values were -0.1, 0.2 and -0.1. With only three values we can do little to check our assumptions. However, we have no reason to believe that our normality or independence assumptions are unreasonable.

Confidence Interval for μ

We return now to our first study objective. We wanted to estimate the average response, μ, of wheat to fungicide seed treatment, for all wheat crops sown in Methven County in the spring of 1986. We have derived a best estimate, $\bar{y} = 1.2$ tonnes/ha, but we would also like to give an indication of the accuracy of this estimate. This is usually achieved by working out a *confidence interval* for μ. For this we make use of the t statistic defined in §3.4.

In building a t statistic we use random variables which are independently distributed as $N(0, \sigma^2)$. Now $Y.U_1 = \sqrt{3}\bar{Y}$ is distributed as $N(\sqrt{3}\mu, \sigma^2)$. By subtracting its mean value we obtain a random variable $W_1 = \sqrt{3}(\bar{Y} - \mu)$ which is distributed as $N(0, \sigma^2)$. Also, the random variables $W_2 = Y.U_2$ and $W_3 = Y.U_3$ are distributed as $N(0, \sigma^2)$. Hence the ratio

$$\frac{\sqrt{3}(\bar{Y} - \mu)}{\sqrt{[(Y.U_2)^2 + (Y.U_3)^2]/2}}$$

is a t_2 statistic. To obtain a 95% confidence interval for μ we simply gamble that the realized value,

$$t = \frac{\sqrt{3}(\bar{y} - \mu)}{\sqrt{[(y.U_2)^2 + (y.U_3)^2]/2}} = \frac{\sqrt{3}(\bar{y} - \mu)}{\sqrt{s^2}} = \frac{\bar{y} - \mu}{\sqrt{s^2/3}}$$

lies between the 2.5 and 97.5 percentiles of the t_2 distribution. That is, our 95% confidence interval is given by

$$t_2(.025) \leq \frac{\bar{y} - \mu}{\sqrt{s^2/3}} \leq t_2(.975)$$

$$\text{or} \quad -4.303 \leq \frac{1.2 - \mu}{\sqrt{0.03/3}} \leq 4.303$$

where $\bar{y} = 1.2$ and 4.303 is the 97.5 percentile of the t_2 distribution obtained from Table T.2. This set of inequalities converts to

$$
\begin{array}{ccccc}
& -0.43 & \leq & 1.2 - \mu \leq & 0.43 \\
\text{that is} & 1.2 - 0.43 \leq & & \mu & \leq \quad 1.2 + 0.43 \\
\text{or} & 0.77 & \leq & \mu & \leq \quad 1.63
\end{array}
$$

In summary we can state, with 95% confidence that we are correct, that the average response, μ, to fungicide seed treatment is between 0.77 and 1.63 tonnes/ha.

Report on Study

For a small study such as this, we would normally present the data, namely responses of 1.1, 1.4 and 1.1 tonnes per hectare on three farms. We could then say we were 95% confident that the average increase in yield due to fungicide seed treatment was between 0.77 and 1.63 tonnes per hectare for wheat sown in Methven County in the spring of 1986. Lastly we could say that there was little chance that the three responses were simply random fluctuations about a true average response of zero.

Farmers in Methven County might well be dissatisfied with such a report since it is based on an analysis of yield increases instead of profit increases. This dissatisfaction could easily be overcome by calculating for each farm the increase in profit due to fungicide seed treatment, then repeating the analysis using the three new data values. This type of economic analysis is discussed further in Saville (1983).

5.2   General Case

Here we summarize the essential steps in the analysis of our illustrative example and generalize these to a sample of arbitrary size. We then relate our geometric results to those given in more traditional texts and link our F test to the t test developed in §1.2.

Setting the Scene

Our interest is in a single population, assumed normally distributed, whose mean μ and variance σ^2 are both unknown. We would like to estimate μ and σ^2, test the hypothesis $H_0 : \mu = 0$, and derive a 95% confidence interval for μ. To this end, we take a random sample $y_1, y_2, \ldots, y_n$ from the population. These observations are values from the n independent sample random variables $Y_1, Y_2, \ldots, Y_n$, all having a $N(\mu, \sigma^2)$ distribution.

Note that we now use the notation n for the sample size, instead of N as in Part I. This is for consistency with subsequent chapters where we shall have several samples of size n which lead to a total sample of size N.

Raw Materials

The model vector is $[\mu, \mu, \ldots, \mu]^T = \mu[1, 1, \ldots, 1]^T$ so the *model space* is $M = \text{span}\{[1, 1, \ldots, 1]^T\}$. This has a single coordinate axis determined by the unit vector $U_1 = [1, 1, \ldots, 1]^T / \sqrt{n}$.

The *direction associated with the hypothesis* $H_0 : \mu = 0$ is $U_1 = [1, 1, \ldots, 1]^T / \sqrt{n}$. To affirm this we calculate the (signed) length of the projection of y onto U_1

$$y.U_1 = \begin{bmatrix} y_1 \\ \vdots \\ y_n \end{bmatrix} \cdot \frac{1}{\sqrt{n}} \begin{bmatrix} 1 \\ \vdots \\ 1 \end{bmatrix} = \frac{1}{\sqrt{n}}(y_1 + \cdots + y_n) = \frac{n\bar{y}}{\sqrt{n}} = \sqrt{n}\bar{y}$$

The corresponding random variable $Y.U_1$ therefore has an expected value of $\sqrt{n}\mu$, a constant multiple of μ. Thus if $\mu = 0$ the projection coefficient $y.U_1$ will be relatively small, whereas if $\mu \neq 0$ the projection coefficient will be relatively large.

The *observation vector* y is just the vector of observations, $[y_1, y_2, \ldots, y_n]^T$.

This completes the collection of all our raw materials: a model space $M = \text{span}\{[1, 1, \ldots, 1]^T\}$, the direction $U_1 = [1, 1, \ldots, 1]^T / \sqrt{n}$ associated with the hypothesis $H_0 : \mu = 0$, and an observation vector y.

Fitting the Model

To fit the model, we project the observation vector y onto the model space M. This yields the fitted model vector $\bar{y} = [\bar{y}, \bar{y}, \ldots, \bar{y}]^T$, the vector in the model space which is closest to our observation vector. Here $\bar{y}$ is termed the *least squares* estimate of the unknown true mean, μ. The associated orthogonal decomposition of the observation vector is

$$\begin{bmatrix} y_1 \\ \vdots \\ y_n \end{bmatrix} = \begin{bmatrix} \bar{y} \\ \vdots \\ \bar{y} \end{bmatrix} + \begin{bmatrix} y_1 - \bar{y} \\ \vdots \\ y_n - \bar{y} \end{bmatrix}$$

or more briefly $y \quad = \quad \bar{y} \quad + \quad (y - \bar{y})$

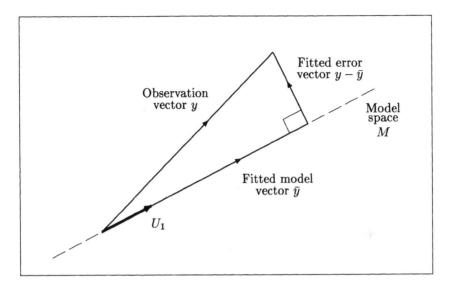

Figure 5.6: The orthogonal decomposition of the observation vector for a single population.

as illustrated in Figure 5.6.

Compare this with Figure 5.3, and note that our analysis still involves a triangular decomposition of the observation vector. The only change is that the error vector now sits in an $n-1$ dimensional error space. One choice of coordinate system for this space is

$$U_2 = \frac{1}{\sqrt{2}} \begin{bmatrix} 1 \\ -1 \\ 0 \\ 0 \\ \vdots \\ 0 \end{bmatrix}, \quad U_3 = \frac{1}{\sqrt{6}} \begin{bmatrix} 1 \\ 1 \\ -2 \\ 0 \\ \vdots \\ 0 \end{bmatrix}, \quad \cdots \quad U_n = \frac{1}{\sqrt{n(n-1)}} \begin{bmatrix} 1 \\ 1 \\ 1 \\ 1 \\ \vdots \\ -(n-1) \end{bmatrix}$$

See the Class Exercise at the end of the chapter for an explanation of how this system was chosen.

Together with $U_1 = [1, 1, \ldots, 1]^T$, $U_2, \ldots, U_n$ make up a coordinate system for n-space which is appropriate for the single population case. Note that just as the fitted model vector, $\bar{y}$, can be written in the form $(y.U_1)U_1$, so too can the fitted error vector, $y - \bar{y}$, be written in the fuller form

$$y - \bar{y} = (y.U_2)U_2 + (y.U_3)U_3 + \cdots + (y.U_n)U_n$$

Testing the Hypothesis

Our test statistic will involve a ratio of squared lengths of projection coefficients, $y.U_i$, so it is now necessary to consider the distributions of the associated random variables, $Y.U_i$.

In §3.2 we demonstrated that the projection coefficients, $Y.U_i$, are normally distributed and always have variance σ^2. It is a straightforward exercise to show that

$$E(Y.U_1) = \sqrt{n}\mu, \quad E(Y.U_2) = 0, \quad \ldots, \quad E(Y.U_n) = 0$$

as summarized in Table 5.3 and illustrated in Figure 5.7. If you refer to §3.2, you will see also that the projection coefficients are independent, since they correspond to projections onto orthogonal directions.

	Average, or expected value	Variance
$Y.U_1$	$\sqrt{n}\mu$	σ^2
$Y.U_2$	0	σ^2
$\vdots$	$\vdots$	$\vdots$
$Y.U_n$	0	σ^2

Table 5.3: Means and variances of the projection coefficients, $Y.U_i$.

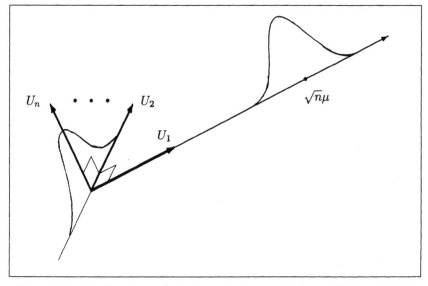

Figure 5.7: The distribution of the projection coefficients, $Y.U_1$ and $Y.U_2$.

Our test of the hypothesis now involves checking whether $(y.U_1)^2$ is acceptably similar to, or substantially larger than, the average of the squared projection coefficients in the error space. The appropriate test statistic is

$$F = \frac{(y.U_1)^2}{\left[(y.U_2)^2 + \cdots + (y.U_n)^2\right]/(n-1)} = \frac{\|\bar{y}\|^2}{\|y - \bar{y}\|^2/(n-1)}$$

If $\mu = 0$ this test statistic follows an $F_{1,n-1}$ distribution, since then $Y.U_1$ as well as $Y.U_2, \ldots, Y.U_n$ follow the $N(0, \sigma^2)$ distribution. If $\mu \neq 0$, then $Y.U_1$ has a non-zero mean and therefore the test statistic is inflated. The ensuing test procedure is to compare the calculated F value with the percentiles of the $F_{1,n-1}$ distribution, and to reject the hypothesis $H_0 : \mu = 0$ if the calculated value is unusually large.

ANOVA Table

The Pythagorean breakup corresponding to $U_1, U_2, \ldots, U_n$ is

$$\|y\|^2 = (y.U_1)^2 + (y.U_2)^2 + \cdots + (y.U_n)^2$$

or, in brief $\|y\|^2 = \|\bar{y}\|^2 + \|y - \bar{y}\|^2$

These ingredients of the test statistic, and the test statistic itself, are most conveniently summarized in the traditional analysis of variance (ANOVA) table, Table 5.4.

Source of Variation	df	SS	MS	F
Mean	1	$\|\bar{y}\|^2$	$\|\bar{y}\|^2$	$\dfrac{\|\bar{y}\|^2}{\|y - \bar{y}\|^2/(n-1)}$
Error	$n-1$	$\|y - \bar{y}\|^2$	$\dfrac{\|y - \bar{y}\|^2}{n-1}$	
Total	n	$\|y\|^2$		

Table 5.4: ANOVA table for a single population.

In the table, normal procedure is to calculate the total and overall mean sums of squares,

$$\|y\|^2 = y_1^2 + \cdots + y_n^2 \quad \text{and} \quad \|\bar{y}\|^2 = \bar{y}^2 + \cdots + \bar{y}^2 = n\bar{y}^2$$

and to obtain the error sum of squares, $\|y - \bar{y}\|^2$, by subtraction as $\|y\|^2 - \|\bar{y}\|^2$.

Estimation of σ^2

To digress briefly, the direction U_1 has been used to show that the least squares estimate of μ is the sample mean $\bar{y}$. The remaining directions, $U_2, \ldots, U_n$ are used to estimate σ^2, as discussed in §3.3. This estimate is known as the *sample variance*, and is given by

$$s^2 = \frac{(y.U_2)^2 + \cdots + (y.U_n)^2}{n-1} = \frac{\|y - \bar{y}\|^2}{n-1} = \frac{\displaystyle\sum_{i=1}^{n}(y_i - \bar{y})^2}{n-1}$$

Confidence Interval

Since $Y.U_1 = \sqrt{n}\,\bar{Y}$ follows the $N(\sqrt{n}\mu, \sigma^2)$ distribution, the random variable $\sqrt{n}(\bar{Y} - \mu)$ must follow the $N(0, \sigma^2)$ distribution. Hence

$$T = \frac{\sqrt{n}(\bar{Y} - \mu)}{\sqrt{\dfrac{(Y.U_2)^2 + \cdots + (Y.U_n)^2}{n-1}}}$$

is a t_{n-1} statistic with $(n-1)$ degrees of freedom. We gamble that the realized value of T lies between the 2.5 and 97.5 percentiles, and obtain a 95% confidence interval for μ of

$$\bar{y} - \frac{s}{\sqrt{n}}t_{n-1}(.975) \ \leq \ \mu \ \leq \ \bar{y} + \frac{s}{\sqrt{n}}t_{n-1}(.975)$$

The t_{n-1} values are given at the end of the book in Table T.2. Note that the random variable T can be written in the form

$$T = \frac{Y.U - E(Y.U)}{\sqrt{\text{Estimator of } \sigma^2}}$$

where U is the direction associated with the hypothesis $H_0 : \mu = 0$. This is the form which will recur through the book.

Geometry and Tradition

We now tie the geometric approach in to the traditional approach. The fitted model vector $(y.U_1)U_1$, we have already noted, is $\bar{y}[1, \ldots, 1]^T$, so the sample mean determines the length of this component of our decomposition. Is the other component, the error vector $(y.U_2)U_2 + \cdots + (y.U_n)U_n$, related to a familiar expression? Yes, since when the error vector is rewritten as $y - \bar{y}$, we have

$$\text{mean squared length of error vector} \ = \ \|y - \bar{y}\|^2/(n-1)$$
$$= \ \sum_{i=1}^{n}(y_i - \bar{y})^2/(n-1)$$
$$= \ s^2, \quad \text{the sample variance}$$

The geometric approach takes the mystery away from the calculation of the sample variance. Why, when we have n squared terms, $(y_1 - \bar{y})^2 + \cdots + (y_n - \bar{y})^2$, do we divide their sum by $n - 1$? The geometric approach tells us immediately that the n dependent squares are really $n - 1$ independent squares, $(y.U_2)^2 + \cdots + (y.U_n)^2$, and these are the ones we are averaging, since each estimates σ^2.

We already know that the larger the sample mean, the further along the equiangular line the model vector lies. Now we can add this fact: the more variable the data, the further the observation vector will lie from the equiangular line. Become familiar with this connection. Picture for yourself the layout when $\bar{y}$ is large and s small, or when $\bar{y}$ is large and s large.

F and t Tests

We can reformulate our test statistic in terms of $\bar{y}$ and s^2 as

$$F = \frac{\|\bar{y}\|^2}{\|y - \bar{y}\|^2/(n-1)} = \frac{n\bar{y}^2}{s^2} = \frac{\bar{y}^2}{(s^2/n)} = \left(\frac{\bar{y}}{s/\sqrt{n}}\right)^2 = t^2$$

where the second-to-last term is the square of the traditional t value, $\bar{y}/(s/\sqrt{n})$. Refer to §3.4 where we showed that the $F_{1,n-1}$ distribution is the square of the t_{n-1} distribution. Thus either the F or t distribution can be used to test our hypothesis. We have preferred to use the $F_{1,n-1}$ distribution in this chapter because it begins the pattern which is followed later.

We now come full circle. How does our test statistic of §1.2, the tea example, relate to the fuller analysis of this chapter? In that situation we had a sample of only 2 drinkers ($n = 2$). For that example we would now quote the test statistic as

$$F = \frac{\|\bar{y}\|^2}{\|y - \bar{y}\|^2/(2-1)} = \left(\frac{A}{B}\right)^2 = t^2$$

using the notation of §1.2. The orthogonal breakup of Figure 1.3 is the same as that of Figure 5.3. The old test statistic, A/B, is actually the square root of the F test statistic used in this chapter, and leads to an equivalent test of the hypothesis.

5.3 Virtues of Our Estimates

Our least squares method has been developed in a natural way, through the cooperation of intuition and vector geometry. But are the resulting estimates for the model parameters, namely the sample mean $\bar{y}$ for μ and the sample variance s^2 for σ^2, also in some sense the best?

Best Linear Unbiased Estimates

Estimation is like archery. A good archer centers his arrows in a tight bunch around the bull's-eye. In the same way it is desirable for our estimates to bunch tightly around the true value, over many repeats of the experiment. We now show that the least squares method does yield the world's best archer.

We shall only consider estimators of μ which are linear combinations of our observations. Such estimators take the form $a_1 Y_1 + a_2 Y_2$, where a_1 and a_2 are real numbers; we deal only with samples of size two, for simplicity of exposition. For example, the sample mean $\overline{Y} = 0.5 Y_1 + 0.5 Y_2$, is of this form.

"Arrows must center on the bull's-eye" we translate into statistical language as "our estimator must have an expectation of μ". That is, we require that

$$E(a_1 Y_1 + a_2 Y_2) = a_1 E(Y_1) + a_2 E(Y_2) = (a_1 + a_2)\mu = \mu$$

or that $a_1 + a_2$ equals one.

"Arrows should be tightly bunched around the origin" we translate into "our estimator must have minimum variance". The variance of our estimator is

$$Var(a_1 Y_1 + a_2 Y_2) = a_1^2 \, Var Y_1 + a_2^2 \, Var Y_2 = (a_1^2 + a_2^2)\sigma^2$$

What pair of values, a_1 and a_2, which sum to one, have their sum of squares least? The answer is $a_1 = a_2 = 1/2$. There is a neat demonstration of this fact using a picture in the plane. A clue: draw all points whose coordinates sum to one, and decide which is nearest the origin.

In summary, the estimator of μ produced by the least squares method, namely the sample mean $\overline{Y} = (Y_1 + Y_2)/2$, has the excellent property that amongst all linear and unbiased estimators it is that with smallest variance.

Maximum Likelihood Estimates

We now demonstrate another important property: our least squares estimator, $\overline{Y}$, is also the *maximum likelihood* estimator of μ.

To help you understand this, imagine for a moment that we know the true population mean μ. Then the probability of drawing the sample $y_1, \ldots, y_n$ is described by the probability density function

$$f(y_1, \ldots, y_n) = \frac{1}{(2\pi\sigma^2)^{\frac{n}{2}}} e^{-\frac{1}{2}\frac{\|y - \mu\|^2}{\sigma^2}}$$

This formula depends solely on the distance $\|y - \mu\|$, so is spherically symmetrical about the true model vector $[\mu, \mu, \ldots, \mu]^T$, as illustrated in Figure 5.8. Think of this as the familiar normal distribution, but spreading in all directions. Observation vectors close to the vector μ are more likely than ones further away. Visualizing many draws of our sample, we can see the observation vector $y = [y_1, y_2, \ldots, y_n]^T$ varying in a cloud of probability which is densest near $[\mu, \mu, \ldots, \mu]^T$, and becomes less and less dense as we move away from $[\mu, \mu, \ldots, \mu]^T$.

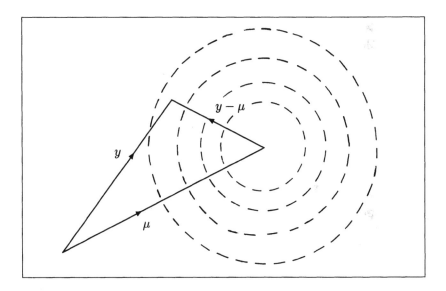

Figure 5.8: Contours of the probability density function centered on the true model vector, $\mu = [\mu, \ldots, \mu]^T$. An observation vector, $y = [y_1, \ldots, y_n]^T$, is also shown.

We now turn the idea around: given that we have a set of data $[y_1, y_2, \ldots, y_n]^T$, and do not know μ, how can we choose μ in such a way that we *maximize the likelihood* of observing these data? The answer is that we should choose μ to minimize the distance $\|y - \mu\|$. This is achieved by projecting y onto the $[1, 1, \ldots, 1]^T$ direction, as illustrated in Figure 5.9. That is, the maximum likelihood estimator of μ is the same as the least squares estimator, $\overline{Y}$.

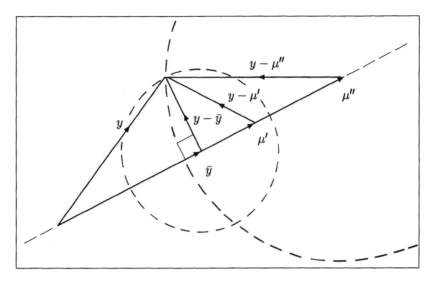

Figure 5.9: An illustration of the fact that the maximum likelihood estimator of μ is the least squares estimator, $\overline{Y}$. The estimate $\hat{\mu} = \bar{y}$ is the *most likely* estimate of μ, since the length of $y - \bar{y}$ is less than the lengths of $y - \mu'$ or $y - \mu''$. The circles are probability density contours passing through the observation vector y for the two scenarios $\mu = \bar{y}$ and $\mu = \mu''$.

5.4   Summary

Our interest in this chapter has been in a population assumed to have a normal distribution, with mean μ and variance σ^2 unknown. Our aim has been to find point and interval estimates of μ, and to test the hypothesis that μ is zero. From the population we have taken a random, hence independent, sample of size n, yielding the observations $y_1, \ldots, y_n$.

Our observation vector y is $[y_1, \ldots, y_n]^T$. Since the expectation of each observation is μ, our model space M is the line in n-space spanned by the unit vector $U_1 = [1, \ldots, 1]^T/\sqrt{n}$. The direction associated with the hypothesis $H_0 : \mu = 0$ is also U_1, since $y.U_1 = \sqrt{n}\bar{y}$, the signed length of the projection of y in the direction U_1, reflects the size of μ. The error space is $n - 1$ dimensional, with axes $U_2, \ldots, U_n$.

In fitting the model we estimate the vector decomposition demanded by the model, namely $y = \mu + (y - \mu)$, by means of the orthogonal decomposition $y = \bar{y} + (y - \bar{y})$, as illustrated in Figure 5.10. This is achieved by projecting y onto M, and yields $\bar{y}$ as an estimate of μ, and $s^2 = \|y - \bar{y}\|^2/(n - 1)$ as our estimate of σ^2.

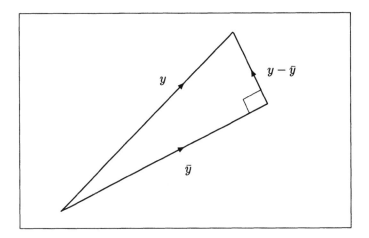

Figure 5.10: The orthogonal decomposition of the observation vector for a single population.

The corresponding Pythagorean decomposition is $\|y\|^2 = \|\bar{y}\|^2 + \|y - \bar{y}\|^2$. Since $\|\bar{y}\|^2 = (y.U_1)^2$ estimates σ^2 only if $\mu = 0$, whereas $\|y - \bar{y}\|^2/(n-1) = \left[(y.U_2)^2 + \cdots + (y.U_n)^2\right]/(n-1)$ always estimates σ^2, we use

$$F = \frac{\|\bar{y}\|^2}{\|y - \bar{y}\|^2/(n-1)} = \frac{n\bar{y}^2}{s^2}$$

as our test statistic. This test statistic follows an $F_{1,n-1}$ distribution if $\mu = 0$, so we reject the hypothesis that $\mu = 0$ if the statistic exceeds the 95 percentile of this distribution.

A 95% confidence interval for μ is given by

$$\bar{y} - \frac{s}{\sqrt{n}} t_{n-1}(0.975) \;\leq\; \mu \;\leq\; \bar{y} + \frac{s}{\sqrt{n}} t_{n-1}(0.975)$$

For additional reading on the geometry of the single population case, refer to Box, Hunter and Hunter (1978), pages 197−203.

Class Exercise

Forty university students were used as guinea pigs to determine whether heartbeat rate is increased by exercise. Each student took his or her heartbeat after ten minutes of sitting, and again after five minutes of running vigorously on the spot. On both occasions the number of beats per 30 seconds was recorded. The changes in heartbeat per 30 seconds are given in Table 5.5.

Student	Change	Student	Change	Student	Change	Student	Change
1.	15	11.	17	21.	12	31.	16
2.	15	12.	13	22.	14	32.	15
3.	13	13.	14	23.	14	33.	13
4.	16	14.	13	24.	12	34.	16
5.	16	15.	16	25.	14	35.	16
6.	17	16.	14	26.	15	36.	15
7.	15	17.	14	27.	12	37.	17
8.	15	18.	15	28.	16	38.	17
9.	14	19.	12	29.	15	39.	17
10.	14	20.	14	30.	15	40.	14

Table 5.5: Changes in the heartbeat rate per 30 seconds.

(a) Each member of the class is asked to use the random numbers given in Table T.1 to select a sample of size four from the data in Table 5.5. The four changes in heartbeat make up the observation vector y. (If the class instructor would like more active class participation, the class can be broken into groups of size four, with each group generating its own data.)

The question to be answered is: Does running affect heartbeat? Assuming no prior knowledge of this subject, we shall carry out a 2-tailed test, $H_0 : \mu = 0$ versus $H_1 : \mu \neq 0$, where μ is the mean change in heartbeat for an infinitely large theoretical population of similar students.

An appropriate coordinate system is

$$
U_1 = \frac{1}{\sqrt{4}} \begin{bmatrix} 1 \\ 1 \\ 1 \\ 1 \end{bmatrix}, \ U_2 = \frac{1}{\sqrt{2}} \begin{bmatrix} 1 \\ -1 \\ 0 \\ 0 \end{bmatrix}, \ U_3 = \frac{1}{\sqrt{6}} \begin{bmatrix} 1 \\ 1 \\ -2 \\ 0 \end{bmatrix}, \ U_4 = \frac{1}{\sqrt{12}} \begin{bmatrix} 1 \\ 1 \\ 1 \\ -3 \end{bmatrix}
$$

Here the first axis direction is the equiangular direction, spanning the model space, M. The other three directions span the error space, and can be chosen in an infinite variety of ways. For our choice, we firstly chose the error axis, U_2, by thinking of a comparison of the first and second observations. Then we chose the second error axis, U_3, to correspond to a comparison of the third observation with the average of the first two observations. Lastly, we chose the remaining error axis, U_4, to correspond to a comparison of the fourth observation with the average of the first three observations. Note that this way of writing down a coordinate system for the error space has been generalized in §5.2.

(b) Each class member is asked to calculate the scalars $y.U_1$, $y.U_2$, $y.U_3$ and $y.U_4$ using his or her own observation vector. The class instructor can then tabulate the results for $y.U_1$ in a histogram, and for $y.U_2$, $y.U_3$ and $y.U_4$ in three further histograms. These histograms will approximate the distributions of the random variables $Y.U_i$, for $i = 1, 2, 3, 4$.

Questions:
1. Which of the random variables appear to have mean zero?
2. Is the variability similar between the histograms?

(c) Each class member is then asked to calculate

$$F = \frac{(y.U_1)^2}{((y.U_2)^2 + (y.U_3)^2 + (y.U_4)^2)/3}$$

and compare his or her answer with the 95 and 99 percentiles of the $F_{1,3}$ distribution.

The instructor can then tabulate the class results in the form of a histogram. This will approximate the distribution of a "noncentral F" statistic. Note: this histogram should be saved for comparison with the class results to be obtained in Chapter 6.

Question:
Does the test have adequate "power" with a sample size of four? That is, was the null hypothesis rejected for most student's data sets?

(d) Each class member can now calculate the projection vectors $(y.U_1)U_1$, $(y.U_2)U_2$, $(y.U_3)U_3$ and $(y.U_4)U_4$. He or she can then confirm that $(y.U_1)U_1 = \bar{y}$, the mean vector, and that $(y.U_2)U_2 + (y.U_3)U_3 + (y.U_4)U_4 = (y - \bar{y})$, the error vector.

(e) Using this new decomposition $y = \bar{y} + (y - \bar{y})$, each class member can now recalculate

$$F = \frac{\|\bar{y}\|^2}{\|y - \bar{y}\|^2/3}$$

to confirm that this formula yields the same answer as in (c).

(f) Lastly, each class member is asked to calculate the 95% confidence interval for the population mean μ, using the formula

$$\bar{y} \ \pm \ \frac{s}{\sqrt{4}} \times 3.182$$

where $\bar{y}$ is his or her sample mean, s is the square root of his or her sample variance $s^2 = \|y - \bar{y}\|^2/3$, "4" is the sample size, and "3.182" is the 97.5 percentile of the t_3 distribution.

A selection of these intervals can be displayed by the instructor. For the class as a whole, what percentage of confidence intervals included the true value of $\mu = 15$?

Exercises

(5.1) An experiment was set up to see whether nitrogen fertilizer was beneficial, detrimental or had no effect on the growth of an alfalfa crop. Five pairs of plots were laid out, with nitrogen applied to just one plot of each pair. The resulting percentage decreases in production of the fertilized, compared to the unfertilised for the five pairs, were

$$8, \ 7, \ 5, \ 6 \ \text{ and } \ 9\%$$

(a) Take these values for the observation vector y and project y onto the coordinate system

$$\frac{1}{\sqrt{5}}\begin{bmatrix} 1 \\ 1 \\ 1 \\ 1 \\ 1 \end{bmatrix}, \ \frac{1}{\sqrt{30}}\begin{bmatrix} 3 \\ 3 \\ -2 \\ -2 \\ -2 \end{bmatrix}, \ \frac{1}{\sqrt{2}}\begin{bmatrix} 1 \\ -1 \\ 0 \\ 0 \\ 0 \end{bmatrix}, \ \frac{1}{\sqrt{2}}\begin{bmatrix} 0 \\ 0 \\ 1 \\ -1 \\ 0 \end{bmatrix}, \ \frac{1}{\sqrt{6}}\begin{bmatrix} 0 \\ 0 \\ 1 \\ 1 \\ -2 \end{bmatrix}$$

writing out each projection vector. Check your working by substituting into the orthogonal decomposition: y = sum of projection vectors.

(b) Calculate the squared lengths of these projections, and check that Pythagoras' Theorem is obeyed.

(c) Assuming that the observations are a sample from a normal population, estimate the true mean percentage decrease in production for the alfalfa crop, μ, and the variance, σ^2.

(d) Use your answers in (b) to calculate the F test of the hypothesis $H_0 : \mu = 0$ against the alternative $H_1 : \mu \neq 0$. What do you conclude?

(e) Recalculate this F test using the breakdown $\|y\|^2 = \|\bar{y}\|^2 + \|y - \bar{y}\|^2$.

(f) Calculate the 95% confidence interval for the true mean percentage decrease in production.

(5.2) In order to monitor the weight changes in a herd of calves, six randomly chosen calves were individually identified with ear tags and weighed monthly. The weights, in kg, on two successive dates were:

Calf tag no.	Weight (22/2/85)	Weight(22/3/85)
635	125	128
123	115	126
715	115	124
817	112	119
125	142	152
347	110	118

We assume that the *changes* in weight come from a normal distribution, $N(\mu, \sigma^2)$, where μ and σ^2 are the mean and variance of the weight changes for the herd of calves.

(a) Write down the observation vector y comprising the six weight changes, a unit vector U_1 spanning the one-dimensional model space, and unit vectors, U_2 to U_6, spanning the error space.

(b) Fit the model by expressing y in the form

$$y = (y.U_1)U_1 + (y.U_2)U_2 + \cdots + (y.U_6)U_6$$

(c) Rewrite the model in the form

$$y = \bar{y} + (y - \bar{y})$$

That is, express the observation vector, y, as a sum of the model vector, $\bar{y}$, together with the error vector, $y - \bar{y}$. Write down the least squares estimator of μ.

(d) Draw a vector diagram showing the relationship of y, $\bar{y}$ and $y - \bar{y}$. Label the vectors with their numerical values.

(e) Calculate the squared lengths of the three vectors, $\|y\|^2$, $\|\bar{y}\|^2$ and $\|y - \bar{y}\|^2$, and set up the standard ANOVA table.

(f) Test the hypothesis $H_0 : \mu = 0$ against the alternative $H_1 : \mu \neq 0$. Is there evidence of any weight change in the herd during the month?

(g) Check the normality assumption by drawing a histogram of the errors, or residuals, $y_i - \bar{y}$. Is there any one value which could be an "outlier"?

(h) Calculate the 95% confidence interval for the true mean weight change of the herd.

(5.3) (a) Write down an orthogonal coordinate system for 8-space which includes the unit vector $[1, 1, 1, 1, 1, 1, 1, 1]^T / \sqrt{8}$. Write down a second such system.

(b) Write down two orthogonal coordinate systems for 5-space which include the unit vector $[1, 1, 1, 1, 1] / \sqrt{5}$. Do not use the system given in Exercise 5.1.

(5.4) A plant breeder has carried out nine trials to evaluate a promising new cultivar of barley when his director queries him as to whether the percentage screenings of the new cultivar is different from the industry average of 10%. The percentage screenings is the percent by weight of grains which are small enough to fall through a standard screen.

The percentage screenings in the nine trials were as follows:

$$S = \text{percentage screenings} = 12, 6, 10, 7, 7, 12, 5, 8, 9$$

Assume these values come from a normal distribution with mean $\mu_s = 10$ and variance σ^2. We wish to test the hypothesis $H_0 : \mu_s = 10$ against the alternative $H_1 : \mu_s \neq 10$.

(a) Transform the problem into one we can solve by calculating $Y = 10 - S$ for each trial. Take the resulting values as your observation vector y.

(b) Test the hypothesis $H_0 : \mu_Y = 0$ against the alternative $H_1 : \mu_Y \neq 0$. This is equivalent to testing $H_0 : \mu_s = 10$ against $H_1 : \mu_s \neq 10$. What can you conclude?

(c) Check the assumption of normality by drawing a histogram of the $y_i - \bar{y}$ values.

(5.5) During the energy crisis in the 1970's the New Zealand Government decided to investigate the feasibility of growing beet as a source of ethanol for use as motor vehicle fuel. One component of this investigation was a research program aimed at determining the fertilizer requirements of beet. Following a search of the international literature to find out which nutrient elements had been shown to be necessary for the growth of beet, several "screening" trials were carried out to determine which of these nutrients were deficient locally.

Six farms were randomly chosen for the study in the spring of 1979. On each farm 30 adjacent plots of beet were sown in five row plots, 5m long and 2.5m wide. The treatments listed in the table were each allocated to two plots in a random manner, with the exception of the "all" and "nil" treatments, which were allocated to six plots.

The treatment design is called a "subtractive" design, since each element in turn is subtracted from the "all fertilizer" mixture. Since treatments 2 to 10 are each to be compared with the "all" treatment, it makes sense to replicate this baseline treatment more heavily than the other treatments. The elements K and Na can substitute for one another, so it was thought necessary to include treatment 6 with neither present in the mixture. Similarly, lime and boron may interact, so treatment 9 with neither present was included. The elements S, Mg, Cu, Zn, Mn and Mo were thought unlikely to be important, so were subtracted from the mixture "in bulk" in treatment 10.

After growing until late autumn the beet in a 4m by 1.5m area in the three center rows of each plot were lifted, cleaned and topped. Roots and tops were weighed separately, and the weights scaled to tonnes per hectare. A subsample of roots was then washed in clean water and a longitudinal section of each root was cut out and grated. The total fermentable sugar concentration was then estimated by laboratory methods using a sample of this grated root. The sugar yield in tonnes per hectare was then calculated for each plot by multiplying the root yield (t/ha) by the percentage of total fermentable sugars. The averages for each experimental treatment for each site are as shown in the table. Mr. R.C. Stephen of the Ministry of Agriculture and Fisheries, Lincoln, New Zealand, is thanked for kindly providing these data.

	Site					
Treatments	1	2	3	4	5	6
1. All	12.0	8.4	14.1	10.3	8.0	10.0
2. All−N	6.8	5.8	12.7	7.5	5.3	10.2
3. All−P	10.5	8.3	13.0	9.2	6.7	9.5
4. All−Na	8.9	7.6	12.1	8.8	6.9	9.4
5. All−K	10.9	8.2	13.6	10.9	7.5	9.9
6. All−K,Na	9.8	8.7	12.7	8.8	7.0	10.5
7. All−Lime	9.8	8.8	14.9	11.1	7.4	10.3
8. All−B	9.8	9.3	13.6	9.7	7.6	10.5
9. All−Lime, B	10.9	7.2	14.3	9.7	7.5	10.4
10. All−S,Mg,Cu,Zn,Mn,Mo	10.4	9.1	12.9	11.2	6.7	10.1
11. Nil	6.7	4.6	10.8	6.2	6.2	10.0
Differences						
N	5.2	2.6	1.4	2.8	2.7	−.2
P	1.5	.1	1.1	1.1	1.3	.5
Na	3.1	.8	2.0	1.5	1.1	.6

The three most important elements appeared to be N, P and Na, so the reductions in sugar yield through withholding nitrogen, phosphate and sodium from the fertilizer mixture were calculated for each site by differencing. The difference of treatments 1 and 2 give the reduction due to withholding nitrogen (for example $12.0 - 6.8 = 5.2$ for the first site), treatments 1 and 3 for phosphate and treatments 1 and 4 for sodium. These differences are shown in the lower part of the table.

In this exercise we shall zero in on the reductions in sugar yield through withholding phosphate from the fertilizer mixture. The underlying population of study is the set of all such reductions for beet crops in Canterbury, New Zealand; we assume this population is normally distributed, with mean μ and variance σ^2. Our observed reductions constitute a sample of size six from this population.

(a) Write down the observation vector, y.

(b) Define the model space, M.

(c) Fit the model in the form $y = \bar{y} + (y - \bar{y})$.

(d) Calculate the Pythagorean breakdown $\|y\|^2 = \|\bar{y}\|^2 + \|y - \bar{y}\|^2$, and use this to test the hypothesis $H_0 : \mu = 0$ against the alternative hypothesis $H_0 : \mu \neq 0$.

(e) Calculate the 95% confidence interval for the mean reduction, μ, through withholding phosphate from the fertilizer mix.

(f) Draw a histogram of the error values, $y_i - \bar{y}$, to check that the normality assumption is reasonable.

(5.6) (a) Make up an observation vector y using the data in Exercise 5.5 on the reduction in sugar yield associated with omitting nitrogen, N, from the sugar beet fertilizer mixture.

(b) Assume the observations come from an $N(\mu, \sigma^2)$ distribution. Use the orthogonal decomposition $y = \bar{y} + (y - \bar{y})$ to decide whether $\mu = 0$ or $\mu \neq 0$. Is nitrogen an important component of the mixture?

(c) Draw a histogram of the errors $y_i - \bar{y}$. Without carrying out any formal tests, decide whether the histogram appears normal.

(d) Compare your results with those of Exercise 5.5, as follows: Is the estimated depression in sugar yield through withholding N greater or less than the estimated depression through withholding P? Which of the calculated F values is greater? Can you explain this apparent contradiction?

Solutions to the Reader Exercises

(5.1) $y.U_1 = \begin{bmatrix} 1.1 \\ 1.4 \\ 1.1 \end{bmatrix} \cdot \frac{1}{\sqrt{3}} \begin{bmatrix} 1 \\ 1 \\ 1 \end{bmatrix} = \frac{1.1 + 1.4 + 1.1}{\sqrt{3}} = \frac{3.6}{\sqrt{3}}$

$y.U_2 = \begin{bmatrix} 1.1 \\ 1.4 \\ 1.1 \end{bmatrix} \cdot \frac{1}{\sqrt{2}} \begin{bmatrix} 1 \\ -1 \\ 0 \end{bmatrix} = \frac{1.1 - 1.4}{\sqrt{2}} = \frac{-0.3}{\sqrt{2}}$

$y.U_3 = \begin{bmatrix} 1.1 \\ 1.4 \\ 1.1 \end{bmatrix} \cdot \frac{1}{\sqrt{6}} \begin{bmatrix} 1 \\ 1 \\ -2 \end{bmatrix} = \frac{1.1 + 1.4 - 2.2}{\sqrt{6}} = \frac{0.3}{\sqrt{6}}$

$$(y.U_1)U_1 = \left(\frac{3.6}{\sqrt{3}}\right)\frac{1}{\sqrt{3}}\begin{bmatrix} 1 \\ 1 \\ 1 \end{bmatrix} = \frac{3.6}{3}\begin{bmatrix} 1 \\ 1 \\ 1 \end{bmatrix} = \begin{bmatrix} 1.2 \\ 1.2 \\ 1.2 \end{bmatrix}$$

$$(y.U_2)U_2 = \left(\frac{-0.3}{\sqrt{2}}\right)\frac{1}{\sqrt{2}}\begin{bmatrix} 1 \\ -1 \\ 0 \end{bmatrix} = \frac{-0.3}{2}\begin{bmatrix} 1 \\ -1 \\ 0 \end{bmatrix} = \begin{bmatrix} -0.15 \\ 0.15 \\ 0 \end{bmatrix}$$

$$(y.U_3)U_3 = \left(\frac{0.3}{\sqrt{6}}\right)\frac{1}{\sqrt{6}}\begin{bmatrix} 1 \\ 1 \\ -2 \end{bmatrix} = \frac{0.3}{6}\begin{bmatrix} 1 \\ 1 \\ -2 \end{bmatrix} = \begin{bmatrix} 0.05 \\ 0.05 \\ -0.10 \end{bmatrix}$$

As a check,

$$(y.U_1)U_1 + (y.U_2)U_2 + (y.U_3)U_3 = \begin{bmatrix} 1.2 \\ 1.2 \\ 1.2 \end{bmatrix} + \begin{bmatrix} -0.15 \\ 0.15 \\ 0 \end{bmatrix} + \begin{bmatrix} 0.05 \\ 0.05 \\ -0.10 \end{bmatrix}$$

$$= \begin{bmatrix} 1.1 \\ 1.4 \\ 1.1 \end{bmatrix} = y, \text{ as required.}$$

(5.2) $y.U_1 = \begin{bmatrix} y_1 \\ y_2 \\ y_3 \end{bmatrix}.\frac{1}{\sqrt{3}}\begin{bmatrix} 1 \\ 1 \\ 1 \end{bmatrix} = \dfrac{y_1 + y_2 + y_3}{\sqrt{3}}$

$$y.U_2 = \begin{bmatrix} y_1 \\ y_2 \\ y_3 \end{bmatrix}.\frac{1}{\sqrt{2}}\begin{bmatrix} 1 \\ -1 \\ 0 \end{bmatrix} = \frac{y_1 - y_2}{\sqrt{2}}$$

$$y.U_3 = \begin{bmatrix} y_1 \\ y_2 \\ y_3 \end{bmatrix}.\frac{1}{\sqrt{6}}\begin{bmatrix} 1 \\ 1 \\ -2 \end{bmatrix} = \frac{y_1 + y_2 - 2y_3}{\sqrt{6}}$$

Hence $Y.U_1 = \dfrac{Y_1 + Y_2 + Y_3}{\sqrt{3}}$, $Y.U_2 = \dfrac{Y_1 - Y_2}{\sqrt{2}}$ and $Y.U_3 = \dfrac{Y_1 + Y_2 - 2Y_3}{\sqrt{6}}$

Therefore

$$E(Y.U_1) = \frac{E(Y_1) + E(Y_2) + E(Y_3)}{\sqrt{3}} = \frac{\mu + \mu + \mu}{\sqrt{3}} = \sqrt{3}\mu$$

$$E(Y.U_2) = \frac{E(Y_1) - E(Y_2)}{\sqrt{2}} = \frac{\mu - \mu}{\sqrt{2}} = 0$$

$$E(Y.U_3) = \frac{E(Y_1) + E(Y_2) - 2E(Y_3)}{\sqrt{6}} = \frac{\mu + \mu - 2\mu}{\sqrt{6}} = 0$$

$$Var(Y.U_1) \quad = \quad \frac{Var(Y_1) + Var(Y_2) + Var(Y_3)}{(\sqrt{3})^2} = \frac{\sigma^2 + \sigma^2 + \sigma^2}{3} = \sigma^2$$

$$Var(Y.U_2) \quad = \quad \frac{Var(Y_1) + Var(Y_2)}{(\sqrt{2})^2} = \frac{\sigma^2 + \sigma^2}{2} = \sigma^2$$

$$Var(Y.U_3) \quad = \quad \frac{Var(Y_1) + Var(Y_2) + 4\,Var(Y_3)}{(\sqrt{6})^2} = \frac{\sigma^2 + \sigma^2 + 4\sigma^2}{6} = \sigma^2$$

$$\textbf{(5.3)} \;\; y.U_2 = \begin{bmatrix} 1.1 \\ 1.4 \\ 1.1 \end{bmatrix} \cdot \frac{1}{\sqrt{14}} \begin{bmatrix} -2 \\ -1 \\ 3 \end{bmatrix} = \frac{-2.2 + -1.4 + 3.3}{\sqrt{14}} = \frac{-0.3}{\sqrt{14}}$$

$$y.U_3 = \begin{bmatrix} 1.1 \\ 1.4 \\ 1.1 \end{bmatrix} \cdot \frac{1}{\sqrt{42}} \begin{bmatrix} 4 \\ -5 \\ 1 \end{bmatrix} = \frac{4.4 - 7 + 1.1}{\sqrt{42}} = \frac{-1.5}{\sqrt{42}}$$

Hence $(y.U_2)^2 + (y.U_3)^2 = \frac{0.09}{14} + \frac{2.25}{42} = 0.06$, as before.

Chapter 6

Questions About
Two Populations

In this chapter we deal with questions concerning two normal populations. The central question to be answered is "Are the two means equal?" If they are deemed unequal then typically we wish to estimate with some precision the difference between them. Situations of this type are common. For example, in designing office furniture it would be of importance to know to what extent the female back is in general shorter than the male back. Furniture designed for each sex should then take into account the difference in back lengths.

We pause to contrast the methods of this chapter with those of the last chapter. There we used *paired* samples and single population methods to answer questions of the type just described. For our back length example, we could take measurements from male-female twins, whence our population would be all back length differences between the males and females in such twin pairs. In this chapter we instead take *independent* samples from the two populations. In our example this would involve taking separate samples of male and female back lengths.

In §1 of this chapter we introduce the appropriate methods by analyzing a problem provided by the wool industry. The general method is succinctly summarized in §2, and computing methods are discussed in §3. A summary is provided in §4, followed by a highly recommended class exercise. In this chapter we consider only the case in which the samples from each population are equal in size. We leave to Appendix A the unequally replicated case. The chapter is lengthy since we set up much of the machinery for handling questions about many normal populations, the material of the next chapter.

6.1   A Case Study

As an introductory example we now consider a study of the effect of length of time in the conditioning room on the bulk of wool.

Background to the Study

In response to an increasing need for objective measurement in the wool industry, the Wool Research Organisation of New Zealand (WRONZ) has developed a prototype "New Zealand Standard" method for measuring the bulk of scoured wool. This involves use of the bulkometer shown in Figure 6.1.

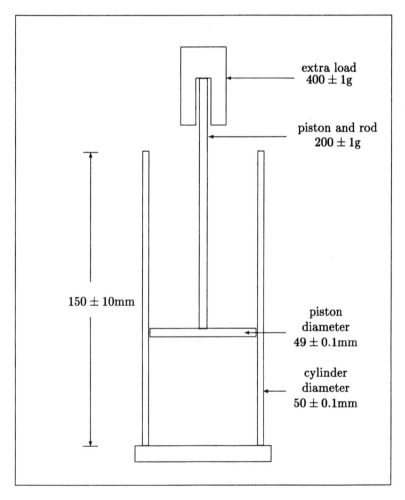

Figure 6.1: Basic bulkometer, used to measure the bulk of wool.

This device exerts a standard pressure on a 2.5gm test specimen of wool, measures the volume occupied by the specimen, and gives a bulk reading in units of cm^3/gm. It is envisaged that the standard method will be adopted by wool testing laboratories.

The draft standard sets out the procedures to be followed in obtaining a sample from a "lot" of wool, and in preparing the sample for measurement. Briefly these are:

1. A representative sample of the lot of wool is obtained by core boring each bale in the lot, and blending these cores.

2. A subsample is taken from the blended sample.

3. The subsample is washed and dried.

4. A card or similar device is used to open up the subsample.

5. The subsample is conditioned in a standard atmosphere.

6. The subsample is split to produce four test specimens.

7. Test specimens are measured in the bulkometer and the results averaged.

Each of these steps has been specified in the Standard as precisely as necessary, on the basis of research undertaken by WRONZ. Of interest here is step five, the conditioning, or bringing to equilibrium, of the cleaned and carded subsamples. Here the Standard specifies that:

> The carded subsamples shall be brought to equilibrium in a standard atmosphere for not less than 12 hours, and not more than 30 hours.

A standard atmosphere is defined as a temperature of 20 ± 2^0C and a relative humidity of $65 \pm 2\%$.

The 12–30 hour length of time for conditioning was decided upon by WRONZ and the commercial Wool Testing Laboratories, thus taking into account commercial practices. Experiments were then carried out to show the appropriateness of the conditioning time. It is one of these experiments which we shall now describe.

Objectives

The objective of the experiment we consider here is to determine whether there is a difference in bulkometer reading between samples conditioned for two days, and those conditioned for three days.

Study Populations

Of interest to us are two populations of mean bulkometer readings: those for subsamples conditioned for two and three days, respectively.

Design

A bulk lot of greasy wool is used for the experiment. Cores are bored into each bale in the lot, and the core borings blended into a single sample. Ten subsamples are taken by hand from this sample, and allocated in a completely random manner to the two treatments:

1. Conditioning for two days.

2. Conditioning for three days.

To assign treatments to subsamples, the subsamples are numbered from 1 to 10. The first treatment is assigned by drawing five random numbers in the range from 1 to 10; the subsample numbers obtained are 2, 4, 5, 6 and 7. The second treatment is then assigned to the remaining subsamples, 1, 3, 8, 9 and 10. This design is called a *completely randomized* design.

During the washing, drying and carding stages procedures are standardized using the prototype standard in its current form. At each stage subsamples are processed either simultaneously or in quick succession in order of their number, from one to ten.

For conditioning, the ten subsamples are placed in the same conditioning room for two or three days, according to the treatment. Upon removal from conditioning, each subsample is split into three test specimens, and bulkometer readings taken. Since subsamples in the first treatment are processed a day earlier than subsamples in the second treatment, care is taken to ensure there are no differences in operator, bulkometer device and procedures for the two batches of test specimens.

Data

The bulkometer reading for each test specimen is given in Table 6.1, together with the average of these readings for each subsample. For the second subsample, one of the test specimens was lost, so the average is a mean of two readings instead of three. This small problem with the data will be ignored.

In order to keep our presentation compact we shall use only the first three subsample means from each treatment, giving an observation vector of

$$y = [29.80, 28.57, 29.97, 30.33, 31.27, 30.37]^T$$

Notation

We use the labels y_{11}, y_{12} and y_{13} for the observations from the first population, and y_{21}, y_{22} and y_{23} for the observations from the second population. In general we shall use the label y_{ij} to denote the jth observation from the ith population, so making the origin of the observation clear.

Building on this, we use the label $\bar{y}_1.$ for the mean of the observations from the first population, and similarly $\bar{y}_2.$ for the second sample mean. In general, $\bar{y}_i.$ will refer to the mean of the observations from the ith population. The dot ($\cdot$) indicates a subscript over which averaging has been performed.

Subsample numbers	Bulkometer readings			Subsample means
Treatment 1 (two day conditioning)				
2	29.4	30.2	–	29.80
4	28.2	28.9	28.6	28.57
5	30.0	29.9	30.0	29.97
6	30.3	29.0	29.3	29.53
7	30.4	28.8	30.5	29.90
Treatment 2 (three day conditioning)				
1	31.2	29.8	30.0	30.33
3	32.0	30.1	31.7	31.27
8	30.5	30.2	30.4	30.37
9	29.8	28.4	29.1	29.10
10	31.7	30.6	30.1	30.80

Table 6.1: Bulkometer readings for individual test specimens together with the average for each subsample, rounded to two decimal places. Data by courtesy of Mr. Steve Ranford, WRONZ, Lincoln, New Zealand.

Model

We assume our two populations are normally distributed, with common variance σ^2. We let μ_1 denote the mean of the population of bulkometer readings from subsamples conditioned for two days, and μ_2 that for subsamples conditioned for three days. For our observations from treatment one, the average over many repetitions of the study is μ_1, while for those from treatment two it is μ_2. Thus model vectors have the form

$$
\begin{bmatrix} \mu_1 \\ \mu_1 \\ \mu_1 \\ \mu_2 \\ \mu_2 \\ \mu_2 \end{bmatrix} = \mu_1 \begin{bmatrix} 1 \\ 1 \\ 1 \\ 0 \\ 0 \\ 0 \end{bmatrix} + \mu_2 \begin{bmatrix} 0 \\ 0 \\ 0 \\ 1 \\ 1 \\ 1 \end{bmatrix}
$$

Our model space M is therefore the 2-dimensional subspace

$$M = \mathrm{span}\{[1,1,1,0,0,0]^T, [0,0,0,1,1,1]^T\}$$

with a ready made coordinate system,

$$U_1 = [1,1,1,0,0,0]^T/\sqrt{3}, \quad U_2 = [0,0,0,1,1,1]^T/\sqrt{3}$$

The model, expressed in vector form, is then

$$
\begin{bmatrix} y_{11} \\ y_{12} \\ y_{13} \\ y_{21} \\ y_{22} \\ y_{23} \end{bmatrix} = \mu_1 \begin{bmatrix} 1 \\ 1 \\ 1 \\ 0 \\ 0 \\ 0 \end{bmatrix} + \mu_2 \begin{bmatrix} 0 \\ 0 \\ 0 \\ 1 \\ 1 \\ 1 \end{bmatrix} + \begin{bmatrix} e_{11} \\ e_{12} \\ e_{13} \\ e_{21} \\ e_{22} \\ e_{23} \end{bmatrix}
$$

where the e_{ij}'s are independent $N[0, \sigma^2]$ values.

Hypothesis

Our study objective is to decide whether two day conditioning affects wool bulk differently to three day conditioning. We translate this into statistical language by saying that we wish to test the null hypothesis $H_0 : \mu_1 = \mu_2$ against the alternative hypothesis $H_1 : \mu_1 \neq \mu_2$.

Where is the direction associated with the hypothesis $H_0 : \mu_1 = \mu_2$? We first rewrite the hypothesis in the equivalent form $H_0 : \mu_1 - \mu_2 = 0$, and then look for a unit vector U, in the model space, for which the projection coefficient has expected value a multiple of $\mu_1 - \mu_2$.

The unit vector $U = [1, 1, 1, -1, -1, -1]^T / \sqrt{6}$ is the one we seek. Why? Clearly it lies in the model space M spanned by $[1, 1, 1, 0, 0, 0]^T$ and $[0, 0, 0, 1, 1, 1]^T$. Also

$$
y.U = \begin{bmatrix} y_{11} \\ y_{12} \\ y_{13} \\ y_{21} \\ y_{22} \\ y_{23} \end{bmatrix} \cdot \frac{1}{\sqrt{6}} \begin{bmatrix} 1 \\ 1 \\ 1 \\ -1 \\ -1 \\ -1 \end{bmatrix} = \frac{y_{11} + y_{12} + y_{13} - y_{21} - y_{22} - y_{23}}{\sqrt{6}}
$$

$$
= \frac{3(\bar{y}_{1.} - \bar{y}_{2.})}{\sqrt{6}} = \frac{\sqrt{3}(\bar{y}_{1.} - \bar{y}_{2.})}{\sqrt{2}}
$$

with an expected value $\sqrt{3}(\mu_1 - \mu_2)/\sqrt{2}$, as required. Hence if $\mu_1 = \mu_2$ the projection coefficient $y.U$ will be "small", averaging to zero over many repetitions of the study, whereas if $\mu_1 \neq \mu_2$ the projection coefficient will be "large", averaging to the non-zero quantity, $\sqrt{3}(\mu_1 - \mu_2)/\sqrt{2}$. This provides us with a method for distinguishing between the cases $\mu_1 = \mu_2$ and $\mu_1 \neq \mu_2$.

The collection of our raw materials is now complete. The model space, direction associated with the hypothesis, and observation vector are:

$$
M = \mathrm{span} \left\{ \begin{bmatrix} 1 \\ 1 \\ 1 \\ 0 \\ 0 \\ 0 \end{bmatrix}, \begin{bmatrix} 0 \\ 0 \\ 0 \\ 1 \\ 1 \\ 1 \end{bmatrix} \right\}, \quad U = \frac{1}{\sqrt{6}} \begin{bmatrix} 1 \\ 1 \\ 1 \\ -1 \\ -1 \\ -1 \end{bmatrix}, \quad \text{and } y = \begin{bmatrix} 29.80 \\ 28.57 \\ 29.97 \\ 30.33 \\ 31.27 \\ 30.37 \end{bmatrix}
$$

Fitting the Model

The model assumed here requires that we express the observation vector y as a point in the 2-dimensional model space, together with an error vector. In vector language, we must estimate the decomposition

$$
\begin{bmatrix} y_{11} \\ y_{12} \\ y_{13} \\ y_{21} \\ y_{22} \\ y_{23} \end{bmatrix}
=
\begin{bmatrix} \mu_1 \\ \mu_1 \\ \mu_1 \\ \mu_2 \\ \mu_2 \\ \mu_2 \end{bmatrix}
+
\begin{bmatrix} e_{11} \\ e_{12} \\ e_{13} \\ e_{21} \\ e_{22} \\ e_{23} \end{bmatrix}
$$

To achieve this we project y onto the model space, yielding $(y.U_1)U_1 + (y.U_2)U_2$ as

$$
29.447 \begin{bmatrix} 1 \\ 1 \\ 1 \\ 0 \\ 0 \\ 0 \end{bmatrix}
+ 30.657 \begin{bmatrix} 0 \\ 0 \\ 0 \\ 1 \\ 1 \\ 1 \end{bmatrix}
=
\begin{bmatrix} 29.447 \\ 29.447 \\ 29.447 \\ 30.657 \\ 30.657 \\ 30.657 \end{bmatrix}
=
\begin{bmatrix} \bar{y}_{1.} \\ \bar{y}_{1.} \\ \bar{y}_{1.} \\ \bar{y}_{2.} \\ \bar{y}_{2.} \\ \bar{y}_{2.} \end{bmatrix}
$$

The fitted model is therefore

$$
\begin{bmatrix} 29.80 \\ 28.57 \\ 29.97 \\ 30.33 \\ 31.27 \\ 30.37 \end{bmatrix}
=
\begin{bmatrix} 29.447 \\ 29.447 \\ 29.447 \\ 30.657 \\ 30.657 \\ 30.657 \end{bmatrix}
+
\begin{bmatrix} .353 \\ -.877 \\ .523 \\ -.327 \\ .613 \\ -.287 \end{bmatrix}
$$

	Observation vector	Fitted model vector	Fitted error vector

or, in brief, y $=$ $\bar{y}_{i.}$ $+$ $(y - \bar{y}_{i.})$

Hence our least squares estimate of μ_1 is $\bar{y}_{1.} = 29.447$, the sample mean for treatment one. Similarly μ_2 is estimated by $\bar{y}_{2.} = 30.657$.

Changing the Coordinate System

A decomposition of y with respect to an orthogonal coordinate system including U_1 and U_2 would permit us to test the hypothesis that $\mu_1 = 0$, and the hypothesis that $\mu_2 = 0$, since the associated projection coefficients reflect the size of μ_1 and μ_2. These hypotheses are not of interest, so we turn now to a more useful coordinate system.

A useful system needs to include U, the direction of the hypothesis of interest, $H_0 : \mu_1 = \mu_2$. Such a system is $U_1, U_2, \ldots, U_6$ as follows:

$$\frac{1}{\sqrt{6}}\begin{bmatrix} 1 \\ 1 \\ 1 \\ 1 \\ 1 \\ 1 \end{bmatrix}, \quad \frac{1}{\sqrt{6}}\begin{bmatrix} 1 \\ 1 \\ 1 \\ -1 \\ -1 \\ -1 \end{bmatrix}, \quad \frac{1}{\sqrt{2}}\begin{bmatrix} 1 \\ -1 \\ 0 \\ 0 \\ 0 \\ 0 \end{bmatrix}, \quad \frac{1}{\sqrt{6}}\begin{bmatrix} 1 \\ 1 \\ -2 \\ 0 \\ 0 \\ 0 \end{bmatrix}, \quad \frac{1}{\sqrt{2}}\begin{bmatrix} 0 \\ 0 \\ 0 \\ 1 \\ -1 \\ 0 \end{bmatrix}, \quad \frac{1}{\sqrt{6}}\begin{bmatrix} 0 \\ 0 \\ 0 \\ 1 \\ 1 \\ -2 \end{bmatrix}$$

Here U_1 and U_2 form a new coordinate system for the model space, while $U_3, \ldots, U_6$ are coordinate axes for the error space. Note that U_2 is the direction associated with the hypothesis of primary interest. Also U_1 reflects the size of the overall mean of our observations, since $y.U_1 = \sqrt{6}\bar{y}_{..}$, where $\bar{y}_{..}$ denotes the mean of all our six observations. Hence $Y.U_1$ has expected value $\sqrt{6}\mu$, where $\mu = (\mu_1 + \mu_2)/2$ is the mean of the population means, and we could use the projection to test the hypothesis that $\mu = 0$. This hypothesis, however, is generally of no interest in our current context.

Note that U_3 and U_4 reflect variation within the first treatment, since $y.U_3 = (y_{11} - y_{12})/\sqrt{2}$ and $y.U_4 = (y_{11} + y_{12} - 2y_{13})/\sqrt{6}$. Similarly, U_5 and U_6 reflect the variation within the second treatment. The distributions of the projection coefficients are pictured in Figure 6.2.

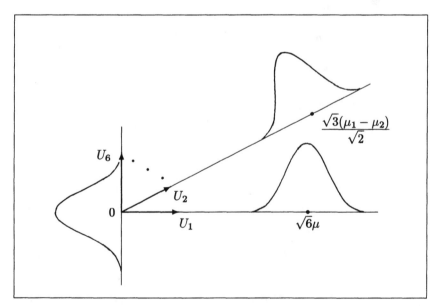

Figure 6.2: Pictorial representation of the distributions of the projection coefficients $Y.U_1, Y.U_2, \ldots, Y.U_6$.

Refitting the Model

Refitting the model using the new U_1 and U_2 vectors yields the model vector $(y.U_1)U_1 + (y.U_2)U_2$ as

$$
\begin{bmatrix} 30.052 \\ 30.052 \\ 30.052 \\ 30.052 \\ 30.052 \\ 30.052 \end{bmatrix}
+
\begin{bmatrix} -.605 \\ -.605 \\ -.605 \\ .605 \\ .605 \\ .605 \end{bmatrix}
=
\begin{bmatrix} \bar{y}_{..} \\ \bar{y}_{..} \\ \bar{y}_{..} \\ \bar{y}_{..} \\ \bar{y}_{..} \\ \bar{y}_{..} \end{bmatrix}
+
\begin{bmatrix} \bar{y}_{1.} - \bar{y}_{..} \\ \bar{y}_{1.} - \bar{y}_{..} \\ \bar{y}_{1.} - \bar{y}_{..} \\ \bar{y}_{2.} - \bar{y}_{..} \\ \bar{y}_{2.} - \bar{y}_{..} \\ \bar{y}_{2.} - \bar{y}_{..} \end{bmatrix}
=
\begin{bmatrix} 29.447 \\ 29.447 \\ 29.447 \\ 30.657 \\ 30.657 \\ 30.657 \end{bmatrix}
$$

the vector $\bar{y}_{i.}$ as previously. The refitted model is therefore

$$
\begin{bmatrix} 29.80 \\ 28.57 \\ 29.97 \\ 30.33 \\ 31.27 \\ 30.37 \end{bmatrix}
=
\begin{bmatrix} 30.052 \\ 30.052 \\ 30.052 \\ 30.052 \\ 30.052 \\ 30.052 \end{bmatrix}
+
\begin{bmatrix} -.605 \\ -.605 \\ -.605 \\ .605 \\ .605 \\ .605 \end{bmatrix}
+
\begin{bmatrix} .353 \\ -.877 \\ .523 \\ -.327 \\ .613 \\ -.287 \end{bmatrix}
$$

	Observation vector	Overall mean vector	Treatment vector	Fitted error vector
or,	y	$\bar{y}_{..}$	$(\bar{y}_{i.} - \bar{y}_{..})$	$(y - \bar{y}_{i.})$

This decomposition is pictured in Figure 6.3.

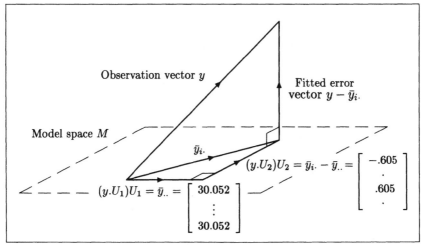

Figure 6.3: The least squares fit of the model, achieved by projecting the observation vector y onto the model space M.

In the decomposition the second component, $\bar{y}_{i.} - \bar{y}_{..}$, reflects the difference between the two treatments. Because of their functions, the first component, $\bar{y}_{..}$, we term the *overall mean vector* and the second $\bar{y}_{i.} - \bar{y}_{..}$ the *treatment vector*. Each lies in its own one-dimensional space. The third component, $y - \bar{y}_{i.}$, is the usual error vector, in this example lying in a four dimensional subspace.

Testing the Hypothesis

We are now in a position to test the hypothesis $H_0 : \mu_1 = \mu_2$ against the alternative hypothesis $H_0 : \mu_1 \neq \mu_2$. This involves checking whether the squared distance $(y.U_2)^2$ is comparable to, or larger than, the average of the corresponding squared distances in the error space. We use the test statistic

$$F \;=\; \frac{(y.U_2)^2}{\left[(y.U_3)^2 + (y.U_4)^2 + (y.U_5)^2 + (y.U_6)^2\right]/4} \;=\; \frac{\|\bar{y}_{i.} - \bar{y}_{..}\|^2}{\|y - \bar{y}_{i.}\|^2/4}$$

The appropriate Pythagorean breakup is

$$\|y\|^2 \;=\; \|\bar{y}_{..}\|^2 \;+\; \|\bar{y}_{i.} - \bar{y}_{..}\|^2 + \|y - \bar{y}_{i.}\|^2$$

$$\text{or,}\quad 5422.5445 \;=\; 5418.6160 + \;\;2.1962\;\; + \;\;1.7324$$

Our test statistic is therefore

$$F \;=\; \frac{2.1962}{1.7324/4} \;=\; \frac{2.1962}{.4331} \;=\; 5.07$$

To decide whether this is large, we compare our value with the 95 and 99 percentiles of the $F_{1,4}$ distribution. These are 7.71 and 21.20 respectively. Since our value of 5.07 is inside the normal range of values for even a 5% level test we declare it to be not significant, and conclude that there is no strong evidence of a difference in bulk between wool conditioned for two and three days.

ANOVA Table

Our analysis so far is conveniently encapsulated in the analysis of variance table, shown in Table 6.2.

Source of Variation	df	SS	MS	F
Overall mean $(H_0 : \mu = 0)$	1	5418.6160	5418.6160	
Treatments $(H_0 : \mu_1 = \mu_2)$	1	2.1962	2.1962	5.07 (ns)
Error	4	1.7324	0.4331	
Total	6	5422.5445		

Table 6.2: ANOVA table for our two treatment bulkometer example.

Simplified Decomposition

The overall mean vector, $\bar{y}_{..}$, does not enter into our calculation of the test statistic. Our orthogonal decomposition can therefore be simplified by subtracting $\bar{y}_{..}$ from both sides of the equation, yielding

$$ y - \bar{y}_{..} \quad = \quad (\bar{y}_{i.} - \bar{y}_{..}) \quad + \quad (y - \bar{y}_{i.}) $$

or,

$$
\begin{bmatrix} -.252 \\ -1.482 \\ -.082 \\ .278 \\ 1.218 \\ .318 \end{bmatrix}
=
\begin{bmatrix} -.605 \\ -.605 \\ -.605 \\ .605 \\ .605 \\ .605 \end{bmatrix}
+
\begin{bmatrix} .353 \\ -.877 \\ .523 \\ -.327 \\ .613 \\ -.287 \end{bmatrix}
$$

as illustrated in Figure 6.4.

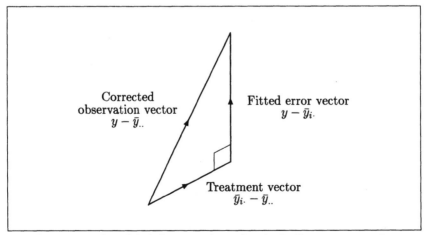

Figure 6.4: Simplified decomposition for the bulkometer example.

The vector $y - \bar{y}_{..}$ is referred to as the observation vector "corrected for the mean", or more simply, the corrected observation vector. The corresponding

Pythagorean breakup is

$$\|y - \bar{y}_{..}\|^2 \;=\; \|\bar{y}_{i.} - \bar{y}_{..}\|^2 \;+\; \|y - \bar{y}_{i.}\|^2$$

$$\text{or,} \quad 3.9285 \;=\; 2.1962 \;+\; 1.7324$$

This simplified breakup is associated with the ANOVA table commonly encountered in statistics texts, here shown in Table 6.3.

Source of Variation	df	SS	MS	F
Treatments $(H_0 : \mu_1 = \mu_2)$	1	2.1962	2.1962	5.07 (ns)
Error	4	1.7324	.4331	
Total (corrected)	5	3.9286		

Table 6.3: The more usual form of the ANOVA table for the two treatment bulkometer example.

Estimation of σ^2

In our example we have transformed our set of independent random variables, $Y_{11}, Y_{12}, Y_{13} \sim N[\mu_1, \sigma^2]$ and $Y_{21}, Y_{22}, Y_{23} \sim N[\mu_2, \sigma^2]$, into a new set of independent random variables

$$Y.U_1 \sim N\left[\sqrt{6}\mu, \sigma^2\right], \qquad Y.U_2 \sim N\left[\sqrt{3}(\mu_1 - \mu_2)/\sqrt{2}, \sigma^2\right],$$
$$Y.U_3, \ \dots \ , Y.U_6 \quad \sim \quad N\left[0, \sigma^2\right]$$

The first and second of these were used to estimate $\mu = (\mu_1 + \mu_2)/2$ and $\mu_1 - \mu_2$ respectively, while the last four were used to estimate σ^2 via

$$s^2 = \frac{(y.U_3)^2 + (y.U_4)^2 + (y.U_5)^2 + (y.U_6)^2}{4} = \frac{\|y - \bar{y}_{i.}\|^2}{4} = \frac{1.7324}{4} = .4331$$

Here $y.U_3$, $y.U_4$, $y.U_5$ and $y.U_6$ are all observations from $N[0, \sigma^2]$ distributions, so their squares all average to σ^2 over many repeats of the experiment, as described in §3.3.

Assumption Checking

Do the assumptions of normality of each population, common variance of the two populations, and independence of our observations appear reasonable?

    If the normality assumption is valid, then a histogram of the errors, or residuals, should be approximately normal. These values are the components of our error vector: the dregs remaining after the observations have been

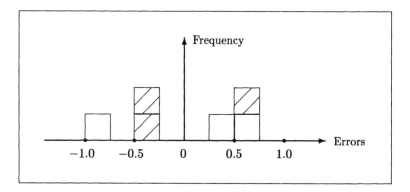

Figure 6.5: Histogram of errors. The errors for the first treatment are un-shaded, while those for the second treatment are shaded.

filtered through the model. The histogram is shown in Figure 6.5, and reveals no obvious non-normality within either treatment.

From the histogram it is also evident that the sample variances for each treatment, s_1^2 and s_2^2, are roughly similar. More formally, we can calculate s_1^2 and s_2^2 and check for equality using the test statistic $F = s_1^2/s_2^2$. Under the null hypothesis of equality of variances this follows an $F_{2,2}$ distribution. Now

$$s_1^2 = \left[(y.U_3)^2 + (y.U_4)^2\right]/2 = \left[.87^2 + (-.641)^2\right]/2 = .5839$$
$$s_2^2 = \left[(y.U_5)^2 + (y.U_6)^2\right]/2 = \left[(-.655)^2 + .351^2\right]/2 = .2827$$

and hence $F = .5839/.2827 = 2.07$. The null hypothesis is rejected at the 5% level of significance if the test statistic lies outside the range .03 to 39.0, the 2.5 and 97.5 percentiles of the $F_{2,2}$ distribution, where the 2.5 percentile can be calculated as the reciprocal of the 97.5 percentile. In our example our test statistic is within the normal range, so our formal test reveals no evidence to suspect that the two population variances, σ_1^2 and σ_2^2, are unequal.

The assumption of independence of the subsamples is checked by examining the basic conduct of the experiment. For example, was the bulk sample adequately mixed between subsamples? In the absence of other information, we must presume that the independence assumption is valid.

In summary, our scrutiny has revealed no reason to be suspicious of our assumptions. Note, however, that with such small sample sizes we would be lucky to detect any violations; for example, the $F_{2,2}$ test of $H_0 : \sigma_1^2 = \sigma_2^2$ is a very weak test.

Confidence Interval for $\mu_1 - \mu_2$

The best estimate for the mean difference in bulkometer reading between samples conditioned for two and three days, $\mu_1 - \mu_2$, is given by $\bar{y}_1. - \bar{y}_2. =$

$29.447 - 30.657 = -1.21$. That is, if asked to gamble on a number, we would say that conditioning samples for three days produced bulkometer readings which were on average 1.21 bulk units higher than readings from subsamples conditioned for only two days. Our statistical analysis, however, cautions us not to rely too heavily on this result, since the data is also consistent with the idea that there is no difference between two and three day conditioning.

How accurate is our estimate of $\mu_1 - \mu_2$? More specifically, for what range of values is $\mu_1 - \mu_2$ consistent with our data? To answer this question we calculate the 95% confidence interval for $\mu_1 - \mu_2$. For this we need the t-statistic as defined in §3.4.

The random variable associated with $\mu_1 - \mu_2$ is $Y.U_2 = \sqrt{3}(\overline{Y}_{1.} - \overline{Y}_{2.})/\sqrt{2}$. This has a $N(\sqrt{3}(\mu_1 - \mu_2)/\sqrt{2}, \sigma^2)$ distribution. By subtracting its mean we obtain the random variable $W_1 = \sqrt{3}\left[(\overline{Y}_{1.} - \overline{Y}_{2.}) - (\mu_1 - \mu_2)\right]\sqrt{2}$ which has a $N(0, \sigma^2)$ distribution. We use W_1 as the numerator of our t-statistic. In the denominator we use $Y.U_3$, $Y.U_4$, $Y.U_5$ and $Y.U_6$ which all have a $N[0, \sigma^2]$ distribution. Hence the ratio

$$\frac{\sqrt{3}\left[(\overline{Y}_{1.} - \overline{Y}_{2.}) - (\mu_1 - \mu_2)\right]/\sqrt{2}}{\sqrt{\left[(Y.U_3)^2 + (Y.U_4)^2 + (Y.U_5)^2 + (Y.U_6)^2\right]/4}}$$

is a t_4 statistic. To obtain a 95% confidence interval for the difference between the two means we gamble that the realized value,

$$t = \frac{\sqrt{3}\left[(\bar{y}_{1.} - \bar{y}_{2.}) - (\mu_1 - \mu_2)\right]/\sqrt{2}}{\sqrt{\left[(y.U_3)^2 + (y.U_4)^2 + (y.U_5)^2 + (y.U_6)^2\right]/4}} = \frac{(\bar{y}_{1.} - \bar{y}_{2.}) - (\mu_1 - \mu_2)}{\sqrt{2s^2/3}}$$

lies between the 2.5 and 97.5 percentiles of the t_4 distribution. Thus our 95% confidence interval for $\mu_1 - \mu_2$ is determined by

$$-2.776 \leq \frac{-1.21 - (\mu_1 - \mu_2)}{\sqrt{2(.4331)/3}} \leq 2.776$$

This can be rearranged in the usual way to yield

$$-0.28 \leq \mu_2 - \mu_1 \leq 2.70$$

In summary, we can state with 95% confidence in our correctness, that the average increase in bulkometer reading, $\mu_2 - \mu_1$, through conditioning for three days rather than two, is between -0.3 and 2.7 units.

Report on Study

In a report on this small study the statistical results can be summarized in the form shown in Table 6.4. Each treatment mean is presented along with the standard error of the difference between the two means (the SED), and the significance of the test of $H_0 : \mu_1 - \mu_2 = 0$. The SED is the estimated standard error of the random variable $\overline{Y}_1. - \overline{Y}_2.$. The latter has a variance of $2\sigma^2/3$, so the SED is $\sqrt{2s^2/3} = \sqrt{2(.4331)/3} = .54$. The SED can be thought of as a typical difference between two treatment means.

Mean bulkometer reading, in cm^3/g	
Treatment	
Conditioned for 2 days	29.5
Conditioned for 3 days	30.7
SED	0.54
Significance of difference	ns

Table 6.4: Results of the bulkometer experiment.

Alternatively, we could state in our report that we are 95% confident that the increase in bulkometer reading through conditioning for three days rather than two days is in the range 1.2 ± 1.5 units.

Exercise for the reader

(6.1) Use the full dataset in Table 6.1, consisting of two samples of size five, to recalculate the simplified decomposition

$$y - \bar{y}.. = (\bar{y}_i. - \bar{y}..) + (y - \bar{y}_i.)$$

Recalculate the test statistic and present your results as shown in Table 6.4. Also recalculate the 95% confidence interval for the difference in bulkometer reading, $\mu_2 - \mu_1$.

6.2   General Case

Here we summarize the essential steps in the analysis of our illustrative example, and generalize these to two samples of arbitrary size. For simplicity of presentation we restrict our attention to problems with equal sample sizes, leaving the case of unequal sample sizes until Appendix A. In this section we also show how our geometric results give meaning to traditional terminology, and lead to the traditional formulae.

Setting the Scene

Our interest is in a pair of populations, assumed normally distributed, and whose means μ_1 and μ_2 and common variance σ^2 are unknown. We would like to estimate μ_1, μ_2 and σ^2, test the hypothesis $H_0 : \mu_1 = \mu_2$, and derive a 95% confidence interval for $\mu_1 - \mu_2$. To this end, we take random samples $y_{11}, \ldots, y_{1n}$ and $y_{21}, \ldots, y_{2n}$ from the two populations. These observations are values from the $2n$ independent sample random variables $Y_{11}, \ldots, Y_{1n} \sim N[\mu_1, \sigma^2]$ and $Y_{21}, \ldots, Y_{2n} \sim N[\mu_2, \sigma^2]$.

Raw Materials

The observation vector y is just the vector of observations, $[y_{11}, \ldots, y_{1n}, y_{21}, \ldots, y_{2n}]^T$.

The model vector is

$$
\begin{bmatrix} \mu_1 \\ \vdots \\ \mu_2 \\ \vdots \end{bmatrix} = \mu_1 \begin{bmatrix} 1 \\ \vdots \\ 0 \\ \vdots \end{bmatrix} + \mu_2 \begin{bmatrix} 0 \\ \vdots \\ 1 \\ \vdots \end{bmatrix}
$$

so the model space is $M = \mathrm{span}\{[1, \ldots, 0, \ldots]^T, [0, \ldots, 1, \ldots]^T\}$. This is a two dimensional subspace of $2n$-space, with coordinate axes determined by the unit vectors

$$
U_1 = \frac{1}{\sqrt{n}} \begin{bmatrix} 1 \\ \vdots \\ 0 \\ \vdots \end{bmatrix} \quad \text{and} \quad U_2 = \frac{1}{\sqrt{n}} \begin{bmatrix} 0 \\ \vdots \\ 1 \\ \vdots \end{bmatrix}
$$

Alternative, more useful coordinate axes, are determined by the unit vectors

$$
U_1 = \frac{1}{\sqrt{2n}} \begin{bmatrix} 1 \\ \vdots \\ 1 \\ \vdots \end{bmatrix} \quad \text{and} \quad U_2 = \frac{1}{\sqrt{2n}} \begin{bmatrix} 1 \\ \vdots \\ -1 \\ \vdots \end{bmatrix}
$$

From now, these are the axes we shall use.

The direction associated with the hypothesis $H_0 : \mu_1 = \mu_2$ is $U_2 = [1, \ldots, -1, \ldots]^T/\sqrt{2n}$. To confirm this we calculate the (signed) length of the projection of y onto U_2,

$$
y.U_2 = \begin{bmatrix} y_{11} \\ \vdots \\ y_{21} \\ \vdots \end{bmatrix} \cdot \frac{1}{\sqrt{2n}} \begin{bmatrix} 1 \\ \vdots \\ -1 \\ \vdots \end{bmatrix} = \frac{1}{\sqrt{2n}} (y_{11} + \cdots + y_{1n} - y_{21} - \cdots - y_{2n})
$$

$$
= \frac{1}{\sqrt{2n}} (n\bar{y}_{1.} - n\bar{y}_{2.}) = \frac{\sqrt{n}}{\sqrt{2}} (\bar{y}_{1.} - \bar{y}_{2.})
$$

The corresponding random variable $Y.U_2$ therefore has an expected value of $\sqrt{n}(\mu_1 - \mu_2)/\sqrt{2}$, a constant multiple of $\mu_1 - \mu_2$. Thus if $\mu_1 = \mu_2$ the projection coefficient $y.U_2$ will be "small", whereas if $\mu_1 = \mu_2$ the projection coefficient will be "big".

This completes the collection of all our raw materials: an observation vector y, a model space M and the direction U_2 associated with the hypothesis $H_0 : \mu_1 = \mu_2$, as follows:

$$
y = \begin{bmatrix} y_{11} \\ \vdots \\ y_{21} \\ \vdots \end{bmatrix}, \quad M = \text{span} \left\{ \begin{bmatrix} 1 \\ \vdots \\ 0 \\ \vdots \end{bmatrix}, \begin{bmatrix} 0 \\ \vdots \\ 1 \\ \vdots \end{bmatrix} \right\}, \quad \text{and} \quad U_2 = \frac{1}{\sqrt{2n}} \begin{bmatrix} 1 \\ \vdots \\ -1 \\ \vdots \end{bmatrix}
$$

Fitting the Model

To fit the model, we project the observation vector y onto the model space M. This yields the fitted model vector $\bar{y}_{i.} = [\bar{y}_{1.}, \ldots, \bar{y}_{1.}, \bar{y}_{2.}, \ldots, \bar{y}_{2.}]^T$, the vector in the model space which is closest to our observation vector. Here $\bar{y}_{1.}$ and $\bar{y}_{2.}$, the treatment means, are termed the "least squares" estimates of μ_1 and μ_2.

Using the alternative coordinate axes for M, we can write $(y.U_1)U_1 + (y.U_2)U_2 = \bar{y}_{..} + (\bar{y}_{i.} - \bar{y}_{..}) = \bar{y}_{i.}$. The resulting orthogonal decomposition of the observation vector is

$$
y \quad = \quad \bar{y}_{..} \quad + \quad (\bar{y}_{i.} - \bar{y}_{..}) \quad + \quad (y - \bar{y}_{i.})
$$

or,

$$
\begin{bmatrix} y_{11} \\ \vdots \\ y_{21} \\ \vdots \end{bmatrix} = \begin{bmatrix} \bar{y}_{..} \\ \vdots \\ \bar{y}_{..} \\ \vdots \end{bmatrix} + \begin{bmatrix} \bar{y}_{1.} - \bar{y}_{..} \\ \vdots \\ \bar{y}_{2.} - \bar{y}_{..} \\ \vdots \end{bmatrix} + \begin{bmatrix} y_{11} - \bar{y}_{1.} \\ \vdots \\ y_{21} - \bar{y}_{2.} \\ \vdots \end{bmatrix}
$$

| | Observation vector | Overall mean vector | Treatment vector | Fitted error vector |

This can be written more simply as a two-way decomposition of the corrected observation vector, into treatment vector plus error vector:

$$y - \bar{y}_{..} = (\bar{y}_{i.} - \bar{y}_{..}) + (y - \bar{y}_{i.})$$

as illustrated in Figure 6.6.

In this triangular decomposition, the corrected observation vector lies in a $2n - 1$ dimensional subspace of $2n$-space, while the treatment vector lies in the one dimensional "treatment space" and the error vector lies in the $2n - 2$ dimensional error space.

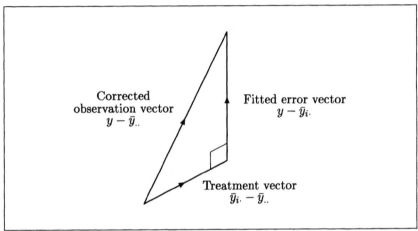

Figure 6.6: Simplified orthogonal decomposition, two treatment case.

A simple choice of coordinate system for the error space is

$$
\begin{array}{cccccccc}
U_3 & U_4 & & U_{n+1} & U_{n+2} & U_{n+3} & & U_{2n} \\
\begin{bmatrix} 1 \\ -1 \\ 0 \\ 0 \\ \vdots \\ 0 \\ 0 \\ 0 \\ 0 \\ \vdots \\ 0 \end{bmatrix} &
\begin{bmatrix} 1 \\ 1 \\ -2 \\ 0 \\ \vdots \\ 0 \\ 0 \\ 0 \\ 0 \\ \vdots \\ 0 \end{bmatrix} & \cdots &
\begin{bmatrix} 1 \\ 1 \\ 1 \\ 1 \\ \vdots \\ -(n-1) \\ 0 \\ 0 \\ 0 \\ \vdots \\ 0 \end{bmatrix} &
\begin{bmatrix} 0 \\ 0 \\ 0 \\ 0 \\ \vdots \\ 0 \\ 1 \\ -1 \\ 0 \\ \vdots \\ 0 \end{bmatrix} &
\begin{bmatrix} 0 \\ 0 \\ 0 \\ 0 \\ \vdots \\ 0 \\ 1 \\ 1 \\ -2 \\ \vdots \\ 0 \end{bmatrix} & \cdots &
\begin{bmatrix} 0 \\ 0 \\ 0 \\ 0 \\ \vdots \\ 0 \\ 1 \\ 1 \\ 1 \\ \vdots \\ -(n-1) \end{bmatrix} \\
\sqrt{2} & \sqrt{6} & & \sqrt{n(n-1)} & \sqrt{2} & \sqrt{6} & & \sqrt{n(n-1)}
\end{array}
$$

The first $n - 1$ of these axes, $U_3, \ldots, U_{n+1}$, are associated with errors within treatment one, while the last $n - 1$ axes, $U_{n+2}, \ldots, U_{2n}$, are associated with errors within treatment two. Together with $U_1 = [1, \ldots, 1]^T / \sqrt{2n}$ and $U_2 = [1, \ldots, -1, \ldots]^T / \sqrt{2n}$, $U_3, \ldots, U_{2n}$ make up a coordinate system for $2n$-space which is appropriate for the two population case. Note that just as the components of the fitted model vector, $\bar{y}_{i \cdot}$, can be written as the projection vectors $\bar{y}_{\cdot \cdot} = (y.U_1)U_1$ and $\bar{y}_{i \cdot} - \bar{y}_{\cdot \cdot} = (y.U_2)U_2$, so too can the fitted error vector, $y - \bar{y}_{i \cdot}$, be written in the fuller form

$$y - \bar{y}_{i \cdot} = (y.U_3)U_3 + (y.U_4)U_4 + \cdots + (y.U_{2n})U_{2n}$$

Testing the Hypothesis

Our test statistic is a ratio of squared lengths of projection coefficients, $y.U_i$, so it is now necessary to consider the distributions of the associated random variables, $Y.U_i$.

From §3.2 we know that the projection coefficients, $Y.U_i$, are independent and normally distributed, with common variance σ^2. It is a straightforward exercise to show that their expected values are

$$E(Y.U_1) = \sqrt{2n}\mu, \quad E(Y.U_2) = \sqrt{n}(\mu_1 - \mu_2)/\sqrt{2},$$
$$E(Y.U_3) = 0, \quad \ldots, \quad E(Y.U_{2n}) = 0$$

as summarized in Table 6.5.

	Average, or expected value	Variance
$Y.U_1$	$\sqrt{2n}\mu$	σ^2
$Y.U_2$	$\sqrt{n}(\mu_1 - \mu_2)/\sqrt{2}$	σ^2
$Y.U_3$	0	σ^2
$\vdots$	$\vdots$	$\vdots$
$Y.U_{2n}$	0	σ^2

Table 6.5: Means and variances of the projection coefficients, $Y.U_i$.

Our test of the hypothesis of equal means, $\mu_1 = \mu_2$, now simply involves checking whether $(y.U_2)^2$ is comparable to, or larger than, the average of the squared projection coefficients in the error space. The appropriate test statistic is

$$F = \frac{(y.U_2)^2}{\left[(y.U_3)^2 + \cdots + (y.U_{2n})^2\right]/(2n-2)} = \frac{\|\bar{y}_{i \cdot} - \bar{y}_{\cdot \cdot}\|^2}{\|y - \bar{y}_{i \cdot}\|^2/(2n-2)}$$

If $\mu_1 = \mu_2$ this test statistic follows an $F_{1, 2n-2}$ distribution, since then $Y.U_2$ as well as $Y.U_3, \ldots, Y.U_{2n}$ follow the $N[0, \sigma^2]$ distribution. If $\mu_1 \neq \mu_2$ then $Y.U_2$ has non-zero mean, so the test statistic is inflated. The ensuing test procedure is to compare the calculated F value with the percentiles of the $F_{1, 2n-2}$ distribution, and to reject the hypothesis $H_0 : \mu_1 = \mu_2$ if the calculated value is unusually large.

ANOVA Table

The Pythagorean breakup of the corrected observation vector is

$$\|y - \bar{y}_{..}\|^2 = (y.U_2)^2 + (y.U_3)^2 + \cdots + (y.U_{2n})^2$$

or, in brief $\|y - \bar{y}_{..}\|^2 = \|\bar{y}_{i.} - \bar{y}_{..}\|^2 + \|y - \bar{y}_{i.}\|^2$

These ingredients of the test statistic, and the test statistic itself, are most conveniently summarized in the traditional analysis of variance table shown here as Table 6.6.

Source of Variation	df	SS	MS	F
Treatments	1	$\|\bar{y}_{i.} - \bar{y}_{..}\|^2$	$\|\bar{y}_{i.} - \bar{y}_{..}\|^2$	$\dfrac{\|\bar{y}_{i.} - \bar{y}_{..}\|^2}{\|y - \bar{y}_{i.}\|^2/(2n-2)}$
Error	$2n - 2$	$\|y - \bar{y}_{i.}\|^2$	$\dfrac{\|y - \bar{y}_{i.}\|^2}{2n-2}$	
Total (corrected)	$2n - 1$	$\|y - \bar{y}_{..}\|^2$		

Table 6.6: ANOVA table for two populations.

In the table, the sums of squares can be expanded to the traditional algebraic expressions as follows:

$$\text{Total SS} = \|y - \bar{y}_{..}\|^2 = \sum_{i=1}^{2}\sum_{j=1}^{n}(y_{ij} - \bar{y}_{..})^2$$

$$= \|y\|^2 - \|\bar{y}_{..}\|^2 = \sum_{i=1}^{2}\sum_{j=1}^{n} y_{ij}^2 - 2n\bar{y}_{..}^{\,2}$$

$$\text{Treatment SS} = \|\bar{y}_{i.} - \bar{y}_{..}\|^2 = n\sum_{i=1}^{2}(\bar{y}_{i.} - \bar{y}_{..})^2$$

$$\text{Error SS} = \sum_{i=1}^{2}\sum_{j=1}^{n}(y_{ij} - \bar{y}_{i.})^2$$

Normal procedure is to calculate the total and treatment sums of squares, and then to obtain the error sum of squares by subtraction.

Estimation of σ^2

The directions U_1 and U_2 have been used in the least squares estimation of the parameters μ_1 and μ_2. The remaining directions, $U_3, \ldots, U_{2n}$ are used to estimate σ^2, as discussed in §3.3. The resulting estimate is

$$
\begin{aligned}
s^2 &= \frac{(y.U_3)^2 + \cdots + (y.U_{2n})^2}{2n-2} = \frac{\|y - \bar{y}_{i\cdot}\|^2}{2n-2} = \frac{\sum\limits_{i=1}^{2}\sum\limits_{j=1}^{n}(y_{ij} - \bar{y}_{i\cdot})^2}{2(n-1)} \\
&= \frac{1}{2(n-1)}\left[\sum_{j=1}^{n}(y_{1j} - \bar{y}_{1\cdot})^2 + \sum_{j=1}^{n}(y_{2j} - \bar{y}_{2\cdot})^2\right] = \frac{s_1^2 + s_2^2}{2}
\end{aligned}
$$

the average of the sample variances, s_1^2 and s_2^2, for treatments one and two. For this reason, s^2 is often termed the *pooled variance estimate*.

Assumption Checking

The assumptions which require checking are those of normality and common variance of the distributions, and independence of the observations, as described in §6.1. A histogram of the errors, such as was shown in Figure 6.5, is the most important diagnostic tool. Normality of distribution can be formally checked by calculating coefficients of skewness or kurtosis. The assumption of common variance, $H_0 : \sigma_1^2 = \sigma_2^2$, can be formally checked by calculating the test statistic $F = s_1^2/s_2^2$. If the assumption holds, this test statistic comes from the $F_{n-1,n-1}$ distribution, since

$$
s_1^2 = \frac{(y.U_3)^2 + \cdots + (y.U_{n+1})^2}{n-1} \quad \text{and} \quad s_2^2 = \frac{(y.U_{n+2})^2 + \cdots + (y.U_{2n})^2}{n-1}
$$

where $U_3, \ldots, U_{2n}$ is the error space coordinate system given earlier in this section. The assumption is rejected if the calculated F value lies outside the interval bounded by the 2.5 and 97.5 percentiles of the $F_{n-1,n-1}$ distribution. The last assumption, that of independence, is checked by considering the details of the conduct of the experiment, and by plotting the errors against the order in which the observations were obtained.

Confidence Interval

Since $Y.U_2 = \sqrt{n}(\overline{Y}_{1.} - \overline{Y}_{2.})/\sqrt{2}$ follows the $N\left[\sqrt{n}(\mu_1 - \mu_2)/\sqrt{2}, \sigma^2\right]$ distribution, the random variable $\sqrt{n}\left[(\overline{Y}_{1.} - \overline{Y}_{2.}) - (\mu_1 - \mu_2)\right]/\sqrt{2}$ must follow the $N[0, \sigma^2]$ distribution. Hence

$$T = \frac{\sqrt{n}\left[(\overline{Y}_{1.} - \overline{Y}_{2.}) - (\mu_1 - \mu_2)\right]/\sqrt{2}}{\sqrt{\dfrac{(Y.U_3)^2 + \cdots + (Y.U_{2n})^2}{2n - 2}}}$$

is a t_{2n-2} statistic with $2n - 2$ degrees of freedom. We gamble that the realized value of T lies between the 2.5 and 97.5 percentiles, and obtain a 95% confidence interval for $\mu_1 - \mu_2$ of

$$(\bar{y}_{1.} - \bar{y}_{2.}) - s\sqrt{\frac{2}{n}}t_{2n-2}(.975) \leq \mu_1 - \mu_2 \leq (\bar{y}_{1.} - \bar{y}_{2.}) + s\sqrt{\frac{2}{n}}t_{2n-2}(.975)$$

The t_{2n-2} values are given at the end of the book in Table T.2. Note that the random variable T can again be written in the form

$$T = \frac{Y.U - E(Y.U)}{\sqrt{\text{Estimator of } \sigma^2}}$$

where U is the direction associated with the hypothesis $H_0 : \mu_1 = \mu_2$.

Equivalence of the F and t Tests

Statistical texts generally use a test statistic of

$$\frac{\bar{y}_{1.} - \bar{y}_{2.}}{\sqrt{2s^2/n}}$$

for this two population, independent samples, problem. Under $H_0 : \mu_1 = \mu_2$ it is said to follow a t_{2n-2} distribution. Is this equivalent to our F test? We can reexpress our test statistic in terms of $\bar{y}_{1.}, \bar{y}_{2.}$ and s^2 as

$$\begin{aligned}
F &= \frac{\|\bar{y}_{i.} - \bar{y}_{..}\|^2}{\|y - \bar{y}_{i.}\|^2/(2n - 2)} = \frac{\|\bar{y}_{i.} - \bar{y}_{..}\|^2}{s^2} = \frac{n(\bar{y}_{1.} - \bar{y}_{..})^2 + n(\bar{y}_{2.} - \bar{y}_{..})^2}{s^2} \\
&= \frac{2n\left[(\bar{y}_{1.} - \bar{y}_{2.})/2\right]^2}{s^2} = \frac{n(\bar{y}_{1.} - \bar{y}_{2.})^2}{2s^2} = \left(\frac{\bar{y}_{1.} - \bar{y}_{2.}}{\sqrt{2s^2/n}}\right)^2 = t^2
\end{aligned}$$

where the second to last term is the square of the traditional t value. Also, recall from §3.4 that the reference distribution $F_{1,2n-2}$ is simply the square of the t_{2n-2}. Hence our F test is equivalent to the t test.

Virtues of Our Estimates

The discussion in §5.3 applies equally to our present two population case.

6.3   Computing

It is only sensible to use a statistical computing package to carry out the arithmetic our analysis involves. Examples of such packages are MINITAB, GENSTAT, SAS, S and BMDP. Because of its widespread distribution we have chosen to use Minitab in this book to illustrate how each design can be analyzed. Any package, however, which includes simple or multiple regression routines could be used in its place.

Most useful in this book is the Minitab REGRESS command, since the word regress is really a synonym for "project". In Table 6.7 we list the commands required to analyze the bulkometer data from §6.1, and in Table 6.8 show the resulting output.

Minitab commands	
List of commands	**Interpretation**
name c1='Bulk' name c2='HypoDirn' name c3='Errors'	Naming the columns for convenience
set c1 29.80 28.57 29.97 30.33 31.27 30.37	Set the observations, y, into column one of the worksheet
set c2 1 1 1 -1 -1 -1 let c2=c2/sqrt(6)	Put the unit vector U_2 into column two
regress c1 1 c2; resids c3.	Project y onto the model space $M = \text{span}\{U_1, U_2\}$, saving errors in column three
histogram c3 stop	Print the error histogram Signifies end of job

Table 6.7: Minitab commands for the bulkometer example.

```
Minitab output

The regression equation is
Bulk = 30.1 - 1.48 HypoDirn

Predictor        Coef    Stdev   t-ratio       p
Constant      30.0517   0.2687   111.86   0.000
HypoDirn      -1.4819   0.6581    -2.25   0.087

s = 0.6581   R-square = 55.9%   R-sq(adj) = 44.9%

Analysis of Variance

SOURCE              DF       SS       MS      F       p
Regression           1   2.1962   2.1962   5.07   0.087
Error                4   1.7323   0.4331
Total                5   3.9285

Histogram of Errors      N = 6

Midpoint        Count
     -0.8          1   *
     -0.6          0
     -0.4          1   *
     -0.2          1   *
      0.0          0
      0.2          0
      0.4          1   *
      0.6          2   **
```

Table 6.8: Selected Minitab output for the bulkometer example.

We now go through the commands in Table 6.7 and the output in Table 6.8, relating these to the concepts discussed in §6.1.

Fitting the Model

The Regress command fits the model by projecting the observation vector y onto the model space $M = \text{span}\{U_1, U_2\}$. Minitab automatically assumes the constant term, corresponding to U_1, is included in the model, so it suffices to specify the projection of y onto U_2. The fitted model is output in the form of a "regression equation". This corresponds to a fitted model vector written in the form

$$\text{Fitted model vector} \;=\; \bar{y}_{..} \;+\; (y.U_2)U_2$$

where $\bar{y}_{..} = 30.1$ and $y.U_2 = -1.48$ in our example.

Testing the Hypothesis

The ANOVA table given in the Minitab output corresponds to the Pythagorean breakup

$$\|y - \bar{y}_{..}\|^2 \;=\; (y.U_2)^2 \;+\; \|y - \bar{y}_{i.}\|^2$$

The test statistic $F = (y.U_2)^2/s^2 = 2.1962/0.4331 = 5.07$ is also given, along with the probability, $p = 0.087$, of observing an F value as large or larger than 5.07 under the null hypothesis $\mu_1 = \mu_2$.

Checking of Assumptions

The error histogram can be obtained by saving the residuals, or errors, from the fitted model. This allows checking of the normality assumption.

ONEWAY Command

In the context of this chapter, the Oneway command of Minitab is even more useful than the Regress command. The appropriate commands are shown in Table 6.9, and the relevant part of the corresponding output in Table 6.10 (for economy of space we have not reprinted the error histogram, or reproduced the confidence intervals for μ_1 and μ_2 as output by Minitab).

The Oneway command produces the ANOVA table corresponding to the Pythagorean breakup

$$\|y - \bar{y}_{..}\|^2 \;=\; \|\bar{y}_{i.} - \bar{y}_{..}\|^2 \;+\; \|y - \bar{y}_{i.}\|^2$$

The test statistic, $F = 5.07$, is again included. The treatment means, $\bar{y}_{1.}$ and $\bar{y}_{2.}$, are also printed, together with the pooled standard deviation, $s = 0.658$, calculated from the pooled variance, $s^2 = 0.433$. These three values, $\bar{y}_{1.}$, $\bar{y}_{2.}$ and s, provide the information necessary for calculating the confidence interval for $\mu_1 - \mu_2$. In addition, the output provides separate estimates of the standard deviations for each population, $s_1 = 0.764$ and $s_2 = 0.532$. These can immediately be used to check the assumption of common variance, using the test statistic $F = s_1^2/s_2^2$.

Minitab commands

List of commands	Interpretation
⌈name c1='Bulk' |name c2='Trtments' ⌊name c3='Errors'	⌈Naming the columns ⌊for convenience
⌈set c1 ⌊29.80 28.57 29.97 30.33 31.27 30.37	⌈Set y into column one ⌊of the worksheet
⌈set c2 ⌊1 1 1 2 2 2	⌈Set treatment numbers ⌊into column two
⌈oneway c1 c2 c3	⌈Perform the ANOVA, saving ⌊errors in column three
⌈histogram c3 ⌈stop	⌈Print error histogram ⌈Signifies end of job

Table 6.9: Using the ONEWAY command of Minitab, bulkometer example.

Minitab output

ANALYSIS OF VARIANCE ON Bulk

SOURCE	DF	SS	MS	F	p
Trtments	1	2.196	2.196	5.07	0.087
ERROR	4	1.732	0.433		
TOTAL	5	3.928			

LEVEL	N	MEAN	STDEV
1	3	29.447	0.764
2	3	30.657	0.532

POOLED STDEV = 0.658

Table 6.10: Part of the corresponding Minitab output.

6.4 Summary

In this chapter we have considered two normally distributed populations with unknown means μ_1 and μ_2, and common unknown variance σ^2. Our aim has been to find point and interval estimates of $\mu_1 - \mu_2$, and to test the hypothesis that $\mu_1 = \mu_2$. From the population we have taken independent, equally sized samples of n observations, $y_{11}, \ldots, y_{1n}$ and $y_{21}, \ldots, y_{2n}$.

Our observation vector y is $[y_{11}, \ldots, y_{1n}, y_{21}, \ldots, y_{2n}]^T$. The model vector of expected values is $[\mu_1, \ldots, \mu_1, \mu_2, \ldots, \mu_2]^T$, so our model space M is the plane in $2n$-space spanned by the unit vectors $U_1 = [1, \ldots, 1, 1, \ldots, 1]^T / \sqrt{2n}$ and $U_2 = [1, \ldots, 1, -1, \ldots, -1]^T / \sqrt{2n}$. The direction associated with the hypothesis $H_0 : \mu_1 = \mu_2$ is the vector U_2, since $y.U_2 = \sqrt{n}(\bar{y}_{1.} - \bar{y}_{2.})/\sqrt{2}$, the signed length of the projection of y in the direction U_2, reflects the size of the difference $\mu_1 - \mu_2$. The error space is $2n - 2$ dimensional, with axes $U_3, \ldots, U_{2n}$.

In fitting the model we estimate the vector decomposition demanded by the model, namely $y = \mu_i + (y - \mu_i)$, by means of the orthogonal decomposition $y = \bar{y}_{i.} + (y - \bar{y}_{i.})$. This is achieved by projecting y onto M, and yields $\bar{y}_{1.}$ and $\bar{y}_{2.}$ as our estimates of μ_1 and μ_2, and $s^2 = \|y - \bar{y}_{i.}\|^2/(2n - 2)$ as our estimate of σ^2.

For hypothesis testing purposes, a more useful form of the orthogonal decomposition is

$$ y - \bar{y}_{..} \quad = \quad (\bar{y}_{i.} - \bar{y}_{..}) \quad + \quad (y - \bar{y}_{i.}) $$

as shown in Figure 6.7.

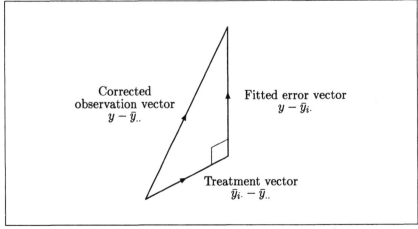

Figure 6.7: Orthogonal decomposition for two populations.

Corresponding to this is the Pythagorean breakup

$$\|y - \bar{y}_{..}\|^2 \;=\; \|\bar{y}_{i\cdot} - \bar{y}_{..}\|^2 \;+\; \|y - \bar{y}_{i\cdot}\|^2$$

Since $\|\bar{y}_{i\cdot} - \bar{y}_{..}\|^2 = (y.U_2)^2$ estimates σ^2 if and only if $\mu_1 = \mu_2$, whereas $s^2 = \|y - \bar{y}_{i\cdot}\|^2/(2n-2) = \left[(y.U_3)^2 + \cdots + (y.U_{2n})^2\right]/(2n-2)$ always estimates σ^2, we use

$$F \;=\; \frac{\|\bar{y}_{i\cdot} - \bar{y}_{..}\|^2}{\|y - \bar{y}_{i\cdot}\|^2/(2n-2)} \;=\; \frac{n(\bar{y}_{1\cdot} - \bar{y}_{2\cdot})^2}{2s^2}$$

as our test statistic. This test statistic follows an $F_{1,2n-2}$ distribution if $\mu_1 = \mu_2$, so we reject the hypothesis that $\mu_1 = \mu_2$ if the test statistic exceeds the 95 percentile of this distribution.

A 95% confidence interval for μ is given by

$$(\bar{y}_{1\cdot} - \bar{y}_{2\cdot}) \;\pm\; s\sqrt{\frac{2}{n}}\,t_{2n-2}(.975)$$

Class Exercise

In the class exercise of Chapter 5 the *change* in heartbeat, before and after exercise, was recorded for each individual student. For the exercise in this chapter we shall consider a different design for the study. Under the new design eighty university students were randomly divided into two groups of forty. In the first group the students recorded their heartbeats, over 30 seconds, after ten minutes of sitting still. In the second group the students recorded their heartbeat immediately after five minutes of running vigorously on the spot. The resulting data are given in the following two tables:

(a) Group 1: students at rest

No.	Beats	No.	Beats	No.	Beats	No.	Beats
1.	31	11.	37	21.	27	31.	35
2.	27	12.	39	22.	31	32.	39
3.	28	13.	38	23.	34	33.	34
4.	37	14.	29	24.	33	34.	29
5.	35	15.	39	25.	42	35.	38
6.	38	16.	36	26.	38	36.	38
7.	41	17.	38	27.	37	37.	40
8.	25	18.	34	28.	27	38.	34
9.	32	19.	33	29.	36	39.	33
10.	42	20.	39	30.	37	40.	31

(b) Group 2: students after exercise

No.	Beats	No.	Beats	No.	Beats	No.	Beats
1.	50	11.	49	21.	55	31.	58
2.	51	12.	52	22.	58	32.	41
3.	52	13.	48	23.	44	33.	52
4.	47	14.	48	24.	52	34.	50
5.	60	15.	60	25.	51	35.	52
6.	50	16.	47	26.	50	36.	49
7.	46	17.	61	27.	59	37.	52
8.	53	18.	62	28.	55	38.	50
9.	53	19.	55	29.	50	39.	46
10.	43	20.	55	30.	50	40.	54

These two groups will for the sake of our exercise be regarded as infinitely large, normally distributed populations with means μ_1 and μ_2, and a common variance σ^2.

(a) Each member of the class is asked to use the random numbers in Table T.1 to select a sample of size four from each of the above two groups. A more active class atmosphere can be created by breaking the class into groups of size eight, and having each group generate its own data, after randomly allocating members to "rest" and "exercise". Either way, the eight heartbeat numbers which are produced make up the observation vector y. The question of interest is: "Does running affect heartbeat?" To answer this question we formally test the hypothesis $H_0 : \mu_1 = \mu_2$ versus $H_1 : \mu_1 \neq \mu_2$.

An appropriate coordinate system is

$$
\begin{array}{cccccccc}
U_1 & U_2 & U_3 & U_4 & U_5 & U_6 & U_7 & U_8 \\
\begin{bmatrix} 1 \\ 1 \\ 1 \\ 1 \\ 1 \\ 1 \\ 1 \\ 1 \end{bmatrix} &
\begin{bmatrix} 1 \\ 1 \\ 1 \\ 1 \\ -1 \\ -1 \\ -1 \\ -1 \end{bmatrix} &
\begin{bmatrix} 1 \\ -1 \\ 0 \\ 0 \\ 0 \\ 0 \\ 0 \\ 0 \end{bmatrix} &
\begin{bmatrix} 1 \\ 1 \\ -2 \\ 0 \\ 0 \\ 0 \\ 0 \\ 0 \end{bmatrix} &
\begin{bmatrix} 1 \\ 1 \\ 1 \\ -3 \\ 0 \\ 0 \\ 0 \\ 0 \end{bmatrix} &
\begin{bmatrix} 0 \\ 0 \\ 0 \\ 0 \\ 1 \\ -1 \\ 0 \\ 0 \end{bmatrix} &
\begin{bmatrix} 0 \\ 0 \\ 0 \\ 0 \\ 1 \\ 1 \\ -2 \\ 0 \end{bmatrix} &
\begin{bmatrix} 0 \\ 0 \\ 0 \\ 0 \\ 1 \\ 1 \\ 1 \\ -3 \end{bmatrix} \\
\sqrt{8} & \sqrt{8} & \sqrt{2} & \sqrt{6} & \sqrt{12} & \sqrt{2} & \sqrt{6} & \sqrt{12}
\end{array}
$$

(b) Each class member is asked to calculate the scalar $y.U_2$, the projection vector $(y.U_2)U_2$, and the treatment sum of squares, $(y.U_2)^2$.

They are also asked to calculate the treatment vector $(\bar{y}_{i.} - \bar{y}_{..})$ and the treatment sum of squares using $\|\bar{y}_{i.} - \bar{y}_{..}\|^2$. These should be the same as $(y.U_2)U_2$ and $(y.U_2)^2$ respectively.

The values of $y.U_2$ for the class as a whole can be plotted in the form of a histogram by the instructor. The histogram will approximate the distribution of the projection coefficient $Y.U_2$. Is it plausible that the standard error of the random variable $Y.U_2$ is $\sigma = 5$?

(c) Class members are next asked to write down their error vector, $y - \bar{y}_{i.}$. This could be obtained more lengthily as $(y.U_3)U_3 + \cdots + (y.U_8)U_8$, but this is not required. The error sum of squares can then be calculated as $\|y - \bar{y}_{i.}\|^2$. Lastly the error mean square, which is an unbiased estimate of σ^2, is obtained by dividing $\|y - \bar{y}_{i.}\|^2$ by 6, the dimension of the error space.

The class instructor can plot the estimates of σ^2 for the class as a whole in the form of a histogram. This histogram should be centered on $\sigma^2 = 25$.

(d) Class members are next asked to write out the orthogonal decomposition

$$y = \bar{y}_{..} + (\bar{y}_{i.} - \bar{y}_{..}) + (y - \bar{y}_{i.})$$

using the vectors they have calculated in (b) and (c). They are also asked to confirm the Pythagorean breakup

$$\|y\|^2 = \|\bar{y}_{..}\|^2 + \|\bar{y}_{i.} - \bar{y}_{..}\|^2 + \|y - \bar{y}_{i.}\|^2$$

using their own vectors.

(e) Each class member is now asked to test the hypothesis $H_0 : \mu_1 = \mu_2$ versus $H_1 : \mu_1 \neq \mu_2$ by calculating the test statistic

$$F = \frac{\|\bar{y}_{i.} - \bar{y}_{..}\|^2}{\|y - \bar{y}_{i.}\|^2/6}$$

and comparing their F value with the tabulated 5% critical value $F_{1,6}(.95)$.

The calculated test statistics can be plotted by the instructor in the form of a histogram. This histogram will approximate a "noncentral" F distribution.

(f) The calculated F values will in general be lower than the F values obtained in the class exercise in Chapter 5. Refer back to the appropriate class histogram to see that this is so. This illustrates the difference between a "paired samples" design (Chapter 5), and an "independent samples" design (Chapter 6). Provided the pairing is effective, the first design will yield a more powerful test than the second design.

(g) If time permits, each class member can also calculate the 95% confidence interval for the difference, $\mu_2 - \mu_1$, using the formula

$$(\bar{y}_{2.} - \bar{y}_{1.}) \pm s\sqrt{\frac{2}{4}} \times 2.447$$

where $\bar{y}_1.$ and $\bar{y}_2.$ are the two sample means, s is the square root of the sample variance $s^2 = \|y - \bar{y}_i.\|^2/6$, "4" is the common sample size, and "2.447" is the 97.5 percentile of the t_6 distribution.

A selection of these intervals can be displayed by the class instructor. For the class as a whole, what percentage of confidence intervals included the true value of $\mu_2 - \mu_1 = 15$?

Exercises

(6.1) An experiment was carried out to determine the tolerance of Golden Queen peach seedlings to the herbicide oxadiazon. Eight plots of seedlings were randomly allocated to the treatments:

> 1. No herbicide (control)
> 2. Oxadiazon at normal rate

All plots were hand weeded throughout the experiment. At the end of the first season the average height of the seedlings in the eight plots were:

1. Control	80, 68, 62, 82 cm
2. Oxadiazon	62, 56, 68, 66 cm

(a) Take the observation vector $y = [80, 68, 62, 82, 62, 56, 68, 66]^T$ and calculate the squared length of its projection onto each of the directions

$$
U_1\ \frac{\begin{bmatrix} 1 \\ 1 \\ 1 \\ 1 \\ 1 \\ 1 \\ 1 \\ 1 \end{bmatrix}}{\sqrt{8}}\quad
U_2\ \frac{\begin{bmatrix} 1 \\ 1 \\ 1 \\ 1 \\ -1 \\ -1 \\ -1 \\ -1 \end{bmatrix}}{\sqrt{8}}\quad
U_3\ \frac{\begin{bmatrix} 1 \\ 1 \\ -1 \\ -1 \\ 0 \\ 0 \\ 0 \\ 0 \end{bmatrix}}{\sqrt{4}}\quad
U_4\ \frac{\begin{bmatrix} 1 \\ -1 \\ 0 \\ 0 \\ 0 \\ 0 \\ 0 \\ 0 \end{bmatrix}}{\sqrt{2}}\quad
U_5\ \frac{\begin{bmatrix} 0 \\ 0 \\ 1 \\ -1 \\ 0 \\ 0 \\ 0 \\ 0 \end{bmatrix}}{\sqrt{2}}\quad
U_6\ \frac{\begin{bmatrix} 0 \\ 0 \\ 0 \\ 0 \\ 1 \\ 1 \\ -1 \\ -1 \end{bmatrix}}{\sqrt{4}}\quad
U_7\ \frac{\begin{bmatrix} 0 \\ 0 \\ 0 \\ 0 \\ 1 \\ -1 \\ 0 \\ 0 \end{bmatrix}}{\sqrt{2}}\quad
U_8\ \frac{\begin{bmatrix} 0 \\ 0 \\ 0 \\ 0 \\ 0 \\ 0 \\ 1 \\ -1 \end{bmatrix}}{\sqrt{2}}
$$

(b) Assume the observations are independent samples from normal populations with true means μ_1 and μ_2, and a common variance σ^2. Test $H_0 : \mu_1 = \mu_2$ against $H_1 : \mu_1 \neq \mu_2$ using an F test obtained by dividing $(y.U_2)^2$ by the average of $(y.U_3)^2, \ldots, (y.U_8)^2$. What is your conclusion?

(c) Recalculate the F value using the vector decomposition $y = \bar{y}.. + (\bar{y}_i. - \bar{y}..) + (y - \bar{y}_i.)$ and the formula

$$
F = \frac{\|\bar{y}_i. - \bar{y}..\|^2}{\|y - \bar{y}_i.\|^2/6}
$$

(d) Calculate the 95% confidence interval for the true difference, $\mu_1 - \mu_2$, between the two populations.

(e) Use the computer package Minitab, or any simple regression program, to recalculate the F value.

(f) Draw a histogram of the errors, $y_{ij} - \bar{y}_{i\cdot}$. These are the components of the error vector, $y - \bar{y}_{i\cdot}$. Are there any obvious problems?

(g) Present your results in a table using the format given in Table 6.4.

(6.2) In Exercise 6.1 the coordinate system $U_3, \ldots, U_8$ for the error space, whilst a perfectly good system, did not follow the pattern given elsewhere in this chapter. An alternative system following this pattern is

$$
\begin{array}{cccccc}
U_3 & U_4 & U_5 & U_6 & U_7 & U_8 \\
\begin{bmatrix} 1 \\ -1 \\ 0 \\ 0 \\ 0 \\ 0 \\ 0 \\ 0 \\ \hline \sqrt{2} \end{bmatrix} &
\begin{bmatrix} 1 \\ 1 \\ -2 \\ 0 \\ 0 \\ 0 \\ 0 \\ 0 \\ \hline \sqrt{6} \end{bmatrix} &
\begin{bmatrix} 1 \\ 1 \\ 1 \\ -3 \\ 0 \\ 0 \\ 0 \\ 0 \\ \hline \sqrt{12} \end{bmatrix} &
\begin{bmatrix} 0 \\ 0 \\ 0 \\ 0 \\ 1 \\ -1 \\ 0 \\ 0 \\ \hline \sqrt{2} \end{bmatrix} &
\begin{bmatrix} 0 \\ 0 \\ 0 \\ 0 \\ 1 \\ 1 \\ -2 \\ 0 \\ \hline \sqrt{6} \end{bmatrix} &
\begin{bmatrix} 0 \\ 0 \\ 0 \\ 0 \\ 1 \\ 1 \\ 1 \\ -3 \\ \hline \sqrt{12} \end{bmatrix}
\end{array}
$$

Calculate $s^2 = \left[(y.U_3)^2 + \ldots + (y.U_8)^2 \right] / 6$ using the new coordinate system to show that the result is the same as that obtained using the old coordinate system.

(6.3) An experiment was carried out to determine the effect of phosphate fertilizer application on the yield of a mixed ryegrass/white clover pasture. A uniform area of pasture was chosen, and six 8m × 3m experimental plots were marked out. These were assigned in a completely random manner to the two treatments

1. No fertilizer application (control)

2. Phosphate (P) at the rate of 40 kg P/ha

During the winter the phosphate was spread uniformly over the appropriate plots. Throughout the following growing season the central 6m × 2m of each plot was cut at monthly intervals, with clippings being weighed and returned to each plot, except for a small sample which was oven dried to determine the percentage of dry matter.

The dry matter yields for each plot, converted to tonnes/ha, were as follows:

1. Control	9.6	7.4	10.0
2. Phosphate	11.3	10.1	12.2

(a) Put these six values into an observation vector y, and calculate the orthogonal decomposition $y - \bar{y}_{..} = (\bar{y}_{i.} - \bar{y}_{..}) + (y - \bar{y}_{i.})$.

(b) Calculate the Pythagorean breakup

$$\|y - \bar{y}_{..}\|^2 = \|\bar{y}_{i.} - \bar{y}_{..}\|^2 + \|y - \bar{y}_{i.}\|^2$$

and the test statistic, F. Summarize your results in an ANOVA table.

(c) Assuming the data are independent values from normal populations with means μ_1 and μ_2 and a common variance σ^2, test the hypothesis $H_0 : \mu_1 = \mu_2$ by comparing the calculated F value with the appropriate critical value from Table T.2. What is your conclusion?

(d) Draw a histogram of the errors, using the error vector, $y - \bar{y}_{i.}$, and visually check for obvious problems.

(e) Use Minitab to check your calculations.

(f) Present your results in a table, similar to Table 6.4.

(6.4) An experiment was established to determine whether boron fertilizer affects the seed yield of alfalfa. The experiment consisted of six plots marked out in a field of alfalfa. Three plots received an application of boron fertilizer and three plots were unfertilized control plots. The two treatments, "control" and "boron", were assigned to the plots in a completely random manner.

The seed yields, in kg/ha, obtained from each plot were as follows:

Treatment	Replicate 1	Replicate 2	Replicate 3
Control	61	65	63
Boron	76	73	70

Test the hypothesis $H_0 : \mu_1 = \mu_2$ by writing out an appropriate vector decomposition and then completing an appropriate statistical analysis. Show your working and state your conclusions clearly.

(6.5) The virtues of tagging

In the majority of experimental trials which involve animals, the experimental animals are given numbered tags. These allow identification of the animal with its treatment group, such as "control", "twice-drenched" or "thrice-drenched". Equally importantly, if weights are being recorded the tags enable the calculation of the weight gain for each animal.

The value of individual identification is highlighted by the following set of data, which corresponds to the weights at two different times of twelve tagged calves from a population of calves.

Calf weights in kg on 22 February 1984	Calf weights in kg on 22 March 1984
172	177
115	125
115	124
112	119
142	153
110	118
152	157
177	185
116	127
125	128
128	136
115	118

The question to be answered is: have the calves in the population as a whole gained weight during the month?

(a) Calculate the weightgain of each individual calf and treat the twelve weightgains as a sample from a normally distributed population with mean μ. Test the hypothesis $H_0 : \mu = 0$ versus $H_1 : \mu \neq 0$ using the methods described in Chapter 5. Use the decomposition $y = \bar{y} + (y - \bar{y})$.

(b) Suppose that the calves were not individually identified and that the above sets of weights were obtained by weighing a different random sample of calves on each occasion. The first column is then a sample from the population of all possible weights for calves weighed on 22 February 1984, and the second column a sample of weights on 22 March 1984. Assuming the two populations of weights are normally distributed with means μ_1 and μ_2 and a common variance σ^2, test the hypothesis $H_0 : \mu_1 = \mu_2$ versus $H_1 : \mu_1 \neq \mu_2$ using the methods described in this chapter. That is, use the decomposition $y = \bar{y} + (\bar{y}_i. - \bar{y}..) + (y - \bar{y}_i.)$.

(c) Which of the two designs is more accurate? Would you agree that individual identification of animals is worth the trouble? The first design is a *paired samples* design and the second is an *independent samples* design.

Solution to the Reader Exercise

(6.1) The simplified orthogonal decomposition for the full data set is

$$
y - \bar{y}_{..} \quad = \quad (\bar{y}_{i\cdot} - \bar{y}_{..}) \quad + \quad (y - \bar{y}_{i\cdot})
$$

$$
\begin{bmatrix}
29.80 - \bar{y}_{..} \\
28.57 - \bar{y}_{..} \\
29.97 - \bar{y}_{..} \\
29.53 - \bar{y}_{..} \\
29.90 - \bar{y}_{..} \\
30.33 - \bar{y}_{..} \\
31.27 - \bar{y}_{..} \\
30.37 - \bar{y}_{..} \\
29.10 - \bar{y}_{..} \\
30.80 - \bar{y}_{..}
\end{bmatrix}
=
\begin{bmatrix}
-.41 \\
-.41 \\
-.41 \\
-.41 \\
-.41 \\
.41 \\
.41 \\
.41 \\
.41 \\
.41
\end{bmatrix}
+
\begin{bmatrix}
.246 \\
-.984 \\
.416 \\
-.024 \\
.346 \\
-.044 \\
.896 \\
-.004 \\
-1.274 \\
.426
\end{bmatrix}
$$

where $\bar{y}_{..} = 29.964$, $\bar{y}_{1\cdot} = 29.554$ and $\bar{y}_{2\cdot} = 30.374$.

The corresponding Pythagorean decomposition is

$$
\|y - \bar{y}_{..}\|^2 \quad = \quad \|\bar{y}_{i\cdot} - \bar{y}_{..}\|^2 \quad + \quad \|y - \bar{y}_{i\cdot}\|^2
$$
$$
5.6214 \quad = \quad 1.6810 \quad + \quad 3.9314
$$

The resulting test statistic is

$$
F \quad = \quad \frac{\|\bar{y}_{i\cdot} - \bar{y}_{..}\|^2}{\|y - \bar{y}_{i\cdot}\|^2/8} \quad = \quad \frac{1.6810}{3.9314/8} \quad = \quad 3.42
$$

This value is not significant when we compare it with the 95 percentile of the $F_{1,8}$ distribution, namely 5.32. Hence we formally "accept" as plausible our null hypothesis $H_0 : \mu_1 = \mu_2$. These results are summarized in the table.

Mean bulkometer reading, in cm^3/g	
Treatment	
Conditioned for 2 days	29.6
Conditioned for 3 days	30.4
SED	0.44
Significance of difference	ns

The revised 95% confidence interval is given by

$$
-2.306 \quad \leq \quad \frac{(\bar{y}_{1\cdot} - \bar{y}_{2\cdot}) - (\mu_1 - \mu_2)}{\sqrt{2s^2/n}} \quad \leq \quad 2.306
$$

where 2.306 is the 95 percentile of the t_8 distribution. Thus

$$
-2.306 \quad \leq \quad \frac{-.82 - (\mu_1 - \mu_2)}{\sqrt{2(.4914)/5}} \quad \leq \quad 2.306
$$

which in turn can be rearranged to give

$$-.20 \quad \leq \quad \mu_2 - \mu_1 \quad \leq \quad 1.84$$

In our report we could therefore say that we are 95% confident that the increase in bulkometer reading through conditioning for three days rather than two days is in the range 0.8 ± 1.0 units.

Chapter 7

Questions About Several Populations

Here we move on to discuss questions concerning an arbitrary number of populations. For simplicity we restrict our attention to the case where we sample equal numbers from each population. The unequal replications case is discussed in Appendix B.

When we discussed two populations we were primarily interested in whether the population means, μ_1 and μ_2, were equal. To put it formally, the question of interest usually involved the *contrast* $\mu_1 - \mu_2$. When we deal with several populations, questions of interest still correspond to contrasts amongst the population means. For example, in §7.1 we shall examine a study of pollution levels where the locations selected are two suburban regions and one urban region. The natural contrast in this study is the contrast of suburban with urban pollution levels, $(\mu_1 + \mu_2)/2 - \mu_3$.

Here is the plan for this chapter. In §1 we discuss a simple example which illustrates the problems of handling more than two populations. We then, in §2, introduce the four main types of contrast which occur in practice, before covering the general situation in §3. The chapter ends in §4 with a short summary.

7.1 A Simple Example

To illustrate the setting up of contrasts among the population means, we shall now consider a simplified study of smoke pollution levels in the South Island of New Zealand.

Example

A study was set up to determine whether winter smoke pollution levels differed from urban to suburban areas in Christchurch, New Zealand. A second objective was to determine whether pollution levels differed between hill suburbs and suburbs in the flat regions of the city.

The Health Department measured smoke pollution hourly, in micrograms per cubic meter, throughout the three months of winter in 1988. These readings were taken in a fixed but randomly chosen selection of locations in each of two randomly selected hill suburbs, two plains suburbs, and two areas in the central city. The mean area readings are recorded in the table.

Population				Means
1. Hill suburb	(Suburban)	124	110	117
2. Plains suburb	(Suburban)	107	115	111
3. Central city	(Urban)	126	138	132

The corresponding observation vector is $y = [124, 110, 107, 115, 126, 138]^T$.

These observations are assumed to come from three normal populations, the winter smoke pollution levels in the respective area types, with true means μ_1, μ_2 and μ_3, and a common variance σ^2. The model vector is therefore

$$
\begin{bmatrix} \mu_1 \\ \mu_1 \\ \mu_2 \\ \mu_2 \\ \mu_3 \\ \mu_3 \end{bmatrix} = \mu_1 \begin{bmatrix} 1 \\ 1 \\ 0 \\ 0 \\ 0 \\ 0 \end{bmatrix} + \mu_2 \begin{bmatrix} 0 \\ 0 \\ 1 \\ 1 \\ 0 \\ 0 \end{bmatrix} + \mu_3 \begin{bmatrix} 0 \\ 0 \\ 0 \\ 0 \\ 1 \\ 1 \end{bmatrix}
$$

so the model space M is a 3-dimensional subspace of 6-space.

The hypotheses of interest relate to a comparison of suburban with urban, $(\mu_1 + \mu_2)/2$ versus μ_3, and a comparison of hill suburbs with those on the plains, μ_1 versus μ_2. That is, we wish to test the hypotheses:

(a) $H_0 : (\mu_1 + \mu_2)/2 = \mu_3$ versus $H_1 : (\mu_1 + \mu_2)/2 \neq \mu_3$
(b) $H_0 : \mu_1 = \mu_2$ versus $H_1 : \mu_1 \neq \mu_2$

These can be rewritten in the alternative forms:

(a) $H_0 : \mu_1 + \mu_2 - 2\mu_3 = 0$ versus $H_1 : \mu_1 + \mu_2 - 2\mu_3 \neq 0$
(b) $H_0 : \mu_1 - \mu_2 = 0$ versus $H_1 : \mu_1 - \mu_2 \neq 0$

Here $(\mu_1 + \mu_2 - 2\mu_3)$ and $(\mu_1 - \mu_2)$ are *contrasts* among the population means.

The directions associated with these hypotheses are

$$U_2 = \frac{1}{\sqrt{12}} \begin{bmatrix} 1 \\ 1 \\ 1 \\ 1 \\ -2 \\ -2 \end{bmatrix} \quad \text{and} \quad U_3 = \frac{1}{\sqrt{4}} \begin{bmatrix} 1 \\ 1 \\ -1 \\ -1 \\ 0 \\ 0 \end{bmatrix}$$

This is because $y.U_2 = 2(\bar{y}_{1.} + \bar{y}_{2.} - 2\bar{y}_{3.})/\sqrt{12}$ has an expected value of $2(\mu_1 + \mu_2 - 2\mu_3)/\sqrt{12}$, so $y.U_2$ is small if the hypothesis is true and large if the hypothesis is false. Similarly $y.U_3 = 2(\bar{y}_{1.} - \bar{y}_{2.})/\sqrt{4} = \bar{y}_{1.} - \bar{y}_{2.}$ has an expected value of $\mu_1 - \mu_2$, so $y.U_3$ is small if the corresponding hypothesis is true, and large if the hypothesis is false.

Since U_2 and U_3 are orthogonal, we say that $(\mu_1 + \mu_2 - 2\mu_3)$ and $(\mu_1 - \mu_2)$ are *orthogonal contrasts*. In this happy situation we can write down a tailor-made coordinate system for the model space M:

$$U_1 = \frac{1}{\sqrt{6}} \begin{bmatrix} 1 \\ 1 \\ 1 \\ 1 \\ 1 \\ 1 \end{bmatrix}, \quad U_2 = \frac{1}{\sqrt{12}} \begin{bmatrix} 1 \\ 1 \\ 1 \\ 1 \\ -2 \\ -2 \end{bmatrix}, \quad U_3 = \frac{1}{\sqrt{4}} \begin{bmatrix} 1 \\ 1 \\ -1 \\ -1 \\ 0 \\ 0 \end{bmatrix}$$

The two hypotheses can then be tested in the usual way, using the test statistics

$$\frac{(y.U_2)^2}{[(y.U_4)^2 + (y.U_5)^2 + (y.U_6)^2]/3} \quad \text{and} \quad \frac{(y.U_3)^2}{[(y.U_4)^2 + (y.U_5)^2 + (y.U_6)^2]/3}$$

where

$$U_4 = \frac{1}{\sqrt{2}} \begin{bmatrix} 1 \\ -1 \\ 0 \\ 0 \\ 0 \\ 0 \end{bmatrix}, \quad U_5 = \frac{1}{\sqrt{2}} \begin{bmatrix} 0 \\ 0 \\ 1 \\ -1 \\ 0 \\ 0 \end{bmatrix}, \quad U_6 = \frac{1}{\sqrt{2}} \begin{bmatrix} 0 \\ 0 \\ 0 \\ 0 \\ 1 \\ -1 \end{bmatrix}$$

are coordinate axes for the error space. Under the null hypotheses, these test values will follow the $F_{1,3}$ distribution.

Exercises for the reader

(7.1) Write out in numerical terms the orthogonal decomposition

$$y = (y.U_1)U_1 + (y.U_2)U_2 + \cdots + (y.U_6)U_6, \quad \text{and show that}$$

(i) $(y.U_1)U_1 = \bar{y}_{..}$

(ii) $(y.U_2)U_2 + (y.U_3)U_3 = \bar{y}_{i.} - \bar{y}_{..}$, and

(iii) $(y.U_4)U_4 + (y.U_5)U_5 + (y.U_6)U_6 = y - \bar{y}_{i.}$

That is, we have the decomposition already met in Chapter 6, of

$$y = \bar{y}_{..} + (\bar{y}_{i.} - \bar{y}_{..}) + (y - \bar{y}_{i.})$$

The only change is that our model space is now 3-dimensional, consisting of a 1-dimensional overall mean space and a 2-dimensional treatment space.

(7.2) Evaluate the test statistics given in the text, and decide whether to accept or reject each of the hypotheses.

7.2   Types of Contrast

Contrasts which occur in practice can be classified into the following four types:

1. Class comparisons

2. Factorial comparisons

3. Polynomial contrasts

4. Pairwise comparisons

As examples, we shall now describe six agricultural experiments, Examples A to F, which involve the various contrast types and combinations of them. The data for Examples A to D have been provided by Mr. C.C.S. McLeod, of the Ministry of Agriculture and Fisheries, Timaru, New Zealand. Example E has been reproduced from "Statistics for Biology, 3rd edition (1980)", by O.N. Bishop, with kind permission from Longmans publishers. Example F is a purely hypothetical data set devised by the authors.

Chapters 8, 9, 10 and 11 look in detail at the four types of contrast, and present full analyses of these datasets. Throughout these chapters analyses will be performed assuming that the field layouts follow the very simplest experimental design, the completely randomized design. This tactic permits us to become familiar with contrasts against the simplest backdrop. In practice such a simple design is rarely used in field experiments, so the authors have been obliged to alter the original designs for Examples A to D. Randomized block designs and split plot designs, necessary for a complete analysis of the original experiments, will be introduced in Chapters 12 and 14.

Example A: Class Comparisons

An experimenter investigating winter greenfeeds set up a trial comparing four mangel cultivars and four fodder beet cultivars.

The trial was laid out in a completely randomized design with 32 plots assigned equally to the eight treatments as follows:

Plot number:	1	2	3	4	5	6	7	8	9	10	11
Treatment number:	5	3	3	1	5	8	7	6	1	4	8

Plot number:	12	13	14	15	16	17	18	19	20	21	22
Treatment number:	7	7	6	3	2	5	2	4	8	6	7

Plot number:	23	24	25	26	27	28	29	30	31	32	
Treatment number:	2	3	1	6	4	4	8	2	1	5	

The field layout is shown in Figure 7.1. Plots were 6m long and 1.8m wide, consisting of three rows sown 60cm apart. Within each row, plants were thinned to 20cm spacing. The 32 field plots were sown in a long row in a fenced-off corner of a farmer's field.

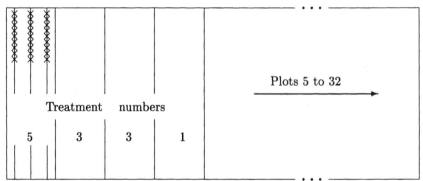

Figure 7.1: Field layout for greenfeed Example A.

The trial was sown on 15 November 1976 and harvested on 9 June 1977. The plants growing in the middle 1.5m of the center row of each plot were harvested, divided into tops and bulbs, and weighed to obtain "wet weights" of tops and bulbs. A sample of each of the tops and bulbs was then weighed, oven dried, and reweighed, to enable estimation of the dry matter percentages. The "dry weights" of tops and bulbs were then estimated by multiplying the respective wet weights and dry matter percentages. These were then scaled up to tonnes per hectare. The data in Table 7.1 is the dry matter of bulbs in tonnes per hectare.

The comparison of particular interest is the contrast of mangel cultivars with fodder beet cultivars. That is, did the mangels in general produce more or less dry matter in the bulbs than the fodder beets?

Treatment	Bulb dry matter				Mean
Brigadier mangels	11.7	11.3	11.5	10.6	11.275
Yellow globe mangels	12.6	13.5	14.8	14.9	13.95
Orange globe mangels	12.5	13.1	11.2	9.8	11.65
Red intermediate mangels	11.0	11.8	13.4	8.1	11.075
Mono rosa fodder beet	13.9	13.0	15.2	17.1	14.8
Mono blanc fodder beet	14.1	12.6	15.6	11.3	13.4
Mono bomba fodder beet	11.9	15.2	13.9	14.7	13.925
Yellow daeno fodder beet	15.5	16.8	22.3	14.4	17.25

Table 7.1: Dry matter of bulbs, in t/ha, for each plot, Example A.

Formally this contrast is

$$ c \;=\; \frac{\mu_1 + \mu_2 + \mu_3 + \mu_4}{4} \;-\; \frac{\mu_5 + \mu_6 + \mu_7 + \mu_8}{4} $$

This is a *class comparison* type contrast. These are discussed in Chapter 8.

Example B: Factorial Contrasts

An experiment was set up in November 1974 to look at the longterm require-ment for nitrolime and superphosphate of Zephyr barley grown for malt-ing. Superphosphate contains phosphorus and sulphur, while nitrolime adds nitrogen. Twenty plots were assigned to four treatments in a completely random manner as follows.

Plot number:	1	2	3	4	5	6	7	8	9	10
Treatment number:	2	1	1	3	2	1	3	4	4	3
Plot number:	11	12	13	14	15	16	17	18	19	20
Treatment number:	1	3	4	2	1	4	3	4	2	2

In this trial, each plot consisted of a single 7-row drill strip. The twenty field plots were sown side by side as shown in Figure 7.1, except that they were much longer and narrower, being 40m long and 1.25m wide. Plots were topdressed with the same fertilizer each year according to the treat-ments given below, and harvested using a combine harvester. Yields in kilo-grams per plot from the eighth harvest, in February 1982, were as shown in Table 7.2.
Questions of interest are:

1. Did nitrolime have any effect on barley yield?

Treatment	Grain yield					Mean
1. Control (no fertilizer)	19.2	18.4	17.0	17.6	17.2	17.88
2. 250 kg/ha superphosphate	18.2	19.8	19.4	19.0	19.8	19.24
3. 250 kg/ha nitrolime	20.0	21.6	22.0	20.8	20.4	20.96
4. 250 super + 250 n/lime	23.6	21.6	23.2	21.4	21.2	22.2

Table 7.2: Barley grain yield, in kg/plot, for the 8th harvest, Example B.

2. Did superphosphate have any effect on barley yield?

3. Did the fertilizers interact? Or, was the response to super in the presence of nitrolime the same as the response to super in the absence of nitrolime?

The corresponding contrasts are:

$$c_1 = \frac{\mu_3 + \mu_4}{2} - \frac{\mu_1 + \mu_2}{2}$$

$$c_2 = \frac{\mu_2 + \mu_4}{2} - \frac{\mu_1 + \mu_3}{2}$$

$$c_3 = (\mu_4 - \mu_3) - (\mu_2 - \mu_1)$$

These are *factorial* contrasts. The factors here are nitrolime and superphosphate. Each factor is at two levels: either it is present or absent. Such contrasts are discussed in Chapter 9.

Example C: Factorial Contrasts Involving Class Comparisons

An experimenter wished to investigate the oversowing into a barley crop of soft turnips for use as winter greenfeed.

A trial with six treatments and four replicates was therefore laid out with the treatments listed below assigned to plots at random. The barley was sown in the spring of 1976. In midsummer two cultivars of turnip were oversown into the barley crop. The barley was harvested six weeks later, and plots either not topdressed or given 125 kg/ha or 250 kg/ha sulphate of ammonia (S/A), a nitrogen fertilizer.

The number of soft turnip plants present in a 1.8m by 0.6m area in the center of each plot in early winter were as given in Table 7.3.

Treatment	Number of plants				Mean
Green globe soft turnips, no S/A	45	69	71	83	67
Green globe soft turnips, 125 kg/ha S/A	27	59	62	64	53
Green globe soft turnips, 250 kg/ha S/A	24	49	55	77	51.25
York globe soft turnips, no S/A	19	17	22	45	25.75
York globe soft turnips, 125 kg/ka S/A	29	13	27	21	22.5
York globe soft turnips, 250 kg/ha S/A	12	20	18	15	16.25

Table 7.3: Number of soft turnip plants per 1.08m^2 area, Example C.

Questions of interest are:

1. Did Green globe turnips establish or survive differently to York globe turnips?

2. Did the nitrogen (S/A) application affect the survival of the turnips?

3. Did the high and low rates of nitrogen differ?

4. Did the effect of the nitrogen application vary with soft turnip cultivar?

5. Did the difference between the high and low rates of nitrogen vary with cultivar?

The corresponding contrasts are:

$$c_1 = \frac{\mu_4 + \mu_5 + \mu_6}{3} - \frac{\mu_1 + \mu_2 + \mu_3}{3}$$
$$c_2 = (\frac{\mu_5 + \mu_6}{2} - \mu_4) + (\frac{\mu_2 + \mu_3}{2} - \mu_1)$$
$$c_3 = (\mu_6 - \mu_5) + (\mu_3 - \mu_2)$$
$$c_4 = (\frac{\mu_5 + \mu_6}{2} - \mu_4) - (\frac{\mu_2 + \mu_3}{2} - \mu_1)$$
$$c_5 = (\mu_6 - \mu_5) - (\mu_3 - \mu_2)$$

These are *factorial* contrasts involving *class comparisons*. The factors are cultivar, at two levels, Green or York globe, and sulphate of ammonia, at three levels, nil, 125 and 250 kg/ha. The first contrast, c_1, corresponds to the only possible comparison within the first factor, namely of Green with York globe. The second and third contrasts, c_2 and c_3, correspond to class comparisons within the second factor, namely, of "none versus some" and "low versus high". The last two contrasts, c_4 and c_5, correspond to the interactions of the first factor contrast, c_1, with the second factor contrasts, c_2 and c_3. These contrasts are discussed further in Chapter 9.

Example D: Polynomial Contrasts

An experimenter set out to determine how the grain yield of spring sown malting barley was affected by seeding rate.

A trial with five treatments and six replicates was therefore laid out in a completely randomized design, with layout similar to that of Example B. Grain yields, in kilograms, harvested from plots of size 40m by 1.25m, were as shown in Table 7.4.

Seeding rate	Grain yield						Mean
50 kg/ha	25.4	22.4	25.2	24.4	24.2	22.0	23.93
75 kg/ha	26.2	26.2	25.2	26.4	25.0	27.8	26.13
100 kg/ha	27.6	27.6	26.0	25.8	26.2	25.8	26.50
125 kg/ha	27.6	28.2	26.8	26.6	28.0	27.8	27.50
150 kg/ha	27.2	28.2	26.8	25.6	27.2	27.6	27.10

Table 7.4: Grain yield of barley per 50m^2, Example D.

Questions of interest are:

1. Did grain yield increase with increasing seeding rate?

2. Did the rate of increase in grain yield drop off with increasing seeding rate?

Corresponding contrasts are:

$$
\begin{aligned}
c_1 &= -2\mu_1 - \mu_2 \qquad\quad + \mu_4 + 2\mu_5 \\
c_2 &= 2\mu_1 - \mu_2 - 2\mu_3 - \mu_4 + 2\mu_5
\end{aligned}
$$

These correspond to the linear and quadratic components of a polynomial response curve, and are called *polynomial* contrasts. These contrasts are discussed in Chapter 10.

Example E: Selected Pairwise Contrasts

An experimenter working with sunflowers set up an experiment in which twelve sunflowers were fed four different fertilizer solutions, labelled A, B, C, and D. Solution A was a solution containing magnesium, nitrogen and trace nutrients, whereas solutions B, C and D each contained only two out of the three ingredients. This treatment design is called a subtractive design since each ingredient in turn is subtracted from the complete solution, A. The experiment was laid out in a completely randomized design, as shown:

| A | C | C | D | A | B | C | A | B | B | D | D |

After nine weeks' growth the sunflowers were pulled, dried and weighed. Their dry weights, in grams, were as shown in Table 7.5.

Treatment	Dry weight			Mean
A. Solution containing magnesium, nitrogen and trace nutrients	1172	750	784	902
B. Solution lacking magnesium	67	95	59	74
C. Solution lacking nitrogen	148	234	92	158
D. Solution lacking trace nutrients	297	243	263	268

Table 7.5: Dry weight of sunflowers in grams, Example E. This data is reproduced from "Statistics for Biology, 3rd edition (1980)", by O.N. Bishop, with kind permission from Longmans publishers.

Questions of interest are:

1. Did omitting magnesium from the solution have any effect on the growth of the sunflowers?

2. Did omitting nitrogen have any effect?

3. Did omitting trace nutrients have any effect?

The corresponding contrasts are:

$$c_1 = \mu_1 - \mu_2$$
$$c_2 = \mu_1 - \mu_3$$
$$c_3 = \mu_1 - \mu_4$$

These are *pairwise* contrasts. Since they all involve μ_1, none are orthogonal to any of the other contrasts. A study of such pairwise comparisons forms the content of Chapter 11.

Example F: All Pairwise Contrasts

Four unreleased chemicals were being evaluated for stripe rust control in wheat. A trial with four treatments and five replicates was therefore laid out in a completely randomized design. Grain yields, in kilograms, from the $50m^2$ plots are shown in Table 7.6.

Chemical	Grain yield					Mean
Product X52	19.0	21.0	23.0	20.0	17.0	20.0
Product B29	23.8	21.8	17.8	23.8	21.8	21.8
Product Z15	21.9	24.9	23.4	18.9	22.9	22.4
Product PP5	24.3	22.3	26.3	26.3	22.3	24.3

Table 7.6: Grain yield of wheat per $50m^2$, Example F.

Questions of interest are:

1. Was product X52 better or worse than product B29?

2. Was product X52 better or worse than product Z15?

3. Was product X52 better or worse than product PP5?

4. Was product B29 better or worse than product Z15?

5. Was product B29 better or worse than product PP5?

6. Was product Z15 better or worse than product PP5?

The corresponding contrasts are:

$$c_1 = \mu_1 - \mu_2$$
$$c_2 = \mu_1 - \mu_3$$
$$c_3 = \mu_1 - \mu_4$$
$$c_4 = \mu_2 - \mu_3$$
$$c_5 = \mu_2 - \mu_4$$
$$c_6 = \mu_3 - \mu_4$$

These are all possible *pairwise* contrasts. Here we need the so-called "multiple comparison procedures" of Chapter 11.

7.3   The Overview

The study of single populations in Chapter 5 we broadened in Chapter 6 to the study of two populations. This chapter completes this first phase of our development by discussing questions about several populations. In this reference section we present the general ideas which are needed for the analysis of such questions.

Our k populations are assumed normal, with means $\mu_1, \ldots, \mu_k$ and common variance σ^2 unknown. A typical question will take the form of a contrast equalling zero; many examples have been presented in the previous section of this book. From each population we draw a random sample of size n, leading to an observation vector of the form

$$y = [y_{11}, \ldots, y_{21}, \ldots, y_{k1}, \ldots]^T$$

Model vectors, the vectors of expected values, must then have the form $[\mu_1, \ldots, \mu_2, \ldots, \mu_k, \ldots]^T$ whence the model space is k-dimensional, being

$$M = \text{span} \left\{ \begin{bmatrix} 1 \\ \cdot \\ 0 \\ \vdots \\ 0 \\ \cdot \end{bmatrix}, \begin{bmatrix} 0 \\ \cdot \\ 1 \\ \vdots \\ 0 \\ \cdot \end{bmatrix}, \cdots, \begin{bmatrix} 0 \\ \cdot \\ 0 \\ \vdots \\ 1 \\ \cdot \end{bmatrix} \right\}$$

If we project y onto M to fit the model in the usual way we produce the familiar breakup

$$y = \bar{y}_{i\cdot} + (y - \bar{y}_{i\cdot})$$

of observation vector into model vector plus error vector, as in Figure 7.2. This leads to least squares estimates $\bar{y}_{i\cdot}$ of the μ_i, the k population means.

Hypotheses of interest can most often be written in the form $H_0 : c = 0$, where $c = c_1\mu_1 + \cdots + c_k\mu_k$ is a contrast, that is, $c_1 + \cdots + c_k = 0$. The direction associated with this hypothesis is

$$U_c = \frac{1}{\sqrt{n \sum_{i=1}^{k} c_i^2}} \begin{bmatrix} c_1 \\ \cdot \\ c_2 \\ \vdots \\ c_k \\ \cdot \end{bmatrix}$$

As usual, we test the hypothesis by comparing the squared length of the projection in this direction with the average squared projection in the error space, s^2. Thus the test statistic will have the form

$$F = \frac{(y.U_c)^2}{\|y - \bar{y}_{i\cdot}\|^2/k(n-1)} = \frac{n(c_1\bar{y}_{1\cdot} + \cdots + c_k\bar{y}_{k\cdot})^2}{(c_1^2 + \cdots + c_k^2)s^2}$$

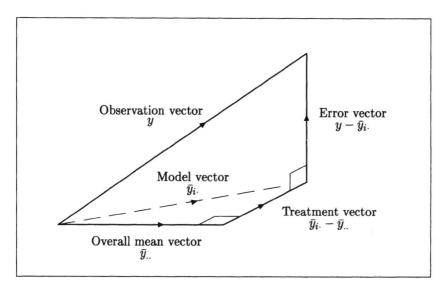

Figure 7.2: Orthogonal decomposition of the observation vector for the case of several populations.

How many such hypotheses can we independently test? Now the vector U_c associated with any contrast lies in the model space, but is orthogonal to the equiangular vector. That is, the vector U_c lies in a $k-1$ dimensional subspace of the model space, termed the treatment space. We can therefore specify at most $k-1$ orthogonal directions of the form U_c, corresponding to at most $k-1$ *orthogonal contrasts*, and thus from §3.2, specify at most $k-1$ independent hypotheses. Note that it is desirable to set up independent hypotheses: a conclusion about one should in no way influence the conclusion about another.

In practice the contrasts corresponding to the hypotheses of interest may not all be orthogonal, or may not number $k-1$. However, if a complete set of $k-1$ orthogonal contrasts is specified, then the $k-1$ corresponding unit vectors, plus the overall mean direction, the equiangular vector, will serve as a very convenient coordinate system for the model space. This was the case in the simple example given in §7.1.

To sum up, if U_1 is the unit vector along the equiangular line, and $U_2, \ldots, U_k$ correspond to $k-1$ orthogonal contrasts of interest, then we have the associated orthogonal decomposition of y as

$$
\begin{aligned}
y &= (y.U_1)U_1 \quad +(y.U_2)U_2 + \cdots + (y.U_k)U_k + (y.U_{k+1})U_{k+1} + \cdots + (y.U_{kn})U_{kn} \\
&= \quad \bar{y}_{..} \quad + \quad (\bar{y}_{i.} - \bar{y}_{..}) \quad + \quad (y - \bar{y}_{i.}) \\
&= \text{Mean vector} + \quad \text{Treatment vector} \quad + \quad \text{Error vector}
\end{aligned}
$$

where $U_{k+1}, \ldots, U_{kn}$ form an orthogonal coordinate system for the error space. This decomposition, summarized in Figure 7.2, leads to the appropriate Pythagorean breakup and test statistics.

A final word or two about contrasts. In our air pollution example of §7.1 notice that the suburban versus urban contrast can be written as either $c = (\mu_1 + \mu_2)/2 - \mu_3$ or $d = \mu_1 + \mu_2 - 2\mu_3$. Both correspond to the same unit vector,

$$U_2 = \frac{1}{\sqrt{3}} \begin{bmatrix} 1/2 \\ 1/2 \\ 1/2 \\ 1/2 \\ -1 \\ -1 \end{bmatrix} = \frac{1}{\sqrt{12}} \begin{bmatrix} 1 \\ 1 \\ 1 \\ 1 \\ -2 \\ -2 \end{bmatrix}$$

In fact there are an infinite number of contrasts which give rise to a particular unit vector such as U_2. Any two, however, are scalar multiples of each other. For hypothesis testing purposes it is immaterial whether we test the hypothesis $H_0 : c = 0$ or the hypothesis $H_0 : d = 0$, so we usually use the contrast expressed with smallest whole number coefficients, here the contrast d. For confidence intervals, however, we prefer to use an easily interpreted form: in this example it would be contrast c, the "average difference in pollution between suburban and urban areas".

For Examples A to F the contrast coefficients expressed in their simplest terms are given in Table 7.7. In the equal replications case contrasts are orthogonal if and only if the vectors shown in Table 7.7 are orthogonal. This allows us to readily check whether the contrasts in the examples form a complete set of $k - 1$. Evidently only for Examples B and C do we have such a complete orthogonal set; Examples A and D have incomplete sets of orthogonal contrasts, while in Examples E and F the contrasts are not all orthogonal to one another.

Exercises for the reader

For these questions, the unit vectors U_c and U_d are those corresponding to the contrasts $c = c_1\mu_1 + \cdots + c_k\mu_k$ and $d = d_1\mu_1 + \cdots + d_k\mu_k$.

(7.3) Show that U_c lies in the model space M. To do this, show that U_c is a linear combination of the spanning set for M.

(7.4) Show that U_c is perpendicular to the equiangular vector.

(7.5) Show that U_c and U_d are orthogonal if and only if $\sum_{i=1}^{k} c_i d_i = 0$.

(A)

Treatment	Contrast c
Brigadier mangels	1
Yellow globe mangels	1
Orange globe mangels	1
Red intermediate mangels	1
Mono rosa fodder beet	−1
Mono blanc fodder beet	−1
Mono bomba fodder beet	−1
Yellow daeno fodder beet	−1

(B)

	c_1	c_2	c_3
Control	−1	−1	1
Super	−1	1	−1
Nitrolime	1	−1	−1
Super + nitrolime	1	1	1

(C)

	c_1	c_2	c_3	c_4	c_5
G. globe, no S/A	−1	−2	0	2	0
G. globe, 125 S/A	−1	1	−1	−1	1
G. globe, 250 S/A	−1	1	1	−1	−1
Y. globe, no S/A	1	−2	0	−2	0
Y. globe, 125 S/A	1	1	−1	1	−1
Y. globe, 250 S/A	1	1	1	1	1

(D)

	c_1	c_2
50 kg/ha	−2	2
75 kg/ha	−1	−1
100 kg/ha	0	−2
125 kg/ha	1	−1
150 kg/ha	2	2

(E)

	c_1	c_2	c_3
Complete solution	1	1	1
Solution − magnesium	−1	0	0
Solution − nitrogen	0	−1	0
Solution − trace nutrients	0	0	−1

(F)

	c_1	c_2	c_3	c_4	c_5	c_6
$X52$	1	1	1	0	0	0
$B29$	−1	0	0	1	1	0
$Z15$	0	−1	0	−1	0	1
$PP5$	0	0	−1	0	−1	−1

Table 7.7: Contrast coefficients in their simplest terms for Examples A to F.

7.4 Summary

In this chapter we have introduced studies designed to answer questions about the means of several populations.

In the case of k populations we are usually interested in testing hypotheses of the form $H_0 : c_1\mu_1 + \cdots + c_k\mu_k = 0$, where $c_1\mu_1 + \cdots + c_k\mu_k$ is a contrast among the population means. If we take samples of size n from each population then the model space is a k-dimensional subspace of kn-space. The coordinate system for the model space can be chosen to be $U_1 = [1, \cdots, 1]^T/\sqrt{kn}$ together with $k-1$ orthogonal unit vectors $U_2, \ldots, U_k$. If the hypotheses of interest can be formulated in terms of $k-1$ orthogonal contrasts, then the unit vectors $U_2, \cdots, U_k$ can be chosen to correspond to these contrasts. This will allow independent testing of $k-1$ hypotheses of the form $H_0 : c_1\mu_1 + \cdots + c_k\mu_k = 0$.

The orthogonal decomposition takes the form

$$y \quad = \quad \bar{y}_{..} \quad + \quad (\bar{y}_{i.} - \bar{y}_{..}) \quad + \quad (y - \bar{y}_{i.})$$

breaking the observation vector into overall mean vector, treatment vector and error vector. This can be simplified to

$$y - \bar{y}_{..} \quad = \quad (\bar{y}_{i.} - \bar{y}_{..}) \quad + \quad (y - \bar{y}_{i.})$$

as shown in Figure 7.3. In turn the treatment vector can be written as

$$\bar{y}_{i.} - \bar{y}_{..} \quad = \quad (y.U_2)U_2 + \cdots + (y.U_k)U_k$$

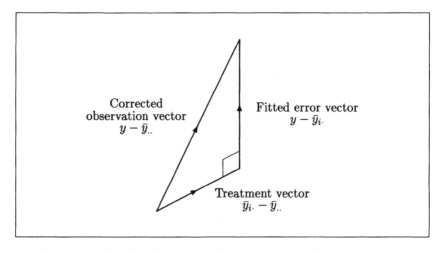

Figure 7.3: Simplified orthogonal decomposition, k treatment case.

We test the hypothesis $H_0 : c_1\mu_1 + \cdots + c_k\mu_k = 0$ by comparing $(y.U_c)^2$ with the error mean square, s^2.

In Chapters 8 to 11 we shall learn how to build up sets of orthogonal contrasts using the four basic types of contrast: class comparisons, factorial comparisons, polynomial comparisons and pairwise comparisons. Examples of these types of contrast have been presented in Examples A to F of this chapter. We shall also learn how to construct confidence intervals for contrasts.

This chapter completes the development laid down in the previous two chapters and sets the scene for the more detailed work of the next four chapters. Table 7.8 presents a succinct summary of our progress to date.

No.	Orthogonal decompositions (plus dimensions)				Parameters
popns	Observation	Mean	Treatment	Error	Parameters
1	y	$=$ $\bar{y}$		$+$ $(y - \bar{y})$	μ
(Ch.5)	(n)	(1)		$(n-1)$	
2	y	$=$ $\bar{y}_{..}$	$+$ $(\bar{y}_{i.} - \bar{y}_{..})$	$+$ $(y - \bar{y}_{i.})$	$\mu, \mu_1 - \mu_2$
(Ch.6)	$(2n)$	(1)	(1)	$(2(n-1))$	
3	y	$=$ $\bar{y}_{..}$	$+$ $(\bar{y}_{i.} - \bar{y}_{..})$	$+$ $(y - \bar{y}_{i.})$	μ, c_1, c_2
(Ch.7)	$(3n)$	(1)	(2)	$(3(n-1))$	
k	y	$=$ $\bar{y}_{..}$	$+$ $(\bar{y}_{i.} - \bar{y}_{..})$	$+$ $(y - \bar{y}_{i.})$	$\mu, c_1 \ldots c_{k-1}$
(Ch.7)	(kn)	(1)	$(k-1)$	$(k(n-1))$	

Table 7.8: The orthogonal decompositions into overall mean vector, treatment vector and error vector, for studies with completely randomized designs. Also shown are the dimensions of the subspaces in which the vectors lie, and the model parameters for the 1, 2, 3 and k population cases.

Solutions to the Reader Exercises

(7.1) The orthogonal decomposition is

$$y = (y.U_1)U_1 + (y.U_2)U_2 + (y.U_3)U_3 + (y.U_4)U_4 + (y.U_5)U_5 + (y.U_6)U_6$$

$$
\begin{bmatrix} 124 \\ 110 \\ 107 \\ 115 \\ 126 \\ 138 \end{bmatrix}
=
\begin{bmatrix} 120 \\ 120 \\ 120 \\ 120 \\ 120 \\ 120 \end{bmatrix}
+
\begin{bmatrix} -6 \\ -6 \\ -6 \\ -6 \\ 12 \\ 12 \end{bmatrix}
+
\begin{bmatrix} 3 \\ 3 \\ -3 \\ -3 \\ 0 \\ 0 \end{bmatrix}
+
\begin{bmatrix} 7 \\ -7 \\ 0 \\ 0 \\ 0 \\ 0 \end{bmatrix}
+
\begin{bmatrix} 0 \\ 0 \\ -4 \\ 4 \\ 0 \\ 0 \end{bmatrix}
+
\begin{bmatrix} 0 \\ 0 \\ 0 \\ 0 \\ -6 \\ 6 \end{bmatrix}
$$

(i) Hence $(y.U_1)U_1 = \bar{y}_{..}$

$$
\text{(ii) } (y.U_2)U_2 + (y.U_3)U_3 =
\begin{bmatrix} -3 \\ -3 \\ -9 \\ -9 \\ 12 \\ 12 \end{bmatrix}
=
\begin{bmatrix} 117 - 120 \\ 117 - 120 \\ 111 - 120 \\ 111 - 120 \\ 132 - 120 \\ 132 - 120 \end{bmatrix}
= \bar{y}_{i.} - \bar{y}_{..}
$$

$$
\text{(iii) } (y.U_4)U_4 + (y.U_5)U_5 + (y.U_6)U_6 =
\begin{bmatrix} 7 \\ -7 \\ -4 \\ 4 \\ -6 \\ 6 \end{bmatrix}
= y - \bar{y}_{i.}
$$

Hence our decomposition reduces to $y = \bar{y}_{..} + (\bar{y}_{i.} - \bar{y}_{..}) + (y - \bar{y}_{i.})$, as required.

(7.2) We can show that $(y.U_2)^2 = 432$ and $(y.U_3)^2 = 36$. Furthermore $(y.U_4)^2 = 98$, $(y.U_5)^2 = 32$ and $(y.U_6)^2 = 72$. Hence the test statistics are:

(a) $\dfrac{432}{(98 + 32 + 72)/3} = \dfrac{432}{67.33} = 6.42$ (b) $\dfrac{36}{(98 + 32 + 72)/3} = \dfrac{36}{67.33} = 0.53$

To test the hypotheses we compare each of these with the distribution of the $F_{1,3}$ statistic. This has a 95 percentile of 10.13. Since neither test statistic exceeds this value we are unable to formally reject either null hypothesis.

(7.3) $U_c = \dfrac{c_1}{\sqrt{\dfrac{k}{n}\sum\limits_{i=1}^{n} c_i^2}} \begin{bmatrix} 1 \\ \vdots \\ 0 \\ \vdots \\ 0 \\ \vdots \end{bmatrix} + \dfrac{c_2}{\sqrt{\dfrac{k}{n}\sum\limits_{i=1}^{n} c_i^2}} \begin{bmatrix} 0 \\ \vdots \\ 1 \\ \vdots \\ 0 \\ \vdots \end{bmatrix} + \dfrac{c_3}{\sqrt{\dfrac{k}{n}\sum\limits_{i=1}^{n} c_i^2}} \begin{bmatrix} 0 \\ \vdots \\ 0 \\ \vdots \\ 1 \\ \vdots \end{bmatrix}$

so the vector U_c lies in the model space, M.

(7.4) $\qquad U_c.[1, 1, \ldots, 1]^T = \dfrac{n}{\sqrt{n \sum\limits_{i=1}^{k} c_i^2}} \sum_{i=1}^{k} c_i = 0$

(7.5)

$$U_c.U_d = 0 \iff \dfrac{1}{\sqrt{n \sum\limits_{i=1}^{k} c_i^2}} \begin{bmatrix} c_1 \\ \cdot \\ c_2 \\ \vdots \\ c_k \\ \cdot \end{bmatrix} . \dfrac{1}{\sqrt{n \sum\limits_{i=1}^{k} d_i^2}} \begin{bmatrix} d_1 \\ \cdot \\ d_2 \\ \vdots \\ d_k \\ \cdot \end{bmatrix} = 0$$

$$\iff \sum_{i=1}^{k} c_i d_i = 0$$

Part III
Orthogonal Contrasts

This part of the book examines the types of questions we can pose about the means of several populations. Questions are phrased in terms of contrasts, of which there are four main types: class comparisons are discussed in Chapter 8, factorial contrasts in Chapter 9, polynomial contrasts in Chapter 10 and pairwise contrasts in Chapter 11. Corresponding to each question there is always a direction in the treatment space, with "independent" questions corresponding to orthogonal directions.

Chapter 8

Class Comparisons

This chapter is devoted to a study of contrasts of the class comparison type. With the exception of pairwise comparisons, this is the easiest type of contrast. A class comparison contrasts the average of the population means in one class of populations, with the average for a second class of populations. For example, in §7.1 the contrast of suburban with urban, $c = (\mu_1 + \mu_2)/2 - \mu_3$, was of this type. Here the class of suburban populations had two members whereas the class of urban populations had just one member.

In this chapter we shall analyze the first case study, Example A, in §1, then go on to describe the procedure for generating a complete set of $k - 1$ orthogonal class contrasts in §2. The chapter ends with a summary in §3, and a class exercise.

8.1 Analyzing Example A

We shall now carry out a statistical analysis of the data given in Example A of §7.2. For ready reference we reprint this here in Table 8.1.

Data

The observation vector is

$$y = [\ y_{11},\ y_{12},\ y_{13},\ y_{14}, \ldots,\ y_{81},\ y_{82},\ y_{83},\ y_{84}]^T$$
$$= [11.7, 11.3, 11.5, 10.6, \ldots, 15.5, 16.8, 22.3, 14.4]^T$$

Our observations are assumed to be drawn in an independent fashion.

Treatment	Bulb dry matter				Mean
Brigadier mangels	11.7	11.3	11.5	10.6	11.275
Yellow globe mangels	12.6	13.5	14.8	14.9	13.95
Orange globe mangels	12.5	13.1	11.2	9.8	11.65
Red intermediate mangels	11.0	11.8	13.4	8.1	11.075
Mono rosa fodder beet	13.9	13.0	15.2	17.1	14.8
Mono blanc fodder beet	14.1	12.6	15.6	11.3	13.4
Mono bomba fodder beet	11.9	15.2	13.9	14.7	13.925
Yellow daeno fodder beet	15.5	16.8	22.3	14.4	17.25

Table 8.1: Dry matter of bulbs, in t/ha, for each plot, Example A.

Model

We assume that each of the eight populations of interest are normally distributed, with means $\mu_1, \ldots, \mu_8$ and common variance σ^2 unknown. The model vector is

$$
\begin{bmatrix} \mu_1 \\ \mu_1 \\ \mu_1 \\ \mu_1 \\ \mu_2 \\ \mu_2 \\ \mu_2 \\ \mu_2 \\ \vdots \\ \mu_8 \\ \mu_8 \\ \mu_8 \\ \mu_8 \end{bmatrix}
= \mu_1 \begin{bmatrix} 1 \\ 1 \\ 1 \\ 1 \\ 0 \\ 0 \\ 0 \\ 0 \\ \vdots \\ 0 \\ 0 \\ 0 \\ 0 \end{bmatrix}
+ \mu_2 \begin{bmatrix} 0 \\ 0 \\ 0 \\ 0 \\ 1 \\ 1 \\ 1 \\ 1 \\ \vdots \\ 0 \\ 0 \\ 0 \\ 0 \end{bmatrix}
+ \cdots + \mu_8 \begin{bmatrix} 0 \\ 0 \\ 0 \\ 0 \\ 0 \\ 0 \\ 0 \\ 0 \\ \vdots \\ 1 \\ 1 \\ 1 \\ 1 \end{bmatrix}
$$

so the model space M is an 8-dimensional subspace of 32-space, with coordinate axis directions $[1, 1, 1, 1, 0, 0, 0, 0, \ldots, 0, 0, 0, 0]^T$ and so on.

Test Hypothesis

We wish to determine whether the mangels and fodder beets included in the study produce, on average, similar quantities of bulb when measured in terms of dry weight. The hypothesis of interest is

$$H_0 : \frac{\mu_1 + \mu_2 + \mu_3 + \mu_4}{4} = \frac{\mu_5 + \mu_6 + \mu_7 + \mu_8}{4}$$

with an alternative of

$$H_1 : \frac{\mu_1 + \mu_2 + \mu_3 + \mu_4}{4} \neq \frac{\mu_5 + \mu_6 + \mu_7 + \mu_8}{4}$$

The corresponding contrast, expressed in its simplest terms, is $c = \mu_1 + \mu_2 + \mu_3 + \mu_4 - \mu_5 - \mu_6 - \mu_7 - \mu_8$. The direction associated with the hypothesis is $U_2 = [1, \ldots, 1, -1, \ldots, -1]^T/\sqrt{32}$, with 16 entries of "1" followed by 16 of "-1". This vector lies in the model space, and has a projection coefficient of

$$y.U_2 = 4(\bar{y}_{1.} + \bar{y}_{2.} + \bar{y}_{3.} + \bar{y}_{4.} - \bar{y}_{5.} - \bar{y}_{6.} - \bar{y}_{7.} - \bar{y}_{8.})/\sqrt{32}$$

Hence the random variable $Y.U_2$ has expected value a constant multiple of $c = \mu_1 + \mu_2 + \mu_3 + \mu_4 - \mu_5 - \mu_6 - \mu_7 - \mu_8$, as required. Thus if the contrast c is zero the projection coefficient $y.U_2$ will be small, averaging to zero over many repetitions of the study. On the other hand, if c is non-zero, the projection coefficient will be large, averaging to the non-zero quantity $4c/\sqrt{32}$. This allows us to distinguish the cases $c = 0$ and $c \neq 0$. This is illustrated in Figure 8.1.

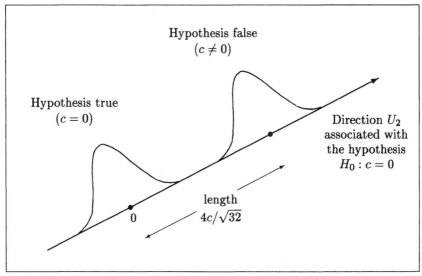

Figure 8.1: Distribution of projection coefficient $Y.U_2$ when hypothesis is true ($c = 0$) and when the hypothesis is false ($c \neq 0$). If $c = 0$ the realized value, $y.U_2$, will be small, whereas if $c \neq 0$ this realized value will be large.

We now have the raw materials for our analysis. The observation vector y, the model space M and the direction U_2 associated with our single hypothesis are as follows:

$$y = \begin{bmatrix} 11.7 \\ \vdots \\ 14.4 \end{bmatrix}, \quad M = \text{span} \left\{ \begin{bmatrix} 1 \\ \vdots \\ 0 \end{bmatrix}, \ldots, \begin{bmatrix} 0 \\ \vdots \\ 1 \end{bmatrix} \right\}, \quad \text{and} \quad U_2 = \frac{1}{\sqrt{32}} \begin{bmatrix} 1 \\ \vdots \\ -1 \end{bmatrix}$$

Fitting the Model

We estimate the model vector, the vector of population means, by projecting the observation vector onto the model space. The resulting fitted model vector is

$$\bar{y}_{i\cdot} = [11.275, 11.275, 11.275, 11.275, \ldots, 17.25, 17.25, 17.25, 17.25]^T$$

the vector of treatment means. The fitted model is therefore

$$
\begin{array}{ccccc}
y & = & \bar{y}_{i\cdot} & + & (y - \bar{y}_{i\cdot}) \\
\begin{bmatrix} 11.7 \\ 11.3 \\ 11.5 \\ 10.6 \\ \vdots \\ 15.5 \\ 16.8 \\ 22.3 \\ 14.4 \end{bmatrix}
& = &
\begin{bmatrix} 11.275 \\ 11.275 \\ 11.275 \\ 11.275 \\ \vdots \\ 17.25 \\ 17.25 \\ 17.25 \\ 17.25 \end{bmatrix}
& + &
\begin{bmatrix} .425 \\ .025 \\ .225 \\ -.675 \\ \vdots \\ -1.75 \\ -.45 \\ 5.05 \\ -2.85 \end{bmatrix} \\
\text{Observation} & = & \text{Model} & + & \text{Fitted error} \\
\text{vector} & & \text{vector} & & \text{vector}
\end{array}
$$

Here the model vector lies in the 8-dimensional model space and the fitted error vector lies in the 24-dimensional error space. The decomposition is an extension of the two population result, and yields $\bar{y}_{1\cdot} = 11.275$, the sample mean for treatment one, as our least squares estimate of μ_1, and so on, up to $\bar{y}_{8\cdot} = 17.25$ as our estimate of μ_8.

Testing the Hypothesis

To test our hypothesis that the contrast c equals zero, we simply check whether the squared distance, $(y.U_2)^2$, is comparable to, or larger than, the average of the corresponding squared distances in the error space. We use the test statistic

$$ F = \frac{(y.U_2)^2}{\left[(y.U_9)^2 + \cdots + (y.U_{32})^2\right]/24} = \frac{(y.U_2)^2}{\|y - \bar{y}_{i\cdot}\|^2/24} $$

where $U_9, \ldots, U_{32}$ are coordinate axes for the error space.

Substituting values for $(y.U_2)^2$ and $\|y - \bar{y}_{i\cdot}\|^2$ we obtain

$$ F = \frac{65.27}{88.67/24} = \frac{65.27}{3.69} = 17.69 $$

Here $(y.U_2)^2$ was calculated using the Regress command of Minitab, as will be described shortly. It could also have been calculated, however, as

$$(y.U_2)^2 = \frac{4^2(11.275 + 13.95 + 11.65 + 11.075 - 14.8 - 13.4 - 13.925 - 17.25)^2}{32}$$

$$= 65.27$$

If $H_0 : \mu_1 + \mu_2 + \mu_3 + \mu_4 = \mu_5 + \mu_6 + \mu_7 + \mu_8$ were true, then the test statistic, F, would be a realized value of the $F_{1,24}$ distribution. This distribution has a 99 percentile of 7.82. Thus our value of 17.69 is well outside the normal range of values. We formally reject H_0 at the "1% level of significance". We conclude that the fodder beet cultivars used in the experiment produced on average a higher dry weight of bulbs than did the mangel cultivars used in the experiment.

Simplified Decomposition

As in the two population case, the unit vector $U_1 = [1, \ldots, 1]^T/\sqrt{32}$ can be used to specify the first axis of a new coordinate system for the model space, with U_2 specifying a second axis. Since $(y.U_1)U_1 = \bar{y}_{..}$ our orthogonal decompostion can now be rewritten in the form

$$
\begin{bmatrix} 11.7 \\ 11.3 \\ 11.5 \\ 10.6 \\ \vdots \\ 15.5 \\ 16.8 \\ 22.3 \\ 14.4 \end{bmatrix}
-
\begin{bmatrix} 13.42 \\ 13.42 \\ 13.42 \\ 13.42 \\ \vdots \\ 13.42 \\ 13.42 \\ 13.42 \\ 13.42 \end{bmatrix}
=
\begin{bmatrix} -2.14 \\ -2.14 \\ -2.14 \\ -2.14 \\ \vdots \\ 3.83 \\ 3.83 \\ 3.83 \\ 3.83 \end{bmatrix}
+
\begin{bmatrix} .425 \\ .025 \\ .225 \\ -.675 \\ \vdots \\ -1.75 \\ -.45 \\ 5.05 \\ -2.85 \end{bmatrix}
$$

Corrected observation vector		Treatment vector	Fitted error vector

$$\text{or} \qquad y \; - \; \bar{y}_{..} \quad = \quad \bar{y}_{i.} - \bar{y}_{..} \quad + \quad y - \bar{y}_{i.}$$

as shown in Figure 8.2. The vector $y - \bar{y}_{..}$ lies in a 31-dimensional subspace of 32-space, the treatment vector lies in the 7-dimensional treatment space, and the error vector lies in the 24-dimensional error space. The projection $(y.U_2)U_2$ lies in a 1-dimensional subspace of the treatment space.

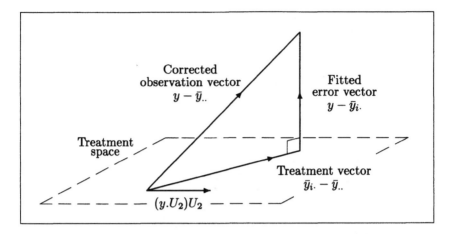

Figure 8.2: Orthogonal decomposition of the corrected observation vector, Example A.

ANOVA Table

The corresponding Pythagorean decomposition is

$$\|y - \bar{y}_{..}\|^2 \;\; = \;\; \|\bar{y}_{i\cdot} - \bar{y}_{..}\|^2 \;\; + \;\; \|y - \bar{y}_{i\cdot}\|^2$$

$$\text{or} \quad 210.04 \;\; = \;\; 121.37 \;\; + \;\; 88.67$$

Here the calculations are based on more accuracy than is shown in the vectors. For example, $\bar{y}_{..} = 13.415625$.

These calculations, plus the all important squared projection length, $(y.U_2)^2 = 65.27$, are summarized in the traditional ANOVA table, shown in Table 8.2.

Source of Variation	df	SS	MS	F
Treatments	7	121.37		
H_0 : mangels=beets	1	65.27	65.27	17.69(**)
Error	24	88.67	3.69	
Total	31	210.04		

Table 8.2: ANOVA table summarizing the calculations involved in a test of the null hypothesis that mangels and fodder beets produce equal quantities of bulb dry matter per hectare.

Estimation of σ^2

In our example we have transformed our set of independent random variables, $Y_{11}, Y_{12}, Y_{13}, Y_{14} \sim N[\mu_1, \sigma^2]$, ..., $Y_{81}, Y_{82}, Y_{83}, Y_{84} \sim N[\mu_8, \sigma^2]$, into a new set of independent random variables

$$Y.U_1 \sim N\left[\sqrt{32}\mu, \sigma^2\right], \quad Y.U_2 \sim N\left[4c/\sqrt{32}, \sigma^2\right], \quad \ldots,$$

$$Y.U_9, \ldots, Y.U_{32} \quad \sim \quad N[0, \sigma^2]$$

The model space directions $U_3, \ldots, U_8$ will not be spelt out until §8.2. The error space directions $U_9, \ldots, U_{32}$ follow the pattern established in Chapters 5 to 7, with the first three directions corresponding to variations within treatment one, and so on.

The first and second random variables, $Y.U_1$ and $Y.U_2$, have been used to estimate $\mu = (\mu_1 + \cdots + \mu_8)/8$ and $c = \mu_1 + \mu_2 + \mu_3 + \mu_4 - \mu_5 - \mu_6 - \mu_7 - \mu_8$ respectively, while the last twenty four were used to estimate σ^2 via

$$s^2 = \frac{(y.U_9)^2 + \cdots + (y.U_{32})^2}{24} = \frac{\|y - \bar{y}_{i.}\|^2}{24} = \frac{88.67}{24} = 3.69$$

Here $y.U_9, \ldots, y.U_{32}$ are observations from the $N[0, \sigma^2]$ distribution, so their squares average to σ^2 over many repeats of the experiment, as described in §3.3.

Checking of Assumptions

The assumptions we need to check are independence, normality of the distributions, and equality of the population variances, σ_1^2 to σ_8^2.

The assumption of independence can be suspect in a field experiment, where variations in the fertility of the soil may cause the errors to be correlated among neighbouring plots. For example, plots 1 to 4 may be on a good patch of soil and plots 21 to 27 may be on a bad patch of soil. In our case the plots 1 to 32 were laid out in a line, so one would normally graph the errors, $y_{ij} - \bar{y}_{i.}$, against the plot number to check this assumption. However, since our layout was altered, we shall not do this here.

To check the assumption of normality, we plot a histogram of the errors in Figure 8.3.

The histogram pools all eight treatments, so the normality of each population is not being checked. Overall, however, the data appears to be approximately normally distributed. The only worry is the value of 5.05 to the right of the distribution. This corresponds to an unusually high value of 22.3 in treatment 8. Our first thought is that this could be due to a transcription or arithmetic error. Upon checking our field books, however, we find that the value is correct.

How unusual is this value of 5.05? In our experiment the estimate of the variance is $s^2 = 3.69$, so the estimate of the standard error, σ, is

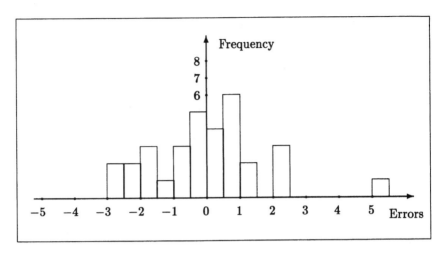

Figure 8.3: Histogram of the errors, $y - \bar{y}_{i.}$, for the greenfeed Example A.

$s = \sqrt{3.69} = 1.92$. Hence when standardized to a standard error of unity, our value is $5.05/1.92 = 2.63$, which lies between the 95 and 99 percentiles of the t_{24} distribution, 2.06 and 2.80 respectively. Since we have 32 observations, it is perhaps not surprising to find this unusual a value. However, the value of 22.3 will probably have increased the standard error, which will serve to reduce the standardized value, so we pause to examine the situation more carefully.

What effect would it have on our analysis if we were to reject this value of 22.3, and substitute a "missing value" estimated as the mean of the other three values, $15.567 = (15.5+16.8+14.4)/3$? Here the reason for substituting a fictitious value is that we preserve orthogonality, so can proceed as before and obtain the revised ANOVA table given in Table 8.3.

Source of Variation	df	SS	MS	F
Treatments	7	79.65		
H_0 : mangels=beets	1	47.45	47.45	19.96(**)
Error	23(1)	54.67	2.38	
Total	30	134.32		

Table 8.3: ANOVA table obtained with one missing value.

Interestingly enough, our test statistic has increased from 17.69 to 19.96, so our conclusions remain unaltered. The reason for this increase is that while the contrast sum of squares $(y.U_2)^2$ has decreased, the estimate of σ^2, namely s^2 the error mean square, has also decreased, and by a proportionately greater amount. It is also interesting to note that if we use our new estimate, $s^2 = 2.38$, then our justification for rejecting the value of 22.3 is improved, since our new estimated standard error is $s = 1.54$, leading to a standardized value of $5.05/1.54 = 3.28$, which is greater than the 99-percentile of the t_{23} distribution, a value of 2.81.

In Table 8.3 the number of error degrees of freedom has decreased from 24 to 23, since in treatment 8 only three independent values are contributing to the error sum of squares. The missing value of 15.567 adds $(y_{83} - \bar{y}_{8.})^2 = (15.567 - 15.567)^2 = 0$ to the sum of squares. The treatment and contrast sums of squares are biased upwards since they are calculated assuming each treatment mean is an average of four values; this is a relatively minor point in our example.

We move on now to consider the validity of the assumption of a common variance, σ^2. If we calculate variance estimates s_i^2 for each treatment we obtain the estimates given in Table 8.4.

Treatment no.	1	2	3	4	5	6	7	8	Average
Variance estimate	0.23	1.22	2.15	4.93	3.17	3.46	2.11	12.30	3.69

Table 8.4: Variance estimates for individual treatments.

We can compare any two of these estimates using the $F_{3,3}$ statistic. For example, we can test the hypothesis $H_0 : \sigma_1^2 = \sigma_2^2$ against the hypothesis $H_1 : \sigma_1^2 \neq \sigma_2^2$ by checking whether the test statistic $s_1^2/s_2^2 = 0.23/1.22 = 0.19$ lies inside or outside the normal range of values of $F_{3,3}$. This is the range .06 to 15.44, bounded by the 2.5 and 97.5 percentiles of the $F_{3,3}$ distribution. That is, exceptionally large or exceptionally small values are significant, since they differ only by virtue of whether the larger or smaller s_i^2 value appears in the numerator of the test statistic. The value of 0.19 is within the normal range, so we would accept as plausible the hypothesis $H_0 : \sigma_1^2 = \sigma_2^2$.

If we slavishly carry out all 28 such pairwise comparisons of variances, we find that treatment 1 had a significantly lower variance than treatments 4 and 8. Treatment 8 has the highest variance estimate, as we would expect, since our unusual value of 22.3 is included. It is unwise to place much emphasis on the results of 28 such 5% level significance tests, so we shall not get too upset that treatment 1 appears to have a lower variance than treatment 4. However, the high variance estimate for treatment 8 serves to reinforce the view that we should declare the value of 22.3 to be "suspect", or "unusual".

In conclusion, our assumption checking has revealed only one problem on which we plan to take action, that of an unusually high value in treatment 8. The action we take is to substitute a value of 15.567 in place of the value of 22.3.

Confidence Interval for the Contrast of Interest

The estimated average difference between mangels and fodder beets in terms of bulb dry matter, in tonnes per hectare, is

$$\frac{11.275 + 13.95 + 11.65 + 11.075}{4} - \frac{14.8 + 13.4 + 13.925 + 15.567}{4} = -2.44$$

excluding the suspicious value of 22.3 from the data. That is, our best estimate is that the fodder beet cultivars outyielded the mangel cultivars by an average of 2.44 tonnes per hectare. We now wish to give an indication of the accuracy of this estimate by calculating a 95% confidence interval for the true value of the contrast

$$c = \frac{\mu_1 + \mu_2 + \mu_3 + \mu_4}{4} - \frac{\mu_5 + \mu_6 + \mu_7 + \mu_8}{4}$$

We follow the procedure outlined in Appendix E. The corresponding unit vector is $U_c = [1, \ldots, -1, \ldots]^T / \sqrt{32}$. The projection coefficient

$$y.U_c = 4\left(\bar{y}_{1.} + \bar{y}_{2.} + \bar{y}_{3.} + \bar{y}_{4.} - \bar{y}_{5.} - \bar{y}_{6.} - \bar{y}_{7.} - \bar{y}_{8.}\right)/\sqrt{32}$$

comes from an

$$N\left[4\left(\mu_1 + \mu_2 + \mu_3 + \mu_4 - \mu_5 - \mu_6 - \mu_7 - \mu_8\right)/\sqrt{32}, \sigma^2\right]$$

distribution. Rewriting this using the shorthand label, $\hat{c}$, for the estimated value of the contrast, c, and rearranging, tells us that the value $4\hat{c}/\sqrt{2}$ comes from an $N(4c/\sqrt{2}, \sigma^2)$ distribution. Hence $4(\hat{c} - c)/\sqrt{2}$ comes from an $N(0, \sigma^2)$ distribution. This will be the numerator for our t statistic. For the denominator we use s, an estimate with 23 degrees of freedom.

To obtain a 95% confidence interval for the contrast c, we gamble that the realized value $t = 4(\hat{c} - c)/\sqrt{2}s$, lies between the 2.5 and 97.5 percentiles of the t_{23} distribution. That is $-2.069 \leq \dfrac{4(\hat{c} - c)}{\sqrt{2}s} \leq 2.069$, or

$$\hat{c} - 2.069\frac{\sqrt{2}s}{4} \quad \leq \quad c \quad \leq \quad \hat{c} + 2.069\frac{\sqrt{2}s}{4}$$

Substituting $\hat{c} = -2.44$ and $s = \sqrt{2.38} = 1.54$, we obtain

$$-2.44 - 1.13 \quad \leq \quad c \quad \leq \quad -2.44 + 1.13$$

That is, our 95% confidence interval is

$$-3.57 \quad \leq \quad \frac{\mu_1 + \mu_2 + \mu_3 + \mu_4}{4} - \frac{\mu_5 + \mu_6 + \mu_7 + \mu_8}{4} \quad \leq \quad -1.31$$

In summary, we are 95% confident that beets outyielded mangels by between 1.3 and 3.6 tonnes per hectare of bulb dry matter.

Report on Study

In a report on this experiment the data can be summarized as shown in Table 8.5.

In this table, the SED is the standard error of the difference between any two treatment means, that is, the standard error of the estimator $\overline{Y}_{i\cdot} - \overline{Y}_{j\cdot}$, given by $\sqrt{2s^2/4} = \sqrt{2(2.38)/4} = 1.09$. We have decided to present the data with the unusual value of 22.3 excluded from the analysis, so a note is included to this effect. The report might also include a 95% confidence interval for the average difference between the four mangel cultivars and the four fodder beet cultivars. This would be an interval of 1.3 to 3.6 tonnes/ha of bulb dry matter.

Treatment	DM bulbs, in tonnes/ha
Brigadier mangels	11.3
Yellow globe mangels	14.0
Orange globe mangels	11.7
Red intermediate mangels	11.1
Mono rosa fodder beet	14.8
Mono blanc fodder beet	13.4
Mono bomba fodder beet	13.9
Yellow daeno fodder beet	15.6
SED	1.09
Significance of contrast	
Mangels vs fodder beets	**

Note: An exceptionally high value of 22.3, recorded for the yellow daeno treatment, was excluded from the analysis

Table 8.5: Summary of results, winter greenfeed trial.

Minitab Analysis

We shall now describe how to calculate the ANOVA table given in Table 8.1 using Minitab. What we require is the Pythagorean decomposition

$$\|y - \bar{y}_{..}\|^2 = \|\bar{y}_{i\cdot} - \bar{y}_{..}\|^2 + \|y - \bar{y}_{i\cdot}\|^2$$

and the square of the projection coefficient, $(y.U_2)^2$. These two requirements can be easily met using the Oneway and Regress commands respectively, as shown in Table 8.6. Note that to calculate the revised ANOVA table in Table 8.3 we simply substitute 15.567 for 22.3 in the data.

List of commands	Interpretation
⌈name c1='Bulbdm' ∣name c2='HypoDirn' ∣name c3='Trtments' ∣name c4='Errors' ⌊name c5='Fitted'	⌈Naming the columns ⌊for convenience
⌈set c1 ⌊11.7 11.3 ... 22.3 14.4	[Set y into column one
⌈set c2 ∣1 1 ... − 1 − 1 ⌊let c2=c2/sqrt(32)	⌈Set the contrast direction U_2 ⌊into column two
[regress c1 1 c2	[Project y onto U_2
⌈set c3 ⌊1 1 ... 8 8	⌈Set the treatment numbers ⌊into column three
[oneway c1 c3 c4 c5	⌈Fit the full model, and save ⌊errors and fitted values
[histogram c4 [print c1 c5 c4 [stop	[Draw the error histogram [Print vector decomposition [Signifies end of job

Table 8.6: Minitab commands for analyzing Example A.

From the output in Table 8.7 we firstly retrieve the contrast sum of squares $(y.U_2)^2 = 65.265$. This appears as the "Regression" sum of squares in the first ANOVA table. The remainder of the output generated by the Regress command can be ignored, since we have projected y onto just a two dimensional subspace of the model space, spanned by U_1 and U_2. The full model is fitted by the Oneway command which gives us the required error mean square, $s^2 = 3.69$.

Also produced is a useful list of treatment means and estimated standard errors within each treatment, $s_1, s_2, \ldots, s_8$. The latter values can be squared to create Table 8.4 in our section on assumption checking. To further facilitate assumption checking we have saved the error vector in column 4 of the worksheet, and produced a histogram of the error components, $y_{ij} - \bar{y}_{i\cdot}$, similar to that in Figure 8.3. For economy of space we have not reproduced this here. To allow us to print out the vector decomposition in the form $y = \bar{y}_{i\cdot} + (y - \bar{y}_{i\cdot})$, we have saved the fitted $\bar{y}_{i\cdot}$ values in column 5 of the worksheet. This allows inspection of individual errors together with the corresponding observed and fitted values. For economy of space we have again omitted this from the table.

Regress output

The regression equation is
Bulbdm = 13.4 − 8.08 HypoDirn

Predictor	Coef	Stdev	t-ratio	p
Constant	13.4156	0.3883	34.55	0.000
HypoDirn	-8.079	2.197	-3.68	0.001

s = 2.197 R-square = 31.1% R-sq(adj) = 28.8%

Analysis of Variance

SOURCE	DF	SS	MS	F	p
Regression	1	65.265	65.265	13.52	0.001
Error	30	144.777	4.826		
Total	31	210.042			

Oneway output

ANALYSIS OF VARIANCE ON Bulbdm

SOURCE	DF	SS	MS	F	p
Trtments	7	121.37	17.34	4.69	0.002
ERROR	24	88.67	3.69		
TOTAL	31	210.04			

LEVEL	N	MEAN	STDEV
1	4	11.275	0.479
2	4	13.950	1.103
3	4	11.650	1.466
4	4	11.075	2.220
5	4	14.800	1.780
6	4	13.400	1.860
7	4	13.925	1.452
8	4	17.250	3.507

POOLED STDEV = 1.922

Table 8.7: Selected Minitab output for Example A.

Fixed Effects Versus Random Effects: An Aside

A point which should be clarified is that in the above analysis, summarized in Table 8.5, the cultivars are treated as *fixed* cultivars, so the contrast is a comparison between the average of a fixed set of four mangel cultivars and another fixed set of four fodder beet cultivars. If any of these cultivars were to change, so too would the true value of the contrast.

This is an important point. We are tempted, but not validly entitled, to say that "fodder beets are *in general* better than mangels" in terms of the quantity of bulb dry matter produced. All we can validly say from Table 8.5 is that the four *particular* fodder beet cultivars on average did better than the four particular mangel cultivars. Of course, we also remember that this was just in one farmer's field, under specific field conditions and under just one pattern of weather conditions.

If we wished to make a more general statement, we would need to know that the mangel cultivars had been chosen *at random* from a larger population of mangel cultivars, such as all cultivars in common use locally, and that the fodder beet cultivars had similarly been chosen at random from a population of fodder beet cultivars, such as all local cultivars. If in fact the cultivars in our experiment had been so chosen, we would have two samples of size four, chosen in a completely random manner from the populations of mangel and fodder beet cultivars. The observation vector would consist of the eight values given in Table 8.5, and the analysis would follow the methods described in Chapter 6. Results are summarized in Table 8.8.

Population	Dry Matter of bulbs (t/ha)
Mangel cultivars	12.0
Fodder beet cultivars	14.4
SED	0.83
Significance of difference	*

Table 8.8: Analysis when cultivars are random selections from larger populations of mangel and fodder beet cultivars.

The words "fixed effects" and "random effects" appear in many textbooks, and we felt that this was a good opportunity to clarify this topic. The fixed effects analysis is given in Table 8.5, and the random effects analysis is given in Table 8.8. The difference in significances between the two analyses (** to *) is typical of results which occur in practice. That is, there is very strong evidence that the four particular mangel cultivars produce less on average than the four particular fodder beet cultivars. However, the variation among the four mangel cultivars, and among the four fodder beet cultivars, is large enough to cast some doubt on whether fodder beets are in general more productive than mangels.

8.2 General Case

In this section we describe the method for generating a complete set of orthogonal class comparisons, discuss the method of analysis using Minitab, and derive a 95% confidence interval for a general contrast, c. En route we briefly digress to discuss the way in which experimental results modify hypotheses.

Generation of Orthogonal Contrasts

In Example A only one contrast, c, was specified before the data was collected, so we have only one special direction, U_2, which can serve as an axis in the treatment space. To illuminate the method for specifying a complete set, however, we shall examine the experimental treatments and invent a further six orthogonal contrasts which could plausibly have been specified by the experimenter. These will correspond to a further six axes, $U_3, \ldots, U_8$, which complete the coordinate system for the treatment space.

The general method for generating a complete set of $k - 1$ orthogonal class comparisons is to initially split the set of treatments into two classes, and to compare one class with the other. Then each of these classes is broken

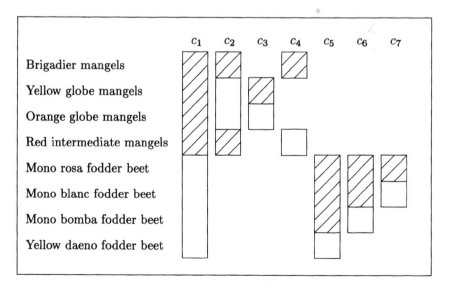

Figure 8.4: The generation of a complete set of orthogonal contrasts for Example A, by splitting treatments into classes, subclasses, and so on. Each vertical column represents a contrast; the split which generates the contrast is depicted by shading one of the classes.

down into a pair of subclasses and the subclasses compared, and so on, until no further splitting is possible.

Applying this method to Example A, we note that the contrast c, henceforth called c_1, has already split the experimental treatments into two classes, the class consisting of treatments 1 to 4 and the class consisting of treatments 5 to 8. This is depicted in the first column of Figure 8.4.

To generate further contrasts, we shall first concentrate on splits within the class of mangel cultivars, treatments 1 to 4. Here we notice that the second and third mangel cultivars are "globe" types of mangel. Lacking any better basis for a split, we decide to contrast the globe cultivars with the non-globe cultivars, generating the contrast $c_2 = \mu_1 - \mu_2 - \mu_3 + \mu_4$. We now have two subclasses of size two. The only way to split these subclasses any further is to compare the two cultivars within each subclass with one another. This generates the contrasts $c_3 = \mu_2 - \mu_3$ and $c_4 = \mu_1 - \mu_4$. The contrasts c_2 to c_4 are also depicted in Figure 8.4.

We now go to the other major class, that containing the fodder beet cultivars, treatments 5 to 8. Here we notice that three of the fodder beet cultivars are "mono" types, so we decide to contrast the class of mono cultivars with the class of the single non-mono cultivar. This generates the contrast $c_5 = \mu_5 + \mu_6 + \mu_7 - 3\mu_8$. We now have a subclass of size three which can be split further and a subclass of size one which clearly cannot be split further. Looking within the subclass of size three, we notice that two of the cultivar names involve colours, rosa and blanc. We decide to contrast these two with the third cultivar, generating the contrast $c_6 = \mu_5 + \mu_6 - 2\mu_7$. Our subclass of size three has been split into a (subsub)class of size two and a class of size one. All that remains is to split our class of size two into two classes of size one. This generates our last remaining contrast $c_7 = \mu_5 - \mu_6$. Figure 8.4 gives a pictorial view of c_5 to c_7.

The coefficients for contrasts c_1 to c_7 are summarized in Table 8.9.

Treatment		c_1	c_2	c_3	c_4	c_5	c_6	c_7
Brigadier mangels	μ_1	1	1	0	1	0	0	0
York globe mangels	μ_2	1	−1	1	0	0	0	0
Orange globe mangels	μ_3	1	−1	−1	0	0	0	0
Red intermediate mangels	μ_4	1	1	0	−1	0	0	0
Mono rosa fodder beet	μ_5	−1	0	0	0	1	1	1
Mono blanc fodder beet	μ_6	−1	0	0	0	1	1	−1
Mono bomba fodder beet	μ_7	−1	0	0	0	1	−2	0
Yellow daeno fodder beet	μ_7	−1	0	0	0	−3	0	0

Table 8.9: The set of orthogonal contrasts for Example A.

Since the treatments are equally replicated the orthogonality of the contrasts can be established by checking the orthogonality of the 8-dimensional vectors given by the columns in the table. If we so desired we could now write down a complete coordinate system for the model space, consisting of U_1, the usual axis for the overall mean space, plus U_2 to U_8, the axes for the treatment space corresponding to the contrasts c_1 to c_7.

MINITAB Analysis

In Minitab, we could set U_2 to U_8 into columns 2 to 8, and project y onto the model space using the command "regress c1 7 c2−c8" . This would fit the model in the form

$$y - \bar{y}_{..} = (y.U_2)U_2 + \cdots + (y.U_8)U_8 + (y - \bar{y}_{i.})$$

with the treatment vector decomposed into projections in the directions associated with the hypotheses $H_0 : c_i = 0$, as depicted in Figure 8.5.

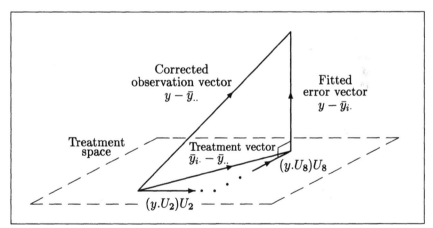

Figure 8.5: The orthogonal decomposition of the corrected observation vector, showing the treatment vector further decomposed into projections in the directions associated with the hypotheses.

The corresponding Pythagorean decomposition is

$$\|y - \bar{y}_{..}\|^2 \quad = \quad (y.U_2)^2 + \cdots + (y.U_8)^2 + \|y - \bar{y}_{i.}\|^2$$

enabling the simultaneous testing of seven independent hypotheses of the form H_0 : contrast = 0.

More Examples

To further illustrate the method of generating orthogonal contrasts of the class comparison type, we have compiled a set of examples in Table 8.10. For the case of two treatments there is clearly only one possible contrast, $\mu_1 - \mu_2$. For the case of three treatments the only possibility is to initially split into classes of size one and two. For four treatments the initial split can be either into classes of size two and two, or into classes of size three and one. For five treatments the initial split can be either into classes of size two and three, or into classes of size four and one.

Two treatments		c_1			
1. Control		−1			
2. Nitrolime		1			

Three treatments		c_1	c_2		
1. Control sheep	(No drench)	−2	0		
2. Sheep drenched once	(Drench)	1	−1		
3. Sheep drenched twice	(Drench)	1	1		

Four treatments			c_1	c_2	c_3
1. Fan heater, brand A	(Convection)		1	1	0
2. Fan heater, brand B	(Convection)		1	−1	0
3. Bar heater, brand P	(Radiation)		−1	0	1
4. Bar heater, brand Q	(Radiation)		−1	0	−1

			c_1	c_2	c_3
1. Lupins	(Legume)		−3	0	0
2. Mustard	(Non legume)	(Noncereal)	1	−2	0
3. Barley	(Non legume)	(Cereal)	1	1	−1
4. Oats	(Non legume)	(Cereal)	1	1	1

Five Treatments			c_1	c_2	c_3	c_4
1. Dacron	(Synthetic Fibre)		3	−1	0	0
2. Terylene	(Synthetic Fibre)		3	1	0	0
3. Cotton	(Natural Fibre)	(Plant fibre)	−2	0	−2	0
4. Angora	(Natural Fibre)	(Animal fibre)	−2	0	1	1
5. Wool	(Natural Fibre)	(Animal fibre)	−2	0	1	−1

		c_1	c_2	c_3	c_4
1. Control	(No herbicide)	−4	0	0	0
2. Systemic herbicide A	(Herbicide)	1	1	1	0
3. Systemic herbicide B	(Herbicide)	1	1	−1	0
4. Contact herbicide X	(Herbicide)	1	−1	0	1
5. Contact herbicide Y	(Herbicide)	1	−1	0	−1

Table 8.10: Examples of sets of contrasts of the class comparison type.

Modifying Ideas

We digress now to dwell for a moment on the way in which hypotheses are formulated and tested. Before any experiment is conducted the experimenter will have strong ideas about the answers to certain questions, and weak ideas about the answers to other questions. These prior expectations, or lack of them, should be recorded at the planning stage. After the data is collected some of these ideas will be held more strongly and some less strongly, depending on whether the data confirms or contradicts the prior ideas. In addition, other new ideas may be formulated in the light of the data. These new ideas become the hypotheses to be investigated in subsequent experimentation. Corresponding to these hypothesis categories are contrast categories which we can label as follows:

1. Strongly prespecified contrasts

2. Weakly prespecified contrasts

3. Contrasts specified from an examination of the data

In Example A, our contrast c_1 falls into the first category. The other contrasts, c_2 to c_7, would fall into the second category if the experimenter possessed some prior knowledge as justification.

When interpreting experimental data and writing reports it is very helpful to first spell out the knowledge possessed prior to the experimentation, then outline the new information discovered, and lastly conclude with a summary of the updated state of knowledge. In this context it is clearly important to distinguish the above three categories.

Confidence Intervals

In §7.3 we gave the direction associated with the contrast $c = c_1\mu_1 + \cdots + c_k\mu_k$ as

$$U_c = \frac{1}{\sqrt{n\sum_{i=1}^{k} c_i^2}} \begin{bmatrix} c_1 \\ \cdot \\ c_2 \\ \vdots \\ c_k \\ \cdot \end{bmatrix}$$

The corresponding projection coefficient, $y.U_c$, is

$$\frac{n(c_1\bar{y}_1. + \cdots + c_k\bar{y}_k.)}{\sqrt{n\sum_{i=1}^{k} c_i^2}} = \frac{\sqrt{n}(c_1\bar{y}_1. + \cdots + c_k\bar{y}_k.)}{\sqrt{\sum_{i=1}^{k} c_i^2}} = \frac{\sqrt{n}\hat{c}}{\sqrt{\sum_{i=1}^{k} c_i^2}}$$

which averages, over many repetitions of the study, to

$$\frac{\sqrt{n}(c_1\mu_1 + \cdots + c_k\mu_k)}{\sqrt{\sum_{i=1}^{k} c_i^2}} \;=\; \frac{\sqrt{n}c}{\sqrt{\sum_{i=1}^{k} c_i^2}}$$

That is, $y.U_c = \sqrt{n}\hat{c}/\sqrt{\sum_{i=1}^{k} c_i^2}$ comes from an $N\left[\sqrt{n}c/\sqrt{\sum_{i=1}^{k} c_i^2}, \sigma^2\right]$ distribution. Hence $\sqrt{n}(c - \hat{c})/\sqrt{\sum_{i=1}^{k} c_i^2}$ comes from an $N[0, \sigma^2]$ distribution. This we use as the numerator for our t statistic. For the denominator we use $\sqrt{s^2}$, where s^2 is an average of the $k(n-1)$ squared projection coefficients, $(y.U_{k+1})^2, \ldots, (y.U_{kn})^2$. The resulting realized value of the t statistic is

$$t \;=\; \frac{\sqrt{n}(\hat{c} - c)/\sqrt{\sum_{i=1}^{k} c_i^2}}{s} \;=\; \frac{\sqrt{n}(\hat{c} - c)}{s\sqrt{\sum_{i=1}^{k} c_i^2}}$$

To obtain a 95% confidence interval for the contrast c, we simply gamble that our realized value lies between the 2.5 and 97.5 percentiles of the $t_{k(n-1)}$ distribution. That is, we gamble that

$$-t_{k(n-1)}(.975) \;\leq\; \frac{\sqrt{n}(\hat{c} - c)}{s\sqrt{\sum_{i=1}^{k} c_i^2}} \;\leq\; t_{k(n-1)}(.975)$$

Upon rearrangement this yields the desired 95% confidence interval for c of

$$\hat{c} - \sqrt{\sum_{i=1}^{k} c_i^2}\,\frac{s}{\sqrt{n}}t_{k(n-1)}(.975) \;\leq\; c \;\leq\; \hat{c} + \sqrt{\sum_{i=1}^{k} c_i^2}\,\frac{s}{\sqrt{n}}t_{k(n-1)}(.975)$$

As an illustration, we can use the contrast in Example A, $c = (\mu_1 + \mu_2 + \mu_3 + \mu_4 - \mu_5 - \mu_6 - \mu_7 - \mu_8)/4$, and substitute $c_1 = c_2 = c_3 = c_4 = 1/4$, $c_5 = c_6 = c_7 = c_8 = -1/4$ together with $\hat{c} = -2.44$, $s = 1.54$, $n = 4$ and $t_{k(n-1)}(.975) = 2.069$. This yields the confidence interval of -2.44 ± 1.13, as obtained previously.

8.3   Summary

Class comparisons readily give rise to orthogonal contrasts. To form a complete set of orthogonal contrasts of this type, simply divide the populations into two classes and contrast the first class with the second class. Then divide the first class into two subclasses and contrast the first subclass with the second subclass. Successively subdivide each subclass until each resulting class contains only one population. Similarly work on the second of the original classes until it has been reduced to classes of size one. This procedure always generates $k - 1$ orthogonal contrasts, where k is the number of populations.

The choice of classes and subclasses is determined by the populations in the study. Of the $k - 1$ orthogonal contrasts which are needed to make up a complete set, a proportion will correspond to hypotheses of real interest to the researcher. The remainder will correspond to nonsensical hypotheses, but may be included to complete the coordinate system for the treatment space.

The full set of population means $\mu_1, ..., \mu_k$ are estimated in fitting the model $y = \bar{y}_{i\cdot} + (y - \bar{y}_{i\cdot})$. When rewritten in corrected form, the fitted model becomes

$$y - \bar{y}_{\cdot\cdot} \;\; = \;\; (\bar{y}_{i\cdot} - \bar{y}_{\cdot\cdot}) + (y - \bar{y}_{i\cdot})$$

The common variance σ^2 is estimated via $s^2 = \|y - \bar{y}_{i\cdot}\|^2/[k(n - 1)]$. This value serves as the baseline for tests of hypotheses such as $H_0 : c_1\mu_1 + \cdots + c_k\mu_k = 0$. Here the relevant test statistic is $F = (y.U_c)^2/s^2$, where U_c is the direction corresponding to the hypothesis $H_0 : c = 0$. Under H_0, the test statistic follows the $F_{1,k(n-1)}$ distribution.

The 95% confidence interval for a contrast $c = c_1\mu_1 + \cdots + c_k\mu_k$ is

$$\hat{c} \;\; \pm \;\; \sqrt{\sum_{i=1}^{k} c_i^2 \frac{s}{\sqrt{n}}} t_{k(n-1)}(.975)$$

where $\hat{c} = c_1\bar{y}_{1\cdot} + \cdots + c_k\bar{y}_{k\cdot}$.

Class Exercise

In this class exercise we shall simulate an experiment in which six lambs are
allocated in a completely random manner to three experimental treatments.
The experiment is designed to determine:

(A) Whether it is necessary to drench lambs for worm control.

(B) Whether a single drench at three months of age is adequate, compared
with an additional drench a month later.

The experimental treatments are as follows:

(1) Control: undrenched lambs

(2) One drench: lambs drenched at three months of age

(3) Two drenches: lambs drenched at three and four months of age

The success of the drench program is judged by the weight gain of the lambs
from three months of age to six months of age. During this three month
experimental period the reader can imagine the six lambs living in a single
group and feeding on identical pasture. Any possible recontamination of the
drenched lambs by the undrenched lambs will be ignored.

(a) For the class exercise each student is asked to use the random numbers
in Table T.1 to select two lambs from each of the populations in Table 8.11.
The six corresponding weightgains form the observation vector, y.

These six observations will be treated as samples of size two from three
infinitely large, normally distributed populations with means μ_1, μ_2 and μ_3,
and a common variance σ^2. The questions of interest can be phrased in
terms of independent hypothesis tests as follows:

(A) $H_0 : \mu_1 = \frac{\mu_2 + \mu_3}{2}$ versus $H_1 : \mu_1 \neq \frac{\mu_2 + \mu_3}{2}$

(B) $H_0 : \mu_2 = \mu_3$ versus $H_1 : \mu_2 \neq \mu_3$

An appropriate coordinate system for 6-space is

$$
\begin{array}{cccccc}
U_1 & U_2 & U_3 & U_4 & U_5 & U_6 \\
\begin{bmatrix} 1 \\ 1 \\ 1 \\ 1 \\ 1 \\ 1 \end{bmatrix} &
\begin{bmatrix} 2 \\ 2 \\ -1 \\ -1 \\ -1 \\ -1 \end{bmatrix} &
\begin{bmatrix} 0 \\ 0 \\ 1 \\ 1 \\ -1 \\ -1 \end{bmatrix} &
\begin{bmatrix} 1 \\ -1 \\ 0 \\ 0 \\ 0 \\ 0 \end{bmatrix} &
\begin{bmatrix} 0 \\ 0 \\ 1 \\ -1 \\ 0 \\ 0 \end{bmatrix} &
\begin{bmatrix} 0 \\ 0 \\ 0 \\ 0 \\ 1 \\ -1 \end{bmatrix} \\
\sqrt{6} & \sqrt{12} & \sqrt{4} & \sqrt{2} & \sqrt{2} & \sqrt{2}
\end{array}
$$

Population one: Control

Lamb no.	Wt. gain	Lamb no.	Wt. gain	Lamb no.	Wt. gain	Lamb no.	Wt. gain	Lamb no.	Wt. gain
1.	11	11.	7	21.	12	31.	10	41.	10
2.	10	12.	9	22.	8	32.	10	42.	12
3.	8	13.	9	23.	9	33.	8	43.	9
4.	11	14.	7	24.	8	34.	11	44.	10
5.	11	15.	9	25.	11	35.	11	45.	11
6.	10	16.	10	26.	9	36.	12	46.	7
7.	12	17.	7	27.	9	37.	10	47.	9
8.	12	18.	11	28.	10	38.	10	48.	10
9.	12	19.	10	29.	7	39.	9	49.	10
10.	9	20.	10	30.	9	40.	9	50.	17

Population two: One drench

1.	18	11.	15	21.	15	31.	17	41.	14
2.	15	12.	14	22.	15	32.	17	42.	16
3.	10	13.	16	23.	16	33.	16	43.	17
4.	16	14.	12	24.	13	34.	14	44.	15
5.	16	15.	18	25.	13	35.	14	45.	14
6.	15	16.	14	26.	17	36.	16	46.	18
7.	14	17.	17	27.	15	37.	16	47.	14
8.	12	18.	12	28.	16	38.	15	48.	14
9.	15	19.	17	29.	13	39.	14	49.	14
10.	15	20.	14	30.	11	40.	16	50.	15

Population three: Two drenches

1.	20	11.	22	21.	19	31.	23	41.	17
2.	19	12.	18	22.	19	32.	20	42.	21
3.	18	13.	15	23.	21	33.	20	43.	21
4.	18	14.	27	24.	23	34.	24	44.	19
5.	22	15.	18	25.	21	35.	20	45.	21
6.	19	16.	23	26.	21	36.	15	46.	19
7.	18	17.	17	27.	20	37.	20	47.	17
8.	18	18.	18	28.	20	38.	15	18.	17
9.	19	19.	25	29.	21	39.	23	49.	20
10.	18	20.	20	30.	21	40.	20	50.	23

Table 8.11: Populations of lamb weightgains, in kilograms, for simulation exercise.

Here U_1 spans the overall mean space, U_2 and U_3 span the treatment space, and U_4, U_5 and U_6 span the error space.

(b) Each class member is now asked to calculate the scalars $y.U_1$, $y.U_2$, $y.U_3$, $y.U_4$, $y.U_5$ and $y.U_6$.

The class instructor is then asked to plot a histogram of the class results for $y.U_1$. This will approximate the distribution of the random variable $Y.U_1$. What is the theoretical mean of this random variable if you assume in god-like manner that $\mu_1 = 10$, $\mu_2 = 15$ and $\mu_3 = 20$? Does your theoretical answer match with the histogram?

The class instructor is also asked to plot five more histograms, on the same axes, one for each of $y.U_2$, $y.U_3$, $y.U_4$, $y.U_5$ and $y.U_6$. Again, what are the theoretical means for these histograms?

(c) The above histograms all have a true variance of σ^2, assuming that $\sigma_1^2 = \sigma_2^2 = \sigma_3^2$. Each class member is now asked to estimate σ^2 by substituting their own values into the formula

$$s^2 = \frac{(y.U_4)^2 + (y.U_5)^2 + (y.U_6)^2}{3}$$

The class instructor is asked to draw a histogram of the class results. How variable are these estimates of σ^2?

(d) Each class member is now asked to test the hypothesis $H_0 : \mu_1 = (\mu_2 + \mu_3)/2$ by calculating the test statistic, $F = (y.U_2)^2/s^2$. If H_0 is true this comes from the $F_{1,3}$ distribution which has a 95 percentile of 10.13. Reject H_0 if your calculated F is greater than 10.13 and accept H_0 if your calculated F is less than 10.13.

The class instructor is asked to draw a histogram of the test statistics, F. This histogram is an approximation to a "noncentral F" distribution. What percentage of the class rejected the null hypothesis?

(e) Each class member can now repeat this last section for the null hypothesis $H_0 : \mu_2 = \mu_3$ by calculating the test statistic $F = (y.U_3)^2/s^2$.

The class instructor is again asked to draw a histogram of the test statistics, F. What percentage of the class rejected the null hypothesis? How does this compare with the results from the last section (d)?

(f) If time permits, each class member can calculate the orthogonal decomposition

$$
\begin{bmatrix} y_{11} \\ y_{12} \\ y_{21} \\ y_{22} \\ y_{31} \\ y_{32} \end{bmatrix}
=
\begin{bmatrix} \bar{y}_{..} \\ \bar{y}_{..} \\ \bar{y}_{..} \\ \bar{y}_{..} \\ \bar{y}_{..} \\ \bar{y}_{..} \end{bmatrix}
+
\begin{bmatrix} \bar{y}_{1.} - \bar{y}_{..} \\ \bar{y}_{1.} - \bar{y}_{..} \\ \bar{y}_{2.} - \bar{y}_{..} \\ \bar{y}_{2.} - \bar{y}_{..} \\ \bar{y}_{3.} - \bar{y}_{..} \\ \bar{y}_{3.} - \bar{y}_{..} \end{bmatrix}
+
\begin{bmatrix} y_{11} - \bar{y}_{1.} \\ y_{12} - \bar{y}_{1.} \\ y_{21} - \bar{y}_{2.} \\ y_{22} - \bar{y}_{2.} \\ y_{31} - \bar{y}_{3.} \\ y_{32} - \bar{y}_{3.} \end{bmatrix}
$$

and confirm that s^2 can also be calculated using the formula

$$s^2 = \frac{\|y - \bar{y}_{i\cdot}\|^2}{3} = \frac{(y_{11} - \bar{y}_{1\cdot}) + \cdots + (y_{32} - \bar{y}_{3\cdot})^2}{3}$$

Exercises

(8.1) An experiment was carried out on a property with a blackberry weed problem to see whether blackberry is controlled more effectively by goats or sheep at equivalent stocking rates. A secondary objective was to determine whether some breeds of sheep or goats are more effective at controlling blackberry than other breeds.

Eight fields with similar amounts of blackberry were allocated in a completely random manner to eight mobs of animals, being two replicates of the treatments listed in the table. At yearly intervals an aerial photograph was taken of the trial, and the percentage of blackberry cover assessed for each field. The data which will be analysed is the reduction in percentage of blackberry cover between the initial photograph and the photograph taken one year later. For example, $44\% = 76\% - 32\%$, where 76% is the initial cover and 32% the cover after one year. This data is given in the table.

Treatments	Reduction in cover	
1. Romney sheep	44	40
2. Merino sheep	42	50
3. Angora goats	51	55
4. Feral goats	58	52

Questions of interest are:

(A) On average, do the two breeds of sheep differ from the two breeds of goat in their effectiveness as blackberry control agents?

(B) Do Romney sheep differ from Merino sheep?

(C) Do angora goats differ from feral goats?

(a) Write down a set of three orthogonal contrasts corresponding to these questions of interest.

(b) Write down the associated unit vectors, U_2, U_3 and U_4.

(c) Confirm that these three unit vectors are orthogonal to one another. They will serve as an orthogonal coordinate system for the treatment space.

(d) Complete the orthogonal coordinate system for 8-space by writing down a coordinate axis, U_1, for the overall mean space and four orthogonal axes, U_5, U_6, U_7 and U_8, for the error space.

(e) Calculate the projection coefficients $y.U_1, \ldots, y.U_8$ in the orthogonal decomposition $y = (y.U_1)U_1 + \cdots + (y.U_8)U_8$.

(f) Test the hypothesis corresponding to question (A) by calculating the test statistic

$$F = \frac{(y.U_2)^2}{[(y.U_5)^2 + \cdots + (y.U_8)^2]/4}$$

What is your conclusion?

(g) Test hypotheses corresponding to questions (B) and (C) similarly, stating conclusions.

(h) Draw up an ANOVA table to summarise your calculations. As a check, your error mean square should be $s^2 = 16.5$.

(i) Present your results in a table similar to Table 8.5.

(8.2) (a) Using the data from Exercise 8.1, calculate the orthogonal decomposition in the form

$$y - \bar{y}_{..} \quad = \quad (\bar{y}_{i.} - \bar{y}_{..}) + (y - \bar{y}_{i.})$$

(b) Calculate the Pythagorean decomposition

$$\|y - \bar{y}_{..}\|^2 = \|\bar{y}_{i.} - \bar{y}_{..}\|^2 + \|y - \bar{y}_{i.}\|^2$$

the breakup of the total sum of squares into treatment and error sums of squares.

(c) Confirm that $\|\bar{y}_{i.} - \bar{y}_{..}\|^2 = (y.U_2)^2 + (y.U_3)^2 + (y.U_4)^2$ using the vectors U_2, U_3 and U_4 from Exercise 8.1.

(d) Also confirm that $\|y - \bar{y}_{i.}\|^2 = (y.U_5)^2 + (y.U_6)^2 + (y.U_7)^2 + (y.U_8)^2$.

(e) Use the Minitab Oneway command to check your answer in (b) above.

(f) Use the Regress command in Minitab to check your values for $(y.U_2)^2$, $(y.U_3)^2$ and $(y.U_4)^2$. Hint: if y is set into column 1, and U_2, U_3 and U_4 are set into columns 2, 3 and 4, the command is "regress c1 3 c2 c3 c4".

(8.3) (a) Calculate a 95% confidence interval for the contrast of sheep versus goats in Exercise 8.1, using the formula given in the summary of §8.3.

(b) Similarly, calculate a 95% confidence interval for the contrast of Romney with Merino sheep.

(c) Which of the above confidence intervals is the narrower? Why?

(8.4) An experiment was conducted to compare four crops which can be grown as green manure for improving the fertility and structure of the soil. The site chosen for the experiment was one of the less fertile fields on an experiment station. The experimental treatments listed below were assigned in a completely random manner to twelve field plots.

1. Peas (cultivar Onward) sown at 300 kg/ha

2. Barley (cultivar Zephyr) sown at 150 kg/ha

3. Lupins (old standard cultivar) sown at 200 kg/ha

4. Lupins (new cultivar, Uniharvest) sown at 200 kg/ha

All plots were sown on the first day of autumn. On the first day of winter, three months later, a 2m by 8m area was harvested from the centre of each plot. Plants were cut off at ground level so that only the above ground portions were included in the harvest samples. These samples were then dried and weighed. The data given in the table are the dry weights of the harvest samples converted to units of tonnes/hectare.

Treatment	Dry matter of "tops"			Mean
Peas	4.7	4.9	4.5	4.7
Barley	3.4	3.9	3.2	3.5
Lupins (old)	5.2	5.1	6.2	5.5
Lupins (new)	5.8	5.0	5.1	5.3

Three out of the four crops are *legumes*, plants which fix nitrogen from the atmosphere. These are the peas and the two lupin cultivars. The fourth crop, barley, is nonleguminous.

Questions of interest are:

- Did the three legume crops produce on average more or less dry matter in the tops than the nonlegume crop?

- Did one legume — peas — produce more or less than the average of the cultivars of the other legume — lupins?

- Was there any difference between the two lupin cultivars?

(a) Write down a complete set of three orthogonal contrasts corresponding to these questions of interest.

(b) Write down the corresponding unit vectors, U_2, U_3 and U_4 which span the treatment space.

(c) Calculate the squared lengths of the projections of y onto these directions. These are $(y.U_2)^2$, $(y.U_3)^2$ and $(y.U_4)^2$.

(d) Write down the fitted model in the form $y = \bar{y}_{i.} + (y - \bar{y}_{i.})$.

(e) Estimate σ^2 using the formula $s^2 = \|y - \bar{y}_{i.}\|^2/8$.

(f) Write down and test hypotheses relating to the questions of interest.

(g) Summarize your calculations in an ANOVA table.

(h) Present your results in a table similar to Table 8.5.

(i) Check your calculations using Minitab.

(8.5) The following experiment was carried out to determine the best method of applying a fertilizer mix to corn. Three treatments, given below, were assigned completely at random to nine plots. The weight of corn, in kilograms, yielded by each plot was as follows:

1. Control (no fertilizer)	45.1	46.7	47.4
2. 300 kg/ha plowed under	56.7	57.3	54.6
3. 300 kg/ha broadcast	53.3	55.0	54.7

The two questions of interest are:

• Did fertilizer increase corn yield?

• Was there a difference due to method of application?

Set up contrasts and unit vectors relevant to this experiment and calculate F tests to allow these two questions to be answered. What are your conclusions? Exhibit your working by listing test hypotheses, a relevant ANOVA table, and a table summarizing your results in the manner of Table 8.5.

(8.6) An experiment is conducted to test the tolerance of wheat to herbicides used in the control of yarrow. A weed free area is divided into twelve plots and six treatments, namely five herbicides and a control, are assigned in a completely random manner to the plots. The treatments and wheat grain yields in t/ha are:

Control (no herbicide)	6.5	6.1
Systemic herbicide A	5.2	5.0
Systemic herbicide B	4.9	5.3
Contact herbicide P, formulation 1	5.8	5.7
Contact herbicide P, formulation 2	5.4	5.6
Contact herbicide Q, formulation 3	5.0	5.3

Questions of interest are:

- Do the herbicides, on average, affect the grain yield?

- Do systemic herbicides differ from contact herbicides?

- Does "systemic A" differ from "systemic B"?

- Does "contact P" differ from "contact Q"?

- Do the formulations of "contact P" differ?

Set up a system of five orthogonal contrasts relating to these questions and write down the unit vectors U_2, U_3, U_4, U_5, and U_6 corresponding to these contrasts. Calculate $(y.U_2)^2, \ldots, (y.U_6)^2$, obtain an estimate, s^2, for σ^2, and perform the five F tests. What are your conclusions? Use hand calculations or Minitab as you please. Summarize your results in a relevant ANOVA table and a table similar to Table 8.5.

(8.7) A completely randomized design experiment was set up to determine the effect of seed treatment by acids on the early growth of rice seedlings. Four experimental treatments, a control, an inorganic and two organic acids, were applied to twelve experimental pots. Results, in mg of shoot dry weight, were as follows:

1. Control	4.23	4.39	4.10
2. HCl acid (inorganic)	3.85	3.78	3.89
3. Propionic acid (organic)	3.75	3.65	3.81
4. Butyric acid (organic)	3.66	3.67	3.62

The questions of interest are:

- Do acid treatments increase/decrease seedling growth?

- Do organic acids differ from inorganic acids?

- Does the first organic acid differ from the second organic acid?

(a) Write down three orthogonal contrasts which are appropriate for testing these questions.

(b) What are the corresponding unit vectors U_2, U_3 and U_4?

(c) Calculate $(y.U_2)^2$, $(y.U_3)^2$ and $(y.U_4)^2$ and s^2.

(d) Present your results in an ANOVA table and results table similar to Table 8.5.

(e) What conclusions have you reached concerning the questions of interest?

(8.8) In an experiment comparing various forms of phosphate, the treatments are:

  1. Rock phosphate from Nauru Island

  2. Rock phosphate from Christmas Island (Quarry no. 1)

  3. Rock phosphate from Christmas Island (Quarry no. 2)

  4. Superphosphate

  5. Control (no phosphate)

Each treatment is replicated twice.

(a) In terms of μ_1, μ_2, μ_3, μ_4, and μ_5 write down four hypotheses which are relevant in this experiment.

(b) Write down the unit vectors corresponding to the hypotheses in (a).

(8.9) Write down an appropriate set of orthogonal contrasts for each of the following treatment lists. Assume each treatment is equally replicated.

(a) Treatments:

  1. Superphosphate applied in the autumn

  2. Superphosphate applied in the spring

  3. Control (no superphosphate)

(b) Treatments:

  1. Protein based diet, brand A

  2. Starch based diet, brand B

  3. Protein based diet, brand C

  4. Starch based diet, brand D

(c) Treatments:

  1. Field peas (legume)

  2. Beans (legume)

  3. Oats (nonlegume)

  4. Mustard (nonlegume)

  5. Garden peas (legume)

(8.10) In a series of experiments in New Zealand the main objective is to compare the sugar yield of fodder beet varieties with the sugar yield of sugar beet varieties.

Seed is available for ten varieties of fodder beet and six varieties of sugar beet. Thirty plots are available per experiment and it is decided to use five replicates of six treatments.

The main interest is in the contrast between sugar beets and fodder beets. Which of the following sets of treatments is best?

(a) 5 sugar beet varieties and 1 fodder beet variety

(b) 4 sugar beet varieties and 2 fodder beet varieties

(c) 3 sugar beets and 3 fodder beets

(d) 2 sugar beets and 4 fodder beets

(e) 1 sugar beet and 5 fodder beets

Give a statistical reason for your answer.

(8.11) An experiment is conducted to compare the honeydew yield of three races of light honey bee and two races of dark honey bee. Fifteen hives, three of each variety, were placed in a single apiary. The table shows the results of the experiment.

Treatment		Honey Yield (in kg)		
Light	Italian	52.6	58.4	49.2
	Caucasian	51.4	56.3	53.1
	Caucasian (new variety)	52.7	53.6	51.2
Dark	German	60.5	58.6	61.3
	African "killer" bee	71.2	76.3	78.4

(a) Write down the contrasts corresponding to

- Light bees versus dark bees

- Caucasian (both races) versus Italian

- Caucasian versus new race of Caucasian

- German versus African "killer" bee

(b) Write down the associated unit vectors, U_2, U_3, U_4 and U_5.

(c) Are these contrasts mutually orthogonal? Give reasons.

(d) Calculate the contrast sums of squares, $(y.U_2)^2$, $(y.U_3)^2$, $(y.U_4)^2$ and

$(y.U_5)^2$.

(e) Sketch the decomposition of y:

$$y = (y.U_1)U_1 + \cdots + (y.U_5)U_5 + \text{error vector}$$

(f) Write out an appropriate ANOVA table, and test the four hypotheses of the form H_0: contrast $= 0$ which correspond to the contrasts you wrote down in (a).

(g) Use Minitab to check your calculations.

(h) Present your results as shown in Table 8.5.

Chapter 9

Factorial Contrasts

This chapter will deal with *factorial* contrasts. These arise when several factors are to be simultaneously investigated in a single experiment. Factorial contrasts can be thought of as clever class comparisons. For example, when we analyze Example B we shall firstly contrast the class of "no superphosphate" treatments with the class of "superphosphate" treatments, then move on to contrast the class of "no nitrolime" treatments with the class of "nitrolime" treatments. Both of these contrasts involve all the observations in the experiment, so in effect we will have investigated two factors for the price of one. When more than two factors are investigated in the same study, even greater economies can be achieved.

The plan for this chapter is to make some further introductory remarks in §1, then carry out statistical analyses of Example B in §2 and Example C in §3. The general method of generating factorial contrasts is then outlined in §4, and a special notation for factorials is discussed in §5. The chapter concludes with a summary in §6, and exercises.

9.1   Introduction

The difference between studies involving class comparisons and those with a factorial design is illustrated in Table 9.1. In Example 1 the only factor is "species", with two levels, sheep and goats. By a *factor* we shall always mean some criterion which partitions the treatments into classes, called the *levels* of the factor. In Example 2 there are two factors, namely species and sex. Factor A is "species" and has two levels, sheep and goats, while factor B is "sex", also with two levels, male and female. All 2×2 combinations of these factor levels appear in the treatment list.

Class Comparisons	Factorial Contrasts
Example 1	**Example 2**
1. Romney sheep	1. Rams (male sheep)
2. Merino sheep	2. Ewes (female sheep)
3. Angora goats	3. Billies (male goats)
4. Feral goats	4. Does (female goats)

Table 9.1: Two treatment lists to illustrate the distinction between class comparisons and factorial contrasts.

In Example 2 each factor splits the treatments into two groups of size two, as illustrated in Figure 9.1. The corresponding factorial contrasts are $c_1 = \mu_1 + \mu_2 - \mu_3 - \mu_4$ and $c_2 = \mu_1 - \mu_2 + \mu_3 - \mu_4$. The important distinction between these contrasts and those of the class comparison type is that c_1 and c_2 both utilize information from all four experimental treatments. We shall shortly see this means that both can be estimated with maximum precision.

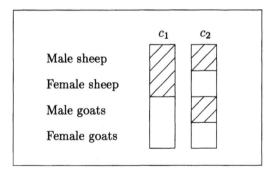

Figure 9.1: Factorial contrasts viewed as clever class comparisons, using Example 2 of Table 9.1. For each contrast one class of treatments is shaded, and the other class left unshaded.

9.2   Analyzing Example B

The treatment design for Example B of §7.2 is of a 2×2 factorial type, as in Example 2 of Table 9.1. The first factor is nitrolime, at two levels, none and 250 kg/ha, and the second factor is superphosphate, also at two levels, none and 250 kg/ha. To illustrate factorial contrasts we now statistically analyze the data for Example B.

Data

For ready reference the data is reproduced in Table 9.2.

Treatment	Grain yield					Mean
1. Control (no fertilizer)	19.2	18.4	17.0	17.6	17.2	17.88
2. 250 kg/ha superphosphate	18.2	19.8	19.4	19.0	19.8	19.24
3. 250 kg/ha nitrolime	20.0	21.6	22.0	20.8	20.4	20.96
4. 250 super + 250 n/lime	23.6	21.6	23.2	21.4	21.2	22.2

Table 9.2: Barley grain yield, in kg/plot, for the 8th harvest, Example B.

The resulting observation vector is $y = \begin{bmatrix} y_{11} \\ y_{12} \\ \vdots \\ y_{44} \\ y_{45} \end{bmatrix} = \begin{bmatrix} 19.2 \\ 18.4 \\ \vdots \\ 21.4 \\ 21.2 \end{bmatrix}$

Model

We have four populations of interest, corresponding to the four combinations of the two factors, each at two levels. We assume each is normally distributed, with means μ_1, μ_2, μ_3 and μ_4, and common variance σ^2. The model vector is

$$\begin{bmatrix} \mu_1 \\ \cdot \\ \mu_2 \\ \cdot \\ \mu_3 \\ \cdot \\ \mu_4 \\ \cdot \end{bmatrix} = \mu_1 \begin{bmatrix} 1 \\ \cdot \\ 0 \\ \cdot \\ 0 \\ \cdot \\ 0 \\ \cdot \end{bmatrix} + \mu_2 \begin{bmatrix} 0 \\ \cdot \\ 1 \\ \cdot \\ 0 \\ \cdot \\ 0 \\ \cdot \end{bmatrix} + \mu_3 \begin{bmatrix} 0 \\ \cdot \\ 0 \\ \cdot \\ 1 \\ \cdot \\ 0 \\ \cdot \end{bmatrix} + \mu_4 \begin{bmatrix} 0 \\ \cdot \\ 0 \\ \cdot \\ 0 \\ \cdot \\ 1 \\ \cdot \end{bmatrix}$$

so the model space M is a four dimensional subspace of 20-space, spanned by $[1, . ., 0, . ., 0, . ., 0, .]^T$ and so on.

Test Hypotheses

The null hypotheses to be tested are:

1. Nitrolime has no effect on barley yield.

2. Superphosphate has no effect on barley yield.

3. There is no interaction between the fertilizers. That is, the response to super in the absence of nitrolime is the same as the response to super in the presence of nitrolime.

In terms of contrasts these hypotheses are equivalent to:

$$H_0 : c_1 = -\mu_1 - \mu_2 + \mu_3 + \mu_4 = 0 \qquad \text{(Nitrolime effect} = 0)$$
$$H_0 : c_2 = -\mu_1 + \mu_2 - \mu_3 + \mu_4 = 0 \qquad \text{(Super effect} = 0)$$
$$H_0 : c_3 = \mu_1 - \mu_2 - \mu_3 + \mu_4 = 0 \qquad \text{(Interaction} = 0)$$

The directions associated with these three hypotheses are

$$U_2 = \frac{1}{\sqrt{20}} \begin{bmatrix} -1 \\ \cdot \\ -1 \\ \cdot \\ 1 \\ \cdot \\ 1 \\ \cdot \end{bmatrix}, \quad U_3 = \frac{1}{\sqrt{20}} \begin{bmatrix} -1 \\ \cdot \\ 1 \\ \cdot \\ -1 \\ \cdot \\ 1 \\ \cdot \end{bmatrix}, \quad U_4 = \frac{1}{\sqrt{20}} \begin{bmatrix} 1 \\ \cdot \\ -1 \\ \cdot \\ -1 \\ \cdot \\ 1 \\ \cdot \end{bmatrix}$$

These make up an orthogonal coordinate system for the treatment space.

We can check the appropriateness of the first of these directions by calculating that the projection coefficient $y.U_2$ has the value

$$5(-\bar{y}_{1.} - \bar{y}_{2.} + \bar{y}_{3.} + \bar{y}_{4.})/\sqrt{20}$$

Hence the random variable $Y.U_2$ has expected value a constant multiple of $c_1 = -\mu_1 - \mu_2 + \mu_3 + \mu_4$. Thus if the contrast c_1 is zero, the projection coefficient $y.U_2$ will be small, averaging to zero over many repetitions of the study. On the other hand, if c_1 is non-zero the projection coefficient $y.U_2$ will be large, averaging to the non-zero quantity $5c_1/\sqrt{20}$. Similar checks can be made for $y.U_3$ and $y.U_4$.

For each hypothesis, the decision as to whether it is true or false hinges on whether the corresponding squared projection length, $(y.U_2)^2$, $(y.U_3)^2$ or $(y.U_4)^2$, is comparable to, or larger than the average of the corresponding squared projection lengths in the error space.

This completes the assembling of the raw materials for our analysis. We have an observation vector, y, a model space, M, and directions U_2, U_3 and U_4 associated with our three hypotheses. Note that the effect of our new treatment design is simply to specify three new directions of interest within the unchanged treatment space. The treatment space is here three dimensional, since we have four treatments.

Fitting the Model

As usual we project the observation vector y onto the model space M and obtain the fitted model vector, $\bar{y}_{i.} = [17.88, 17.88, \ldots, 22.2, 22.2]^T$, the vector of treatment means.

Hence the fitted model is $y = \bar{y}_{i.} + (y - \bar{y}_{i.})$, a familiar decomposition. When the overall mean vector is subtracted from both sides of this equation we arrive at the simplified form of the decomposition:

$$y \;-\; \bar{y}_{..} \;=\; (\bar{y}_{i.} - \bar{y}_{..}) \;+\; (y - \bar{y}_{i.})$$

$$\begin{array}{ccc} \text{Corrected} & = & \text{Treatment} & + & \text{Error} \\ \text{observation vector} & & \text{vector} & & \text{vector} \end{array}$$

$$\begin{bmatrix} 19.2 \\ \cdot \\ 18.2 \\ \cdot \\ 20.0 \\ \cdot \\ 23.6 \\ \cdot \end{bmatrix} - \begin{bmatrix} 20.07 \\ \cdot \\ \cdot \\ \cdot \\ \cdot \\ \cdot \\ \cdot \\ 20.07 \end{bmatrix} = \begin{bmatrix} -2.19 \\ \cdot \\ -.83 \\ \cdot \\ .89 \\ \cdot \\ 2.13 \\ \cdot \end{bmatrix} + \begin{bmatrix} 1.32 \\ \cdot \\ \cdot \\ \cdot \\ \cdot \\ \cdot \\ \cdot \\ -1.0 \end{bmatrix}$$

as illustrated in the top half of Figure 9.2. Here the corrected observation vector, $y - \bar{y}_{..}$, lies in a 19 dimensional subspace of 20-space, the treatment vector $\bar{y}_{i.} - \bar{y}_{..}$ lies in a 3 dimensional subspace, and the error vector $y - \bar{y}_{i.}$ lies in a 16 dimensional subspace.

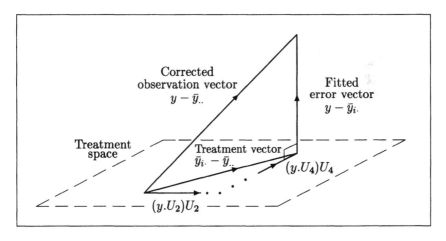

Figure 9.2: The orthogonal decomposition of $y - \bar{y}_{..}$, showing the treatment vector further decomposed into projections in the directions associated with the hypotheses.

The treatment vector, $\bar{y}_{i.} - \bar{y}_{..}$, is more usefully written out in terms of the projections onto U_2, U_3 and U_4, as also illustrated in Figure 9.2. Substituting the numerical values of the projection coefficients, we obtain the decomposition

$$y - \bar{y}_{..} \;=\; 6.75U_2 \;+\; 2.91U_3 \;-\; 0.13U_4 \;+\; (y - \bar{y}_{i.})$$

Testing the Hypotheses

Our first hypothesis is that nitrolime has no effect on barley yield, or equivalently that the contrast $c_1 = -\mu_1 - \mu_2 + \mu_3 + \mu_4$ is zero. In order to test this hypothesis we must check whether the squared distance, $(y.U_2)^2$, is comparable to, or larger than, the average of the corresponding squared distances in the error space. The appropriate test statistic is

$$F = \frac{(y.U_2)^2}{\left[(y.U_5)^2 + \cdots + (y.U_{20})^2\right]/16} = \frac{(y.U_2)^2}{\|y - \bar{y}_{i\cdot}\|^2/16} = \frac{45.602}{0.802} = 56.86$$

where $U_5, \ldots, U_{20}$ are coordinate axes for the error space. Since $F = 56.86$ exceeds the 99 percentile of the $F_{1,16}$ distribution, 8.53, we reject the hypothesis at the 1% level of significance. We conclude that the application of nitrolime has increased the yield of the barley.

The second hypothesis, that superphosphate has no effect, or equivalently that $c_2 = -\mu_1 + \mu_2 - \mu_3 + \mu_4 = 0$, is tested using the test statistic

$$F = \frac{(y.U_3)^2}{\|y - \bar{y}_{i\cdot}\|^2/16} = \frac{8.450}{0.802} = 10.54$$

This again exceeds the 99 percentile of the $F_{1,16}$ distribution, so we again reject the hypothesis at the 1% level of significance. We conclude that the application of superphosphate has also increased the yield of the barley.

The third hypothesis, that the fertilizers do not interact, or $c_3 = \mu_1 - \mu_2 - \mu_3 + \mu_4 = 0$, is tested using the statistic

$$F = \frac{(y.U_4)^2}{\|y - \bar{y}_{i\cdot}\|^2/16} = \frac{0.018}{0.802} = 0.02$$

This is not statistically significant: there is no evidence of an interaction between the two fertilizers. The two fertilizers appear to act independently of one another, in that the response to one fertilizer is unaffected by whether the other is applied.

ANOVA Table

The above calculations are usually summarized in an ANOVA table such as that given in Table 9.3.

Source of Variation	df	SS	MS	F
Treatments	3	54.070		
Nitrolime	1	45.602	45.602	56.86(**)
Superphosphate	1	8.450	8.450	10.54(**)
Super $\times$ nitrolime	1	.018	.018	.02
Error	16	12.832	.802	
Total	19	66.902		

Table 9.3: ANOVA table for the 2×2 factorial Example B.

In the table the entries in the sums of squares column are the squared lengths of the vectors displayed in Figure 9.2. The overall Pythagorean breakup is

$$\|y - \bar{y}_{..}\|^2 = \|\bar{y}_{i.} - \bar{y}_{..}\|^2 + \|y - \bar{y}_{i.}\|^2$$
$$66.902 = 54.070 + 12.832$$

In the more detailed breakup, the treatment sum of squares is written as the sum of the three contrast sums of squares

$$\|\bar{y}_{i.} - \bar{y}_{..}\|^2 = (y.U_2)^2 + (y.U_3)^2 + (y.U_4)^2$$
$$54.070 = 8.450 + 45.602 + .018$$

Estimation of σ^2

In our analysis we have transformed our original set of independent random variables $Y_{11}, \ldots, Y_{15} \sim N[\mu_1, \sigma^2]$, $\ldots$, $Y_{41}, \ldots, Y_{45} \sim N[\mu_4, \sigma^2]$ into a new set of independent, normal random variables, $Y.U_1, \ldots, Y.U_{20}$, with means and variances as given in Table 9.4. The first four of these random variables are used to estimate the overall mean $\mu = (\mu_1 + \mu_2 + \mu_3 + \mu_4)/4$ and the contrasts c_1, c_2 and c_3; equivalently we can think of these model space projection coefficients as providing estimates of μ_1, μ_2, μ_3 and μ_4. The last sixteen are used to estimate σ^2 via

$$s^2 = \frac{(y.U_5)^2 + \cdots + (y.U_{20})^2}{16} = \frac{\|y - \bar{y}_{i.}\|^2}{16} = \frac{12.832}{16} = 0.802$$

Here $y.U_5, \ldots, y.U_{20}$ are all observations from a $N[0, \sigma^2]$ distribution, so their squares average to σ^2 over many repetitions of the experiment.

	Mean	Variance
$Y.U_1$	$\sqrt{20}\mu$	σ^2
$Y.U_2$	$\dfrac{\sqrt{5}(-\mu_1 - \mu_2 + \mu_3 + \mu_4)}{\sqrt{4}}$	σ^2
$Y.U_3$	$\dfrac{\sqrt{5}(-\mu_1 + \mu_2 - \mu_3 + \mu_4)}{\sqrt{4}}$	σ^2
$Y.U_4$	$\dfrac{\sqrt{5}(\mu_1 - \mu_2 - \mu_3 + \mu_4)}{\sqrt{4}}$	σ^2
$Y.U_5$	0	σ^2
$\vdots$	$\vdots$	$\vdots$
$Y.U_{20}$	0	σ^2

Table 9.4: Means and variances of the projection coefficients, $Y.U_i$.

Checking of Assumptions

We need to check our assumptions of independence, normality of the distributions, and equality of variances.

To check independence we would usually plot the errors against the plot number, since plots were laid out consecutively from 1 to 20. Since we falsified the randomization, however, there is nothing to be gained from this process here.

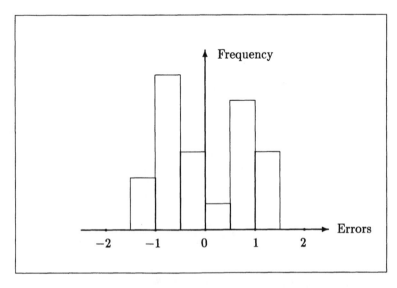

Figure 9.3: Histogram of the errors for Example B.

To check our assumption of normality, we draw a histogram of the errors $y_{ij} - \bar{y}_{i\cdot}$ as shown in Figure 9.3. The histogram reveals no reason to question our assumption of normality, so we shall dispense with more formal checks.

Variance estimates for each treatment are given in Table 9.5, obtained here by squaring the standard deviations produced by Minitab. We could conduct formal tests of hypotheses such as $H_0 : \sigma_1^2 = \sigma_2^2$ using test statistics such as $F = s_1^2/s_2^2$ which is distributed as $F_{4,4}$ under H_0. However, the variances are sufficiently similar for us to accept the assumption of common variances without going through any formal procedure.

Treatment number	1	2	3	4	Average
Variance estimate	.832	.448	.687	1.241	.802

Table 9.5: Variance estimates for each treatment in Example B.

Confidence Intervals

We would now like to obtain 95% confidence intervals for the average response to nitrolime, $c_1 = (-\mu_1 - \mu_2 + \mu_3 + \mu_4)/2$ and the average response to superphosphate, $c_2 = (-\mu_1 + \mu_2 - \mu_3 + \mu_4)/2$. These are of interest because it appears from the absence of any significant interaction that the fertilizers acted independently of one another.

From §8.2 the formula for a 95% confidence interval for a contrast $c = c_1\mu_1 + \cdots + c_k\mu_k$ is

$$(c_1\bar{y}_{1\cdot} + \cdots + c_k\bar{y}_{k\cdot}) \ \pm \ \sqrt{\sum_{i=1}^{k} c_i^2 \frac{s}{\sqrt{n}}} t_{k(n-1)}(.975)$$

where $c_1\bar{y}_{1\cdot} + \cdots + c_k\bar{y}_{k\cdot} = \hat{c}$ is the estimate and $\sqrt{\sum_{i=1}^{k} c_i^2 \times s^2/n} = $ s.e.$(\hat{C})$ is the standard error of the estimator $\hat{C} = c_1\overline{Y}_{1\cdot} + \cdots + c_k\overline{Y}_{k\cdot}$.

In the case of nitrolime, the estimate of the average response is

$$\hat{c}_1 = \frac{(\bar{y}_{3\cdot} - \bar{y}_{1\cdot}) + (\bar{y}_{4\cdot} - \bar{y}_{2\cdot})}{2} = \frac{(20.96 - 17.88) + (22.2 - 19.24)}{2} = 3.02$$

Also the standard error of the estimator $\hat{C}_1$ is

$$\text{s.e.}(\hat{C}_1) = \sqrt{\left[(\tfrac{1}{2})^2 + (-\tfrac{1}{2})^2 + (\tfrac{1}{2})^2 + (-\tfrac{1}{2})^2\right] \times .802/5} = 0.4005$$

Hence the 95% confidence interval for c_1 is $3.02 \pm .4005 \times 2.120$, where $2.120 = t_{16}(.975)$. This is 3.02 ± 0.85. That is, we are 95% sure of ourselves when we state that the true average response to nitrolime was in the range 3.02 ± 0.85 kilograms per plot.

Similarly, the estimate of the average response to superphosphate is

$$\hat{c}_2 = \frac{(\bar{y}_2. - \bar{y}_1.) + (\bar{y}_4. - \bar{y}_3.)}{2} = \frac{(19.24 - 17.88) + (22.2 - 20.96)}{2} = 1.30$$

Also, as above, s.e.$(\hat{C}_2)$ = .4005.   Hence the 95% confidence interval for c_2 is $1.30 \pm .4005 \times 2.120$. Hence we can be 95% confident of ourselves if we state that the average response to superphosphate was in the range 1.30 ± 0.85 kg/plot.

Report on Study

Our conclusions are straightforward. Nitrolime and superphosphate have both been proven to increase the grain yield, with nitrolime giving the bigger increase. Also, the fertilizers appear to act independently of one another. The 95% confidence interval for the response to nitrolime is 3.02 ± 0.85 kg/plot, and for the response to superphosphate is 1.30 ± 0.85 kg/plot. Our statistical analysis can be summarized either by presenting these confidence intervals, or by using one of the methods shown in Table 9.6.

Method one		Method two	
	Grain yield		**Grain yield**
Treatment	(kg/plot)	**Main effects**	(kg/plot)
Control	17.9	**Nitrolime**	
Nitrolime	21.0	None	18.6
Super	19.2	Some	21.6
Nitrolime+super	22.2	SED	.40
SED	.57	Significance	**
Contrasts		**Super**	
Nitrolime	**	None	19.4
Super	**	Some	20.7
Interaction	ns	SED	.40
		Significance	**
		Interaction	ns

Table 9.6: Two ways of presenting the results for Example B.

Method one in the table follows the format used in Chapter 8, with a presentation of individual treatment means, SED $= \sqrt{2s^2/5}$, and the significance of the three contrasts. Method two presents the mean of the ten plots receiving no nitrolime, followed by the mean of the ten plots receiving nitrolime; then the mean of the ten plots receiving no superphosphate, followed by the mean of the ten plots receiving superphosphate. These are known as the *main effect* means. The standard error of the difference between two such means is SED $= \sqrt{2s^2/10}$. After the presentation of the main effects, the significance of the interaction is presented.

A small comment on units. Notice that we have neglected to convert from a unit of kg/plot to a more universal unit such as tonnes/ha. This is no worry. The constant multiplier is $(10,000/50)/1000 = 0.2$, since the harvest area was 50 square metres. Hence to convert to tonnes/ha we can simply multiply the means and SED's in Table 9.6 by 0.2; the significances are unchanged. Similarly for our confidence intervals we multiply the numerical values by 0.2.

Minitab Analysis

To compute the sums of squares for the ANOVA table given in Table 9.3 it suffices to Regress the observation vector y onto U_2, U_3 and U_4 as these form a complete set of coordinate axes for the treatment space. As a useful addition, however, the Oneway command can be used to calculate treatment means and other essentials. The commands and output are summarized in Tables 9.7 and 9.8.

Notice that the s^2 value of .802 is produced by both the Regress command and the Oneway command, since the full model is fitted in both cases. In our Minitab job we decided to print the errors $y_{ij} - \bar{y}_{i.}$, so had these printed alongside the corresponding observations and treatment means in the form of the fitted model $y = \bar{y}_{i.} + (y - \bar{y}_{i.})$. We also decided to produce a histogram of the errors. The relevant commands are included in Table 9.7, but the output has been excluded from Table 9.8 for economy of space.

List of commands	Interpretation
name c1 = 'GrainYld' name c2 = 'Nitro' name c3 = 'Super' name c4 = 'Intn' name c5 = 'Trtments' name c6 = 'Errors' name c7 = 'Fitted'	Naming the columns for convenience
set c1 19.2 18.4 17.0 ...	Set y into column one
set c2 $(-1, -1, 1, 1)5$ let c2=c2/sqrt(20)	Set U_2 into column two using Minitab shorthand
set c3 $(-1, 1, -1, 1)5$ let c3=c3/sqrt(20)	Set U_3 into column three
set c4 $(1, -1, -1, 1)5$ let c4=c4/sqrt(20)	Set U_4 into column four
regress c1 3 c2−c4	Project y onto U_2, U_3 and U_4
set c5 $(1, 2, 3, 4)5$	Set treatment numbers into column five
oneway c1 c5 c6 c7	Fit the model, saving errors in column six, and fitted values in column seven
print c1 c7 c6	Display the model $y = \bar{y}_{i\cdot} + (y - \bar{y}_{i\cdot})$
histogram c6	Produce histogram of errors
stop	Signifies end of job

Table 9.7: Minitab commands for Example B.

Regress output

The regression equation is
GrainYld = 20.1 + 6.75 Nitro + 2.91 Super − 0.134 Intn

Predictor	Coef	Stdev	t-ratio	p
Constant	20.0700	0.2002	100.22	0.000
Nitro	6.7529	0.8955	7.54	0.000
Super	2.9069	0.8955	3.25	0.005
Intn	−0.1342	0.8955	−0.15	0.883

s = 0.8955 R-sq = 80.8% R-sq(adj) = 77.2%

Analysis of Variance

SOURCE	DF	SS	MS	F	p
Regression	3	54.070	18.023	22.47	0.000
Error	16	12.832	0.802		
Total	19	66.902			

SOURCE	DF	SEQ SS
Nitro	1	45.602
Super	1	8.450
Intn	1	0.018

Oneway output

ANALYSIS OF VARIANCE ON GrainYld

SOURCE	DF	SS	MS	F	p
Trtments	3	54.070	18.023	22.47	0.000
ERROR	16	12.832	0.802		
TOTAL	19	66.902			

LEVEL	N	MEAN	STDEV
1	5	17.880	0.912
2	5	19.240	0.669
3	5	20.960	0.829
4	5	22.200	1.114

POOLED STDEV = 0.896

Table 9.8: Selected Minitab output for Example B.

Interpretation of Interaction

The interaction contrast, c_3, was of little interest in our example, the estimate of -0.12 being small. But exactly what do we mean by the word "interaction"? We clarify this point using a 2×2 table of treatment means, as shown in Table 9.9.

		Superphosphate −	Superphosphate +	Superphosphate responses
Nitrolime	−	17.88 $(\bar{y}_{1.})$	19.24 $(\bar{y}_{2.})$	+1.36
	+	20.96 $(\bar{y}_{3.})$	22.20 $(\bar{y}_{4.})$	+1.24
Nitrolime responses		+3.08	+2.96	−0.12 The interaction

Table 9.9: Treatment means and responses for Example B.

In Table 9.9 the two responses to nitrolime, 3.08 and 2.96, are similar: superphosphate application has little effect on the response to nitrolime. Also, the responses to superphosphate, of 1.36 and 1.24, are similar regardless of whether nitrolime has been applied.

The interaction contrast, $\mu_1 - \mu_2 - \mu_3 + \mu_4$, can be thought of in three equivalent ways:

1. $(\mu_4 - \mu_3) - (\mu_2 - \mu_1)$, being the difference between the response to super when nitrolime is present and the response to super when nitrolime is absent. In Table 9.9 this is estimated as $1.24 - 1.36 = -0.12$.

2. $(\mu_4 - \mu_2) - (\mu_3 - \mu_1)$, being the difference between the response to nitrolime when super is present, and the response to nitrolime when super is absent. In Table 9.9 this is estimated as $2.96 - 3.08 = -0.12$.

3. $(\mu_4 - \mu_1) - [(\mu_2 - \mu_1) + (\mu_3 - \mu_1)]$, being the difference between the combined effect of applying both fertilizers and the sum of the individual effects. In Table 9.9 this is estimated as $(22.2 - 17.88) - [1.36 + 3.08] = 4.32 - [1.36 + 3.08] = -0.12$.

Consider now Table 9.10 where we present examples of experiments in which there are (a) positive and (b) negative interactions.

In (a) the growth of the alfalfa appears to have been limited by a shortage of both phosphorus, P, and sulphur, S, so that the plants did not grow any better until both nutrients were applied.

In (b) the pasture appears to have suffered from a molybdenum, Mo, deficiency, which could be remedied directly by applying molybdenum as fertilizer, or indirectly by applying lime to raise the pH of the soil and release trace elements including molybdenum. In case (a) there are two "limiting factors", and in case (b) the factors "substitute" for one another.

(a) Yield of alfalfa in t/ha			(b) Pasture yield in t/ha		
	No Sulphur	Sulphur		No lime	Lime
No P	6.7	6.9	No Mo	6.9	10.8
P	6.7	10.5	Mo	10.9	11.1
	Interaction = 3.6			Interaction = −3.7	

Table 9.10: Examples of positive and negative interactions.

When the interaction is believed to be nonzero, the discussion of the results is centered on the 2 × 2 table of treatment means, so the results are best presented either in a 2 × 2 table or using method one in Table 9.6. If the interaction is believed to be zero, method two of Table 9.6 can be used.

The Efficiency of the Factorial Design

The factorial design is one of the most important statistical designs. In the absence of interaction, replication is greatly increased compared to using the same resources in single factor experiments. To understand this, suppose that the twenty plots in our barley fertilizer trial, shown in Figure 9.4(a), had been used for two single factor trials of ten plots looking at the effect of nitrolime and superphosphate separately, as illustrated in Figure 9.4(b).

From the single factor trial on nitrolime, the response to nitrolime would be calculated as a difference between two means which are each an average of five observations. With the factorial design, however, the response to nitrolime is calculated as a difference between two main effect means, which are each an average of ten observations. This means that if the trials produce the same s^2 value, the variance of the estimated response to nitrolime is halved by using the 2 × 2 factorial design instead of two single factor trials.

If the factors A and B interact the added replication of the factorial design is lost. Nevertheless, the design still allows the experimenter to find out that the factors do not operate independently.

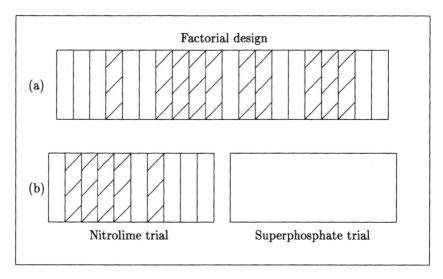

Figure 9.4: Using twenty field plots (a) as five replications of a 2×2 factorial design, and (b) as two single factor experiments, each with five replications of fertilized and control treatments. Shaded plots received nitrolime, and unshaded plots did not. Superphosphate applications are not shown.

A Cautionary Tale

There is a very important point to keep in mind when analyzing factorial designs. In our example, the contrasts c_1 and c_2 which we specified as the nitrolime and super effects have easy interpretations only in the absence of interaction of these two factors. This point is easily overlooked: it is the price we pay for the efficiency of this design.

To put it another way, only if the factors operate independently will the magnitude of the responses estimated from a factorial design in the long run equal those obtained from single factor trials. To help understand why this is so, we introduce the notation:

$$
\begin{aligned}
1 &= \text{mean of the control} \\
a &= \text{response to factor A in absence of B} \\
b &= \text{response to factor B in absence of A} \\
ab &= \text{response to applying factors A and B together,} \\
&\quad \text{over and above the individual responses} \\
&= \text{interaction contrast, } c_3
\end{aligned}
$$

Here a and b are the responses obtained in the long run from *single factor* trials.

		Factor B		Factor A
		Level one	Level two	means
Factor	Level one	$1\ (\mu_1)$	$1+b\ (\mu_2)$	$1+b/2$
A	Level two	$1+a\ (\mu_3)$	$1+a+b+ab\ (\mu_4)$	$1+a+b/2+ab/2$
Factor B means		$1+a/2$	$1+a/2+b+ab/2$	

Table 9.11: A new notation for our 2×2 table of means.

For a trial with a 2×2 factorial design, $1 + b$ is the mean of population two, $1 + a$ is the mean of population three, and $1 + a + b + ab$ is the mean of population four, as shown in Table 9.11. Therefore the factor A main effect, or the nitrolime effect in our example, is

$$c_1 = \frac{(\mu_3 + \mu_4) - (\mu_1 + \mu_2)}{2} = (1 + a + b/2 + ab/2) - (1 + b/2) = a + ab/2$$

which equals the single factor response, a, only when $ab = 0$, or the interaction contrast, c_3, is zero. That is, c_1 can be interpreted as a single factor response only in the absence of interaction. Otherwise it must be interpreted as the average of the responses to factor A in the presence and absence of factor B. This line of reasoning applies equally to the factor B, or superphosphate, main effect.

9.3 Analyzing Example C

In the previous section we illustrated the basic ideas concerning factorial contrasts using a 2×2 factorial design. We now enrich these ideas using a 2×3 design, Example C of §7.2. We interweave class comparisons with factorial contrasts, introduce the need for data transformations and complete the setting for the analysis of general factorial designs.

In Example C we set up an experiment with six treatments: all combinations of two turnip varieties (Green globe and York globe) with three nitrogen fertilizer levels (none, 125 kg/ha and 250 kg/ha). The aim is to investigate the effect of these factors on turnip survival.

Data

For convenience we reprint the data in Table 9.12. The observation vector for Example C is $y = [45, 69, \ldots, 18, 15]^T$.

Treatment	Number of plants				Mean
Green globe soft turnips, no S/A	45	69	71	83	67
Green globe soft turnips, 125 kg/ha S/A	27	59	62	64	53
Green globe soft turnips, 250 kg/ha S/A	24	49	55	77	51.25
York globe soft turnips, no S/A	19	17	22	45	25.75
York globe soft turnips, 125 kg/ka S/A	29	13	27	21	22.5
York globe soft turnips, 250 kg/ha S/A	12	20	18	15	16.25

Table 9.12: Number of soft turnip plants per $1.08m^2$ area, Example C.

Model

Our experiment involves six populations, which as usual we assume normally distributed with common variance. The situation is summarized in Table 9.13.

		Nitrogen		
		None	125 kg/ha	250 kg/ha
Cultivar	Green globe	μ_1	μ_2	μ_3
	York globe	μ_4	μ_5	μ_6

Table 9.13: The six populations of Example C, with their means.

The model vector is $[\mu_1, \mu_1, \ldots, \mu_6, \mu_6]^T$, so the model space, M, is a 6-dimensional subspace of 24-space.

Test Hypotheses

The null hypotheses to be tested are:

1. Green globe turnips were identical to York globe turnips in terms of establishment and survival.

2. Nitrogen application did not affect turnip survival overall.

3. The high and low rates of nitrogen did not differ in their effect.

4. The overall effect of nitrogen did not vary between the two soft turnip cultivars.

5. The difference between high and low rates of nitrogen did not vary with cultivar.

In terms of contrasts these hypotheses can be written most simply as:

$$
\begin{aligned}
c_1 &= -\mu_1 - \mu_2 - \mu_3 + \mu_4 + \mu_5 + \mu_6 = 0 \quad \text{(cultivar)} \\
c_2 &= -2\mu_1 + \mu_2 + \mu_3 - 2\mu_4 + \mu_5 + \mu_6 = 0 \quad \text{(no N vs. N)} \\
c_3 &= \quad\;\; - \mu_2 + \mu_3 \quad\quad\;\; - \mu_5 + \mu_6 = 0 \quad \text{(125 vs. 250 N)} \\
c_4 &= 2\mu_1 - \mu_2 - \mu_3 - 2\mu_4 + \mu_5 + \mu_6 = 0 \quad \text{(cultivar} \times \text{(no N,N))} \\
c_5 &= \quad\quad\;\; \mu_2 - \mu_3 \quad\quad\;\; - \mu_5 + \mu_6 = 0 \quad \text{(cultivar} \times \text{(125,250))}
\end{aligned}
$$

Here the first contrast, c_1, is the main effect contrast for factor A, cultivar. The second and third contrasts, c_2 and c_3, are the main effect contrasts for factor B, nitrogen. The last two contrasts, c_4 and c_5, are interaction contrasts, checking for dependence between the factors.

The directions associated with these five hypotheses are

$$
\begin{array}{ccccc}
U_2 & U_3 & U_4 & U_5 & U_6 \\
\begin{bmatrix} -1 \\ \cdot \\ -1 \\ \cdot \\ -1 \\ \cdot \\ 1 \\ \cdot \\ 1 \\ \cdot \\ 1 \end{bmatrix} &
\begin{bmatrix} -2 \\ \cdot \\ 1 \\ \cdot \\ 1 \\ \cdot \\ -2 \\ \cdot \\ 1 \\ \cdot \\ 1 \end{bmatrix} &
\begin{bmatrix} 0 \\ \cdot \\ -1 \\ \cdot \\ 1 \\ \cdot \\ 0 \\ \cdot \\ -1 \\ \cdot \\ 1 \end{bmatrix} &
\begin{bmatrix} 2 \\ \cdot \\ -1 \\ \cdot \\ -1 \\ \cdot \\ -2 \\ \cdot \\ 1 \\ \cdot \\ 1 \end{bmatrix} &
\begin{bmatrix} 0 \\ \cdot \\ 1 \\ \cdot \\ -1 \\ \cdot \\ 0 \\ \cdot \\ -1 \\ \cdot \\ 1 \end{bmatrix} \\
\sqrt{24} & \sqrt{48} & \sqrt{16} & \sqrt{48} & \sqrt{16}
\end{array}
$$

These make up an orthogonal coordinate system for the treatment space.

For each hypothesis, the decision on whether it is true or false hinges on whether the appropriate squared projection coefficient, $(y.U_2)^2, \ldots, (y.U_6)^2$, is comparable to or larger than the average of the corresponding squared projection lengths in the error space.

The collection of the raw materials for our analysis is now complete. We have an observation vector, y, a model space, M, and directions U_2, U_3, U_4, U_5 and U_6 associated with our five hypotheses.

Fitting the Model

As always we project the observation vector y onto the model space M and obtain the fitted model vector, $\bar{y}_{i.} = [67, 67, \ldots, 16.25, 16.25]^T$. Hence the fitted model is the usual $y = \bar{y}_{i.} + (y - \bar{y}_{i.})$. When the overall mean is subtracted from both sides, we arrive at the simplified decomposition:

$$y \quad - \quad \bar{y}_{..} \qquad = \qquad (\bar{y}_{i.} - \bar{y}_{..}) \quad + \quad (y - \bar{y}_{i.})$$

| Corrected observation vector | = | Treatment vector | + | Error vector |

$$\begin{bmatrix} 45 \\ \cdot \\ 27 \\ \cdot \\ 24 \\ \cdot \\ 19 \\ \cdot \\ 29 \\ \cdot \\ 12 \\ \cdot \end{bmatrix} - \begin{bmatrix} 39.3 \\ \cdot \\ 39.3 \\ \cdot \\ 39.3 \\ \cdot \\ 39.3 \\ \cdot \\ 39.3 \\ \cdot \\ 39.3 \\ \cdot \end{bmatrix} = \begin{bmatrix} 27.7 \\ \cdot \\ 13.7 \\ \cdot \\ 12.0 \\ \cdot \\ -13.5 \\ \cdot \\ -16.8 \\ \cdot \\ -23.0 \\ \cdot \end{bmatrix} + \begin{bmatrix} -22 \\ \cdot \\ -26 \\ \cdot \\ -27.25 \\ \cdot \\ -6.75 \\ \cdot \\ 6.5 \\ \cdot \\ -4.25 \\ \cdot \end{bmatrix}$$

Here the corrected observation vector, $y - \bar{y}_{..}$, lies in a 23 dimensional subspace, the treatment vector $\bar{y}_{i.} - \bar{y}_{..}$ lies in a 5 dimensional subspace, and the error vector $y - \bar{y}_{i.}$ lies in an 18 dimensional subspace of 24-space.

When the treatment vector is written out in terms of the projections onto $U_2, \ldots, U_6$, this decomposition becomes

$$y - \bar{y}_{..} \quad = \quad -87.2U_2 - 24.5U_3 - 8.0U_4 + 9.8U_5 - 4.5U_6 \quad + \quad (y - \bar{y}_{i.})$$

as illustrated in Figure 9.5.

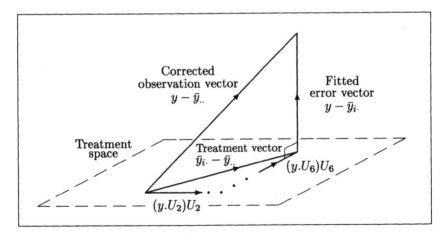

Figure 9.5: The orthogonal decomposition of $y - \bar{y}_{..}$, showing the treatment vector further decomposed into projections in the directions associated with the hypotheses.

Testing the Hypotheses

In order to test our hypotheses, each of the form H_0 : contrast $= 0$, we as usual must check whether the associated squared distance, $(y.U_c)^2$, is comparable to or larger than the average of the corresponding squared distances in the error space. For example, the test statistic for the first hypothesis is

$$F = \frac{(y.U_2)^2}{\left[(y.U_7)^2 + \cdots + (y.U_{24})^2\right]/18} = \frac{(y.U_2)^2}{\|y - \bar{y}_{i\cdot}\|^2/18} = \frac{7597}{211} = 36.00$$

where $U_7, \ldots, U_{24}$ are coordinate axes for the error space.

The most useful Pythagorean breakup is

$$\|y - \bar{y}_{..}\|^2 = (y.U_2)^2 + (y.U_3)^2 + (y.U_4)^2 + (y.U_5)^2 + (y.U_6)^2 + \|y - \bar{y}_{i\cdot}\|^2$$
$$12177 = 7597 + 602 + 64 + 96 + 20 + 3797$$

Here the squared projection lengths in the treatment space sum to $\|\bar{y}_{i\cdot} - \bar{y}_{..}\|^2 = 8380$, the treatment sum of squares. These calculations, and the resulting test statistics, are summarized in the ANOVA table, presented here in Table 9.14.

Source of Variation	df	SS	MS	F
Treatments	5	8380		
Cultivar	1	7597	7597	36.00 (**)
no N v. N	1	602	602	2.85 (ns)
125 v. 250 N	1	64	64	0.30 (ns)
Cult. × (no N,N)	1	96	96	0.46 (ns)
Cult. × (125,250)	1	20	20	0.10 (ns)
Error	18	3797	211	
Total	23	12177		

Table 9.14: ANOVA table for Example C, a 2×3 factorial design.

The only significant F value in the table is the one corresponding to the comparison of the two cultivars, Green globe and York globe. All other F values are nonsignificant when compared with the 95 percentile of the $F_{1,18}$ distribution, 4.41.

Estimation of σ^2

In this analysis we have transformed the original set of random variables, $Y_{11}, \ldots, Y_{14} \sim N\left[\mu_1, \sigma^2\right]$, $\ldots$ $Y_{61}, \ldots, Y_{64} \sim N\left[\mu_6, \sigma^2\right]$ into a new set of independent random variables, $Y.U_1, \ldots, Y.U_{24}$ with means and variances as given in Table 9.15. The first six of these random variables are used to estimate μ, $c_1, \ldots, c_5$, or equivalently $\mu_1, \ldots, \mu_6$, while the last eighteen are used to estimate σ^2 via

$$s^2 = \frac{(y.U_7)^2 + \cdots + (y.U_{24})^2}{18} = \frac{\|y - \bar{y}_{i\cdot}\|^2}{18} = \frac{3797}{18} = 211$$

	Mean	Variance
$Y.U_1$	$\sqrt{24}\mu$	σ^2
$Y.U_2$	$\dfrac{\sqrt{4}(-\mu_1 - \mu_2 - \mu_3 + \mu_4 + \mu_5 + \mu_6)}{\sqrt{6}}$	σ^2
$Y.U_3$	$\dfrac{\sqrt{4}(-2\mu_1 + \mu_2 + \mu_3 - 2\mu_4 + \mu_5 + \mu_6)}{\sqrt{12}}$	σ^2
$Y.U_4$	$\dfrac{\sqrt{4}(-\mu_2 + \mu_3 - \mu_5 + \mu_6)}{\sqrt{4}}$	σ^2
$Y.U_5$	$\dfrac{\sqrt{4}(2\mu_1 - \mu_2 - \mu_3 - 2\mu_4 + \mu_5 + \mu_6)}{\sqrt{12}}$	σ^2
$Y.U_6$	$\dfrac{\sqrt{4}(\mu_2 - \mu_3 - \mu_5 + \mu_6)}{\sqrt{4}}$	σ^2
$Y.U_7$	0	σ^2
.	.	.
$Y.U_{24}$	0	σ^2

Table 9.15: Means and variances of the projection coefficients, $Y.U_i$.

Checking of Assumptions

The assumptions we can check are those of normality of distribution and equality of variances. Further checks will be delayed until we see the correct field plan in Chapter 14; at this point we shall need to revise our analysis.

To check on the normality assumption, we inspect the histogram of errors given in Figure 9.6. This appears fairly normal in distribution, so we accept the assumption of normality.

Variance estimates for individual treatments are given in Table 9.16. We appear to have problems. As one might have anticipated, the lower plant numbers for the York globe treatments seem to have been associated with a lower level of variability. We examine $H_0 : (\sigma_1^2 + \sigma_2^2 + \sigma_3^2)/3 = (\sigma_4^2 + \sigma_5^2 + \sigma_6^2)/3$

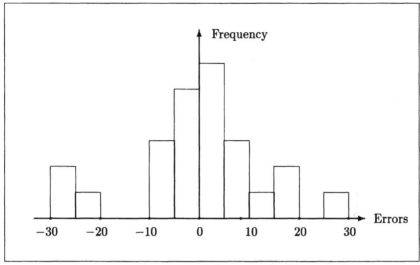

Figure 9.6: Histogram of the errors for Example C.

by calculating the test statistic

$$F = \frac{(s_1^2 + s_2^2 + s_3^2)/3}{(s_4^2 + s_5^2 + s_6^2)/3} = \frac{344.3}{77.7} = 4.43$$

If the hypothesis is true, F comes from the $F_{9,9}$ distribution which has a 97.5 percentile of 4.03. We therefore reject the hypothesis, and conclude that the Green globe treatments are more variable than the York globe treatments.

Treatment number	1	2	3	4	5	6	Average
Variance estimate	253	305	475	169	52	12	211

Table 9.16: Variance estimates for the individual treatments.

What do we do now? One solution is to "transform" the data in order to equalize the within treatment variation. With *counts* of plants, insects, freckles on a human face, etc., the traditional advice is to use the square root transformation. This means we take the square root of each observation and carry out the same statistical analysis as before, using the new values. In Minitab this is easy; we simply add the command: let c1 = sqrt(c1) .

When we rework the analysis we find that the above F value of 4.43 is reduced to 2.42, which is nonsignificant. Apart from this, little else changes. The F values for testing the five hypotheses of interest are virtually unchanged, and the histogram of errors is little different. For simplicity of presentation we shall therefore stick with the original analysis.

Confidence Interval

The 95% confidence interval for the average difference in plant numbers between the two cultivars, $c = (\mu_1 + \mu_2 + \mu_3)/3 - (\mu_4 + \mu_5 + \mu_6)/3$, is

$$\hat{c} \quad \pm \quad \text{s.e.}(\hat{C}) \times t\text{-value}$$

$$\text{where} \quad \hat{c} \ = \ \frac{67 + 53 + 51.25}{3} - \frac{25.75 + 22.5 + 16.25}{3} \ = \ 35.6, \ \text{and}$$

$$\text{s.e.}(\hat{C}) \ = \ \sqrt{\left[\left(\tfrac{1}{3}\right)^2 + \left(\tfrac{1}{3}\right)^2 + \left(\tfrac{1}{3}\right)^2 + \left(-\tfrac{1}{3}\right)^2 + \left(-\tfrac{1}{3}\right)^2 + \left(-\tfrac{1}{3}\right)^2\right] \times s^2/n}$$

$$= \ 5.93$$

Using $t_{18}(.975) = 2.101$, we find that a 95% confidence interval for the difference between Green globe and York globe soft turnip plant numbers is 35.6 ± 12.5 plants per plot. Harvested plot areas were 1.08 square metres, so this could be converted to plants/m^2 by dividing each figure by 1.08.

Report on Study

The results can be presented using confidence intervals, or in tabular form using either of the ways shown in Table 9.17. As in our previous example, there is little obvious advantage in either method. Both give the significance of the five contrasts plus an estimate of the standard error. In all cases SED $= \sqrt{2s^2/m}$, where m is the number of observations going into each of the displayed means, with $m = 4$, 8 and 12 respectively. Method two condenses the information more than method one, so that the results are more easily digested. This condensing, however, results in the loss of information about possible interactions.

A summary of the statistical evidence is simple. The Green globe cultivar of soft turnip had significantly higher plant numbers in early winter than the York globe cultivar. There was no definite proof of differences in plant number due to nitrogen application. There was also no evidence of any interaction.

It may puzzle the reader as to why plant numbers tended to decrease with nitrogen instead of the more usual increase. The reason, based on field observations, is as follows. During the harvest of the barley, some grain fell to the ground and subsequently germinated to compete with the young soft turnip plants. In the plots where nitrogen was applied, the young barley plants competed more vigorously with the soft turnips than in the plots where no nitrogen was applied, so that fewer soft turnip plants survived in the nitrogen plots than in the no nitrogen plots.

Method 1		Method 2	
	Plant nos early winter	**Main effect means**	Plant nos early winter
Treatment		**Turnip cultivar**	
G.g., no S/A	67	Green globe	57.1
G.g., 125 kg/ha S/A	53	York globe	21.5
G.g., 250 kg/ha S/A	51	SED	5.9
Y.g., no S/A	26	Significance	**
Y.g., 125 kg/ha S/A	23		
Y.g., 250 kg/ha S/A	16		
SED	10.3	**Nitrogen application**	
		no S/A	46.4
		125 kg/ha S/A	37.8
Contrasts		250 kg/ha S/A	33.8
Cultivar	**	SED	7.3
no N v. N	ns	Contrasts:	
125 v.250 N	ns	no N v. N	ns
Cultivar×(no N,N)	ns	125 v. 250 N	ns
Cultivar×(125,250)	ns		
		Interaction contrasts	
		Cultivar×(no N,N)	ns
		Cultivar×(125,250)	ns

Table 9.17: Presentation of results by two methods, for Example C. The plant numbers are per $1.08m^2$.

Interpretation of Contrasts

The five contrasts divide naturally into main effect and interaction contrasts. The contrast c_1 is a main effect contrast for factor A, cultivar, and c_2 and c_3 are main effect contrasts for factor B, nitrogen. The other two contrasts, c_4 and c_5, are A × B, cultivar × nitrogen, interaction contrasts.

The interpretation of main effect contrasts is straightforward. We simply examine the main effect means and contrasts as presented by Method two in Table 9.17 for each factor in turn, acting as though the other factor was not present in the experiment. For the cultivar factor, we see that c_1, the contrast of the Green globe treatments with the York globe treatments, is estimated as $\hat{c}_1 = 21.5 - 57.1 = -35.6$ plants/plot. That is, the Green globe treatments averaged 35.6 more plants per plot than the York globe treatments. Note that we are using the original definition of c_1, from §7.2. Our F test tells us that $\hat{c}_1$ is significantly different from zero at the 1% level of significance. In this case, then, we are confident that the true value, c_1, is different from zero.

For the nitrogen factor, our best estimate of the average nitrogen response is $\hat{c}_2 = 35.8 - 46.4 = -10.6$ plants/plot, where 35.8 is the average of the 37.8 and 33.8 values in Table 9.17. Similarly, the difference between the low and high rates of nitrogen is $\hat{c}_3 = 33.8 - 37.8 = -4.0$ plants/plot. Our F tests tell us that neither of these contrast estimates is significantly different from zero. In other words, while $\hat{c}_2 = -10.6$ and $\hat{c}_3 = -4.0$ are our best estimates, we remain uncertain about whether the true values, c_2 and c_3, are really different from zero.

Turning to the cultivar × nitrogen interaction contrasts, our attention passes to the 2×3 table of treatment means displayed in Table 9.18. To interpret the interaction contrasts, however, we need to make up the two further 2×2 tables shown in Table 9.19.

	no S/A	125 S/A	250 S/A
Green globe	67	53	51
York globe	26	23	16

Table 9.18: Treatment means for Example C, displayed in a 2×3 table.

(a)	no N	N	Response
Green globe	67	52	-15
York globe	26	19	-7

(b)	125 S/A	250 S/A	Difference
Green globe	53	51	-2
York globe	23	16	-7

Table 9.19: Tables for interpreting interaction contrasts, Example C. The numbers have been excessively rounded, for ease of explanation.

The cultivar × (no N,N) interaction table is shown in Table 9.19(a). The contrast c_4 is estimated as the difference between the nitrogen responses for Green globe and York globe cultivars; that is, $\hat{c}_4 = (19 - 26) - (52 - 67) = 8$ to our level of accuracy. The corresponding F test indicates that $\hat{c}_4 = 8$ is not significantly different from zero. In other words, the true value, c_4, has not been proven to be nonzero: the average nitrogen responses have not been proven to differ between the cultivars.

The cultivar × (125,250N) interaction table is shown in Table 9.19(b). For each cultivar there is a difference between low and high rates of nitrogen. The contrast c_5 is estimated as the difference between these two differences.

That is, $\hat{c}_5 = (16 - 23) - (51 - 53) = -5$ to our level of accuracy. Again the corresponding F test indicates that $\hat{c}_5$ is not significantly different from zero. Hence c_5 may in reality be zero: the response to additional nitrogen may be the same for the two cultivars.

Minitab Analysis

The analysis of Example C using Minitab echoes the analysis described in Tables 9.7 and 9.8, so we shall only outline it here. In brief, we set y, $U_2, \ldots, U_6$ into columns 1 to 6 of the worksheet, then Regress y onto U_2 to U_6. We can then calculate treatment means, estimates of variance for individual treatments, and an error histogram using the Oneway and Histogram commands.

9.4 Generating Factorial Contrasts

The complete sets of orthogonal factorial contrasts used in Examples B and C are generated in a common fashion which may already be clear to the reader. The simple procedure which generates a complete set of factorial contrasts is as follows:

1. *Write down*, for each factor, a complete set of orthogonal contrasts in terms of the levels of that factor.

2. *Expand* each of these contrasts to a contrast involving all the treatments by repeating coefficients. This will generate all the main effect contrasts.

3. *Cross* each main effect contrast for factor A with each main effect contrast for factor B by multiplying corresponding coefficients. This will generate all the interaction contrasts. If there is a third factor C, cross all pairs of A and B contrasts, all pairs of A and C contrasts, all pairs of B and C contrasts, and all triplets of A, B and C contrasts.

We first illustrate this procedure using Example C. Here factor A, cultivar, has two levels, so we have just one main effect contrast, given by the contrast coefficients $(-1, 1)$. Factor B, nitrogen, has three levels, so we have two main effect contrasts, with coefficients $(-2, 1, 1)$ and $(0, -1, 1)$. Expanding these three contrasts in terms of the full list of treatments generates the main effect contrasts c_1, c_2 and c_3 shown in Table 9.20(a). Crossing the factor A contrast, c_1, with each of the factor B contrasts, c_2 and c_3, generates the interaction contrasts c_4 and c_5 shown in Table 9.20(b).

	(a) Main effect contrasts			(b) Interaction contrasts	
	Factor A	Factor B		$A \times B$	
	c_1	c_2	c_3	$c_4 = c_1 \times c_2$	$c_5 = c_1 \times c_3$
μ_1	-1	-2	0	2	0
μ_2	-1	1	-1	-1	1
μ_3	-1	1	1	-1	-1
μ_4	1	-2	0	-2	0
μ_5	1	1	-1	1	-1
μ_6	1	1	1	1	1

Table 9.20: Generating the factorial contrasts in Example C.

In general, if factor A has i levels and factor B has j levels, the above procedure will generate $i - 1$ main effect contrasts for factor A, $j - 1$ main effect contrasts for factor B, and $(i - 1) \times (j - 1)$ interaction contrasts. The resulting $(i - 1) + (j - 1) + (i - 1)(j - 1) = ij - 1$ contrasts make up a complete set of orthogonal contrasts among the ij treatments.

Example

To illustrate the procedure using a more complicated example, we shall consider an experiment which is carried out to compare leguminous and nonleguminous crops grown for green manure under varying levels of moisture availability. The experiment is laid down in a 3×4 factorial design. Factor A consists of three levels of moisture availability: unirrigated, and irrigated when the plant available soil moisture (a.s.m.) falls to levels of either 25% or 50% of saturation. Factor B consists of four crops, lupins, peas, mustard and barley, of which the first two are legumes.

The complete set of treatments and orthogonal contrasts for each factor are shown in Table 9.21. For factor A the first contrast compares the "dry" treatment with the average of the two "irrigated" treatments and the second contrast compares the "low" level of irrigation with the "high" level of irrigation. For factor B the legumes are compared with the nonlegumes, then comparisons are made within each of the two categories of plant species.

The second stage of the procedure is to expand these contrasts into the main effect contrasts c_1, c_2, c_3, c_4 and c_5 as shown in Table 9.22. The last stage is to cross the two factor A contrasts, c_1 and c_2 with the three factor B contrasts, c_3, c_4 and c_5. This generates the six interaction contrasts $c_6 = c_1 \times c_3$, $c_7 = c_1 \times c_4$, $c_8 = c_1 \times c_5$, $c_9 = c_2 \times c_3$, $c_{10} = c_2 \times c_4$ and $c_{11} = c_2 \times c_5$ as shown in Table 9.22.

		Factor B (crop)				A contrasts	
		Legumes		Nonlegumes			
		Lupins	Peas	Mustard	Barley		
Factor	Dry	μ_1	μ_2	μ_3	μ_4	-2	0
A	25% asm	μ_5	μ_6	μ_7	μ_8	1	-1
(moisture)	50% asm	μ_9	μ_{10}	μ_{11}	μ_{12}	1	1
		-1	-1	1	1		
B contrasts		-1	1	0	0		
		0	0	-1	1		

Table 9.21: Treatment list and orthogonal contrasts for factors A and B.

	Main effect contrasts					Interaction contrasts					
	A		**B**			$A \times B$					
	c_1	c_2	c_3	c_4	c_5	c_6	c_7	c_8	c_9	c_{10}	c_{11}
μ_1	-2	0	-1	-1	0	2	2	0	0	0	0
μ_2	-2	0	-1	1	0	2	-2	0	0	0	0
μ_3	-2	0	1	0	-1	-2	0	2	0	0	0
μ_4	-2	0	1	0	1	-2	0	-2	0	0	0
μ_5	1	-1	-1	-1	0	-1	-1	0	1	1	0
μ_6	1	-1	-1	1	0	-1	1	0	1	-1	0
μ_7	1	-1	1	0	-1	1	0	-1	-1	0	1
μ_8	1	-1	1	0	1	1	0	1	-1	0	-1
μ_9	1	1	-1	-1	0	-1	-1	0	-1	-1	0
μ_{10}	1	1	-1	1	0	-1	1	0	-1	1	0
μ_{11}	1	1	1	0	-1	1	0	-1	1	0	-1
μ_{12}	1	1	1	0	1	1	0	1	1	0	1

Table 9.22: Complete set of factorial contrasts generated using the basic contrasts given in the preceding table.

Computing Shortcuts

Setting up of such sets of orthogonal contrasts, and feeding their corresponding unit vectors into Minitab, will quickly become tedious. In practice researchers use packages such as GENSTAT (Alvey, Galwey and Lane, 1982) which will automatically generate the full set of factorial contrasts from the contrasts for factors A and B as given in Table 9.21. This is clearly the neatest solution to the computing problem. However, since we wish to continue using Minitab in this text, we shall outline a shortcut which will make our work considerably easier. The procedure, applied to Table 9.22, is as follows:

1. SET the main effect contrast vectors into columns 2 to 6 of the worksheet, but refrain from making these into unit vectors.

2. Generate the interaction contrast vectors by cross multiplying columns, putting the resulting vectors into columns 7 to 12.

3. Make columns 2 to 12 into unit vectors.

4. REGRESS c1 11 c2-c12 as usual.

The corresponding Minitab commands are given in Table 9.23, assuming that there are three replicates of each treatment.

Alternative Factorial Notation

Appendix C describes an alternative and more traditional notation for factorial experiments. In this system, which we do not use in the main body of this text, the symbol μ_{ij} is used to denote a population mean, and y_{ijk} to denote an observation. Any reader having problems relating our results to those in certain other texts may find this appendix useful.

List of commands	Interpretation
$\begin{bmatrix} \text{set c1} \\ \text{data} \dots \end{bmatrix}$	$[$Set y into column one
$\begin{bmatrix} \text{set c2} \\ (-2,\ 1,\ 1)12 \end{bmatrix}$	$\begin{bmatrix} \text{Set } U_2 \text{ into column two} \\ \text{without dividing by } \sqrt{72} \end{bmatrix}$
$\begin{bmatrix} \text{set c3} \\ (\ 0, -1,\ 1)12 \\ \text{set c4} \\ 3(-1, -1,\ 1,\ 1)3 \\ \text{set c5} \\ 3(-1,\ 1,\ 0,\ 0)3 \\ \text{set c6} \\ 3(\ 0,\ 0, -1,\ 1)3 \end{bmatrix}$	$[$Similarly for U_3 to U_6
$\begin{bmatrix} \text{let \ c7=c2*c4} \\ \text{let \ c8=c2*c5} \\ \text{let \ c9=c2*c6} \\ \text{let c10=c3*c4} \\ \text{let c11=c3*c5} \\ \text{let c12=c3*c6} \end{bmatrix}$	$\begin{bmatrix} \text{Generate the unit vectors} \\ \text{corresponding to the} \\ \text{interaction contrasts,} \\ \text{without divisors} \end{bmatrix}$
$\begin{bmatrix} \text{let \ c2=c2/sqrt(72)} \\ \text{let \ c3=c3/sqrt(24)} \\ \vdots \\ \text{let c12=c12/sqrt(12)} \end{bmatrix}$	$[$Convert to unit vectors
$[$regress c1 11 c2$-$c12	$[$Project y onto U_2 to U_{12}
Other commands as before: oneway, histogram etc.	

Table 9.23: Simplified Minitab commands for factorial experiments.

9.5 Summary

Factorial contrasts are appropriate when several factors are being studied in the one experiment. To generate these contrasts the procedure is to firstly specify a complete set of orthogonal contrasts for each factor in turn. For example, if factor A has three levels, namely a control and two diets, then there are two orthogonal contrasts, $c_1 = \mu_2 + \mu_3 - 2\mu_1$ and $c_2 = \mu_2 - \mu_3$. Similarly if factor B has four levels there are three orthogonal contrasts. Secondly, the factorial "main effect" contrasts are written out in terms of the complete list of treatments, of which there are $3 \times 4 = 12$ in our example, simply by repeating the appropriate coefficients. Lastly, the "interaction" contrasts are generated by multiplying the corresponding coefficients in the main effect contrasts.

If the factors operate independently of one another, a factorial design is extremely efficient. For example, in a two factor experiment the full set of experimental observations is used to analyze the effect of factor A, and reused to analyze the effect of factor B, so performing two jobs for the price of one. Similarly, in a four factor experiment four jobs are done for the price of one.

If the factors do not operate independently, valuable knowledge of how they interact is obtained from the data analysis. For these reasons factorial designs are extremely popular and useful.

For additional reading on factorial designs, refer to Cochran and Cox (1957), pages 148–156 and 161–175; Snedecor and Cochran (1968), pages 339–349 and 359–364; and Steel and Torrie (1980), pages 336–355.

Exercises

(9.1) A survey of household cats was carried out to determine

  (i) whether, on average, male cats differ from females in weight,

 (ii) whether, on average, Siamese cats differ in weight from ordinary cats,

(iii) whether the sex weight difference is the same for both Siamese and ordinary cats.

The local veterinary association was approached for lists of all known Siamese and ordinary fully grown cats in the area. From each of these lists three males and three females were chosen at random. The populations of study and the resulting cat weights were as follows:

Population	Cat weights (kg)		
1. Siamese male	4.1	5.2	3.6
2. Ordinary male	6.3	7.5	8.7
3. Siamese female	3.0	4.7	3.4
4. Ordinary female	5.7	4.7	6.1

(a) Write down the observation vector y and the model space M.

(b) Write down hypotheses of interest in terms of a complete set of orthogonal contrasts. Also write down the corresponding unit vectors U_i.

(c) Calculate the projection coefficients $y.U_i$ corresponding to the hypotheses of interest, either by hand or using Minitab.

(d) Write out the error vector $y - \bar{y}_{i.}$ and calculate s^2.

(e) Test the hypotheses of interest. What are your conclusions?

(f) Write out the appropriate ANOVA table and present your results using method one of Table 9.6.

(g) Calculate the 99% confidence interval for the average difference in weight between Siamese and ordinary cats.

(9.2) In a completely randomized design experiment involving sixteen rats housed in individual cages, average daily feed intake, in grams, was measured for four diets treated with two solutions, saline and "DMH". Treatments thus had a 2 × 4 factorial structure. There were two replicates for each treatment. The data is given in the following table.

		Factor B (diet)			
		Fibre free	Cellulose fibre	Pectin fibre	Guar fibre
Factor A	Saline	24.0, 19.0	26.8, 29.0	21.6, 28.6	21.5, 23.9
(solution)	DMH	20.1, 26.5	23.6, 31.6	19.3, 26.7	20.7, 29.5

(a) Think of an appropriate set of three orthogonal contrasts for factor B, diet. Use these and the factor A, solution, contrast to generate a set of seven orthogonal contrasts among the eight treatment means. Say which are main effect contrasts and which are interaction contrasts.

(b) Write down the corresponding unit vectors $U_2, U_3, \ldots, U_8$.

(c) Analyze the data using the Regress command of Minitab.

(d) Present an ANOVA table showing the seven F tests and their significances.

(e) Use the Oneway command of Minitab to calculate the eight treatment means. Present your results using method two of Table 9.17.

(f) Draw a histogram of the errors to check the normality assumption.

(9.3) An experiment was carried out to compare two chemicals used for controlling the weed yarrow. Each chemical was applied at the rate recommended by the manufacturer and at three times this rate. A control treatment was also included in the experiment.
Treatments were:

    1. Chemical A at normal rate

    2. Chemical A at 3× normal rate

    3. Chemical B at normal rate

    4. Chemical B at 3× normal rate

    5. No chemical treatment (control).

Write down the four orthogonal contrasts among the treatment means μ_1, μ_2, μ_3, μ_4 and μ_5 which are appropriate for an analysis of this experiment as a 2×2 factorial + 1 design.

(9.4) A completely randomized design field experiment was set up to investigate the response of wheat to two forms of phosphate (P) fertiliser and two forms of potassium (K) fertiliser. The treatments formed a 3×3 factorial structure and there were three replicates. The plot grain yields converted to tonnes/ha are given in the following table.

	Control (no K)			Soluble K			Insoluble K		
Control (no P)	4.6	4.9	4.3	6.3	6.1	6.4	6.6	6.7	6.9
Rock phosphate	5.4	5.6	5.2	6.8	7.5	6.7	7.5	8.0	7.3
Superphosphate	5.3	5.7	5.1	7.5	7.0	7.2	7.1	7.4	8.1

(a) Decide on two orthogonal contrasts for the phosphate factor and another two for the potassium factor. Use these to generate a set of eight orthogonal contrasts among the nine treatment means.

(b) Write down the corresponding orthogonal unit vectors $U_2, \ldots U_9$ and use the Regress command of Minitab to analyze the dataset. If you are so inclined include the Oneway command in your job as a check.

(c) Summarize your results in an appropriate ANOVA table and an appropriate table of means, standard errors and contrast significances.

(d) Calculate the 95% confidence interval for the main effect contrast for the "average response to K".

(9.5) An investigation into the effects of fungicide and anti-desiccant treatments on the amount of rot in sugar beet stored in a "clamp" (large heap) over winter, was carried out in the winter of 1983. Twenty eight bags of treated or untreated sugar beet were randomly positioned inside the clamp when it was being constructed, and removed three months later. The table following lists the seven treatments which were used, and the percentage rot by volume in each of the four replicate bags for each treatment.

Treatment	Rot percentage			
1. Untreated (control 1)	5.6	6.7	7.3	8.2
2. Water (control 2)	6.2	3.6	4.0	4.6
3. Wax	4.3	3.3	4.7	6.5
4. Wiltpruf	6.3	5.3	5.1	3.7
5. Thiabendazole	2.5	4.2	2.1	3.2
6. Wax + Thiabendazole	2.3	3.4	2.4	4.2
7. Wiltpruf + Thiabendazole	3.4	4.4	5.8	4.0

Treatment one was completely untreated. Wax and Wiltpruf are antidesiccants and Thiabendazole is a fungicide, all applied in a solution of water. Treatments 2 to 7 therefore make up a class of treatments which all require the application of water to the beets. These six treatments form a 2×3 factorial, with factors fungicide(nil, Thiabendazole) and antidesiccant (nil, wax or Wiltpruf).

(a) Given these facts, set up an appropriate set of six orthogonal contrasts.

(b) Use the corresponding unit vectors $U_2, \ldots, U_7$ to analyze the data set using Minitab.

(c) Summarize your calculations in an ANOVA table. What can you conclude?

(d) Check the normality assumption by drawing a histogram of the errors.

(e) Summarize your results in a table using method one of Table 9.17.

(9.6) Some farmers conjecture that "ill thrift" (poor growth rate) in lambs is caused by a dietary deficiency of certain trace elements. One such claim was checked out on a farm with an illthrift problem by setting up a $2 \times 2 \times 2$ factorial experiment involving the trace elements cobalt (Co), copper (Cu) and iodine (I). The eight treatments shown below were each allocated at

random to three lambs, which were drenched at two weekly intervals according to treatment. The 24 lambs were run together as one mob on a common
pasture. The data below are the weightgains of the individual lambs over
the three month trial period.

Treatment	Weightgain (kg)		
Control	17	18	14
I	16	15	18
Cu	18	19	16
I+Cu	14	19	16
Co	20	18	17
I+Co	19	17	16
Cu+Co	15	17	19
I+Cu+Co	16	21	18

(a) Write down the contrasts, and the associated unit vectors (U_2, U_3
and U_4) corresponding to the following three hypotheses:

  1. Cobalt has no effect on weight gain.

  2. Copper has no effect on weight gain.

  3. Iodine has no effect on weight gain.

(b) Use Minitab to calculate $(y.U_2)^2$, $(y.U_3)^2$ and $(y.U_4)^2$ using the Regress
command.

(c) Use the Oneway command to fit the full model and hence estimate σ^2
as $s^2 = \|y - \bar{y}_{i.}\|^2/16$.

(d) Summarize your calculations in an ANOVA table.

(e) Present your results in a table of means, etc, using the style of method
two in Table 9.6.

(9.7) A 2×2 factorial experiment was set up to look at the effects of boron
and copper on pasture production. Three replicates were laid down in a
completely randomized design.
  The treatment mean yields in t/ha from the first season of growth were
as follows:

Treatment	Boron	Copper	Mean Yields
1	−	−	5.8
2	−	+	6.2
3	+	−	6.7
4	+	+	8.3

(a) Write down the contrast among the means μ_1, μ_2, μ_3 and μ_4 for just the boron main effect.

(b) Write down the corresponding unit vector U assuming the data vector y is in the order Treatment 1: rep 1, rep 2, rep 3, followed by Treatment 2: rep 1, rep 2, rep 3, and so on.

(c) Calculate $(y.U)^2$, the sum of squares for the boron main effect.

(d) Given that $s^2 = 2$, calculate the F test for the hypothesis H_0 : boron has no effect on pasture yield.

(e) Calculate the 95% confidence interval for the boron main effect contrast, written as an average response to boron.

Chapter 10

Polynomial Contrasts

The experiments we have considered so far have always involved treatments which are *qualitative* in nature: for example, variety of lupin or breed of sheep. Consequently the contrasts of interest were comparisons of one class of treatment with another (Chapter 8). In the case of factorial experiments (Chapter 9) the contrasts of interest also included the interactions between factors. Frequently in an experimental situation, however, the treatments are *quantitative* in nature. Such is the case when treatments correspond to the seeding rate of barley in kg/ha, or the dose rate of a medicine in a clinical trial in mg/person. In the first example, it is then natural to ask a question of the type "Does the yield of barley increase as the seeding rate increases?" In the second example, the natural question may be "Does blood pressure go down as dose rate goes up?" In this chapter we find out how to answer such questions. The mechanism will be essentially that of the previous two chapters: the effects we wish to study correspond to contrasts, and testing is carried out using the appropriate unit vector in the treatment space.

The plan for this chapter is to firstly analyze Example D in §1, then discuss the general case in §2, expanding upon important aspects. We then analyze a factorial experiment in which one of the factors involves polynomial contrasts, in §3. The chapter closes with a summary in §4, and exercises.

10.1 Analyzing Example D

We shall now statistically analyze the data given in Example D of §7.2. Recall that the aim of this experiment was to determine the effect of seeding rate on the yield of spring sown malting barley. A trial with five treatments, corresponding to seeding rates of 50, 75, 100, 125 and 150 kg/ha, each with six replicates, was laid out in a completely randomized design.

Specifically, the questions of interest are

1. Did grain yield increase with increasing seeding rate?

2. Did the rate of increase in grain yield drop off with increasing seeding rate?

With five treatments we have, as usual, a four dimensional treatment space. Where are the unit vectors in this treatment space which enable us to test these questions? In order to find them we shall need to consider the fitting of a sequence of polynomials of increasing order.

Data

We firstly convert our grain yield data from units of kg/plot to units of kg/ha. Since each plot was 50m^2 in area, the conversion factor is $10000/50 = 200$. The resulting grain yields in kg/ha are given in Table 10.1.

Seeding rate	Grain yield						Mean
50 kg/ha	5080	4480	5040	4880	4840	4400	4787
75 "	5240	5240	5040	5280	5000	5560	5227
100 "	5520	5520	5200	5160	5240	5160	5300
125 "	5520	5640	5360	5320	5600	5560	5500
150 "	5440	5640	5360	5120	5440	5520	5420

Table 10.1: Grain yields of Example D converted to kg/ha.

The corresponding observation vector is $y = \begin{bmatrix} y_{11} \\ y_{12} \\ \vdots \\ y_{56} \end{bmatrix} = \begin{bmatrix} 5080 \\ 4480 \\ \vdots \\ 5520 \end{bmatrix}.$

To see the relationship between grain yield and seeding rate we plot a "scattergram" of yield against seeding rate, as shown in Figure 10.1.

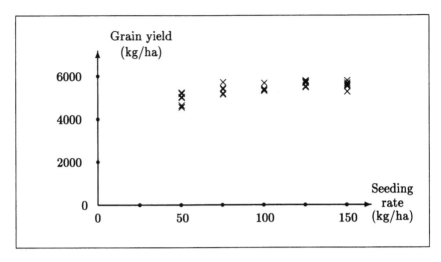

Figure 10.1: Scattergram of yield against seeding rate, Example D.

Model

We assume that each observation y_{ij} is independently drawn from a normal population with mean μ_i and a common variance σ^2. In Figure 10.1 this means that for each seeding rate we assume the vertical spread of yields is normally distributed with a common variance.

The resulting *model vector* is

$$
\begin{bmatrix} \mu_1 \\ \cdot \\ \mu_2 \\ \cdot \\ \mu_3 \\ \cdot \\ \mu_4 \\ \cdot \\ \mu_5 \\ \cdot \end{bmatrix}
= \mu_1 \begin{bmatrix} 1 \\ \cdot \\ 0 \\ \cdot \\ 0 \\ \cdot \\ 0 \\ \cdot \\ 0 \\ \cdot \end{bmatrix}
+ \mu_2 \begin{bmatrix} 0 \\ \cdot \\ 1 \\ \cdot \\ 0 \\ \cdot \\ 0 \\ \cdot \\ 0 \\ \cdot \end{bmatrix}
+ \mu_3 \begin{bmatrix} 0 \\ \cdot \\ 0 \\ \cdot \\ 1 \\ \cdot \\ 0 \\ \cdot \\ 0 \\ \cdot \end{bmatrix}
+ \mu_4 \begin{bmatrix} 0 \\ \cdot \\ 0 \\ \cdot \\ 0 \\ \cdot \\ 1 \\ \cdot \\ 0 \\ \cdot \end{bmatrix}
+ \mu_5 \begin{bmatrix} 0 \\ \cdot \\ 0 \\ \cdot \\ 0 \\ \cdot \\ 0 \\ \cdot \\ 1 \\ \cdot \end{bmatrix}
$$

so the model space M is a five dimensional subspace of 30-space.

Of interest to us is the curve through the unknown population means μ_i. Recall that a straight line is uniquely determined by two points, a quadratic by three points, a cubic by four points, and a quartic by five points. In our case, then, the five points $(50, \mu_1), \ldots, (150, \mu_5)$ uniquely determine a quartic, or fourth order polynomial,

$$ y = \alpha_0 + \alpha_1 x + \alpha_2 x^2 + \alpha_3 x^3 + \alpha_4 x^4 $$

where y denotes mean yield and x denotes seeding rate. This can be written alternatively as

$$y = \alpha_0 + \alpha_1(x - \bar{x}) + \alpha_2(x - \bar{x})^2 + \alpha_3(x - \bar{x})^3 + \alpha_4(x - \bar{x})^4$$
$$= \alpha_0 + \alpha_1(x - 100) + \alpha_2(x - 100)^2 + \alpha_3(x - 100)^3 + \alpha_4(x - 100)^4$$

where the α_i values may differ between the two forms. For example, the polynomial $y = 9801 - 198x + x^2$ is exactly the same as the polynomial $y = 1 + 2(x - 100) + (x - 100)^2$. For us, however, the second form of the equation will prove more convenient.

Figure 10.1 and our geometric breakup, $y = \mu_i + (y - \mu_i)$, are connected: the *heights* of the true curve at the seeding rates of $50, \ldots, 150$ kg/ha, namely $\mu_1, \ldots, \mu_5$, are the *entries* of our model vector. Since we have six replicates at each seeding rate, these heights are repeated in the model vector. Thus our model vector can usefully be thought of as a discrete version of the true curve.

The simultaneous equations linking the heights $\mu_1, \ldots, \mu_5$ to the coefficients $\alpha_0, \ldots, \alpha_4$ are

$$\mu_1 = \alpha_0 + \alpha_1(50 - 100) + \alpha_2(50 - 100)^2 + \alpha_3(50 - 100)^3 + \alpha_4(50 - 100)^4$$
$$\mu_2 = \alpha_0 + \alpha_1(75 - 100) + \alpha_2(75 - 100)^2 + \alpha_3(75 - 100)^3 + \alpha_4(75 - 100)^4$$
$$\mu_3 = \alpha_0 + \alpha_1(100 - 100) + \alpha_2(100 - 100)^2 + \alpha_3(100 - 100)^3 + \alpha_4(100 - 100)^4$$
$$\mu_4 = \alpha_0 + \alpha_1(125 - 100) + \alpha_2(125 - 100)^2 + \alpha_3(125 - 100)^3 + \alpha_4(125 - 100)^4$$
$$\mu_5 = \alpha_0 + \alpha_1(150 - 100) + \alpha_2(150 - 100)^2 + \alpha_3(150 - 100)^3 + \alpha_4(150 - 100)^4$$

since $y = \mu_1$ when $x = 50$, and so on.
We can view this as writing our model vector alternatively as

$$
\begin{bmatrix} \mu_1 \\ \cdot \\ \mu_2 \\ \cdot \\ \mu_3 \\ \cdot \\ \mu_4 \\ \cdot \\ \mu_5 \\ \cdot \end{bmatrix}
= \alpha_0 \begin{bmatrix} 1 \\ \cdot \\ 1 \\ \cdot \\ 1 \\ \cdot \\ 1 \\ \cdot \\ 1 \\ \cdot \end{bmatrix}
+ \alpha_1 \begin{bmatrix} -50 \\ \cdot \\ -25 \\ \cdot \\ 0 \\ \cdot \\ 25 \\ \cdot \\ 50 \\ \cdot \end{bmatrix}
+ \alpha_2 \begin{bmatrix} 50^2 \\ \cdot \\ 25^2 \\ \cdot \\ 0^2 \\ \cdot \\ 25^2 \\ \cdot \\ 50^2 \\ \cdot \end{bmatrix}
+ \alpha_3 \begin{bmatrix} -50^3 \\ \cdot \\ -25^3 \\ \cdot \\ 0^3 \\ \cdot \\ 25^3 \\ \cdot \\ 50^3 \\ \cdot \end{bmatrix}
+ \alpha_4 \begin{bmatrix} 50^4 \\ \cdot \\ 25^4 \\ \cdot \\ 0^4 \\ \cdot \\ 25^4 \\ \cdot \\ 50^4 \\ \cdot \end{bmatrix}
$$

We have just re-expressed the model vector in terms of an oblique, or nonorthogonal coordinate system for the model space:

$$M = \text{span} \left\{ \begin{bmatrix} 1 \\ \cdot \\ \cdot \\ \cdot \\ \cdot \\ \cdot \\ \cdot \\ \cdot \\ 1 \end{bmatrix}, \begin{bmatrix} -50 \\ \cdot \\ -25 \\ \cdot \\ 0 \\ \cdot \\ 25 \\ \cdot \\ 50 \\ \cdot \end{bmatrix}, \begin{bmatrix} (-50)^2 \\ \cdot \\ (-25)^2 \\ \cdot \\ 0^2 \\ \cdot \\ 25^2 \\ \cdot \\ 50^2 \\ \cdot \end{bmatrix}, \begin{bmatrix} (-50)^3 \\ \cdot \\ (-25)^3 \\ \cdot \\ 0^3 \\ \cdot \\ 25^3 \\ \cdot \\ 50^3 \\ \cdot \end{bmatrix}, \begin{bmatrix} (-50)^4 \\ \cdot \\ (-25)^4 \\ \cdot \\ 0^4 \\ \cdot \\ 25^4 \\ \cdot \\ 50^4 \\ \cdot \end{bmatrix} \right\}$$

For future reference we label these vectors X_1, X_2, X_3, X_4 and X_5.

The corresponding orthogonal coordinate system

To enable us to use our standard method for model fitting, we shall now orthogonalize the coordinate system $X_1, \ldots, X_5$, using the "Gram-Schmidt" method. In brief, the method involves taking each vector X_i in turn, and making it orthogonal to its predecessors by subtracting its projection onto the space which they span. The resulting orthogonal vectors $T_1, \ldots, T_5$ are then made into unit vectors $U_1, \ldots, U_5$ by dividing each by its length.

The direction of the first coordinate axis is simply chosen to be $T_1 = X_1$. Hence

$$U_1 = \frac{T_1}{\|T_1\|} = \frac{1}{\sqrt{N}} \begin{bmatrix} 1 \\ \vdots \\ 1 \end{bmatrix} = \frac{1}{\sqrt{30}} \begin{bmatrix} 1 \\ \vdots \\ 1 \end{bmatrix}$$

Note that U_1 is just the familiar equiangular vector.

The second coordinate axis is also easily produced, since X_2 is already orthogonal to X_1. Hence its direction is just $T_2 = X_2$ and

$$U_2 = \frac{T_2}{\|T_2\|} = \frac{x - \bar{x}}{\|x - \bar{x}\|} = \frac{1}{\sqrt{37500}} \begin{bmatrix} -50 \\ \cdot \\ -25 \\ \cdot \\ 0 \\ \cdot \\ 25 \\ \cdot \\ 50 \end{bmatrix} = \frac{1}{\sqrt{60}} \begin{bmatrix} -2 \\ \cdot \\ -1 \\ \cdot \\ 0 \\ \cdot \\ 1 \\ \cdot \\ 2 \end{bmatrix}$$

where x is the vector of x-values and $\bar{x}$ the overall mean x vector.

Moving on to the third coordinate axis, U_3, we require that the span of U_1, U_2 and U_3 be the same as the span of X_1, X_2 and X_3. Figure 10.2 tells us what we must do. To find a vector in the span of X_1, X_2 and X_3 which is orthogonal to both U_1 and U_2, we simply subtract the projection of X_3

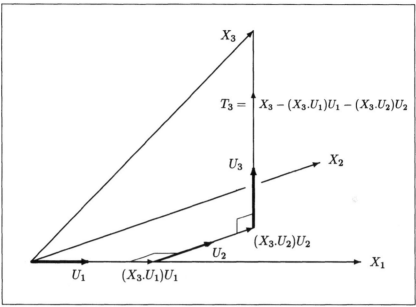

Figure 10.2: Finding the third orthogonal unit vector, U_3, by subtracting from X_3 its projection onto the subspace spanned by X_1 and X_2.

onto U_1 and U_2, from the vector X_3. The resulting vector is

$$
\begin{aligned}
T_3 &= X_3 - (X_3.U_1)U_1 - (X_3.U_2)U_2 \\
&= \begin{bmatrix} (x_1 - \bar{x})^2 \\ \vdots \\ (x_k - \bar{x})^2 \end{bmatrix} - \left\{ \begin{bmatrix} (x_1 - \bar{x})^2 \\ \vdots \\ (x_k - \bar{x})^2 \end{bmatrix} \cdot \frac{1}{\sqrt{N}} \begin{bmatrix} 1 \\ \vdots \\ 1 \end{bmatrix} \right\} \frac{1}{\sqrt{N}} \begin{bmatrix} 1 \\ \vdots \\ 1 \end{bmatrix} \\
&\quad - \left\{ \begin{bmatrix} (x_1 - \bar{x})^2 \\ \vdots \\ (x_k - \bar{x})^2 \end{bmatrix} \cdot \frac{1}{\sqrt{M_2}} \begin{bmatrix} x_1 - \bar{x} \\ \vdots \\ x_k - \bar{x} \end{bmatrix} \right\} \frac{1}{\sqrt{M_2}} \begin{bmatrix} x_1 - \bar{x} \\ \vdots \\ x_k - \bar{x} \end{bmatrix} \\
&= \begin{bmatrix} (x_1 - \bar{x})^2 \\ \vdots \\ (x_k - \bar{x})^2 \end{bmatrix} - \frac{M_2}{N} \begin{bmatrix} 1 \\ \vdots \\ 1 \end{bmatrix} - \frac{M_3}{M_2} \begin{bmatrix} x_1 - \bar{x} \\ \vdots \\ x_k - \bar{x} \end{bmatrix} \\
&= \begin{bmatrix} (x_1 - \bar{x})^2 - \frac{M_3}{M_2}(x_1 - \bar{x}) - \frac{M_2}{N} \\ \vdots \\ (x_k - \bar{x})^2 - \frac{M_3}{M_2}(x_k - \bar{x}) - \frac{M_2}{N} \end{bmatrix}
\end{aligned}
$$

where $M_2 = \|x - \bar{x}\|^2 = \sum_{i=1}^{k} \sum_{j=1}^{n} (x_i - \bar{x})^2 = n \sum_{i=1}^{k} (x_i - \bar{x})^2$ and $M_3 = n \sum_{i=1}^{k} (x_i - \bar{x})^3$. To convert T_3 to a unit vector we simply divide it by its length:

$$U_3 = \frac{T_3}{\|T_3\|} = \frac{X_3 - (X_3.U_1)U_1 - (X_3.U_2)U_2}{\text{length of numerator}}$$

In our example $N = 30,$

$$M_2 = 6 \times 25^2[(-2)^2 + (-1)^2 + 0^2 + 1^2 + 2^2] = 37500,$$

and $\qquad M_3 = 6 \times 25^3[(-2)^3 + (-1)^3 + 0^3 + 1^3 + 2^3] = 0.$

$$\text{Hence } T_3 = \begin{bmatrix} (-50)^2 - 1250 \\ \cdot \\ (-25)^2 - 1250 \\ \cdot \\ 0^2 - 1250 \\ \cdot \\ 25^2 - 1250 \\ \cdot \\ 50^2 - 1250 \\ \cdot \end{bmatrix} = \begin{bmatrix} 1250 \\ \cdot \\ -625 \\ \cdot \\ -1250 \\ \cdot \\ -625 \\ \cdot \\ 1250 \\ \cdot \end{bmatrix}, \quad \text{so } U_3 = \frac{1}{\sqrt{84}} \begin{bmatrix} 2 \\ \cdot \\ -1 \\ \cdot \\ -2 \\ \cdot \\ -1 \\ \cdot \\ 2 \\ \cdot \end{bmatrix}$$

To obtain U_4 and U_5 we simply carry on this process using the expressions

$$U_4 = \frac{T_4}{\|T_4\|} = \frac{X_4 - (X_4.U_1)U_1 - (X_4.U_2)U_2 - (X_4.U_3)U_3}{\text{length of numerator}}$$

$$U_5 = \frac{T_5}{\|T_5\|} = \frac{X_5 - (X_5.U_1)U_1 - (X_5.U_2)U_2 - (X_5.U_3)U_3 - (X_5.U_4)U_4}{\text{length of numerator}}$$

Exercises for the reader

(10.1) For our example, calculate T_3 directly by substituting into the expression $T_3 = X_3 - (X_3.U_1)U_1 - (X_3.U_2)U_2$ the values $X_3 = [(-50)^2, \ldots, 50^2]^T$, $U_1 = [1, \ldots, 1]^T/\sqrt{30}$ and $U_2 = [-2, \ldots, 2]^T/\sqrt{60}$.

(10.2) For our example, calculate T_4 directly by substituting into the above expression $X_4 = [(-50)^3, \ldots, 50^3]^T$. Hence calculate U_4.

(10.3) Similarly, calculate T_5 by substituting $X_5 = [(-50)^4, \ldots, 50^4]^T$. Hence calculate U_5.

The resulting orthogonal coordinate system for the model space M is

$$
\begin{array}{ccccc}
U_1 & U_2 & U_3 & U_4 & U_5 \\[4pt]
\begin{bmatrix} 1 \\ \cdot \\ \cdot \\ \cdot \\ \cdot \\ \cdot \\ 1 \end{bmatrix} &
\begin{bmatrix} -2 \\ \cdot \\ -1 \\ \cdot \\ 0 \\ \cdot \\ 1 \\ \cdot \\ 2 \end{bmatrix} &
\begin{bmatrix} 2 \\ \cdot \\ -1 \\ \cdot \\ -2 \\ \cdot \\ -1 \\ \cdot \\ 2 \end{bmatrix} &
\begin{bmatrix} -1 \\ \cdot \\ 2 \\ \cdot \\ 0 \\ \cdot \\ -2 \\ \cdot \\ 1 \end{bmatrix} &
\begin{bmatrix} 1 \\ \cdot \\ -4 \\ \cdot \\ 6 \\ \cdot \\ -4 \\ \cdot \\ 1 \end{bmatrix} \\[4pt]
\dfrac{}{\sqrt{30}} & \sqrt{60} & \sqrt{84} & \sqrt{60} & \sqrt{420}
\end{array}
$$

Orthogonal polynomial components

We can now rewrite our polynomial breakup of the model vector in orthogonal form using the orthogonal vectors

$$
\begin{aligned}
T_1 &= X_1 \\
T_2 &= X_2 \\
T_3 &= X_3 - (X_3.U_1)U_1 - (X_3.U_2)U_2 \\
T_4 &= X_4 - (X_4.U_1)U_1 - (X_4.U_2)U_2 - (X_4.U_3)U_3 \\
T_5 &= X_5 - (X_5.U_1)U_1 - (X_5.U_2)U_2 - (X_5.U_3)U_3 - (X_5.U_4)U_4
\end{aligned}
$$

We have worked out the first three of these:

$$
T_1 = \begin{bmatrix} 1 \\ \vdots \\ 1 \end{bmatrix}, \quad
T_2 = \begin{bmatrix} 50 - 100 \\ \vdots \\ 150 - 100 \end{bmatrix}, \quad
T_3 = \begin{bmatrix} (50 - 100)^2 - 1250 \\ \vdots \\ (150 - 100)^2 - 1250 \end{bmatrix}
$$

At the start of this section we used the non-orthogonal polynomial components of our curve, α_0, $\alpha_1(x - \bar{x})$ and so on, to break up our model vector in terms of the nonorthogonal vectors $X_1, \ldots, X_5$. Having orthogonalized these to $T_1, \ldots, T_5$, we can now reverse the process and use them to write down the *orthogonal polynomial components* of the true curve. From T_1, T_2 and T_3 we can see that the first three such components must be:

$$
\begin{aligned}
&\text{Constant component:} \quad && p_0(x) = 1 \\
&\text{Linear component:} \quad && p_1(x) = x - 100 \\
&\text{Quadratic component:} \quad && p_2(x) = (x - 100)^2 - 1250
\end{aligned}
$$

We shall not bother to work out the cubic and quartic components, $p_3(x)$ and $p_4(x)$. The resulting *orthogonal polynomial* form of the true curve is

$$
\begin{aligned}
y &= \beta_0 && + \beta_1(x - 100) + \beta_2[(x - 100)^2 - 1250] + \beta_3 p_3(x) + \beta_4 p_4(x) \\
&= \beta_0 p_0(x) + && \beta_1 p_1(x) + \qquad\quad \beta_2 p_2(x) \qquad\qquad + \beta_3 p_3(x) + \beta_4 p_4(x)
\end{aligned}
$$

When written as an orthogonal sum our model vector is now

$$
\begin{bmatrix} \mu_1 \\ \cdot \\ \mu_2 \\ \cdot \\ \mu_3 \\ \cdot \\ \mu_4 \\ \cdot \\ \mu_5 \\ \cdot \end{bmatrix}
= \beta_0 \begin{bmatrix} 1 \\ \cdot \\ 1 \\ \cdot \\ 1 \\ \cdot \\ 1 \\ \cdot \\ 1 \\ \cdot \end{bmatrix}
+ \beta_1 \begin{bmatrix} 50 - 100 \\ \cdot \\ 75 - 100 \\ \cdot \\ 100 - 100 \\ \cdot \\ 125 - 100 \\ \cdot \\ 150 - 100 \\ \cdot \end{bmatrix}
+ \beta_2 \begin{bmatrix} (50 - 100)^2 - 1250 \\ \cdot \\ (75 - 100)^2 - 1250 \\ \cdot \\ (100 - 100)^2 - 1250 \\ \cdot \\ (125 - 100)^2 - 1250 \\ \cdot \\ (150 - 100)^2 - 1250 \end{bmatrix}
+ \cdots
$$

$$
= \beta_0 \begin{bmatrix} p_0(50) \\ \cdot \\ p_0(75) \\ \cdot \\ \vdots \end{bmatrix}
+ \beta_1 \begin{bmatrix} p_1(50) \\ \cdot \\ p_1(75) \\ \cdot \\ \vdots \end{bmatrix}
+ \beta_2 \begin{bmatrix} p_2(50) \\ \cdot \\ p_2(75) \\ \cdot \\ \vdots \end{bmatrix}
+ \cdots
$$

$$
= \quad \beta_0 T_1 \quad + \quad \beta_1 T_2 \quad + \quad \beta_2 T_3 \quad + \cdots
$$

To clarify this further, we have drawn the first three "building block" components in Figure 10.3.

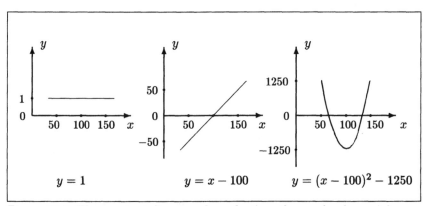

Figure 10.3: The constant, linear and quadratic orthogonal polynomial components, the "building blocks", for Example D.

The constant component, $p_0(x) = 1$, is used to approximate the *elevation* of the true curve. The linear component, $p_1(x) = x - 100$, is used to approximate the *tilt* of the true curve, and the quadratic component, $p_2(x) = (x - 100)^2 - 1250$, is used to approximate the *curvature* of the true curve. The corresponding fitted coefficients, b_0, b_1 and b_2, will estimate the elevation, the degree of tilt and the amount of curvature in the unknown true curve.

Successive polynomial approximations

The usefulness of these orthogonal components can be seen from Table 10.2, in which we consider polynomials of increasing order as possible curves with which to approximate the unknown true curve.

Polynomial order	Fitted model vector	Equation for fitted model in orthogonal form
0	$(y.U_1)U_1 = b_0 T_1$	b_0
1	$\sum_{i=1}^{2}(y.U_i)U_i = \sum_{i=1}^{2} b_{i-1}T_i$	$b_0 + b_1(x - 100)$
2	$\sum_{i=1}^{3}(y.U_i)U_i = \sum_{i=1}^{3} b_{i-1}T_i$	$b_0 + b_1(x - 100)$ $+b_2[(x - 100)^2 - 1250]$
3	$\sum_{i=1}^{4}(y.U_i)U_i = \sum_{i=1}^{4} b_{i-1}T_i$	$b_0 + b_1(x - 100)$ $+b_2[(x - 100)^2 - 1250]$ $+b_3 p_3(x)$
4	$\sum_{i=1}^{5}(y.U_i)U_i = \sum_{i=1}^{5} b_{i-1}T_i$	$b_0 + b_1(x - 100)$ $+b_2[(x - 100)^2 - 1250]$ $+b_3 p_3(x) + b_4 p_4(x)$

Table 10.2: Polynomial approximations of increasing order, Example D.

When the polynomial equations are written in orthogonal form, the estimated coefficients, $b_0, \ldots, b_4$, do not change from one order of polynomial to the next. The corresponding fitted values are appropriate sums of projection vectors, of the form $\sum(y.U_i)U_i = \sum b_{i-1}T_i$. For each increase in the order of polynomial the last projection, $(y.U_i)U_i = b_{i-1}T_i$, gives the change in the fitted values due to the addition of the extra term. Hence:

$(y.U_2)U_2$ is the "purely linear" part of the fitted model

$(y.U_3)U_3$ is the "purely quadratic" part of the fitted model

$(y.U_4)U_4$ is the "purely cubic" part of the fitted model

$(y.U_5)U_5$ is the "purely quartic" part of the fitted model

In the case of our example the relevant vectors work out to be:

$$
\begin{array}{ccccc}
(y.U_1)U_1 & (y.U_2)U_2 & (y.U_3)U_3 & (y.U_4)U_4 & (y.U_5)U_5 \\
\begin{bmatrix} 5247 \\ \cdot \\ \cdot \\ \cdot \\ \cdot \\ \cdot \\ \cdot \\ 5247 \end{bmatrix} &
\begin{bmatrix} -308 \\ \cdot \\ -154 \\ \cdot \\ 0 \\ \cdot \\ 154 \\ \cdot \\ 308 \\ \cdot \end{bmatrix} &
\begin{bmatrix} -130 \\ \cdot \\ 65 \\ \cdot \\ 130 \\ \cdot \\ 65 \\ \cdot \\ -130 \\ \cdot \end{bmatrix} &
\begin{bmatrix} -9 \\ \cdot \\ 17 \\ \cdot \\ 0 \\ \cdot \\ -17 \\ \cdot \\ 9 \\ \cdot \end{bmatrix} &
\begin{bmatrix} -13 \\ \cdot \\ 51 \\ \cdot \\ -77 \\ \cdot \\ 51 \\ \cdot \\ -13 \\ \cdot \end{bmatrix}
\end{array}
$$

Already we start to see which of the polynomial components are important.

Polynomial Contrasts

In previous chapters we have always written down the contrast first, then the corresponding unit vector. This is because the contrast was the basic entity of interest. In our present context the situation is different. We have derived the unit vectors corresponding to the hypotheses of interest, without any reference to contrasts. Here we do not even need to write down any contrasts. The reason we sometimes do write down "polynomial contrasts" is that they prove useful when *only some* of the experimental treatments are rates of some quantity, such as seeding rate or medical dose rate. Then we shall be in the happy position of being able to mix polynomial contrasts with contrasts of other types. An example of this type will be given in §10.2. From the unit vectors U_2, U_3, U_4, U_5 then, we see that the corresponding contrasts are:

$$
\begin{aligned}
c_1 &= -2\mu_1 - \mu_2 && + \mu_4 + 2\mu_5 && \text{(linear contrast)} \\
c_2 &= 2\mu_1 - \mu_2 - 2\mu_3 - \mu_4 + 2\mu_5 && \text{(quadratic contrast)} \\
c_3 &= -\mu_1 + 2\mu_2 && - 2\mu_4 + \mu_5 && \text{(cubic contrast)} \\
c_4 &= \mu_1 - 4\mu_2 + 6\mu_3 - 4\mu_4 + \mu_5 && \text{(quartic contrast)}
\end{aligned}
$$

Test Hypotheses

We can now return to the study objectives. Whether grain yield increases with increasing seeding rate is now seen to translate into testing

$$H_0 : \beta_1 = 0 \quad \text{versus} \quad H_1 : \beta_1 \neq 0$$

Informally, the null hypothesis states that the tilt of the curve is zero, while the alternative states that it is non-zero.

The second question, whether the rate of increase in grain yield drops off with increasing seeding rate, is tested using

$$H_0 : \beta_2 = 0 \quad \text{versus} \quad H_1 : \beta_2 \neq 0$$

Here the null hypothesis is that the quadratic curvature is zero.

Of lesser interest are the null hypotheses:

$$H_0 : \beta_3 = 0 \text{ versus } H_1 : \beta_3 \neq 0 \quad \text{and} \quad H_0 : \beta_4 = 0 \text{ versus } H_1 : \beta_4 \neq 0$$

These correspond to cubic and quartic components. For the sake of completeness these hypotheses will also be tested.

The directions associated with these four hypotheses are U_2, U_3, U_4 and U_5. To see that this is the case we firstly show that the projection coefficient $y.U_2$ averages to a scalar multiple of β_1. Then if $\beta_1 = 0$ the projection coefficient $y.U_2$ will be small, averaging to zero over many repetitions of the study, whereas if $\beta_1 \neq 0$ the projection coefficient will be large.

$$\text{Now} \quad y.U_2 = \begin{bmatrix} y_{11} \\ \cdot \\ y_{21} \\ \vdots \end{bmatrix} \cdot \begin{bmatrix} -2 \\ \cdot \\ -1 \\ \vdots \end{bmatrix} / \sqrt{60} \quad = \quad 6 \left[-2\bar{y}_{1.} - \bar{y}_{2.} + \bar{y}_{4.} + 2\bar{y}_{5.} \right] / \sqrt{60},$$

so the random variable $Y.U_2$ averages to $6 \left[-2\mu_1 - \mu_2 + \mu_4 + 2\mu_5 \right] / \sqrt{60}$ which can be written alternatively as $\mu_i.U_2$, where μ_i denotes the model vector $[\mu_1, \; . \; , \mu_2, \ldots, \mu_5]^T$. If we then expand the model vector in the form $\beta_0 T_1 + \beta_1 T_2 + \beta_2 T_3 + \beta_3 T_4 + \beta_4 T_5$, where the T_i's are mutually orthogonal, we find that

$$\begin{aligned} \mu_i.U_2 &= (\beta_0 T_1 + \beta_1 T_2 + \beta_2 T_3 + \beta_3 T_4 + \beta_4 T_5).U_2 \\ &= \beta_1(T_2.U_2) = \beta_1 \|T_2\| = \beta_1 \|x - \bar{x}\| = 25\sqrt{60}\beta_1 \end{aligned}$$

which is a scalar multiple of β_1, as required.

Similarly, the random variables $Y.U_3$, $Y.U_4$ and $Y.U_5$ have as expected values scalar multiples of β_2, β_3 and β_4 respectively. These are $\beta_2 \|T_3\| = 625\sqrt{84}\beta_2$, $\beta_3 \|T_4\| = 18750\sqrt{60}\beta_3$ and $\beta_4 \|T_5\| = 133929\sqrt{420}\beta_4$, using the solutions to the reader exercises for the last two equalities. These results are summarized in Table 10.3, together with the expected values of the projection coefficients in terms of the contrasts $c_1, \ldots, c_4$; these were calculated using the same methods as were used in Chapters 8 and 9. Clearly the contrasts $c_1, \ldots, c_4$ are simply constant multiples of $\beta_1, \ldots, \beta_4$ respectively, so a test of $H_0 : \beta_1 = 0$ is equivalent to a test of $H_0 : c_1 = 0$, and so on.

We now have all the raw materials for our analysis. We have a model space M, directions associated with our hypotheses, and an observation vector y.

	Mean, or expected, value		Variance
	(a) Using β_i values	(b) Using μ_i values	
$Y.U_1$	$\sqrt{30}\beta_0$	$\sqrt{30}\mu$	σ^2
$Y.U_2$	$25\sqrt{60}\beta_1$	$\dfrac{\sqrt{6}(-2\mu_1 - \mu_2 + \mu_4 + 2\mu_5)}{\sqrt{10}}$	σ^2
$Y.U_3$	$625\sqrt{84}\beta_2$	$\dfrac{\sqrt{6}(2\mu_1 - \mu_2 - 2\mu_3 - \mu_4 + 2\mu_5)}{\sqrt{14}}$	σ^2
$Y.U_4$	$18750\sqrt{60}\beta_3$	$\dfrac{\sqrt{6}(-\mu_1 + 2\mu_2 - 2\mu_4 + \mu_5)}{\sqrt{10}}$	σ^2
$Y.U_5$	$133929\sqrt{420}\beta_4$	$\dfrac{\sqrt{6}(\mu_1 - 4\mu_2 + 6\mu_3 - 4\mu_4 + \mu_5)}{\sqrt{70}}$	σ^2
$Y.U_6$	0	0	σ^2
$\vdots$	$\vdots$	$\vdots$	$\vdots$
$Y.U_{30}$	0	0	σ^2

Table 10.3: Distribution of the projection coefficients, $Y.U_i$, showing the relationship between the polynomial coefficients $\beta_1, \ldots, \beta_4$ and the polynomial contrasts $c_1, \ldots, c_4$.

Fitting the Model

As usual we project the observation vector y onto the model space M and obtain the *fitted model vector*, $\bar{y}_{i.} = [4787, 4787, \ldots, 5420, 5420]^T$, the vector of treatment means. The fitted model is $y = \bar{y}_{i.} + (y - \bar{y}_{i.})$. Written in the more usual "corrected" form this is

$$
\begin{array}{ccccccc}
y & - & \bar{y}_{..} & = & (\bar{y}_{i.} - \bar{y}_{..}) & + & (y - \bar{y}_{i.})
\end{array}
$$

$$
\begin{bmatrix} 5080 \\ \cdot \\ \cdot \\ \cdot \\ \cdot \\ \cdot \\ \cdot \\ \cdot \\ 5520 \end{bmatrix} - \begin{bmatrix} 5247 \\ \cdot \\ \cdot \\ \cdot \\ \cdot \\ \cdot \\ \cdot \\ \cdot \\ 5247 \end{bmatrix} = \begin{bmatrix} -460 \\ \cdot \\ -20 \\ \cdot \\ 53 \\ \cdot \\ 253 \\ \cdot \\ 173 \\ \cdot \end{bmatrix} + \begin{bmatrix} 293 \\ \cdot \\ \cdot \\ \cdot \\ \cdot \\ \cdot \\ \cdot \\ \cdot \\ 100 \end{bmatrix}
$$

Observation vector corrected for mean $=$ Treatment vector $+$ Error vector

Here the treatment vector lies in the four dimensional subspace of 30-space spanned by U_2, U_3, U_4 and U_5, and the error vector lies in a 25 dimensional subspace.

When the treatment vector is written in terms of its projections onto U_2, U_3, U_4 and U_5, we obtain the orthogonal decomposition

$$y - \bar{y}_{..} = 1193U_2 - 598U_3 + 67U_4 - 264U_5 + (y - \bar{y}_{i.})$$

as illustrated in Figure 10.4.

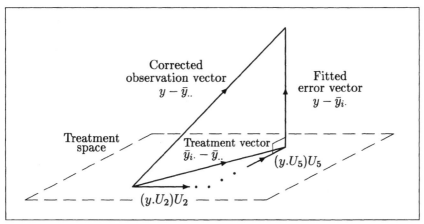

Figure 10.4: Orthogonal decomposition for polynomial contrasts case, showing the treatment vector broken into linear, quadratic, cubic and quartic vectors.

Testing the Hypotheses

As usual we test each hypothesis by checking whether the appropriate squared distance, $(y.U_i)^2$, is comparable to, or greater than, the average of the corresponding squared distances in the error space. For example, we test our first hypothesis, that grain yield does not change with seeding rate, by comparing the test statistic

$$F = \frac{(y.U_2)^2}{\left[(y.U_6)^2 + \cdots + (y.U_{30})^2\right]/25} = \frac{(y.U_2)^2}{\|y - \bar{y}_{i.}\|^2/25} = \frac{1422960}{39755} = 35.79$$

to the percentiles of the $F_{1,25}$ distribution. Here our hypothesis is rejected at the 1% level of significance, so we conclude that grain yield does change with seeding rate.

The appropriate Pythagorean decomposition is

$$\begin{aligned}\|y - \bar{y}_{..}\|^2 &= (y.U_2)^2 + (y.U_3)^2 + (y.U_4)^2 + (y.U_5)^2 + \|y - \bar{y}_{i.}\|^2 \\ 2848267 &= 1422960 + 357505 + 4507 + 69429 + 993867\end{aligned}$$

This decomposition, and the test statistics for our four hypotheses, are summarized in the ANOVA table presented in Table 10.4.

Source of Variation	df	SS	MS	F
Treatments	4	1854400		
$H_0 : \beta_1 = 0$ (linear)	1	1422960	1422960	35.79 (**)
$H_0 : \beta_2 = 0$ (quadratic)	1	357505	357505	8.99 (**)
$H_0 : \beta_3 = 0$ (cubic)	1	4507	4507	.11
$H_0 : \beta_4 = 0$ (quartic)	1	69429	69429	1.75
Error	25	993867	39755	
Total	29	2848267		

Table 10.4: ANOVA table for Example D (polynomial approximation).

From Table 10.4 we conclude that the coefficients β_1 and β_2 for the linear and quadratic terms in the polynomial are nonzero. However, we have no proof that the cubic and quartic coefficients (β_3 and β_4) are nonzero.

In terms of our original questions of interest, we conclude that (1) grain yield increases with increasing seeding rate, and (2) the rate of increase in grain yield drops off with increasing seeding rate.

Fitted Equation

In our example we decide that a quadratic polynomial is sufficient to approximate the unknown true curve relating grain yield to seeding rate. How do we find the equation of this fitted quadratic? From Table 10.2 we have the following equalities:

$$
\begin{aligned}
\text{Constant vector} &= (y.U_1)U_1 = b_0 T_1 = b_0.1 \\
\text{Linear vector} &= (y.U_2)U_2 = b_1 T_2 = b_1(x - 100) \\
\text{Quadratic vector} &= (y.U_3)U_3 = b_2 T_3 = b_2[(x - 100)^2 - 1250] \\
\text{Cubic vector} &= (y.U_4)U_4 = b_3 T_4 = b_3 p_3(x) \\
\text{Quartic vector} &= (y.U_5)U_5 = b_4 T_5 = b_4 p_4(x)
\end{aligned}
$$

where "1", $(x - 100)$ and so on denote vectors in 30-space. Here we know the $y.U_i$ values: $y.U_1 = 28737$, $y.U_2 = 1193$, $y.U_3 = -598$, $y.U_4 = 67$ and $y.U_5 = -264$. Writing out the first three terms as vectors enables us to calculate b_0, b_1 and b_2 and hence write down the required equation.

Firstly we examine the constant vector

$$
(y.U_1)U_1 = \bar{y}.. = \begin{bmatrix} 5247 \\ \vdots \\ 5247 \end{bmatrix} = \begin{bmatrix} b_0 \\ \vdots \\ b_0 \end{bmatrix}
$$

which tells us that $b_0 = \bar{y}_{..} = 5247$.

Secondly we examine the linear vector

$$(y.U_2)U_2 = 1193 \begin{bmatrix} -2 \\ \vdots \\ 2 \end{bmatrix} / \sqrt{60} = b_1 \begin{bmatrix} -50 \\ \vdots \\ 50 \end{bmatrix}$$

and equate coefficients to give $b_1 = (y.U_2/\sqrt{60})/25 = 6.16$.

Thirdly we examine the quadratic vector

$$(y.U_3)U_3 = -598 \begin{bmatrix} 2 \\ \vdots \\ 2 \end{bmatrix} / \sqrt{84} = b_2 \begin{bmatrix} 1250 \\ \vdots \\ 1250 \end{bmatrix}$$

and equate coefficients to give $b_2 = (y.U_3/\sqrt{84})/625 = -0.104$. Note that in general $b_{i-1} = y.U_i/\|T_i\|$, for $i = 1, \ldots, 5$.

The resulting fitted quadratic equation is

$$
\begin{aligned}
y &= b_0 + b_1(x - 100) + b_2\left[(x - 100)^2 - 1250\right] \\
&= 5250 + 6.16(x - 100) - 0.104\left[(x - 100)^2 - 1250\right]
\end{aligned}
$$

This is superimposed on the original scattergram in Figure 10.5.

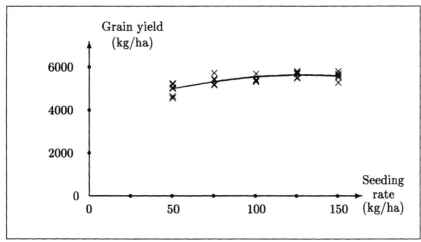

Figure 10.5: Fitted quadratic added to the scattergram, Example D.

Estimation of σ^2

As in previous chapters, we have transformed our original random variables $Y_{ij} \sim N\left[\mu_i, \sigma^2\right]$ into a new set of independent normal random variables

$Y.U_1, \ldots, Y.U_{30}$ with means and variances as given in Table 10.4. The first five of these are used to estimate the polynomial coefficients β_0, β_1, β_2, β_3 and β_4; or equivalently the overall mean μ and the contrasts c_1, c_2, c_3 and c_4; or equivalently the population means μ_1, μ_2, μ_3, μ_4 and μ_5. The remaining 25 random variables are used to estimate σ^2 via

$$s^2 = \frac{(y.U_6)^2 + \cdots + (y.U_{30})^2}{25} = \frac{\|y - \bar{y}_i.\|^2}{25} = \frac{993867}{25} = 39755$$

Note that even though we have decided to approximate the true curve with a quadratic, we do not use $(y.U_4)^2$ and $(y.U_5)^2$ for error estimation. We adopt this conservative strategy, since should the true curve not be a polynomial these squared projection coefficients will not yield unbiased estimates of the variance, σ^2.

Checking of Assumptions

The assumptions we need to check are those of the independence and normality of our observations, and equality of variance between our populations. As in Examples A to C, however, we cannot check our assumption of independence since our randomization order was simplified.

An inspection of the data as displayed in Figure 10.1 reveals no reason to be suspicious of the assumptions of normality of distribution or equality of variances. To be thorough, we would plot the histogram of errors and see whether it appears "fairly normal". We would also estimate the variances for the individual treatments, as given in Table 10.5.

Treatment	Variance estimate
50 kg/ha	81054
75 "	40120
100 "	29929
125 "	17109
150 "	30555
Average	39755

Table 10.5: Variance estimates for individual treatments.

These appear to exhibit a slight trend, decreasing as the seeding rate increases. As a rough test of this, we could compare the two lowest seeding rate treatments with the two highest seeding rate treatments, testing the hypothesis $H_0 : (\sigma_1^2 + \sigma_2^2)/2 = (\sigma_4^2 + \sigma_5^2)/2$. The test statistic is

$$F = \frac{(s_1^2 + s_2^2)/2}{(s_4^2 + s_5^2)/2} = \frac{60587}{23832} = 2.54$$

This is not significant when compared with 3.72, the 97.5 percentile of the $F_{10,10}$ distribution. We therefore accept the assumption of common variance.

Confidence Intervals

We shall now obtain the 95% confidence intervals for the linear and quadratic coefficients, β_1 and β_2, using the results in Table 10.3.

For the linear coefficient, β_1, we use the fact that $y.U_2 = 25\sqrt{60}b_1$ comes from the $N(25\sqrt{60}\beta_1, \sigma^2)$ distribution. Thus the quantity $25\sqrt{60}(b_1 - \beta_1)$ comes from the $N(0, \sigma^2)$ distribution, whence $t = 25\sqrt{60}(b_1 - \beta_1)/s$ comes from the t_{25} distribution. To obtain the 95% confidence interval for β_1 we gamble that our observed value t lies between the 2.5 and 97.5 percentiles of the t_{25} distribution. That is,

$$-2.060 \quad \leq \quad \frac{25\sqrt{60}(b_1 - \beta_1)}{s} \quad \leq \quad 2.060$$

This translates to the confidence interval

$$b_1 - \frac{s}{25\sqrt{60}}2.060 \quad \leq \quad \beta_1 \quad \leq \quad b_1 + \frac{s}{25\sqrt{60}}2.060$$

Here $b_1 = 6.16$ and $s = \sqrt{39755} = 199$. Therefore the 95% confidence interval for β_1 is 6.16 ± 2.12, in units of kg grain/ha per kg seed/ha.

Similarly for the quadratic coefficient β_2 we gamble that the value of $t = 25^2\sqrt{84}(b_2 - \beta_2)/s$ lies between the 2.5 and 97.5 percentiles of the t_{25} distribution. The resulting confidence interval is

$$b_2 - \frac{s}{25^2\sqrt{84}}2.060 \quad \leq \quad \beta_2 \quad \leq \quad b_2 + \frac{s}{25^2\sqrt{84}}2.060$$

Here $b_2 = -.104$. Hence the 95% confidence interval for β_2 is $-.104 \pm .072$.

The confidence interval for β_1 can be interpreted as saying that increasing the seeding rate by a kilogram per hectare increased the grain yield by 6.16 ± 2.12 kg/ha, as an average over the range from 50 to 150 kg seed/ha. However, the confidence interval for β_2 qualifies this last sentence by saying that the "return rate", averaging 6.16, dropped off as the seeding rate increased. We shall return to this point in the next section.

Report on Study

In order to report on the results of the experiment the most enlightening presentation is the scattergram of the data, with the addition of an approximating polynomial curve, as shown in Figure 10.5. A less illuminating method of presenting the results is by way of a table of treatment means, etc., as shown in Table 10.6.

Treatment	Grain yield (kg/ha)
50 kg seed/ha	4790
75 "	5230
100 "	5300
125 "	5500
150 "	5420
SED	115
Significance of contrasts	
Linear	**
Quadratic	**

Table 10.6: Presentation of results, Example D.

For interest's sake, we shall carry this a little further. To a farmer, the obvious question now is "what is the optimum economic seeding rate?" Or, rephrased, "what is the seeding rate which will return the greatest profit?" The key to answering this question is to determine, for each seeding rate, the amount of extra grain produced per hectare in response to an extra kilogram of seed per hectare. For this purpose, we assume that the true curve is a quadratic:

$$y = \beta_0 + \beta_1(x - 100) + \beta_2[(x - 100)^2 - 1250]$$

To obtain the rate of change of yield y with increasing seeding rate x, we simply differentiate:

$$\frac{dy}{dx} = \beta_1 + 2\beta_2(x - 100)$$

This is approximated by

$$\frac{dy}{dx} = 6.16 - .209(x - 100)$$

as illustrated in Figure 10.6.

Now if the net cost to the farmer of a kilogram of seed is double the net payment for a kilogram of grain, the farmer "breaks even" if an extra kilogram of seed produces two extra kilograms of grain. That is, the farmer breaks even on an extra kilogram of seed when dy/dx equals two, and makes a loss on an extra kilogram of seed when dy/dx falls below two. Hence the optimum economic seeding rate in this experiment is estimated from the equation $dy/dx = 2$, that is, $6.16 - .209(x - 100) = 2$. The optimum economic seeding rate is therefore estimated as $x = 120$ kg/ha.

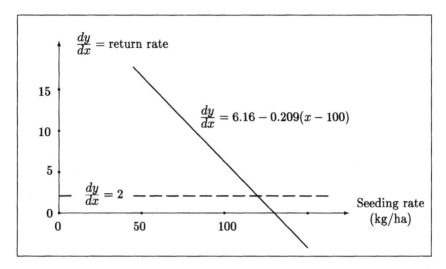

Figure 10.6: Showing the decline in return rate as the seeding rate increases. The farmer breaks even when the lines intersect, at 120 kg/ha.

This experiment was actually one of a series of eight trials spread over four growing seasons and several sites throughout the wheat growing district. The final recommendation to farmers, based on this series of trials, was for a seeding rate of 75−100 kg/ha, our experiment being more responsive than average. This is described in McLeod (1982). Further details of the economic analysis are given in Saville (1983).

Minitab Analysis

The analysis using Minitab follows the pattern of preceding chapters for the basic analysis. Commands are as summarized in Table 10.7.

To compute the fitted quadratic equation we can include further commands in our job, as listed in Table 10.8, for both nonorthogonal and orthogonal forms of the model. These commands should be inserted immediately before the stop in Table 10.7.

In the resulting Minitab output the variance estimate, s^2, will be slightly different from the estimate in Table 10.4 of 39755 since the directions corresponding to the cubic and quartic components have been included in the error space.

For the more interested reader, in Table 10.9 we now give the Minitab commands required to fit both the nonorthogonal and orthogonal sequences of polynomials. In this program we calculate X_2, X_3, X_4, X_5 as well as T_2, T_3, T_4, T_5 and U_2, U_3, U_4, U_5 for a general dataset. This program will also be useful when we discuss polynomial regression in Chapter 16.

List of commands	Interpretation
$\lceil$set c1 $\lfloor$25.4 22.4 25.2 ...	$\lceil$Set y into column one
$\lceil$let c1 = c1*10000/50	$\lceil$Convert to kg/ha
$\lceil$set c2 $\lvert(-2, -1, 0, 1, 2)6$ $\lfloor$let c2 = c2/sqrt(60)	$\lceil$Set U_2 into column two
$\lceil\ .\ .\ .$ $\lfloor\ .\ .\ .$	$\lceil$Set U_3, U_4, U_5 into $\lfloor$columns three, four and five
$\lceil$regress c1 4 c2−c5	$\lceil$Project y onto $U_2, \ldots, U_5$
$\lceil$set c6 $\lfloor(\ 1, \ 2, \ 3, \ 4, \ 5)6$	$\lceil$Set treatment numbers $\lfloor$into column six
$\lceil$oneway c1 c6 c7 c8	$\lceil$Do the ANOVA
$\lceil$print c1 c8 c7	$\lceil$Display $y = \bar{y}_{i\cdot} + (y - \bar{y}_{i\cdot})$
$\lceil$histogram c7	$\lceil$Draw the error histogram
$\lceil$stop	$\lceil$End of the job

Table 10.7: Minitab commands for basic analysis, Example D.

List of commands	Interpretation
$\lceil$set c12 $\lfloor(-50, -25, 0, 25, 50)6$	$\lceil$Set $X_2 = (x - \bar{x})$ $\lfloor$into column twelve
$\lceil$let c13 = c12**2	$\lceil$Calculate $X_3 = (x - \bar{x})^2$
$\lceil$regress c1 2 c12−c13	$\lceil$Fit the quadratic $\lfloor$in non-orthogonal form
$\lceil$let c22=c12	$\lceil$Calculate $T_2 = X_2$
$\lceil$let c23=c13−mean(c13)−sum(c13*c2)*c2	
	$\lceil$Calculate $T_3 = X_3 - (X_3.U_1)U_1 - (X_3.U_2)U_2$
$\lceil$regress c1 2 c22−c23	$\lceil$Fit the quadratic $\lfloor$in orthogonal form

Table 10.8: Additional Minitab commands required to find the equation of the quadratic, both in nonorthogonal and orthogonal form.

List of commands	Interpretation
⌈set c1 |25.4 22.4 25.2 ... ⌊let c1 = c1*10000/50	[Set y into column one
⌈set c2 |(50,75,100,125,150)6 [plot c1 c2	[Set x into column two [Draw a scattergram of y on x
[let c2=c2−mean(c2)	[Calculate $X_2 = (x - \bar{x})$
⌈let c3=c2**2 |let c4=c2**3 ⌊let c5=c2**4	⌈Calculate $X_3 = (x - \bar{x})^2$ |Calculate $X_4 = (x - \bar{x})^3$ ⌊Calculate $X_5 = (x - \bar{x})^4$
⌈regress c1 1 c2 |regress c1 2 c2−c3 |regress c1 3 c2−c4 ⌊regress c1 4 c2−c5	⌈Fit a sequence of polynomials ⌊in nonorthogonal form
[let c12=c2	[Calculate $T_2 = X_2$
[let c22=c12/sqrt(sum(c12**2))	[Calculate $U_2 = T_2/\|T_2\|$
[let c13=c3−mean(c3)−sum(c3*c22)*c22	
	[$T_3 = X_3 - (X_3.U_1)U_1 - (X_3.U_2)U_2$
[let c23=c13/sqrt(sum(c13**2))	[Calculate $U_3 = T_3/\|T_3\|$
[let c14=c4−mean(c4)−sum(c4*c22)*c22−sum(c4*c23)*c23	
	[Calculate $T_4 = X_4 - (X_4.U_1)U_1 - \cdots$
[let c24=c14/sqrt(sum(c14**2))	[Calculate $U_4 = T_4/\|T_4\|$
⌈let c15=c5−mean(c5)−sum(c5*c22)*c22−sum(c5*c23)*c23 −sum(c5*c24)*c24 ⌊	
	[Calculate $T_5 = X_5 - (X_5.U_1)U_1 - \cdots$
[let c25=c15/sqrt(sum(c15**2))	[Calculate $U_5 = T_5/\|T_5\|$
[print c22-c25	[Print U_2 to U_5 to check all is well
⌈regress c1 1 c12 |regress c1 2 c12−c13 |regress c1 3 c12−c14 ⌊regress c1 4 c12−c15	⌈Fit a sequence of polynomials ⌊in orthogonal form
[regress c1 4 c22−c25	[Calculate coefficients, $y.U_i$
[stop	[End of job

Table 10.9: Minitab commands for fitting the nonorthogonal and orthogonal sequences listed in Table 10.2. Using the notation established earlier, X_2, X_3, X_4, X_5 are stored in columns 2−5 of the worksheet; T_2, T_3, T_4, X_5 are stored in columns 12−15; U_2, U_3, U_4, U_5 are stored in columns 22−25.

Exercises for the reader

(10.4) Run the Minitab job as described in Tables 10.7 and 10.8. Compare your output with the results given in the text (for example, the orthogonal decomposition in Figure 10.3, the ANOVA table in Table 10.4, and the equation of the fitted quadratic in Figure 10.5).

(10.5) Run the Minitab job described in Table 10.9. Using the output, rewrite Table 10.2, substituting fitted values for the projection coefficients $y.U_1, \ldots, y.U_5$ and for the polynomial coefficients $b_0, \ldots, b_4$. As a check, does the equation for the quadratic agree with that calculated earlier in this section? Using a similar format, also write out the polynomial approximations of increasing order for the nonorthogonal fits. Notice that the coefficients change from one order of fit to the next.

Problems with the Field Technique

As an illustrative example the barley experiment has served our purposes admirably. We have found that a quadratic describes the data quite tidily, and we have even sidetracked to do an interesting economic analysis. The full story, however, is not quite as tidy. We shall now briefly consider a problem with field technique encountered in this series of experiments.

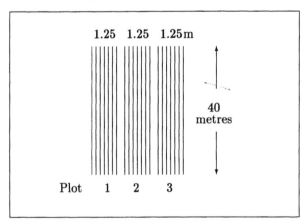

Figure 10.7: Diagram of field layout, not to scale. Each plot consisted of 7 rows of barley sown at 18cm spacing. The gap between plots averaged 36cm.

Figure 10.7 shows the layout of the experimental plots. Each plot was sown at the appropriate seeding rate using one pass of a seven coulter agricultural drill. This resulted in seven rows of barley per plot, with close to 18cm between the rows. Plots were separated by a gap for ease of sowing, identification and harvesting. The gap averaged 36cm between adjacent rows of neighbouring plots.

The edge effect caused by this gap was very noticeable, and is a well documented phenomenon. Basically the plants in the outer rows of each plot yield considerably more grain than the plants in the inner rows, since they have access to the extra moisture and nutrients in the gap. However, it was assumed that the edge effect would affect all treatments equally.

The validity of the assumption was questioned towards the end of the series of trials when discrepancies were noticed between the grain yield data obtained by harvesting entire plots, and other data collected from the three centre rows of each plot. As a result, in the last experiment of the series, not Example D, a four metre length of each individual row in each plot was hand harvested and threshed. The resulting data is graphed in Figure 10.8.

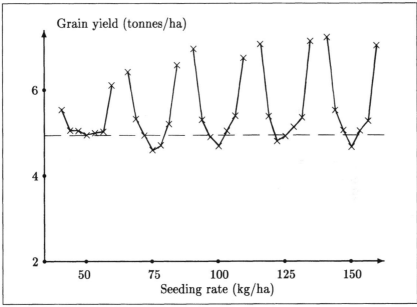

Figure 10.8: Hand sampled yield, in tonnes per hectare, of the individual rows, for each seeding rate, averaged over the six replicate plots. The horizontal line represents the average, over all seeding rates, of the yield of the three center rows.

The figure shows that the edge effect was less pronounced at the lower rates of seeding than at the higher rates of seeding. This trend was statistically significant at the 1% level when analyzed as described in Saville (1984). Presumably there were insufficient plants at the lower rates of seeding to fully exploit the extra resources available in the gap between the plots.

Since a typical farmer would aim to have no gaps between adjacent passes of the drill, we must disregard the grain yields obtained from harvesting entire plots. Turning our attention to the average grain yields from the three centre rows of each plot, anticipated to be free of edge effects, we see from Figure 10.8 that there is little difference between the seeding rates. In fact, a constant polynomial, as shown in the figure, gives a good description of this data when it is statistically analyzed. This means that the estimated optimum economic seeding rate is reduced to 50 kg/ha.

The moral of the story is that one should pay as much attention to the assumptions inherent in the techniques of experimentation or surveying, as to the mathematical assumptions inherent in the analysis. Violations of either of these sets of assumptions can have important effects on the conclusions. Our detailed conclusions for Example D have been invalidated by the problems just described.

10.2   Consolidating the Ideas

This is a catch-all section. We begin with a detailed summary of the routine for fitting polynomial models. This routine is fully general in that it applies even when the x-values are unevenly spaced, or when the replications are unequal. We then summarize the much simpler routine that can be used when the x-values are evenly spaced and equally replicated, as in Example D; this routine uses precalculated sets of orthogonal contrasts. This is followed by a closer look at the linear and quadratic contrasts, including a simple rule for deriving the linear contrast in all circumstances. An example is then given to introduce you to the art of mixing contrasts of the various basic types. We conclude this section by deriving general expressions for the orthogonal polynomial components and the confidence intervals for the coefficients β_0, β_1, ...

Fitting Polynomial Models: A General Routine

Here is a step-by-step summary of what happened when we analyzed the barley example of the previous section. We place things in a general setting, so that it will be clear where difficulties will lie in handling other problems of a similar type. In general, we deal with k populations with means μ_i and unknown variance σ^2. Each population is associated with an accurately measurable quantity, x_i. Our aim is to express the relationship between the means, μ_i, and the quantities, x_i.

1. Given a k-dimensional model space we have room to fit a $(k-1)^{th}$ *degree polynomial model* of the form

$$y = \alpha_0 + \alpha_1 x + \alpha_2 x^2 + \cdots + \alpha_{k-1} x^{k-1}$$

This can also be written in the form

$$y = \alpha_0 + \alpha_1 (x - \bar{x}) + \alpha_2 (x - \bar{x})^2 + \cdots + \alpha_{k-1}(x - \bar{x})^{k-1}$$

2. By putting the model in vector form we find that a *non-orthogonal coordinate system for the model space* is

$$X_1 = \begin{bmatrix} 1 \\ \vdots \\ 1 \end{bmatrix}, \quad X_2 = \begin{bmatrix} x_1 - \bar{x} \\ \vdots \\ x_k - \bar{x} \end{bmatrix}, \quad \ldots, \quad X_k = \begin{bmatrix} (x_1 - \bar{x})^{k-1} \\ \vdots \\ (x_k - \bar{x})^{k-1} \end{bmatrix}$$

where $x_1, \ldots, x_k$ are the k distinct x-values.

3. These vectors can be *orthogonalized* using the Gram-Schmidt method to yield

$$
\begin{array}{cccc}
T_1 & T_2 & T_3 & T_k \\[4pt]
\begin{bmatrix} 1 \\ \vdots \\ 1 \end{bmatrix} &
\begin{bmatrix} x_1 - \bar{x} \\ \vdots \\ x_k - \bar{x} \end{bmatrix} &
\begin{bmatrix} (x_1 - \bar{x})^2 - \frac{M_3}{M_2}(x_1 - \bar{x}) - \frac{M_2}{N} \\ \vdots \\ (x_k - \bar{x})^2 - \frac{M_3}{M_2}(x_k - \bar{x}) - \frac{M_2}{N} \end{bmatrix} &
\begin{bmatrix} p_{k-1}(x_1) \\ \vdots \\ p_{k-1}(x_k) \end{bmatrix}
\end{array}
$$

4. To form an *appropriate orthogonal coordinate system* $U_1, \ldots, U_k$ for the model space, we make each of $T_1, \ldots, T_k$ into a unit vector:

$$U_1 = \frac{T_1}{\|T_1\|}, \quad U_2 = \frac{T_2}{\|T_2\|}, \quad \ldots, \quad U_k = \frac{T_k}{\|T_k\|}$$

If required the corresponding *polynomial contrasts* can now be written down.

5. The resulting *orthogonal polynomial components* are

$$
\begin{array}{lll}
p_0(x) &= 1 & \text{(constant)} \\
p_1(x) &= x - \bar{x} & \text{(linear)} \\
p_2(x) &= (x - \bar{x})^2 - \frac{M_3}{M_2}(x - \bar{x}) - \frac{M_2}{N} & \text{(quadratic)} \\
&\vdots & \\
p_{k-1}(x) &= (x - \bar{x})^{k-1} - (\text{terms in } x^{k-2}, \ldots, x, 1) & ((k-1)^{th} \text{ order})
\end{array}
$$

6. The *orthogonal polynomial* model is

$$y = \beta_0 + \beta_1(x - \bar{x}) + \beta_2 p_2(x) + \cdots + \beta_{k-1} p_{k-1}(x)$$

7. We *fit the model* by forming

$$\bar{y}_{i\cdot} = (y.U_1)U_1 + (y.U_2)U_2 + (y.U_3)U_3 + \cdots + (y.U_k)U_k$$

a sum of constant, linear, quadratic up to $(k-1)^{th}$ order terms. This can also be written in the form

$$\bar{y}_{i\cdot} = b_0 T_1 + b_1 T_2 + b_2 T_3 + \cdots + b_{k-1} T_k$$

where $T_1, \ldots, T_k$ are vectors of polynomial component values; for example, T_2 is the vector of $p_1(x_i) = (x_i - \bar{x})$ values.

8. The equation of the fitted polynomial model is therefore

$$y = b_0 + b_1(x - \bar{x}) + b_2 p_2(x) + \cdots + b_{k-1} p_{k-1}(x)$$

Here $b_0 = (y.U_1)/\|T_1\|, \quad \ldots, \quad b_{k-1} = (y.U_k)/\|T_k\|$

9. The variance estimate is as usual

$$s^2 = \frac{(y.U_{k+1})^2 + \cdots + (y.U_N)^2}{N - k} = \frac{\|y - \bar{y}_{i\cdot}\|^2}{N - k}$$

10. In order to test whether the lth order polynomial coefficient β_l is zero we use the ratio $(y.U_{l+1})^2/s^2$ which comes from an $F_{1,N-k}$ distribution under the null hypothesis that the coefficient $\beta_l = 0$.

Simplified Routine

To the non-mathematical reader, the general routine will appear quite formidable. The good news is that many designed experiments, such as Example D, involve equally spaced and equally replicated x-values. In such cases the necessary sets of orthogonal contrasts are easily found in tables, which means our routine can be simplified to the following:

1. Look up the appropriate set of orthogonal polynomial contrasts in published tables, such as Table 10.10.

2. Write down the corresponding unit vectors, $U_2, \ldots, U_k$. Use these to fit the model by projecting the observation vector onto the model space M spanned by $U_1, \ldots, U_k$. Produce the usual ANOVA table and test hypotheses using the customary test statistics; for example, $H_0 : \beta_1 = 0$ is tested using $(y.U_2)^2/s^2$. This is readily accomplished using the Regress command in Minitab.

3. The last step will tell you whether the relationship between y and x is best approximated by a constant, linear, quadratic or higher order polynomial. If you want the equation of the fitted polynomial, the simplest method is to refit the model in non-orthogonal form

$$y = a_0 + a_1(x - \bar{x}) + a_2(x - \bar{x})^2 + \cdots$$

using the first half of the Minitab program given in Table 10.9.

Table 10.10 gives the sets of orthogonal polynomial contrasts for $k = 3$ up to $k = 6$ treatments. For more extensive tables the reader is referred to Pearson and Hartley (1966) or Little and Hills (1978). The latter text also gives sets of contrasts for some unequally spaced, equally replicated cases.

Number of equally spaced x-values	Polynomial contrasts				
	Linear	Quad	Cubic	Quartic	Quintic
3	$\begin{bmatrix} -1 \\ 0 \\ 1 \end{bmatrix}$	$\begin{bmatrix} 1 \\ -2 \\ 1 \end{bmatrix}$			
4	$\begin{bmatrix} -3 \\ -1 \\ 1 \\ 3 \end{bmatrix}$	$\begin{bmatrix} 1 \\ -1 \\ -1 \\ 1 \end{bmatrix}$	$\begin{bmatrix} -1 \\ 3 \\ -3 \\ 1 \end{bmatrix}$		
5	$\begin{bmatrix} -2 \\ -1 \\ 0 \\ 1 \\ 2 \end{bmatrix}$	$\begin{bmatrix} 2 \\ -1 \\ -2 \\ -1 \\ 2 \end{bmatrix}$	$\begin{bmatrix} -1 \\ 2 \\ 0 \\ -2 \\ 1 \end{bmatrix}$	$\begin{bmatrix} 1 \\ -4 \\ 6 \\ -4 \\ 1 \end{bmatrix}$	
6	$\begin{bmatrix} -5 \\ -3 \\ -1 \\ 1 \\ 3 \\ 5 \end{bmatrix}$	$\begin{bmatrix} 5 \\ -1 \\ -4 \\ -4 \\ -1 \\ 5 \end{bmatrix}$	$\begin{bmatrix} -5 \\ 7 \\ 4 \\ -4 \\ -7 \\ 5 \end{bmatrix}$	$\begin{bmatrix} 1 \\ -3 \\ 2 \\ 2 \\ -3 \\ 1 \end{bmatrix}$	$\begin{bmatrix} -1 \\ 5 \\ -10 \\ 10 \\ -5 \\ 1 \end{bmatrix}$

Table 10.10: Standard sets of orthogonal polynomial contrasts for equally replicated, equally spaced x-values.

Rule of Thumb for Deriving the Linear Contrast

The linear component unit vector always has the simple form

$$U_2 = \frac{1}{\|x - \bar{x}\|} \begin{bmatrix} x_1 - \bar{x} \\ \vdots \\ x_k - \bar{x} \end{bmatrix}$$

This means that in the equal replications case the linear contrast is any scalar multiple of

$$c = (x_1 - \bar{x})\mu_1 + \cdots + (x_k - \bar{x})\mu_k$$

This expression is analogous to the formula for the bending moment about a fulcrum in physics. In Example D, for instance, we can think of weights of $\mu_1, \mu_2, \mu_3, \mu_4$ and μ_5 positioned on a seesaw, or teeter-totter, at distances of $-50, -25, 0, 25$ and 50 from the fulcrum at $\bar{x} = 100$, as shown in Figure 10.9.

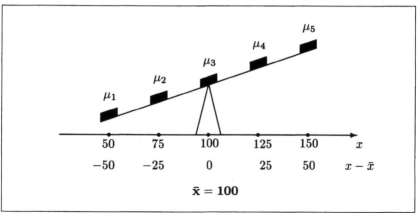

Figure 10.9: Seesaw analogy for linear contrast, Example D.

The bending moment about the fulcrum at $\bar{x} = 100$ is

$$\sum_{i=1}^{5} \text{weight at } x_i \times \text{distance from fulcrum} = \sum_{i=1}^{5} \mu_i \times (x_i - \bar{x})$$
$$= 25(-2\mu_1 - \mu_2 + \mu_4 + 2\mu_5)$$

which is the linear contrast.

As a second example, we shall consider an unequally-spaced experiment with a single replicate of four rates of boron (B) fertilizer: 25, 50, 100 and 175 kg B/ha. Here the relevant picture is Figure 10.10. This shows that the linear contrast is

$$
\begin{aligned}
c &= -62.5\mu_1 - 37.5\mu_2 + 12.5\mu_3 + 87.5\mu_4 \\
&= 12.5(-5\mu_1 - 3\mu_2 + \mu_3 + 7\mu_4)
\end{aligned}
$$

This checks out with the linear component unit vector,

$$
U_2 = \frac{1}{\|x - \bar{x}\|}
\begin{bmatrix} x_1 - \bar{x} \\ \cdot \\ \cdot \\ x_k - \bar{x} \end{bmatrix}
= \frac{1}{\sqrt{13125}}
\begin{bmatrix} -62.5 \\ -37.5 \\ 12.5 \\ 87.5 \end{bmatrix}
= \frac{1}{\sqrt{84}}
\begin{bmatrix} -5 \\ -3 \\ 1 \\ 7 \end{bmatrix}
$$

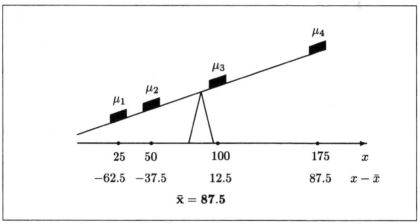

Figure 10.10: Seesaw analogy for an unequally spaced example. The fulcrum is at $\bar{x} = 87.5$.

The seesaw analogy can also be used in the unequal replications case. The necessary modifications are:

1. Be careful to average over all the x-values when calculating $\bar{x}$. For example, if the experiment has 37 experimental units, average over the 37 x-values.

2. On the seesaw, place as many weights μ_i as there are replicates at each x value.

Understanding the Quadratic Contrast

In the last section we saw that the linear contrast is basically a comparison of *low rates* of seeding, for example, versus *high rates* of seeding, with most emphasis placed on the lowest and highest rates. The quadratic contrast,

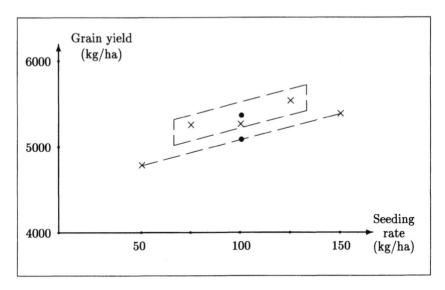

Figure 10.11: The quadratic contrast compares the average of the middle values with the average of the extreme values. The values are denoted by "×" and the averages by "•".

on the other hand, is roughly explained as a comparison of the *middle rates* with the *extreme rates*, as illustrated for Example D in Figure 10.11.

In Figure 10.11 a straight line response, with no curvature, would mean that the average of the middle rates would be equal to the average of the extreme rates. A difference between these averages indicates curvature. In our case, the quadratic contrast is

$$c \;=\; 2\mu_1 - \mu_2 - 2\mu_3 - \mu_4 + 2\mu_5$$

This can be rewritten as

$$c \;=\; \frac{\mu_1 + \mu_5}{2} \;-\; \frac{\mu_2 + 2\mu_3 + \mu_4}{4}$$

which is the simple average of the extreme rates minus a weighted average of the middle rates. This is estimated as

$$
\begin{aligned}
\hat{c} &= \frac{4787 + 5420}{2} - \frac{5227 + 2(5300) + 5500}{4} \\
&= 5103 - 5332 \\
&= -229
\end{aligned}
$$

That is, in Example D we observed *negative curvature*.

To examine another example, refer back to Table 10.10. This tells us that for four equally spaced, equally replicated x-values, the quadratic contrast is

$$c = \mu_1 - \mu_2 - \mu_3 + \mu_4$$

This can be rewritten as

$$c = \frac{\mu_1 + \mu_4}{2} - \frac{\mu_2 + \mu_3}{2}$$

Hence the quadratic contrast is just the difference between the simple average of the two extreme rates and the simple average of the two middle rates.

Mixing Contrasts of Various Types: An Example

We now illustrate how to write down a set of orthogonal contrasts when the treatments involving a range of x-values are only a part of the treatment structure. For example, in an investigation of the effects on strawberry production of mulching with straw or polythene, the experimental treatments could be as given in Table 10.11. Here it would make sense firstly to contrast the mulched treatments, 2 to 5, with the unmulched treatment, 1, via the contrast

$$c_1 = -4\mu_1 + \mu_2 + \mu_3 + \mu_4 + \mu_5$$

Secondly one could contrast the average of the straw mulched treatments, 2 to 4, with the polythene mulched treatment, 5, via the contrast

$$c_2 = \mu_2 + \mu_3 + \mu_4 - 3\mu_5$$

These contrasts are both of the class comparison type.

Number	Treatment name
1	No mulching (control)
2	Straw mulching at half the recommended rate
3	Straw mulching at the recommended rate
4	Straw mulching at twice the recommended rate
5	Polythene mulching

Table 10.11: Experiment in which only treatments 2 to 4 constitute a range of x-values.

The last two contrasts, c_3 and c_4, which can be orthogonally specified are the linear and quadratic contrasts within the straw mulched class of treatments, 2 to 4. The rates of straw mulching are in the proportions 1,2 and 4, with $\bar{x} = 2\frac{1}{3}$, so the linear contrast is

$$c_3 = (-1\tfrac{1}{3})\mu_2 + (-\tfrac{1}{3})\mu_3 + (1\tfrac{2}{3})\mu_4$$
$$\text{or} \quad c_3 = -4\mu_2 - \mu_3 + 5\mu_4$$

The quadratic contrast is

$$c_4 = 2\mu_2 - 3\mu_3 + \mu_4$$

This could be derived using Gram-Schmidt orthogonalization, or by looking up the standard tables published by Little and Hills (1978).

Notice that the linear contrast basically contrasts low with high rates of straw mulching, whereas the quadratic contrast compares the middle rate with the two extreme rates. The full list of four orthogonal contrasts is:

$$
\begin{aligned}
c_1 &= -4\mu_1 + & \mu_2 + & \mu_3 + & \mu_4 + & \mu_5 & \text{(mulch vs. no mulch)}\\
c_2 &= & \mu_2 + & \mu_3 + & \mu_4 - & 3\mu_5 & \text{(straw vs. polythene)}\\
c_3 &= & -4\mu_2 - & \mu_3 + 5\mu_4 & & & \text{(linear trend in straw)}\\
c_4 &= & 2\mu_2 - & 3\mu_3 + & \mu_4 & & \text{(quadratic curvature in straw)}
\end{aligned}
$$

Orthogonal Polynomial Components

For the reader who is interested in writing down the polynomial components of order three or more, the recursive pattern is as follows:

Constant component:
$$p_0(x) = 1$$

Linear component:
$$p_1(x) = (x - \bar{x}) - \frac{X_2.T_1}{\|T_1\|^2}p_0(x)$$

Quadratic component:
$$p_2(x) = (x - \bar{x})^2 - \frac{X_3.T_2}{\|T_2\|^2}p_1(x) - \frac{X_3.T_1}{\|T_1\|^2}p_0(x)$$

Cubic component:
$$p_3(x) = (x - \bar{x})^3 - \frac{X_4.T_3}{\|T_3\|^2}p_2(x) - \frac{X_4.T_2}{\|T_2\|^2}p_1(x) - \frac{X_4.T_1}{\|T_1\|^2}p_0(x)$$

Quartic component:
$$p_4(x) = (x - \bar{x})^4 - \frac{X_5.T_4}{\|T_4\|^2}p_3(x) - \frac{X_5.T_3}{\|T_3\|^2}p_2(x) - \frac{X_5.T_2}{\|T_2\|^2}p_1(x) - \frac{X_5.T_1}{\|T_1\|^2}p_0(x)$$

This pattern is revealed by a simple rearrangement of the Gram-Schmidt orthogonalized vectors. For example, the vector of $p_2(x_i)$ values, $T_3 = X_3 - (X_3.U_2)U_2 - (X_3.U_1).U_1$, is rearranged by substituting $U_1 = T_1/\|T_1\|$ and $U_2 = T_2/\|T_2\|$. The resulting coefficients, $X_i.T_j/\|T_j\|^2$, can be calculated readily in Minitab, using commands similar to those given in Table 10.10.

In the display we have written each expression in full. In the linear component, however, $X_2.T_1$ is always zero, so the second term drops out. Similarly, in Example D we found that for each polynomial component every second term dropped out, thanks to the equality of replication and spacing.

Confidence Intervals for β_i Values

We now work out 95% confidence intervals for the coefficients $\beta_0, \beta_1, \ldots, \beta_{k-1}$ in the orthogonal polynomial

$$y = \beta_0 + \beta_1(x - \bar{x}) + \beta_2 p_2(x) + \cdots + \beta_{k-1} p_{k-1}(x)$$

Recall from Table 10.2 and the discussion preceding Figure 10.5 that the estimates of the coefficients are

$$b_0 = (y.U_1)/\|T_1\| = y.U_1/\sqrt{N} = \bar{y}_{..}$$

$$b_1 = (y.U_2)/\|T_2\| = (y.U_2)/\|x - \bar{x}\| = \sum_{i=1}^{N} y_i(x_i - \bar{x})/\|x - \bar{x}\|$$

$$\vdots$$

$$b_{k-1} = (y.U_k)/\|T_k\|$$

In general,

$$b_l \|T_{l+1}\| = y.U_{l+1}$$

which comes from an $N(\beta_l \|T_{l+1}\|, \sigma^2)$ distribution. Thus the random variable $(B_l - \beta_l)\|T_{l+1}\|$ comes from an $N(0, \sigma^2)$ distribution. This will be the numerator for our t statistic.

The denominator is the usual $\sqrt{S^2}$, where S^2 is the average of $N - k$ squares of $N(0, \sigma^2)$ random variables. Our t statistic is therefore $t = (B_l - \beta_l)\|T_{l+1}\|/S$. To obtain our 95% confidence interval we gamble that the realized value is within the limits

$$t_{N-k}(.025) \leq \frac{(b_l - \beta_l)\|T_{l+1}\|}{s} \leq t_{N-k}(.975)$$

This leads to the 95% confidence interval

$$b_l - \frac{s}{\|T_{l+1}\|} t_{N-k}(.975) \leq \beta_l \leq b_l + \frac{s}{\|T_{l+1}\|} t_{N-k}(.975)$$

For $l = 0$ this becomes

$$\bar{y} - \frac{s}{\sqrt{N}} t_{N-k}(.975) \leq \beta_0 \leq \bar{y} + \frac{s}{\sqrt{N}} t_{N-k}(.975)$$

For $l = 1$ this becomes

$$b_1 - \frac{s}{\|x - \bar{x}\|} t_{N-k}(.975) \leq \beta_1 \leq b_1 + \frac{s}{\|x - \bar{x}\|} t_{N-k}(.975)$$

An example has already been calculated in §10.1.

10.3 A Case Study

To further consolidate the methods of this chapter we now analyze a typical factorial experiment involving polynomial contrasts. The data is by courtesy of John Eiseman of Lincoln College, Canterbury, New Zealand.

Description

The aim of the experiment was to investigate the effect of two factors, treatment by fungicide and sowing time, on the germination rate of Golden Queen peach stones. Stones were planted at five-weekly intervals, giving four sowing times of April 15, May 22, June 29 and August 5. At each of these times some stones were planted after being treated with the fungicide *thiram*, while the remainder were planted untreated, so requiring a 4×2 factorial design with $k = 8$ treatments. When the experiment was laid out five replicates of each treatment were included, requiring a total of $5 \times 8 = 40$ plots. Treatments were assigned at random to these plots, and 100 peach stones planted in each plot. From each plot the number, hence percentage, germinating was recorded.

Questions of Interest

 (i) What is the relationship between the percentage germination and the time of planting?

 (ii) Does the fungicide affect germination?

(iii) Does the presence or absence of fungicide affect the relationship between germination and time of planting?

Results

The results of the experiment are summarized in Table 10.12.

Treatment		Germination					Treatment
Month	Fungicide	percentage					means
April	−	45	56	50	41	58	50.0
May	−	59	52	60	62	57	58.0
June	−	47	51	65	49	41	50.6
August	−	34	46	37	31	44	38.4
April	+	54	49	54	48	48	50.6
May	+	67	63	51	54	66	60.2
June	+	45	56	44	32	40	43.4
August	+	29	37	41	50	44	40.2

Table 10.12: Results of the Golden Queen peach seedling experiment.

Analysis

A quick assessment of the results is provided by the scattergram of Figure 10.12. At a glance we see that May appears to be a good time to sow, with a reduction in germination as the Southern Hemisphere's winter approaches. Also, fungicide appears to have little effect.

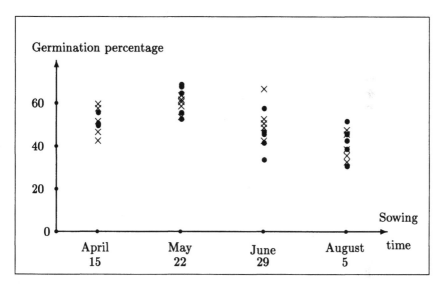

Figure 10.12: Scattergram showing the manner in which the percentage germination depends on the time of sowing of the peach stone. Untreated plots are coded as "×", fungicide treated plots as "•".

For a proper statistical analysis we shall construct an orthogonal set of contrasts. For factor A, fungicide, there is only one possible contrast, that of no fungicide versus fungicide. For factor B, the time of sowing, we shall specify linear, quadratic and cubic contrasts as follows:

$$
\begin{aligned}
\text{Linear contrast} &= -3\mu_1 - \mu_2 + \mu_3 + 3\mu_4 \\
\text{Quadratic contrast} &= \mu_1 - \mu_2 - \mu_3 + \mu_4 \\
\text{Cubic contrast} &= -\mu_1 + 3\mu_2 - 3\mu_3 + \mu_4
\end{aligned}
$$

These coefficients were obtained from Table 10.10. The full set of main effect and interaction contrasts can now be generated as described in Chapter 9. The resulting contrasts are given in Table 10.13.

	Main effect contrasts			Interaction contrasts		
	A	B			A × B		
		Lin	Quad	Cub	Fung. × lin	Fung. × quad	Fung. × cub
	c_1	c_2	c_3	c_4	c_5	c_6	c_7
μ_1	-1	-3	1	-1	3	-1	1
μ_2	-1	-1	-1	3	1	1	-3
μ_3	-1	1	-1	-3	-1	1	3
μ_4	-1	3	1	1	-3	-1	-1
μ_5	1	-3	1	-1	-3	1	-1
μ_6	1	-1	-1	3	-1	-1	3
μ_7	1	1	-1	-3	1	-1	-3
μ_8	1	3	1	1	3	1	1

Table 10.13: Factorial contrasts for peach seedling data.

The analysis now proceeds in the usual manner. The resulting ANOVA table is given in Table 10.14.

Source of Variation	df	SS	MS	F
Treatments	7	2168.38		
Fungicide	1	4.23	4.23	.09(ns)
Time (linear)	1	1017.01	1017.01	21.07(**)
Time (quadratic)	1	680.62	680.62	14.10 (**)
Time (cubic)	1	320.04	320.04	6.63 (*)
Fung. × time (lin)	1	4.21	4.21	.09 (ns)
Fung. × time (quad)	1	34.22	34.22	.71 (ns)
Fung. × time (cubic)	1	108.04	108.04	2.23 (ns)
Error	32	1544.40	48.26	
Total	39	3712.78		

Table 10.14: ANOVA table for peach seedling experiment.

Conclusions

It came as a surprise to the researchers to find that the fungicide had no influence on germination. However, this eliminated any possibility of interactions, so effectively doubled the replication of the sowing time treatments. The relationship between percentage germination and sowing time turned out to be relatively complex, with a polynomial of order three required for its description.

Report on Study

A good way to report the experimental results is to present the scattergram shown in Figure 10.12 with the addition of a cubic fitted to the main effect means for sowing time. The equation of the cubic is

$$y = -4.7 + 86.5x - 35.75x^2 + 4.217x^3,$$

assuming the times of sowing, x, are coded as $1, 2, 3, 4$.

An alternative is to present either a table of treatment means or main effect means. We shall choose the latter option since the fungicide effect has been minimal; this is given here in Table 10.15.

Our results indicate that August, in late winter, was too late for successful germination, and that May seemed to be the optimum time. The

	Percentage Germination
Time of sowing	
April 15	50.3
May 22	59.1
June 29	47.0
August 5	39.3
SED	3.15
Contrasts:	
Linear	**
Quadratic	**
Cubic	*
Fungicide	
Nil	49.3
Fungicide	48.6
SED	2.23
Significance	ns
Interaction contrasts	None significant

Table 10.15: Main effect means, peach seedling experiment.

difference in germination percentage between April and May was 8.8, with a standard error of 3.15, so this appeared to be a genuine difference.

10.4 Summary

In this chapter we have described how to derive and use orthogonal polynomial contrasts and orthogonal polynomial components. The need for such methods arises when all or some of the treatments in an experiment are associated with an accurately measured continuous variable, which we call x. The treatments, for example, might be determined by the intensity, x, of a laser beam being used to remove port wine skin disease. Here the observations would comprise the reduction in area of the complaint following treatment. Of interest is the curve which relates the treatment means to the measured x-values.

With k treatments we can fit a polynomial curve in the form

$$y = \alpha_0 + \alpha_1(x - \bar{x}) + \alpha_2(x - \bar{x})^2 + \cdots + \alpha_{k-1}(x - \bar{x})^{k-1}$$

By evaluating the component polynomials, 1, , $x - \bar{x}$, $(x - \bar{x})^2$ and so on at the successive x-values, we can form a non-orthogonal coordinate system for the model space, $X_1, X_2, \ldots, X_k$.

We orthogonalize these vectors to form $T_1, T_2, \ldots, T_k$ which in turn correspond to an *orthogonal* breakup of the curve of true mean values as

$$y = \beta_0 + \beta_1(x - \bar{x}) + \beta_2 p_2(x) + \cdots + \beta_{k-1} p_{k-1}(x)$$

Here $p_0(x) = 1$ is the *constant* component,

                $p_1(x) = x - \bar{x}$ is the *linear* component,

                $p_2(x)$ is the *quadratic* component, and so on.

The appropriate orthogonal coordinate system $U_1, \ldots, U_k$ for the model space is formed by dividing each of $T_1, \ldots, T_k$ by its length. The contrasts which correspond to each of $U_2, \ldots, U_k$ are called *polynomial contrasts*: linear, quadratic, cubic, and so on. For standard spacings of the x-values these contrasts are given in Table 10.10.

The fitted model is

$$y \quad = \quad (y.U_1)U_1 + (y.U_2)U_2 + \cdots + (y.U_k)U_k + (y - \bar{y}_{i.})$$

where the model vector, $\bar{y}_{i.}$, is broken into constant, linear, ... and $(k-1)^{th}$ order terms. To test whether the lth order polynomial coefficient β_l is zero, we use the test statistic $(y.U_{l+1})^2/s^2$, which comes from an $F_{1,N-k}$ distribution under the null hypothesis that the coefficient $\beta_l = 0$.

The 95% confidence interval for the coefficient β_l is

$$b_l \quad \pm \quad \frac{s}{\|T_{l+1}\|} t_{N-k}(.975)$$

For a more detailed summary of this chapter we refer the reader to the start of §10.2.

Exercises

(10.1) A long term pasture trial was set up on a leached pakihi soil to determine the phosphate "maintenance" requirement under sheep grazing. Seven treatments, 0, 15, 30, 45, 60, 75 and 120 kg of phosphate/ha, were allocated in a completely random manner to 21 plots, with three replicates per treatment. Pasture yields, in kg dry matter/ha, from the first year of the trial were as follows:

Treatment (kg P/ha)	Yield (kg DM/ha)		
0	6521	5470	6090
15	5871	6114	6551
30	6976	7137	8103
45	8459	6795	6947
60	7041	7244	6753
75	6854	8163	10089
120	8619	8508	7627

(a) Graph the basic data and treatment means, with yield on the y-axis and rate of phosphate on the x-axis.

(b) Work out the unit vector U_2 corresponding to the linear component, using the rates given. Express the vector with integer entries. [Hint: $\bar{x} = 15 \times \frac{23}{7}$]

(c) Calculate the corresponding linear contrast.

(d) The linear component of the treatment sum of squares is $(y.U_2)^2$. Find this using the Regress command in Minitab, or by hand calculation.

(e) Use the Oneway command in Minitab to obtain an estimate of s^2, the residual (or error) mean square.

(f) What is the F value for testing whether the linear component is zero? Is the linear component significant?

(g) Summarize your sums of squares and F value in an ANOVA table.

(h) Calculate the equation of the regression line and add this to your graph. This can be done either by hand calculation or by putting the rates of P, one for each plot, into Minitab and Regressing the observation vector y onto this vector of x-values.

(i) Do the assumptions for the analysis appear reasonable from a visual check of your scattergram in (a)? Did the application of phosphate increase pasture yield in the first year of the trial?

(j) Calculate the 95% confidence interval for the true slope, β_1, of the line.

(10.2) Consider an experiment in which the four treatments are four equally-spaced, equally-replicated *rates*, such as dose rates of a compound used in the treatment of arthritis. We'll code the rates to be 1, 2, 3, 4 for ease of calculation and work with just one replicate for simplicity. Hence our vector of x-values is $[x_1, x_2, x_3, x_4]^T = [1, 2, 3, 4]^T$.

(a) Calculate the vectors X_1, X_2, X_3 and X_4 which are

$$
\begin{bmatrix} 1 \\ 1 \\ 1 \\ 1 \end{bmatrix}
\quad
\begin{bmatrix} x_1 - \bar{x} \\ \cdot \\ \cdot \\ x_4 - \bar{x} \end{bmatrix}
\quad
\begin{bmatrix} (x_1 - \bar{x})^2 \\ \cdot \\ \cdot \\ (x_4 - \bar{x})^2 \end{bmatrix}
\quad
\begin{bmatrix} (x_1 - \bar{x})^3 \\ \cdot \\ \cdot \\ (x_4 - \bar{x})^3 \end{bmatrix}
$$

These span the 4-space but are not all mutually orthogonal.

(b) Follow the orthogonalization method described in §10.1 and calculate the orthogonal vectors T_1, T_2, T_3 and T_4 using the notation of this chapter.

(c) Convert these to an orthogonal set of unit vectors U_1, U_2, U_3 and U_4, expressed in their simplest terms.

(d) Write down the corresponding polynomial contrasts. Do your answers match those given in Table 10.10?

(10.3) An experiment was carried out to investigate the effect of seeding rate on grain yield for two cultivars of wheat, Kopara and Karamu. The treatments had a 2×3 factorial structure, and were numbered one to six as follows:

| | Seeding rate | | |
Cultivar	50 kg/ha	100 kg/ha	150 kg/ha
Kopara	μ_1	μ_2	μ_3
Karamu	μ_4	μ_5	μ_6

(a) Write down a complete set of orthogonal contrasts involving $\mu_1, \ldots, \mu_6$ which is appropriate for this set of treatments.

(b) The experiment was laid down in as a completely randomized design with four replicates. The resulting grain yields, converted to tonnes/ha, were as follows:

Kopara, 50 kg/ha	3.64	3.74	4.02	3.72
Kopara, 100 kg/ha	4.14	3.84	3.94	3.97
Kopara, 150 kg/ha	4.08	4.14	4.40	4.17
Karamu, 50 kg/ha	3.80	3.50	4.34	3.79
Karamu, 100 kg/ha	4.43	4.25	4.65	4.43
Karamu, 150 kg/ha	4.42	4.92	4.34	4.52

Draw a scattergram of the yields y against the seeding rates x, using different symbols to distinguish Kopara from Karamu (for example, use "×" for Kopara and "•" for Karamu).

(c) Carry out an analysis of variance involving the contrasts listed in (a), using Minitab. Summarize your results in an ANOVA table.

(d) Are the assumptions for the analysis reasonable on the basis of a quick visual check of your scattergram?

(e) Without any formal calculation draw an approximate line of best fit through the data for Kopara, and again for Karamu.

(f) Visually, are these lines different in elevation? Do they differ in slope?

(g) From the ANOVA table in (c), extract the F tests for the differences in elevation and the differences in slope. Are the results significant? If so, at what level of significance, 5% or 1%?

(10.4) An experiment was carried out to test the effectiveness of a new chemical treatment, Code X, for the control of stripe rust disease in wheat. The optimum rate for the new chemical was thought to be 100g of active ingredient (a.i.) per hectare, so the rates included were one half, one, and two times this rate. For comparative purposes three standard chemicals were included at their recommended rates. An untreated control was also included to measure the baseline level of disease in the wheat crop. These seven treatments were set out in the field as a completely randomized design with four replicates.

Just prior to harvest an assessment was made of the severity of stripe rust on the ears of wheat in each plot. The resulting data were as follows:

Treatments	Ear rust severity			
Code X, 50g a.i./ha	10.6	11.3	26.2	11.4
Code X, 100g a.i./ha	9.7	10.3	7.6	15.8
Code X, 200g a.i./ha	10.4	12.3	4.1	16.2
Bayleton	11.3	16.9	17.3	15.9
Tilt	1.4	6.9	22.1	13.2
Corbel	7.0	13.3	15.3	19.7
Untreated (control)	20.6	25.7	33.3	24.3

(a) Write down the three orthogonal contrasts corresponding to the following comparisons:

1. Treated versus untreated.

2. Code X (average of 3 rates) versus Standards (average of 3 standards).

3. Linear trend within Code X chemical.

(b) Fit the full model using the Oneway command in Minitab. Write down s^2, your estimate of σ^2, from the printout.

(c) Calculate the contrast sums of squares for the contrasts in (a), using the Regress command in Minitab. Calculate the corresponding F values, summarizing your calculations in an ANOVA table.

(d) Summarize your results in a table which includes treatment means, SED and significance of contrasts.

(e) Was the new chemical any more or less effective than the standard chemicals? As an average over all the treated plots, what was the percentage reduction in stripe rust severity, compared to the control, in this experiment?

(10.5) An experiment was carried out to investigate the best method and best time of grafting a new improved peach cultivar onto Golden Queen rootstock. Three methods were tried at three different times, so the treatment design was a 3×3 factorial. The experiment was laid out in a completely random manner with four replicates. Each plot consisted of twenty stems of rootstock, each of which was grafted with a bud of the new cultivar. The numbers of buds which successfully "took" in the various plots were as follows:

Method	Time of grafting		
	8 February	4 March	28 March
Chip	20 19 19 19	16 12 19 15	12 16 13 15
T, with wood	17 18 19 20	19 13 19 13	17 16 17 17
T, minus wood	16 20 20 20	14 19 15 17	17 19 14 17

(a) Write down an appropriate set of eight orthogonal contrasts among the treatment means.

(b) Convert the data to percentages and statistically analyze, using Minitab for the conversion and analysis. Write out the ANOVA table.

(c) Present your results in a table of main effect means and so on, as shown in Table 10.14.

(d) What are your conclusions?

(10.6) (a) Calculate the linear and quadratic contrasts when the treatments are rates, spaced as 1, 2, 6. Show your working.

(b) Calculate the linear contrast when the spacings are:

(i) 1, 2, 7, 8

(ii) 2, 4, 10, 15, 34

(10.7) An experiment investigating the optimum depth of water in rice fields is being planned. The main question of interest is "will the rice yield increase if the depth of water is increased from four to eight inches?" Two experimental designs, each requiring 12 plots, are under consideration:

Design 1   Two treatments (four and eight inches of water) with six replicates laid down in a completely randomized design.

Design 2   Three treatments (four, six and eight inches of water) with four replicates laid down in a completely randomized design.

For the first design the mean rate of change of yield with depth of water is $(\mu_2 - \mu_1)/4$. For the second design the mean rate of change of yield with depth of water is $(\mu_3 - \mu_1)/4$.

In thinking about which design is better, imagine that the following data is observed:

Design 1		Design 2	
Treatment	Mean yield	Treatment	Mean yield
4 inches	6.4	4 inches	6.4
8 inches	8.4	6 inches	7.1
		8 inches	8.4
$s^2 = 2.0$			
		$s^2 = 2.0$	

(a) Estimate the true slope $\beta_1 = (\mu_2 - \mu_1)/4$ and $\beta_1 = (\mu_3 - \mu_1)/4$ for the two designs.

(b) Estimate the standard error of the estimator of the true slope, β_1, for each of the two designs.

(c) Calculate the 95% confidence interval for the true slope β_1 for each design.

(d) Which design is superior in terms of the accuracy with which the true slope is estimated?

Solutions to the Reader Exercises

(10.1)

$$
\begin{aligned}
T_3 &= X_3 - (X_3.U_1)U_1 - (X_3.U_2)U_2 \\
&= X_3 - (X_3.U_1)U_1 \quad \text{since } X_3 \text{ is orthogonal to } U_2
\end{aligned}
$$

$$
=
\begin{bmatrix}
(-50)^2 \\
\cdot \\
(-25)^2 \\
\cdot \\
0^2 \\
\cdot \\
25^2 \\
\cdot \\
50^2 \\
\cdot
\end{bmatrix}
-
\left\{
\left(
\begin{bmatrix}
(-50)^2 \\
\cdot \\
(-25)^2 \\
\cdot \\
0^2 \\
\cdot \\
25^2 \\
\cdot \\
50^2 \\
\cdot
\end{bmatrix}
\cdot \frac{1}{\sqrt{30}}
\begin{bmatrix}
1 \\
\cdot \\
1 \\
\cdot \\
1 \\
\cdot \\
1 \\
\cdot \\
1 \\
\cdot
\end{bmatrix}
\right)
\right\}
\frac{1}{\sqrt{30}}
\begin{bmatrix}
1 \\
\cdot \\
1 \\
\cdot \\
1 \\
\cdot \\
1 \\
\cdot \\
1 \\
\cdot
\end{bmatrix}
$$

$$
=
\begin{bmatrix}
2500 \\
\cdot \\
625 \\
\cdot \\
0 \\
\cdot \\
625 \\
\cdot \\
2500 \\
\cdot
\end{bmatrix}
-
\begin{bmatrix}
1250 \\
\cdot \\
1250 \\
\cdot \\
1250 \\
\cdot \\
1250 \\
\cdot \\
1250 \\
\cdot
\end{bmatrix}
=
\begin{bmatrix}
1250 \\
\cdot \\
-625 \\
\cdot \\
-1250 \\
\cdot \\
-625 \\
\cdot \\
1250 \\
\cdot
\end{bmatrix}
=
625
\begin{bmatrix}
2 \\
\cdot \\
-1 \\
\cdot \\
-2 \\
\cdot \\
-1 \\
\cdot \\
2 \\
\cdot
\end{bmatrix}
$$

(10.2)

$$
\begin{aligned}
T_4 &= X_4 - (X_4 . U_1)U_1 - (X_4 . U_2)U_2 - (X_4 . U_3)U_3 \\
&= X_4 - (X_4 . U_2)U_2 \quad \text{since } X_4 \text{ is orthogonal to } U_1 \text{ and } U_3
\end{aligned}
$$

$$
=
\begin{bmatrix}
(-50)^3 \\
\cdot \\
(-25)^3 \\
\cdot \\
0^3 \\
\cdot \\
25^3 \\
\cdot \\
50^3 \\
\cdot
\end{bmatrix}
-
\frac{3187500}{\sqrt{60}} \frac{1}{\sqrt{60}}
\begin{bmatrix}
-2 \\
\cdot \\
-1 \\
\cdot \\
0 \\
\cdot \\
1 \\
\cdot \\
2 \\
\cdot
\end{bmatrix}
=
\begin{bmatrix}
-125000 \\
\cdot \\
-15625 \\
\cdot \\
0 \\
\cdot \\
15625 \\
\cdot \\
125000 \\
\cdot
\end{bmatrix}
-
\begin{bmatrix}
-106250 \\
\cdot \\
-53125 \\
\cdot \\
0 \\
\cdot \\
53125 \\
\cdot \\
106250 \\
\cdot
\end{bmatrix}
$$

$$= \begin{bmatrix} -18750 \\ \cdot \\ 37500 \\ \cdot \\ 0 \\ \cdot \\ -37500 \\ \cdot \\ 18750 \\ \cdot \end{bmatrix} = 18750 \begin{bmatrix} -1 \\ \cdot \\ 2 \\ \cdot \\ 0 \\ \cdot \\ -2 \\ \cdot \\ 1 \\ \cdot \end{bmatrix}. \quad \text{Hence} \quad U_4 = \frac{T_4}{\|T_4\|} = \frac{1}{\sqrt{60}} \begin{bmatrix} -1 \\ \cdot \\ 2 \\ \cdot \\ 0 \\ \cdot \\ -2 \\ \cdot \\ 1 \\ \cdot \end{bmatrix}$$

(10.3)

$$T_5 = X_5 - (X_5.U_1)U_1 - (X_5.U_2)U_2 - (X_5.U_3)U_3 - (X_5.U_4)U_4$$
$$= X_5 - (X_5.U_1)U_1 - (X_5.U_3)U_3 \quad \text{since } X_5 \text{ is orthogonal to } U_2 \text{ and } U_4$$

$$= \begin{bmatrix} (-50)^4 \\ \cdot \\ (-25)^4 \\ \cdot \\ 0^4 \\ \cdot \\ 25^4 \\ \cdot \\ 50^4 \\ \cdot \end{bmatrix} - \frac{79687500}{\sqrt{30}} \frac{1}{\sqrt{30}} \begin{bmatrix} 1 \\ \cdot \\ 1 \\ \cdot \\ 1 \\ \cdot \\ 1 \\ \cdot \\ 1 \\ \cdot \end{bmatrix} - \frac{145312500}{\sqrt{84}} \frac{1}{\sqrt{84}} \begin{bmatrix} 2 \\ \cdot \\ -1 \\ \cdot \\ -2 \\ \cdot \\ -1 \\ \cdot \\ 2 \\ \cdot \end{bmatrix}$$

$$= \begin{bmatrix} 133929 \\ \cdot \\ -535714 \\ \cdot \\ 803571 \\ \cdot \\ -535714 \\ \cdot \\ 133929 \\ \cdot \end{bmatrix} = 133929 \begin{bmatrix} 1 \\ \cdot \\ -4 \\ \cdot \\ 6 \\ \cdot \\ -4 \\ \cdot \\ 1 \\ \cdot \end{bmatrix}, \quad \text{so} \quad U_5 = \frac{T_5}{\|T_5\|} = \frac{1}{\sqrt{420}} \begin{bmatrix} 1 \\ \cdot \\ -4 \\ \cdot \\ 6 \\ \cdot \\ -4 \\ \cdot \\ 1 \\ \cdot \end{bmatrix}$$

(10.5) The Minitab output enables us to rewrite Table 10.2 with estimated parameters as follows:

Fitted model vector	Equation for fitted model in orthogonal form
$28737U_1 = 5247T_1$	5247
$28737U_1 + 1193U_2$ $= 5247T_1 + 6.16T_2$	$5247 + 6.16(x - 100)$
$28737U_1 + 1193U_2 - 598U_3$ $= 5247T_1 + 6.16T_2 - .104T_3$	$5247 + 6.16(x - 100) - .104\left[(x - 100)^2 - 1250\right]$
$28737U_1 + 1193U_2 - 598U_3$ $+67U_4 = 5247T_1 + \cdots$	$5247 + 6.16(x - 100) - .104\left[(x - 100)^2 - 1250\right]$ $+ .00046p_3(x)$
$28737U_1 + 1193U_2 - 598U_3$ $+67U_4 - 263U_5 = \cdots$	$5247 + 6.16(x - 100) - .104\left[(x - 100)^2 - 1250\right]$ $+ .00046p_3(x) - .000096p_4(x)$

Note that the quadratic equation is the same as that given earlier.

The corresponding equations for the nonorthogonal fits are as follows:

Equation for fitted model in nonorthogonal form
5247
$5247 + 6.16(x - 100)$
$5377 + 6.16(x - 100) - .104(x - 100)^2$
$5377 + 5.18(x - 100) - .104(x - 100)^2 + .00046(x - 100)^3$
$5300 + 5.18(x - 100) + .161(x - 100)^2 + .00046(x - 100)^3 - .000096(x - 100)^4$

Chapter 11

Pairwise Comparisons

The simplest type of contrast occurs when we compare just one population mean with another. This chapter is devoted to such contrasts, termed *pairwise comparisons*. In Example E of §7.2, which involved four fertilizer treatments for sunflowers, such contrasts occur naturally. We can assess, for example, the effect of omitting magnesium from the fertilizer solution by comparing treatment one, the complete solution, with treatment two, the solution lacking magnesium.

When used to test prespecified hypotheses, pairwise contrasts are no different from any other type. However, pairwise contrasts are occasionally used in an extravagant manner to test all possible hypotheses of the form $H_0 : \mu_i = \mu_j$. Such overuse of pairwise contrasts for testing poorly specified hypotheses has made statisticians uneasy, and has led to a search for a more conservative procedure for dealing with the case of all pairwise contrasts when no firm hypotheses are prespecified. To date, dozens of more conservative procedures, commonly referred to as *multiple comparison procedures*, have been devised. However, none of these procedures is commonly accepted as the best.

We begin the chapter with an analysis in §1 of Example E, a study in which three pairwise contrasts have been prespecified. We then introduce the Least Significant Difference (LSD) test in §2, a test equivalent to the pairwise contrast F test. In §3 we analyze Example F, a study in which all pairwise comparisons are required, and as well discuss multiple comparison procedures. The chapter closes with a summary in §4, a class exercise, and general exercises.

11.1   Analyzing Example E

We now statistically analyze Example E of §7.2. In that experiment twelve sunflowers were used to test four fertilizer solutions. The first solution included magnesium, nitrogen and trace nutrients, the second omitted magnesium, the third solution omitted nitrogen and the fourth omitted the trace nutrients. The dry matter of the sunflowers was measured after nine weeks growth. The aim of the experiment is to determine whether a lack of any one constituent affected sunflower growth.

Data

The data is reproduced in Table 11.1. The corresponding observation vector is

$$y = [y_{11}, y_{12}, \ldots, y_{42}, y_{43}]^T = [1172, 750, \ldots, 243, 263]^T$$

Treatment				Mean
A. Solution containing magnesium,				
nitrogen and trace nutrients	1172	750	784	902
B. Solution lacking magnesium	67	95	59	74
C. Solution lacking nitrogen	148	234	92	158
D. Solution lacking trace nutrients	297	243	263	268

Table 11.1: Weight of sunflowers in grams, Example E.

Model

We assume that our four populations are normally distributed, with means μ_1 to μ_4 and common variance σ^2. The model vector is $[\mu_1, \ldots, \mu_4]^T$, and the model space M is a 4 dimensional subspace of 12-space.

Test Hypotheses

The hypotheses of interest are:

$$H_0 : c_1 = \mu_1 - \mu_2 = 0 \quad \text{(magnesium effect} = 0)$$

$$H_0 : c_2 = \mu_1 - \mu_3 = 0 \quad \text{(nitrogen effect} = 0)$$

$$H_0 : c_3 = \mu_1 - \mu_4 = 0 \quad \text{(trace nutrients effect} = 0)$$

The directions associated with these hypotheses are:

$$
U_2 = \frac{1}{\sqrt{6}}
\begin{bmatrix}
1 \\ 1 \\ 1 \\ -1 \\ -1 \\ -1 \\ 0 \\ 0 \\ 0 \\ 0 \\ 0
\end{bmatrix}
, \quad
U_3 = \frac{1}{\sqrt{6}}
\begin{bmatrix}
1 \\ 1 \\ 1 \\ 0 \\ 0 \\ 0 \\ -1 \\ -1 \\ -1 \\ 0 \\ 0
\end{bmatrix}
, \quad
U_4 = \frac{1}{\sqrt{6}}
\begin{bmatrix}
1 \\ 1 \\ 1 \\ 0 \\ 0 \\ 0 \\ 0 \\ 0 \\ -1 \\ -1 \\ -1
\end{bmatrix}
$$

The angle θ between any two of these vectors is 60^0, since for example, $\cos\theta = U_2.U_3 = (1+1+1)/(\sqrt{6}\sqrt{6}) = 0.5$. This is illustrated in Figure 11.1.

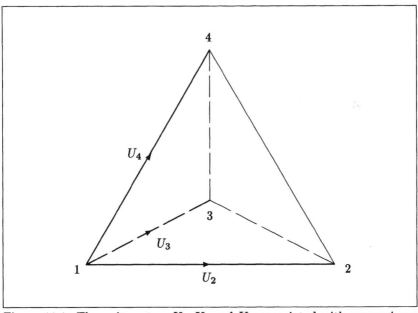

Figure 11.1: The unit vectors U_2, U_3 and U_4, associated with comparisons of treatment 1 with treatments 2, 3 and 4, comprise three of the edges of a tetrahedron, with vertices corresponding to treatments. The remaining three edges correspond to comparisons among treatments 2, 3 and 4.

Hence U_2, U_3 and U_4 do not make up an orthogonal coordinate system for the treatment space, although they do make up an oblique system. Consequently c_1, c_2 and c_3 are a set of nonorthogonal contrasts.

We now have all the raw materials for our analysis. We have a model space M, directions associated with our hypotheses, U_2, U_3 and U_4, and an observation vector, y.

Fitting the Model

In the usual manner we project the observation vector y onto the model space M and obtain the fitted model vector,

$$\bar{y}_{i.} = [902, \ldots, 74, \ldots, 158, \ldots, 268, \ldots]^T$$

the vector of treatment means.

When written in corrected form the fitted model is

$$
\begin{array}{ccccc}
y - \bar{y}_{..} & = & (\bar{y}_{i.} - \bar{y}_{..}) & + & (y - \bar{y}_{i.}) \\
\text{Corrected} & & \text{Treatment} & & \text{Error} \\
\text{observation vector} & & \text{vector} & & \text{vector}
\end{array}
$$

$$
\begin{bmatrix}
1172 - 350 \\
750 - 350 \\
784 - 350 \\
67 - 350 \\
95 - 350 \\
59 - 350 \\
148 - 350 \\
234 - 350 \\
92 - 350 \\
297 - 350 \\
243 - 350 \\
263 - 350
\end{bmatrix}
=
\begin{bmatrix}
552 \\
\cdot \\
\cdot \\
-277 \\
\cdot \\
\cdot \\
-192 \\
\cdot \\
\cdot \\
-82 \\
\cdot \\
\cdot
\end{bmatrix}
+
\begin{bmatrix}
270 \\
-152 \\
-118 \\
-7 \\
21 \\
-15 \\
-10 \\
76 \\
-66 \\
29 \\
-25 \\
-5
\end{bmatrix}
$$

These three vectors lie in 11, 3 and 8 dimensional subspaces respectively, as illustrated in Figure 11.2.

Testing the Hypotheses

Our first hypothesis was that the omitting of magnesium has no effect on the growth of sunflowers, or equivalently, that the contrast $c_1 = \mu_1 - \mu_2$ is zero. In order to confirm or deny this hypothesis, we check whether the squared distance $(y.U_2)^2$ is comparable to, or larger than, the average of the corresponding squared distances in the error space. The test statistic is

$$F = \frac{(y.U_2)^2}{\left[(y.U_5)^2 + \cdots + (y.U_{12})^2\right]/8} = \frac{(y.U_2)^2}{\|y - \bar{y}_{i.}\|^2/8} = \frac{1029204}{15296} = 67.29$$

where $U_5, \ldots, U_{12}$ are coordinate axes for the error space.

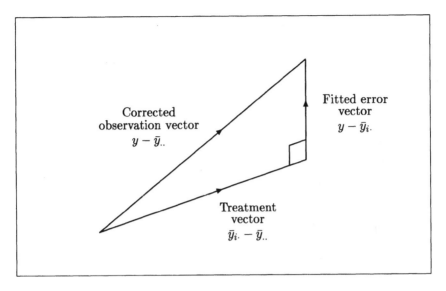

Figure 11.2: The usual orthogonal decomposition.

For our other two hypotheses, the test statistics are obtained by substituting $(y.U_3)^2$ and $(y.U_4)^2$ into the numerator of the F ratio. Results are summarized in the ANOVA table shown in Table 11.2.

Source of Variation	df	SS	MS	F
Treatments	3	1274119		
$H_0 : \mu_1 = \mu_2$	1	1029204	1029204	67.29(**)
$H_0 : \mu_1 = \mu_3$	1	830304	830304	54.28(**)
$H_0 : \mu_1 = \mu_4$	1	603568	603568	39.46(**)
Error	8	122365	15296	
Total	11	1396485		

Table 11.2: ANOVA table for Example E, a subtractive design.

Our F values are all so large that we have no difficulty in rejecting all three hypotheses. We conclude that magnesium, nitrogen and trace nutrients are all essential components of the sunflower fertilizer solution.

The three hypothesis tests are not independent, since U_2, U_3 and U_4 are not orthogonal, so the three corresponding sums of squares do not add up to the treatment sum of squares. In fact, in our example they add to almost twice the treatment sum of squares! The dependence among the tests is due to the fact that the first treatment is involved in all three hypothesis tests. This means that the errors in estimating μ_1 affect all three hypothesis tests. For example, we could stretch our imaginations and argue that the first treatment may have been lucky in that it was allocated the only three good sunflowers in the batch. If so, this good luck on the part of treatment one would have inflated the F values for all three hypothesis tests. Hence the "dependency" among the tests.

Estimation of σ^2

As usual, we estimate σ^2 via

$$s^2 = \frac{(y.U_5)^2 + \cdots + (y.U_{12})^2}{8} = \frac{\|y - \bar{y}_{i.}\|^2}{8} = \frac{122365}{8} = 15296$$

Here $y.U_5, \ldots, y.U_{12}$ are all observations from an $N\left[0, \sigma^2\right]$ distribution, so their squares average to σ^2 over many repetitions of the experiment.

In previous chapters U_2, U_3 and U_4 have usually been orthogonal, so that the random variables $Y.U_2$, $Y.U_3$ and $Y.U_4$ have been independent. In our present example this independence is lost, as mentioned in the last section. The expected values and variances of these random variables are summarized in Table 11.3.

	Average, or expected value	Variance
$Y.U_2$	$\dfrac{\sqrt{3}(\mu_1 - \mu_2)}{\sqrt{2}}$	σ^2
$Y.U_3$	$\dfrac{\sqrt{3}(\mu_1 - \mu_3)}{\sqrt{2}}$	σ^2
$Y.U_4$	$\dfrac{\sqrt{3}(\mu_1 - \mu_4)}{\sqrt{2}}$	σ^2

Table 11.3: Distribution of the projection coefficients, $Y.U_2$, $Y.U_3$ and $Y.U_4$, in Example E.

Checking of Assumptions

In plant growth studies it is usual for the standard error of the plant weights to be roughly proportional to the treatment mean. This is only common sense, since in absolute terms small plants have much less scope for varying in weight than do larger plants. On a relative basis, however, the *percentage* differences in size are often similar between treatment groups of varying general sizes. For example, plants may weigh 10, 15 and 20 grams in one group and 100, 150 and 200 grams in a second group.

In our study we are therefore immediately suspicious of the assumption of a common variance σ^2. To check it out we plot the errors against the fitted values in Figure 11.3, and calculate separate variance estimates for each treatment, shown in Table 11.4.

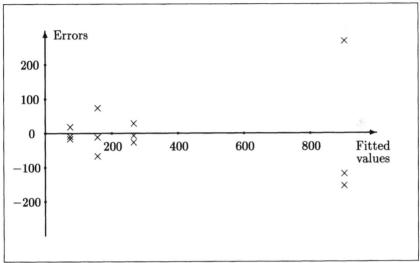

Figure 11.3: Plot of errors against fitted values, Example E.

Treatment	Variance estimate
Complete solution	54964
Minus magnesium	357
Minus nitrogen	5116
Minus trace nutrients	745
Average	15296

Table 11.4: Variance estimates for individual treatments.

In particular we would expect that the first treatment would be considerably more variable than the other treatments, since the sunflowers in this treatment are much bigger than the sunflowers in the other treatments. This is visually confirmed in Figure 11.3. For a more formal confirmation, we can test the hypothesis $H_0 : \sigma_1^2 = (\sigma_2^2 + \sigma_3^2 + \sigma_4^2)/3$ by comparing the variance estimate for treatment one with the average of the variance estimates for treatments two to four, as given in Table 11.4. The relevant test statistic is

$$F = \frac{s_1^2}{(s_2^2 + s_3^2 + s_4^2)/3} = \frac{54964}{2073} = 26.5$$

Under the null hypothesis this comes from an $F_{2,6}$ distribution, which has a 99.5 percentile of 14.5. We therefore reject the hypothesis at the 1% level of significance, and conclude that the first treatment is indeed more variable than the other treatments.

Revising the Model

A more appropriate model involves the use of *percentage* differences rather than absolute differences. This can be achieved by transforming the data using the *logarithm* transformation. Then differences become, for example, $\log y_{11} - \log y_{12} = \log(y_{11}/y_{12})$, involving the ratio of the two weights. The resulting data is shown in Table 11.5.

Treatment	log(weights)			Mean
1. Complete solution	7.066	6.620	6.664	6.784
2. Solution lacking magnesium	4.205	4.554	4.078	4.279
3. Solution lacking nitrogen	4.997	5.455	4.522	4.991
4. Solution lacking trace nutrients	5.694	5.493	5.572	5.586

Table 11.5: The log transformed weights for the Example E sunflower data.

We again assume that each observation, which we call y_{ij} again, is independently drawn from a normal population with mean μ_i and a common variance σ^2. We then reanalyze the data as before. The resulting ANOVA table is presented in Table 11.6.

Our conclusions are again that magnesium, nitrogen and trace nutrients are all essential components of the sunflower fertilizer solution.

When we check our assumptions we find that the variances are more similar between treatments, being .0605, .0608, .2179 and .0102 respectively. We conclude that our revised model is more appropriate than our original model. From here on we shall therefore work only with the revised model.

Source of Variation	df	SS	MS	F
Treatments	3	10.1191		
$H_0 : \mu_1 = \mu_2$	1	9.4121	9.4121	107.74(**)
$H_0 : \mu_1 = \mu_3$	1	4.8180	4.8180	55.15(**)
$H_0 : \mu_1 = \mu_4$	1	2.1504	2.1504	24.61(**)
Error	8	.6989	.0874	
Total	11	10.8179		

Table 11.6: ANOVA table for Example E, using $\log_e$(weights) as observations.

Confidence Intervals

As an example we shall calculate the 95% confidence interval for the effect of omitting magnesium from the fertilizer solution, using the logarithm transformed data.

Table 11.3 tells us that the value $y.U_2 = \sqrt{3}(\bar{y}_{1.} - \bar{y}_{2.})/\sqrt{2}$ comes from an $N[\sqrt{3}(\mu_1 - \mu_2)/\sqrt{2}, \sigma^2]$ distribution. Rearranging this, we see that $w_1 = \sqrt{3}[(\bar{y}_{1.} - \bar{y}_{2.}) - (\mu_1 - \mu_2)]/\sqrt{2}$ comes from an $N(0, \sigma^2)$ distribution. Hence w_1/s comes from a t_8 distribution. Our resulting 95% level gamble is

$$-2.306 \quad \leq \quad \frac{\sqrt{3}[\bar{y}_{1.} - \bar{y}_{2.}) - (\mu_1 - \mu_2)]}{\sqrt{2}s} \quad \leq \quad 2.306$$

Rearranging gives us the 95% confidence interval

$$(\bar{y}_{1.} - \bar{y}_{2.}) - \frac{\sqrt{2}s}{\sqrt{3}}2.306 \quad \leq \quad \mu_1 - \mu_2 \quad \leq \quad (\bar{y}_{1.} - \bar{y}_{2.}) + \frac{\sqrt{2}s}{\sqrt{3}}2.306$$

Substituting $\bar{y}_{1.} = 6.784$, $\bar{y}_{2.} = 4.279$ and $s = \sqrt{.0874}$ leads to

$$2.505 - .557 \quad \leq \quad \mu_1 - \mu_2 \quad \leq \quad 2.505 + .557$$
$$\text{or} \quad 1.948 \quad \leq \quad \mu_1 - \mu_2 \quad \leq \quad 3.062$$

This confidence band is on the scale of $\log_e$(weights). To convert it to something which is easier to interpret, we write $\mu_1 = \log \mu_1'$ and $\mu_2 = \log \mu_2'$, where μ_1' and μ_2' are in units of grams. Then $\mu_1 - \mu_2 = \log \mu_1' - \log \mu_2' = \log(\mu_1'/\mu_2')$. The confidence band is then

$$1.948 \quad \leq \quad \log\left(\frac{\mu_1'}{\mu_2'}\right) \quad \leq \quad 3.062$$

We now *back transform* this confidence band by taking antilogarithms, that
is, using the exponential transformation:

$$e^{1.948} \leq \frac{\mu_1'}{\mu_2'} \leq e^{3.062}$$

$$\text{or} \qquad 7.0 \leq \frac{\mu_1'}{\mu_2'} \leq 21.4$$

This says that we are 95% confident that sunflowers grown with the complete
solution are on average between 7.0 and 21.4 times heavier than sunflowers
grown in the same solution without magnesium.

Note that the estimated mean difference, $e^{2.505} = 12.2$, is not the mid-
point of this band, since the exponential transformation is nonlinear. If
asked to gamble on a particular estimate, we would gamble on this figure of
12.2 since it represents the estimate which is "most likely".

Report on Study

A method for presenting the results of the experiment is given in Table
11.7. The SED and contrasts relate to the means of the log(weights), so are
aligned with these values in the table. Back transformed means are often
presented alongside so that the reader of the report can more readily see
the approximate size of the plants in each treatment. The back transformed
means can also be used to calculate the multiplicative differences between
treatments; for example, the 12.2 in the last paragraph equals 884/72.

Treatment	Means of $\log_e$(dry weights),with back transformed means in brackets	
Complete solution	6.784	(884)
Solution lacking magnesium	4.279	(72)
Solution lacking nitrogen	4.991	(147)
Solution lacking trace nutrients	5.586	(267)
SED	.241	
Significance of contrasts		
Omitting magnesium	**	
Omitting nitrogen	**	
Omitting trace nutrients	**	

Table 11.7: Summary of sunflower results, Example E.

In Table 11.7 the back transformed mean is the geometric mean of the
plant weights. For example, for the complete solution the back transformed
mean is $884 = (1172 \times 750 \times 784)^{\frac{1}{3}}$. This usually differs somewhat from the
arithmetic mean, here equal to 902.

Minitab Analysis

The analysis of the log transformed values using Minitab is summarized in Table 11.8. The pattern is similar to that of previous chapters; firstly $(y.U_2)^2$, $(y.U_3)^2$ and $(y.U_4)^2$ are calculated using the Regress command, then the baseline s^2 is calculated using the Oneway command.

List of commands	Interpretation
name c1 = 'Weights' name c2 = 'Mg' name c3 = 'N' name c4 = 'TrNut' name c5 = 'Trtments' name c6 = 'Errors' name c7 = 'Fitted'	[Naming the columns
set c1 1172 750 ... 243 263 let c1=log(c1)	[Set y into column one and log transform
set c2 (1, −1, 0, 0)3 let c2=c2/sqrt(6) ⋮	[Set U_2, U_3 and U_4 into columns two to four
regress c1 1 c2 regress c1 1 c3 regress c1 1 c4	Calculate $(y.U_2)^2$ Calculate $(y.U_3)^2$ Calculate $(y.U_4)^2$
set c5 (1, 2, 3, 4)3	[Set treatment numbers into column five
[oneway c1 c5 c6 c7	[Fit full model
[print c1 c7 c6	[Display $y = \bar{y}_{i\cdot} + (y - \bar{y}_{i\cdot})$
[histogram c6	[Do a histogram of the errors
[plot c6 c7	[Plot errors versus fitted values
[stop	[End of job

Table 11.8: Minitab commands for log analysis, Example E.

Optimal Design

Our experiment provides an example of a *subtractive* design, since treatments two to four are each generated by subtracting one chemical ingredient from the complete solution. In this situation the contrasts of interest all involve comparisons with the mean of treatment one. This suggests it would be sensible to replicate treatment one more heavily than the other treatments, so that we produce a better estimate of the baseline mean of treatment one.

Scheffé (1959) has addressed this problem. He found that for the best use of a given number of experimental units, such as field plots, animals or human patients, the optimal strategy is to allocate $\sqrt{k}$ times as many experimental units to the baseline treatment as to the other k treatments. In our case this means that we should have $\sqrt{3} = 1.7$ times as many sunflowers allocated to the complete solution as to the other three solutions. That is, if fourteen sunflowers were available, we should have three replicates of treatments two to four, and five ($\approx \sqrt{3} \times 3$) replicates of treatment one.

Using the theory for unequal replications as discussed in Appendix B, the directions corresponding to the hypotheses of interest would in this latter case be

$$
U_2 = \frac{1}{\sqrt{\frac{1}{5}+\frac{1}{3}}} \begin{bmatrix} 1/5 \\ 1/5 \\ 1/5 \\ 1/5 \\ 1/5 \\ -1/3 \\ -1/3 \\ -1/3 \\ 0 \\ 0 \\ 0 \\ 0 \\ 0 \\ 0 \end{bmatrix}, \quad
U_3 = \frac{1}{\sqrt{\frac{1}{5}+\frac{1}{3}}} \begin{bmatrix} 1/5 \\ 1/5 \\ 1/5 \\ 1/5 \\ 1/5 \\ 0 \\ 0 \\ 0 \\ -1/3 \\ -1/3 \\ -1/3 \\ 0 \\ 0 \\ 0 \end{bmatrix}, \quad
U_4 = \frac{1}{\sqrt{\frac{1}{5}+\frac{1}{3}}} \begin{bmatrix} 1/5 \\ 1/5 \\ 1/5 \\ 1/5 \\ 1/5 \\ 0 \\ 0 \\ 0 \\ 0 \\ 0 \\ 0 \\ -1/3 \\ -1/3 \\ -1/3 \end{bmatrix}
$$

The angle θ between any two of these vectors is given by

$$
\cos \theta \quad = \quad \frac{(1/5)^2+(1/5)^2+(1/5)^2+(1/5)^2+(1/5)^2}{1/5 \; + \; 1/3} \quad = \quad \frac{3}{8}
$$

That is, $\theta = 68°$, as compared to $60°$ when all four treatments were equally replicated. This shows that the dependence among the three hypothesis tests is reduced when treatment one is more heavily replicated.

With the new design the SED for a contrast of interest would be $s\sqrt{1/5 + 1/3} = .730s$. This is lower than the SED obtained from allocating sunflowers equally to the treatments; that is, $14/4 = 3.5$ replicates per treatment, giving an SED of $s\sqrt{1/3.5 + 1/3.5} = .756s$. Readers who dislike thinking of 3.5 replicates can double the experiment's size to 28 units.

11.2 Least Significant Difference

Here we use Example E to demonstrate the equivalence of the F test, the t test, and the least significant difference test for pairwise contrasts.

In Table 11.3 we saw that the value $y.U_2 = \sqrt{3}(\bar{y}_{1.} - \bar{y}_{2.})/\sqrt{2}$ comes from a $N[\sqrt{3}(\mu_1 - \mu_2)/\sqrt{2}, \sigma^2]$ distribution. We then tested $H_0 : \mu_1 = \mu_2$ by calculating the ratio,

$$F = \frac{(y.U_2)^2}{s^2} = \frac{3(\bar{y}_{1.} - \bar{y}_{2.})^2}{2s^2}$$

This value was compared with the percentiles of the $F_{1,8}$ distribution.

An alternative, equivalent, way of testing $H_0 : \mu_1 = \mu_2$ is to calculate

$$t = \frac{\sqrt{3}(\bar{y}_{1.} - \bar{y}_{2.})}{\sqrt{2}s}$$

This value is then compared with the percentile of the t_8 distribution.

In this t test a 5% significant result is observed if the calculated t value is greater than $t_8(.975)$, ignoring signs. In other words, the result is 5% significant if

$$\frac{\sqrt{3}|\bar{y}_{1.} - \bar{y}_{2.}|}{\sqrt{2}s} > t_8(.975)$$

This can be rearranged as

$$|\bar{y}_{1.} - \bar{y}_{2.}| > \frac{\sqrt{2}s}{\sqrt{3}} \times t_8(.975) = \text{SED} \times t_8(.975)$$

Here the expression on the right hand side is called the 5% least significant difference, the LSD(5%), since it is the *least difference* which is *significant* at the 5% level. For our log transformed data as summarized in Table 11.7,

$$\begin{aligned} \text{LSD}(5\%) &= \text{SED} \times t_8(.975) \\ &= .241 \times 2.306 \\ &= .557 \end{aligned}$$

while
$$\begin{aligned} \text{LSD}(1\%) &= \text{SED} \times t_8(.995) \\ &= .241 \times 3.355 \\ &= .809 \end{aligned}$$

Hence pairwise contrast differences greater than .557 will be 5% significant, and differences greater than .809 will be 1% significant.

These LSD(5%) and LSD(1%) values are often added to tables such as Table 11.7. The procedure of calculating the LSD and comparing differences with the LSD, is called the "least significant difference" test. Clearly it is equivalent to the F and t tests.

11.3   Multiple Comparison Procedures

In this section we analyze Example F from §7.2, using the simplest and most natural multiple comparison procedure, the LSD test as outlined in the last section. We shall then describe the test for the overall null hypothesis $H_0 : \mu_1 = \mu_2 = \mu_3 = \mu_4$. Lastly, we shall describe an alternative multiple comparison procedure and point out the "inconsistency" of such alternatives.

Analyzing Example F

We now use the LSD test to statistically analyze the hypothetical data given in Example F. This experiment was designed to evaluate four new chemicals for stripe rust control in wheat. The four treatments were laid out with five replications in a completely randomized design.

The grain yields in Table 7.6 are in units of kilograms per $50m^2$. We therefore convert to more standard units of kg per hectare by multiplying each data value by 200 (one hectare equals $10,000m^2$) as shown in Table 11.9. The resulting observation vector is $y = [3800, \ldots, 4460]^T$.

Chemical	Grain yield					Mean
Product X52	3800	4200	4600	4000	3400	4000
Product B29	4760	4360	3560	4760	4360	4360
Product Z15	4380	4980	4680	3780	4580	4480
Product PP5	4860	4460	5260	5260	4460	4860

Table 11.9: Grain yield of wheat in kg/ha, Example F.

We assume, as usual, that the populations are normally distributed with means μ_1, μ_2, μ_3 and μ_4, and a common variance σ^2. The model vector is $[\mu_1, \ldots, \mu_4]^T$, and the model space M is a four dimensional subspace of 20-space.

The treatment space is a three dimensional subspace. The six hypotheses corresponding to all pairwise contrasts among the population means give rise to the following directions in the treatment space:

$$
\begin{array}{cccccc}
U_2 & U_3 & U_4 & U_5 & U_6 & U_7 \\
\begin{bmatrix} 1 \\ \cdot \\ -1 \\ \cdot \\ 0 \\ \cdot \\ 0 \\ \cdot \\ \hline \sqrt{10} \end{bmatrix} &
\begin{bmatrix} 1 \\ \cdot \\ 0 \\ \cdot \\ -1 \\ \cdot \\ 0 \\ \cdot \\ \hline \sqrt{10} \end{bmatrix} &
\begin{bmatrix} 1 \\ \cdot \\ 0 \\ \cdot \\ 0 \\ \cdot \\ -1 \\ \cdot \\ \hline \sqrt{10} \end{bmatrix} &
\begin{bmatrix} 0 \\ \cdot \\ 1 \\ \cdot \\ -1 \\ \cdot \\ 0 \\ \cdot \\ \hline \sqrt{10} \end{bmatrix} &
\begin{bmatrix} 0 \\ \cdot \\ 1 \\ \cdot \\ 0 \\ \cdot \\ -1 \\ \cdot \\ \hline \sqrt{10} \end{bmatrix} &
\begin{bmatrix} 0 \\ \cdot \\ 0 \\ \cdot \\ 1 \\ \cdot \\ -1 \\ \cdot \\ \hline \sqrt{10} \end{bmatrix}
\end{array}
$$

Clearly these six directions are not all mutually orthogonal.

As usual, we fit the model by projecting the observation vector, y, onto the model space, M, producing $\bar{y}_i. = [4000, ., 4360, ., 4480, ., 4860, .]^T$, the fitted model vector of treatment means. The resulting fitted model is the familiar

$$y - \bar{y}.. = (\bar{y}_i. - \bar{y}..) + (y - \bar{y}_i.)$$

Our estimate of the common variance σ^2 is

$$s^2 = \frac{\|y - \bar{y}_i.\|^2}{16} = \frac{3200000}{16} = 200000$$

This gives rise to a 5% least significant difference of

$$\text{LSD}(5\%) = \text{SED} \times t_{16}(.975) = \sqrt{\frac{2 \times 200000}{5}} \times 2.120 = 283 \times 2.120 = 600$$

and a 1% least significant difference of

$$\text{LSD}(1\%) = \text{SED} \times t_{16}(.995) = 283 \times 2.921 = 826$$

The experimental results can now be summarized as shown in Table 11.10.

Treatment(chemical)	Grain yield (kg/ha)
Product X52	4000
Product B29	4360
Product Z15	4480
Product PP5	4860
SED	283
LSD(5%)	600
LSD(1%)	830

Table 11.10: Summary of results for Example F.

We shall not check our assumptions since we contrived the data set, and know that the assumptions are not violated. For analysis using Minitab, we simply fit the model using the Oneway command as in Table 11.8, and use the resulting s^2 to calculate our LSD values.

Of the six pairwise comparisons which are possible, only one is significant. This is the comparison of product X52 with product PP5. Hence the only conclusion which we can draw from our study is that treatment with product PP5 resulted in a significantly higher grain yield than treatment with product X52. Note that this difference was significant at the 1% level.

Recall from §11.2 that the LSD test is equivalent to the F test. This means that we could have obtained the same conclusions using the ANOVA table shown in Table 11.11.

Source of Variation	df	SS	MS	F
Treatments	3	1885500		
$H_0 : \mu_1 = \mu_2$	1	324000	324000	1.62 (ns)
$H_0 : \mu_1 = \mu_3$	1	576000	576000	2.88 (ns)
$H_0 : \mu_1 = \mu_4$	1	1849000	1849000	9.25 (**)
$H_0 : \mu_2 = \mu_3$	1	36000	36000	.18 (ns)
$H_0 : \mu_2 = \mu_4$	1	625000	625000	3.13 (ns)
$H_0 : \mu_3 = \mu_4$	1	361000	361000	1.81 (ns)
Error	16	3200000	200000	
Total	19	5085500		

Table 11.11: ANOVA table for Example F.

Testing the Overall Null Hypothesis

In Example F we are in the unusual position of having no prior expectations concerning the outcome of the study. In this case even the overall null hypothesis $H_0 : \mu_1 = \mu_2 = \mu_3 = \mu_4$ may be plausible. How strong, then, is the evidence against this hypothesis in our example?

To enable us to derive a test of this hypothesis, we must first write down a complete set of orthogonal contrasts. Any set will do. One such set is

$$
\begin{aligned}
c_1 &= \mu_1 - \mu_2 - \mu_3 + \mu_4 \\
c_2 &= \mu_1 \qquad\qquad\; - \mu_4 \\
c_3 &= \qquad\;\; \mu_2 - \mu_3
\end{aligned}
$$

Notice now that $\mu_1 = \mu_2 = \mu_3 = \mu_4$ if and only if all three contrasts are zero.

Let U_2, U_3 and U_4 denote the directions associated with c_1, c_2 and c_3. Then the appropriate statistic for testing $H_0 : \mu_1 = \mu_2 = \mu_3 = \mu_4$ is

$$
F = \frac{[(y.U_2)^2 + (y.U_3)^2 + (y.U_4)^2]/3}{s^2} = \frac{\|\bar{y}_{i\cdot} - \bar{y}_{\cdot\cdot}\|^2/3}{s^2}
$$

If H_0 is true then $Y.U_i \sim N(0, \sigma^2)$ for $i = 2, 3, 4$, hence F comes from an $F_{3,16}$ distribution. If H_0 is false then the numerator is inflated.

With our data the test statistic is

$$
F = \frac{[500 + 1849000 + 36000]/3}{200000} = \frac{1885500/3}{200000} = 3.14
$$

This is less than 3.24, the 95 percentile of the $F_{3,16}$ distribution. We therefore accept as plausible the overall null hypothesis $H_0 : \mu_1 = \mu_2 = \mu_3 = \mu_4$.

In the general case the test statistic is

$$F = \frac{\|\bar{y}_{i\cdot} - \bar{y}_{\cdot\cdot}\|^2/(k-1)}{\|y - \bar{y}_{i\cdot}\|^2/[k(n-1)]} = \frac{\text{treatment mean square}}{\text{error mean square}} = \frac{628500}{200000}$$

which can be easily calculated from Table 11.11.

Our data has been contrived to illustrate the dilution effect which arises with this overall F test. In our data, $\bar{y}_1. = 4000$ and $\bar{y}_4. = 4860$ are distinct, and $\bar{y}_2. = 4360$ and $\bar{y}_3. = 4480$ are about midway between them. This means that of the three contrasts, only $\hat{c}_2 = 4000 - 4860 = -860$ is large ($\hat{c}_1 = 20$ and $\hat{c}_3 = -120$). If we were interested in these contrasts in their own right, we would have calculated individual test statistics,

$$
\begin{aligned}
F_1 &= (y.U_2)^2/s^2 &=& \quad 500/s^2 &=& \quad .0025 \\
F_2 &= (y.U_3)^2/s^2 &=& \quad 1849000/s^2 &=& \quad 9.25 \\
F_3 &= (y.U_4)^2/s^2 &=& \quad 36000/s^2 &=& \quad .18
\end{aligned}
$$

Our overall F value is the average of F_1, F_2 and F_3. In our example, the large F value associated with the second contrast has been diluted by the small F values associated with the other two contrasts, so that the overall F value is not significant.

Interestingly, the overall F value (3.14 here) can be calculated as the average of the complete set of pairwise contrast F values, as listed in Table 11.11 for our example. This means the overall F value can be crudely thought of as an average pairwise F value. Hence if a set of population means are all equal except for one or two differences, the overall F test may be too diluted to detect differences which would be detected by the individual pairwise F tests. On the other hand, if a set of population means are in reality all equal, the overall F test will spuriously declare otherwise in only 5% of *studies*, whereas the pairwise F tests will spuriously declare otherwise in 5% of *comparisons*.

This conflict, between a nonsignificant result from a test of the overall hypothesis $H_0 : \mu_1 = \mu_2 = \mu_3 = \mu_4$, and a 1% significant result from a test of the particular hypothesis $H_0 : \mu_1 = \mu_2$, causes a dilemma for the researcher. Which is to be believed? Fortunately, the conflict seldom arises in practice, since in most studies the population means are quite different and the overall F test is significant as well as particular pairwise differences. When the conflict does arise, however, the researcher must use information from related studies to decide which test to believe. This is the situation in Example F, where the researcher must decide whether to explore the apparent difference between PP5 and X52 in subsequent experimentation.

Alternative Multiple Comparison Procedures

We now briefly digress to discuss alternative multiple comparison procedures. Our main objective will be to point out the inconsistent behaviour of these alternatives.

The simplest alternative to the LSD test is *Fisher's restricted* LSD procedure. The procedure consists of two stages, as follows (e.g. at P=.05):

1. Use an overall F value to test the hypothesis $H_0 : \mu_1 = \mu_2 = \cdots = \mu_k$. If the F value is not significant, declare all population means to be equal. If the F value is 5% significant, carry on to stage two.

2. Compare all pairs of population means using $\text{LSD}(5\%) = \text{SED} \times t\text{-value}$. Note that the significance level is tied to that used in stage 1.

This procedure is more conservative than the unrestricted LSD procedure since it includes a preliminary overall F test. More importantly, however, it is quite anomalous in its behaviour, as we now explain.

In Figure 11.4 we pretend that three researchers, Ron, Mary and Bob, have carried out experiments. Each experiment has four treatments, including by chance a common pair, A and B. All experiments were laid out as completely randomized designs, with five replicates, so the number of error degrees of freedom was 16 in each case. The standard error of the mean, SEM, was 200 and the observed difference between A and B was 860 in all cases.

In our hypothetical example a statistician has advised the researchers to use the restricted LSD procedure to analyse their datasets. Results are displayed in Figure 11.4. Imagine the statistician's embarrassment when the researchers compare notes and discover that the significance of the difference between treatments A and B has varied from "not significant", to "5% significant", to "1% significant". This means that the evidence concerning treatments A and B has varied from "no proof of a difference", to "suggestive of an effect", to "convincing evidence of a difference", to use the words of O'Brien (1983).

This "inconsistency" in the decisions returned by Fisher's restricted LSD procedure is in practice unacceptable. The phenomenon arises because the significance of the difference between A and B is tied to the significance of the overall F test, with the latter varying from 5.62 (**) to 4.21 (*) to 3.14 (n.s.) in Figure 11.4.

In fact, all of the alternatives to the unrestricted LSD procedure are *inconsistent* (Saville 1990). Here a procedure is called inconsistent if the probability that two populations will be declared significantly different is dependent not only on the difference between the two observed means, the

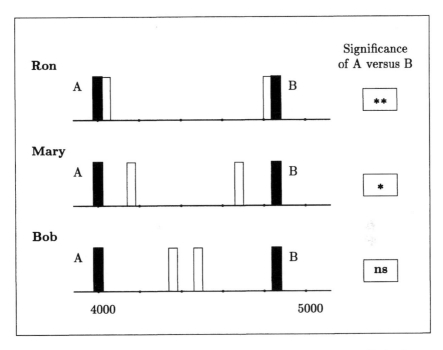

Figure 11.4: Results obtained by the three researchers using Fisher's restricted LSD procedure. Note that the significance of the difference between treatments A and B is dependent on the values of the other two treatment means. This figure follows Fig. 2 in "Multiple Comparison Procedures: The Practical Solution", D.J. Saville, *The American Statistician* 44: 174–180.

SED, and the error degrees of freedom, but also on the number of populations or the values of the other observed means. This means that in practice, the only multiple comparison procedure which is suitable for routine use is the unrestricted LSD procedure as described in §11.2.

In this text we shall therefore use the same formal procedure for examining unplanned pairwise contrasts as we would use for examining planned pairwise contrasts. In both cases we shall use the pairwise contrast F test or the equivalent t test, or the equivalent unrestricted LSD test. However, when it comes to interpretation of the results, we shall make a large distinction between the unplanned and planned cases. If a pairwise comparison is planned, in that the researcher has a definite hypothesis to check, the analysis will serve the purpose of *confirming* that the prior hypothesis is true, or otherwise. If a pairwise comparison is unplanned, the analysis serves simply to *generate* a hypothesis; this can be confirmed by subsequent research if it is of sufficient interest to the researcher.

11.4   Summary

In this chapter we have discussed contrasts of one population mean with another. This type of contrast is very simple, and the usual methods apply. The only complication is that of nonorthogonality, and because of this care must be taken with the use of computing packages.

The pairwise contrast F value for testing $H_0 : \mu_1 = \mu_2$ can be algebraically simplified to

$$F = \frac{n(\bar{y}_{1\cdot} - \bar{y}_{2\cdot})^2}{2s^2}$$

where n is the number of observations used in calculating each mean. This comes from an $F_{1,k(n-1)}$ distribution under the null hypothesis. The corresponding t value of $\sqrt{n}(\bar{y}_{1\cdot} - \bar{y}_{2\cdot})/(\sqrt{2}s)$ follows a $t_{k(n-1)}$ distribution under the null hypothesis. Rearranging the t test gives us the least difference which will be 5% significant. This is called the least significant difference, or

$$\text{LSD}(5\%) = \frac{\sqrt{2}s}{\sqrt{n}} \times t_{k(n-1)}(.975) = \text{SED} \times t_{k(n-1)}(.975)$$

The LSD test can be used to examine pairwise differences in a planned or unplanned manner. Planned usage corresponds to confirming hypotheses, whereas unplanned usage corresponds to generating hypotheses which will require subsequent confirmation. When used in the latter way, the LSD test is being used as a multiple comparison procedure.

In this chapter we have also described the test of the overall null hypothesis $H_0 : \mu_1 = \mu_2 = \cdots = \mu_k$. Here the test statistic is

$$F = \frac{\|\bar{y}_{i\cdot} - \bar{y}_{\cdot\cdot}\|^2/(k-1)}{s^2}$$

This follows the $F_{k-1,k(n-1)}$ distribution if H_0 is true, since $\bar{y}_{i\cdot} - \bar{y}_{\cdot\cdot}$, the treatment vector, is the sum of the projections of y onto the $k-1$ axes in the treatment space. In practice this overall test is seldom of any interest to researchers.

The 95% confidence interval for the difference $\mu_i - \mu_j$ between any two population means is given by the formula

$$(\bar{y}_{i\cdot} - \bar{y}_{j\cdot}) \quad \pm \quad \sqrt{\frac{2s^2}{n}} \times t_{k(n-1)}(.975)$$

This can be written more simply as {observed difference $\pm$ LSD(5%)}.

Class Exercise

To get a feel for how the LSD test behaves when used as a multiple comparison procedure, we shall use the data from Table 5.5 to simulate two situations:

- the case when all populations means are equal

- the case when only two of the population means are equal

For the exercise, we shall pretend that each class member has carried out an experiment involving ten students. The students were allocated in a completely random manner, two students to each of five unrelated diets, A, B, C, D and E. After three months on the diet the change in heartbeat induced by jogging for ten minutes were measured to ascertain whether the general level of fitness was affected by diet.

We first simulate the case of all population means being equal.

(a) Each class member is asked to use random numbers to draw five samples of size two from Table 5.5. These can be regarded as the data for diets A, B, C, D and E.

(b) Using this data, each class member is then asked to calculate $s^2 = \|y - \bar{y}_{i.}\|^2/5$, and hence the LSD(5%).

(c) Each class member can then draw up a table of means, plus an LSD(5%), and compare each pair of diets (there are ten such pairs). How many pairwise differences are significant?

(d) The class instructor is now asked to draw up a histogram of the number of spurious significances. That is, how many class members observed no significant differences, how many observed one, how many two, ..., up to ten?

Is there anything interesting about this histogram? There is a 5% chance of a spurious significance for each pairwise comparison, so the class values will in the long run average to $.05 \times 10 = .5$. Does this seem plausible for your class?

We now simulate the case when only two of the population means are equal.

(e) Each class member is now asked to use a new choice of random numbers to draw five samples of size two from Table 5.5, for diets A, B, C, D and E. For diet A, 5 beats per half minute should be added to each of the observations. For diet D, 5 beats per half minute should be subtracted from each observation. For diet E, 10 beats per half minute should be subtracted. This means that only diets B and C are truly equal in our simulation.

(f) The LSD(5%) can now be recalculated, and a new table of means plus

LSD can be drawn up. Which of the ten pairwise differences are significant?

(g) The class instructor can now find out how many class members observed a spurious significance between diets B and C. The class average should be about 5%, or 1 in 20.

Diets A and B truly differed by 5 beats per half minute. What percentage of class members observed a significant difference between diets A and B?

Diets C and E truly differed by 10 beats per half minute. What percentage of the class picked up this difference?

Diets A and E truly differed by 15 beats per half minute. What percentage of the class picked up this difference?

The probability of detecting a genuine difference is called the *power* of the test. These last three questions should illustrate the fact that the power increases with the size of the difference.

Exercises

(11.1) In §8.1 four fodder beet and four mangel cultivars were evaluated for use as winter greenfeed. The results were summarized in Table 8.5, after adjustment for an unusually high data value.

(a) Calculate the LSD(5%).

(b) Use this LSD(5%) to compare each mangel cultivar with each other mangel cultivar. Which mangel cultivars differ significantly?

(c) Repeat (b) for the fodder beet cultivars.

(d) In (b) and (c) are any two comparisons, both of which you have found to be significant, dependent on one another?

(11.2) An experiment was carried out at Winchmore Irrigation Research Station, near Ashburton, New Zealand, to compare the productivity of fourteen different alfalfa cultivars under irrigation. Four replicate plots of each cultivar were sown. The total dry matter production of alfalfa, excluding weeds, from the three cuts made in the second year of the experiment is given in the table for each plot.

Cultivar	Total alfalfa production			
	(DM kg/ha) for 77/78 season			
Wairau	1611	1757	1964	2444
Caliverde	1586	3071	2565	1547
Ranger	1041	3857	934	856
Vernal	1079	2172	529	1310
Dawson	989	1922	455	474
Kanza	2923	5947	3713	2319
Lahontan	3887	5092	4543	5538
Washoe	6543	7315	8215	5375
Iroquois	797	3286	1919	2764
Saranac	1066	2549	1049	864
520	1074	4532	1412	1062
530	1224	2135	670	2459
Atra 55	1531	2638	2333	1052
Mesilla	1976	4113	1827	1710

(a) Use the Oneway command of Minitab to fit the model $(y - \bar{y}_{..})$ $= (\bar{y}_{i.} - \bar{y}_{..}) + (y - \bar{y}_{i.})$ in the usual manner. Hence obtain s^2, the estimate of σ^2. Note: save your Minitab computer file for reuse in Exercise 11.3.

(b) Use s^2 to calculate an LSD(5%) and an LSD(1%).

(c) Present your results in a table of means, the SED and LSD values, such as shown in Table 11.10, rounding the treatment means to the nearest ten and sorting them into order of size, with the largest at the top.

(d) Did any one cultivar outyield the others in this experiment? If you were asked to choose between two and five cultivars for further evaluation, which cultivars would you choose? Explain the reasons for your choice.

(11.3) In Exercise 11.2 the treatment means were spread over a wide range, with the highest mean being about nine times the lowest mean. A logarithm transformation is therefore likely to be necessary.

(a) Include the line "let c1 = log(c1)" in your Minitab job, and rerun the analysis of Exercise 11.2(a). This provides a new s^2 value.

(b) Recalculate the LSD(5%) and LSD(1%).

(c) Present your results as in Exercise 11.2(c), but with backtransformed means in brackets. Use three decimal places for your log means.

(d) Do the conclusions reached in Exercise 11.2(d) change? What are your revised conclusions?

(11.4) Seed from eight cultivars of barley was imported into New Zealand from throughout the world for local evaluation. The seed was sown in four replicate plots for each of the eight cultivars, using a completely randomized design.

Mean grain yields for each cultivar, in kg/ha, were as follows:

Cultivar	A	B	C	D	E	F	G	H
Grain yield	4910	5120	4440	4910	5030	5560	4870	5160

The variance estimate was $s^2 = 176400$, with 24 degrees of freedom.

(a) Calculate the LSD(5%), LSD(1%) and LSD(0.1%). For the latter, the percentile is $t_{24}(.9995) = 3.745$.

(b) Which pairs of cultivars appear to differ from one another? Present your results in a table of sorted means and LSDs, and in words.

Aside: In this hypothetical example the overall F value is $\left(\|\bar{y}_{i\cdot} - \bar{y}_{\cdot\cdot}\|^2/7\right)/s^2$ $= 2.27$ which is not significant. The restricted LSD procedure would therefore have declared all differences to be nonsignificant.

(11.5) A local carnation grower experienced a severe infestation of the disease ringspot in his polythene tunnel houses. An experiment was therefore set up in one tunnel house, to try out four chemicals which could have some effect on the disease. Three cultivars of carnation were available in the tunnel house, so two 1m by 0.3m trays of each cultivar were allocated to each of the chemicals plus a control treatment. The treatment design was therefore a 5×3 factorial design; the layout was completely randomized, with two replicates.

The appropriate chemical treatment was applied initially, then again at the end of the first and second weeks. At the end of the third week of the experiment the percentage of carnation leaves affected by ringspot was estimated by assessing the leaves on ten stems from each tray. The resulting data, by courtesy of Mark Braithwaite, Ministry of Agriculture and Fisheries, Lincoln, is given in the table.

Cultivar:	Mei Cheng		Mei Boa		Mei Sol	
Chemical	Rep 1	Rep 2	Rep 1	Rep 2	Rep 1	Rep 2
Baycor	85.0	99.0	87.0	91.4	89.2	92.8
Sportak	69.2	57.0	76.1	67.2	79.9	83.2
Saprol	43.2	37.8	40.4	37.7	47.0	52.5
Bravo	62.0	51.3	46.5	69.1	71.7	52.1
Control	89.4	99.1	89.7	93.6	88.0	95.3

(a) Use Minitab and the Oneway command to calculate s^2. You will need to fit 15 treatment means.

(b) Calculate the five main effect means for the chemical factor. Each of these is a mean of six observations.

(c) Calculate the LSD(5%) and LSD(1%) which apply to pairwise comparisons between the main effect means in (b). Hint: Use $n = 6$ as the divisor for s^2.

(d) Present the chemical main effect means and LSD's in a table.

(e) Which of the chemicals had some effect on the disease ringspot, and which appeared to have no effect? Which chemical was the most effective?

Part IV
Introducing Blocking

In Part III we described how to capture the variation between treatments using one dimensional subspaces corresponding to contrasts of interest. Here in Part IV we use similar methods to capture some of the variation between the experimental units. We do this by grouping similar experimental units into blocks, then contrasting the blocks with each other. In order to illustrate the ideas, three designs will be covered: the randomized block design, the latin square design, and the split plot design.

Chapter 12

Randomized Block Design

In preceding chapters our measurements have largely arisen through applying experimental treatments to experimental units. For example, we measured the "bulk" of two and three day conditioned samples of wool. When the experimental units are entirely uniform, the only major influence on the measured value is the treatment. In practice, however the experimental units are seldom entirely uniform; for example, the nature of the wool may vary from sample to sample. This variation in the experimental units will in turn influence our measurements, and disguise the variation between treatments, the issue of central interest. Intelligent design, which recognizes the variation between experimental units, can help overcome this problem.

Just as we created a treatment space to isolate the treatment differences, so now we create a block space to isolate the differences between blocks, or groupings, of experimental units. These blocks are chosen so that each comprises units of similar type. For example, in an animal trial the first block might consist of the heaviest animals, the second block the next heaviest animals, and so on.

In this chapter we introduce the simplest of the block designs, the *Randomized Block Design*. In this design we choose blocks of k experimental units, where k is the number of treatments, then within each block we randomly allocate the k units to the k treatments. This enables us to account for some of the variation between experimental units, and to see more clearly the variation between treatments. We note that the design used in earlier chapters is termed the *Completely Randomized Design*, since units were assigned to treatments in a completely random fashion.

The plan for the chapter is to work through an illustrative example in §1, discuss the general case in §2, work through a real dataset in §3, and discuss why and how to block in §4. The chapter closes with a summary in §5, a class exercise, and general exercises.

12.1   Illustrative Example

As a simple introductory example we shall consider a comparison of two diets for overweight hospital patients.

Design

A hospital dietitian wishes to compare two diets, A and B, which have been designed for overweight hospital patients. The dietitian decides to carry out the study in just one ward of the hospital in which there were female patients recovering from leg fractures. She therefore approached the ten patients in the ward who were more than 20 kg overweight and more than eight weeks from their expected date of dehospitalization; six out of the ten volunteered to participate in the study. These six patients were overweight by 33, 47, 29, 42, 24 and 41 kg respectively.

Since the patients who were most overweight were likely to lose most weight during the study, the dietician decided to allocate the patients to the diets within overweight "blocks". The first block consisted of the two most overweight patients, 47 and 42 kg overweight, the second block the two next most overweight patients, 41 and 33 kg overweight, and the last block the two least overweight patients, 29 and 24 kg overweight. Within each block the two patients were then randomly allocated to the two diets.

Data

The study was intended to last for eight weeks. However, after six weeks two of the patients were ready for release from hospital, so it was decided to terminate the official data collection at this point. The data given in Table 12.1 is the weight loss of each patient over the six week period.

Block	1	2	3
Diet A	12.4	11.2	8.5
Diet B	9.2	9.8	6.5

Table 12.1: Weight loss, in kg, of patients over a six week period.

The resulting observation vector is

$$y = \begin{bmatrix} y_{11} \\ y_{12} \\ y_{13} \\ y_{21} \\ y_{22} \\ y_{23} \end{bmatrix} = \begin{bmatrix} 12.4 \\ 11.2 \\ 8.5 \\ 9.2 \\ 9.8 \\ 6.5 \end{bmatrix}$$

Here y_{ij} denotes the observation from the ith treatment and j^{th} block; for example, y_{12} is the weight loss of the patient receiving diet A in the second block.

Model

With our randomized block design we have six distinct populations, corresponding to the six possible combinations of the two treatments and three blocks. To model the means of these six populations we assume that the mean of each observation random variable Y_{ij} is the sum of the usual i^{th} treatment (diet) mean, μ_i, and a block effect, β_j. This block effect is just the difference between the mean of the jth block and the overall mean, μ. Table 12.2 summarizes these six means. Note that $\beta_1 + \beta_2 + \beta_3 = 0$, since the block effects are defined as deviations from their average. Finally, as usual the six populations are assumed normally distributed, with common variance σ^2.

Block	1	2	3	Average
Diet A	$\mu_1 + \beta_1$	$\mu_1 + \beta_2$	$\mu_1 + \beta_3$	μ_1
Diet B	$\mu_2 + \beta_1$	$\mu_2 + \beta_2$	$\mu_2 + \beta_3$	μ_2
Average	$\mu + \beta_1$	$\mu + \beta_2$	$\mu + \beta_3$	μ

Table 12.2: Long term averages or expected values for each observation, treatment mean, block mean and overall mean.

The resulting *model vector* is

$$
\begin{bmatrix} \mu_1 + \beta_1 \\ \mu_1 + \beta_2 \\ \mu_1 + \beta_3 \\ \mu_2 + \beta_1 \\ \mu_2 + \beta_2 \\ \mu_2 + \beta_3 \end{bmatrix} = \mu_1 \begin{bmatrix} 1 \\ 1 \\ 1 \\ 0 \\ 0 \\ 0 \end{bmatrix} + \mu_2 \begin{bmatrix} 0 \\ 0 \\ 0 \\ 1 \\ 1 \\ 1 \end{bmatrix} + \beta_1 \begin{bmatrix} 1 \\ 0 \\ 0 \\ 1 \\ 0 \\ 0 \end{bmatrix} + \beta_2 \begin{bmatrix} 0 \\ 1 \\ 0 \\ 0 \\ 1 \\ 0 \end{bmatrix} + \beta_3 \begin{bmatrix} 0 \\ 0 \\ 1 \\ 0 \\ 0 \\ 1 \end{bmatrix}
$$

This vector lies in the model space, $M = \text{span}\{V_1, V_2, V_3, V_4, V_5\}$, where $V_1, \ldots, V_5$ are the vectors on the right hand side of the expression. Notice, however, that these vectors are not all orthogonal, since $V_1 + V_2 = V_3 + V_4 + V_5 = [1, 1, 1, 1, 1, 1]^T$. Appropriate coordinate axes for M can be easily found, however, as follows:

$$
U_1 = \frac{1}{\sqrt{6}} \begin{bmatrix} 1 \\ 1 \\ 1 \\ 1 \\ 1 \\ 1 \end{bmatrix}, \quad U_2 = \frac{1}{\sqrt{6}} \begin{bmatrix} -1 \\ -1 \\ -1 \\ 1 \\ 1 \\ 1 \end{bmatrix}, \quad U_3 = \frac{1}{\sqrt{4}} \begin{bmatrix} -1 \\ 1 \\ 0 \\ -1 \\ 1 \\ 0 \end{bmatrix}, \quad U_4 = \frac{1}{\sqrt{12}} \begin{bmatrix} -1 \\ -1 \\ 2 \\ -1 \\ -1 \\ 2 \end{bmatrix}
$$

Here U_1 spans the one-dimensional overall mean space, U_2 spans the one-dimensional treatment space, and U_3, U_4 span the two-dimensional *block space*. Just as U_2 corresponds to the contrast between the treatment means, $\mu_2 - \mu_1$, so U_3 corresponds to the contrast between the block effects, $\beta_2 - \beta_1$, and U_4 corresponds to the contrast among the block effects $\beta_3 - (\beta_1 + \beta_2)/2$. In summary, the model space is a four dimensional subspace consisting of an overall mean space, a treatment space and a block space.

Test Hypothesis

Our objective is to find out whether diets A and B are equally effective. That is, we wish to test the hypothesis $H_0 : \mu_1 = \mu_2$ against the alternative hypothesis $H_0 : \mu_1 \neq \mu_2$.

The associated direction is $U_2 = [-1, -1, -1, 1, 1, 1]^T/\sqrt{6}$. To check this is correct we calculate

$$
y.U_2 = \begin{bmatrix} y_{11} \\ y_{12} \\ y_{13} \\ y_{21} \\ y_{22} \\ y_{23} \end{bmatrix} \cdot \frac{1}{\sqrt{6}} \begin{bmatrix} -1 \\ -1 \\ -1 \\ 1 \\ 1 \\ 1 \end{bmatrix} = \frac{1}{\sqrt{6}}(-y_{11} - y_{12} - y_{13} + y_{21} + y_{22} + y_{23})
$$

From Table 12.2 the expected value of the random variable $Y.U_2$ is

$$
\frac{1}{\sqrt{6}}[-(\mu_1 + \beta_1) - (\mu_1 + \beta_2) - (\mu_1 + \beta_3) + (\mu_2 + \beta_1) + (\mu_2 + \beta_2) + (\mu_2 + \beta_3)]
$$

$$
= \frac{1}{\sqrt{6}}[-3\mu_1 + 3\mu_2] = \frac{\sqrt{3}(\mu_2 - \mu_1)}{\sqrt{2}}
$$

Hence in general the projection coefficient $y.U_2$ will be small if $\mu_1 = \mu_2$, and large if $\mu_1 \neq \mu_2$. This is the desired state of affairs.

The collection of our raw materials is now complete. The observation vector is $y = [12.4, 11.2, 8.5, 9.2, 9.8, 6.5]^T$, the model space is $M = span\{U_1, U_2, U_3, U_4\}$ and the direction associated with the hypothesis $H_0 : \mu_1 = \mu_2$ is $U_2 = [-1, -1, -1, 1, 1, 1]^T/\sqrt{6}$.

Fitting the Model

To fit the model we project the observation vector y onto the model space M. The resulting fitted model vector is

$$(y.U_1)U_1 \quad + \quad (y.U_2)U_2 + (y.U_3)U_3 \quad + \quad (y.U_4)U_4$$

$$= \begin{bmatrix} 9.6 \\ 9.6 \\ 9.6 \\ 9.6 \\ 9.6 \\ 9.6 \end{bmatrix} + \begin{bmatrix} 1.1 \\ 1.1 \\ 1.1 \\ -1.1 \\ -1.1 \\ -1.1 \end{bmatrix} + \begin{bmatrix} .15 \\ -.15 \\ 0 \\ .15 \\ -.15 \\ 0 \end{bmatrix} + \begin{bmatrix} 1.05 \\ 1.05 \\ -2.10 \\ 1.05 \\ 1.05 \\ -2.10 \end{bmatrix}$$

$$= \quad \bar{y}.. \quad + \quad (\bar{y}_{i\cdot} - \bar{y}..) + (y.U_3)U_3 \quad + \quad (y.U_4)U_4$$

$$\text{Overall mean} \quad \text{Treatment} \quad \text{Block contrast}$$
$$\text{vector} \quad \quad \text{vector} \quad \quad \text{vectors}$$

$$= \begin{bmatrix} 10.7 \\ 10.7 \\ 10.7 \\ 8.5 \\ 8.5 \\ 8.5 \end{bmatrix} + \begin{bmatrix} 1.2 \\ .9 \\ -2.1 \\ 1.2 \\ .9 \\ -2.1 \end{bmatrix} = \begin{bmatrix} 11.9 \\ 11.6 \\ 8.6 \\ 9.7 \\ 9.4 \\ 6.4 \end{bmatrix}$$

$$= \quad \bar{y}_{i\cdot} \quad + \quad (\bar{y}._j - \bar{y}..) \quad = \quad \bar{y}_{i\cdot} + (\bar{y}._j - \bar{y}..)$$

$$\text{Treatment mean} \quad \quad \text{Block} \quad \quad \text{Vector of}$$
$$\text{vector} \quad \quad \text{vector} \quad \quad \text{fitted values}$$

The resulting fitted model is

$$\begin{bmatrix} 12.4 \\ 11.2 \\ 8.5 \\ 9.2 \\ 9.8 \\ 6.5 \end{bmatrix} = \begin{bmatrix} 11.9 \\ 11.6 \\ 8.6 \\ 9.7 \\ 9.4 \\ 6.4 \end{bmatrix} + \begin{bmatrix} .5 \\ -.4 \\ -.1 \\ -.5 \\ .4 \\ .1 \end{bmatrix}$$

$$\text{Observation} \quad = \quad \text{Model} \quad + \quad \text{Error}$$
$$\text{vector} \quad \quad \text{vector} \quad \quad \text{vector}$$

For computation, the easiest form of the fitted model vector is $\bar{y}_{i\cdot} + (\bar{y}._j - \bar{y}..)$, the sum of the treatment mean vector and the block vector. Here $\bar{y}._j$ is the mean of all observations in the jth block, analogous to the treatment mean, $\bar{y}_{i\cdot}$. The computations can be set out as shown in Table 12.3. Firstly the data in Table 12.1 are used to calculate treatment means, block means, and block effects, then the fitted values are calculated by adding the block effects to the treatment means.

Block	1	2	3	Trt means	Fitted values		
Diet A	12.4	11.2	8.5	10.7	11.9	11.6	8.6
Diet B	9.2	9.8	6.5	8.5	9.7	9.4	6.4
Block means	10.8	10.5	7.5	9.6			
Block effects	1.2	.9	−2.1		1.2	.9	−2.1

Table 12.3: Calculation of treatment means, block effects and fitted values.

In fitting the model we have obtained estimates $\hat{\mu}_1 = 10.7$, $\hat{\mu}_2 = 8.5$, $\hat{\beta}_1 = 1.2$, $\hat{\beta}_2 = .9$ and $\hat{\beta}_3 = -2.1$. Since the model space is only four dimensional, these five estimates are dependent. In fact, $\hat{\beta}_1 + \hat{\beta}_2 + \hat{\beta}_3$ is zero, so from any two estimated block effects we can calculate the third. If independent estimates are preferred, we can calculate the estimated block contrasts, $\hat{\beta}_2 - \hat{\beta}_1 = -.3$ and $\hat{\beta}_3 - (\hat{\beta}_1 + \hat{\beta}_2)/2 = -3.15$.

Error Space

Since the model space is a four dimensional subspace of six space, the error space is a two dimensional subspace. If we so desire, we can explicitly write down a set of coordinate axes for the error space by "crossing" the coordinate axis for the treatment space, U_2, with each coordinate axis in the block space, U_3 and U_4. The resulting axes are

$$U_5 = \text{``}U_2 \times U_3\text{''} = \frac{1}{\sqrt{4}} \begin{bmatrix} 1 \\ -1 \\ 0 \\ -1 \\ 1 \\ 0 \end{bmatrix}, \quad \text{and} \quad U_6 = \text{``}U_2 \times U_4\text{''} = \frac{1}{\sqrt{12}} \begin{bmatrix} 1 \\ 1 \\ -2 \\ -1 \\ -1 \\ 2 \end{bmatrix}$$

This method is the same as that used to generate interaction vectors in Chapter 9. In fact, some readers may find it helpful to think of the error space as the subspace corresponding to treatment by block interaction contrasts. It follows that the number of error space coordinate axes is $(k - 1) \times (n - 1)$, where k is the number of treatments and n is the number of replicate blocks.

Testing the Hypothesis

We are now in a position to test the hypothesis $H_0 : \mu_1 = \mu_2$ against $H_1 : \mu_1 \neq \mu_2$, using the direction U_2 associated with this hypothesis. As usual, the routine is to compare the squared distance, $(y.U_2)^2$, with the average of the squared distances in the error space. The appropriate test statistic is

$$F = \frac{(y.U_2)^2}{\left[(y.U_5)^2 + (y.U_6)^2\right]/2} = \frac{(y.U_2)^2}{\|\text{error vector}\|^2/2} = \frac{7.26}{.42} = 17.29$$

If $H_0 : \mu_1 = \mu_2$ were true then this test statistic would come from the $F_{1,2}$ distribution which has a 95 percentile of 18.51. Our value of 17.29 is therefore not significant at the 5% level, although it would be significant at the 10% level. Our data is not good enough, therefore, for us to formally reject the hypothesis $H_0 : \mu_1 = \mu_2$. We conclude that there is insufficient data for us to say with any certainty that diet A is superior to diet B.

ANOVA Table

The usual analysis of variance table is based on the orthogonal decomposition

$$(y - \bar{y}_{..}) = (\bar{y}_{i\cdot} - \bar{y}_{..}) + (\bar{y}_{\cdot j} - \bar{y}_{..}) + (y - \bar{y}_{i\cdot} - \bar{y}_{\cdot j} + \bar{y}_{..})$$

$$\begin{bmatrix} 12.4 \\ 11.2 \\ 8.5 \\ 9.2 \\ 9.8 \\ 6.5 \end{bmatrix} - \begin{bmatrix} 9.6 \\ 9.6 \\ 9.6 \\ 9.6 \\ 9.6 \\ 9.6 \end{bmatrix} = \begin{bmatrix} 1.1 \\ 1.1 \\ 1.1 \\ -1.1 \\ -1.1 \\ -1.1 \end{bmatrix} + \begin{bmatrix} 1.2 \\ .9 \\ -2.1 \\ 1.2 \\ .9 \\ -2.1 \end{bmatrix} + \begin{bmatrix} .5 \\ -.4 \\ -.1 \\ -.5 \\ .4 \\ .1 \end{bmatrix}$$

Observation vector   =   Treatment +   Block   +   Error
corrected for mean   vector   vector   vector

This decomposition is illustrated in Figure 12.1.

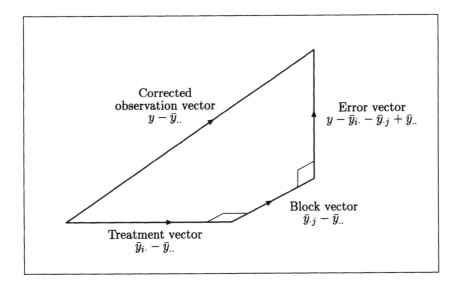

Figure 12.1: Orthogonal decomposition of the corrected observation vector, randomized block design.

The resulting ANOVA table is given in Table 12.4. In the general case of more than two treatments, the treatment sum of squares can be decomposed into contrast sums of squares in the usual manner.

Source of Variation	df	SS	MS	F
Block	2	13.32	6.66	
Treatment	1	7.26	7.26	17.29 (ns)
Error	2	.84	.42	
Total	5	21.42		

Table 12.4: ANOVA table for randomized block design.

Estimation of σ^2

As in previous chapters, we have transformed our original set of random variables, $Y_{ij} \sim N[\mu_i + \beta_j, \sigma^2]$, into a new set of random variables, $Y.U_1, \ldots, Y.U_6$, with means and variances shown in Table 12.5. The first of these is used to estimate the overall mean, μ, while the second is used to estimate the treatment contrast, $\mu_2 - \mu_1$. The third and fourth allow us to estimate the block contrasts, $\beta_2 - \beta_1$ and $\beta_3 - (\beta_1 + \beta_2)/2$. The remaining pair of random variables are used to estimate σ^2 via

$$s^2 = \frac{(y.U_5)^2 + (y.U_6)^2}{2} = \frac{\|\text{error vector}\|^2}{2} = \frac{.84}{2} = .42$$

Here $y.U_5$ and $y.U_6$ are observations from an $N[0, \sigma^2]$ distribution, so their squares average to σ^2 over many repetitions of the experiment.

	Average, or expected value	Variance
$Y.U_1$	$\sqrt{6}\mu$	σ^2
$Y.U_2$	$\sqrt{3}(\mu_2 - \mu_1)/\sqrt{2}$	σ^2
$Y.U_3$	$\sqrt{2}(\beta_2 - \beta_1)/\sqrt{2}$	σ^2
$Y.U_4$	$\sqrt{2}(2\beta_3 - \beta_1 - \beta_2)/\sqrt{6}$	σ^2
$Y.U_5$	0	σ^2
$Y.U_6$	0	σ^2

Table 12.5: Means and variances of the projection coefficients, $Y.U_i$.

In Table 12.5 the expected values of the random variables $Y.U_i$ were calculated as in previous chapters by expanding each $y.U_i$ in terms of the observations, then substituting the expected values given in Table 12.2. The pattern in the expected values for the block contrasts mimics that for the treatment contrasts. For a treatment contrast, $c_1\mu_1 + \cdots + c_k\mu_k$, the expected value of the corresponding $Y.U_i$ is $\sqrt{n}(c_1\mu_1 + \cdots + c_k\mu_k)/\sqrt{\sum_{i=1}^{k} c_i^2}$ while for a block contrast, $d_1\beta_1 + \cdots + d_n\beta_n$, the corresponding expected value is $\sqrt{k}(d_1\beta_1 + \cdots + d_n\beta_n)/\sqrt{\sum_{j=1}^{n} d_j^2}$.

Assumption Checking

Since our data is contrived, there is little point in drawing a histogram of the residuals or checking that the assumptions of independence and equal variance are reasonable. However, we should note that another assumption has been made. This is the assumption that the treatment and block effects are *additive*. That is, we have assumed that diets A and B differ by a constant amount over all blocks. This model seems reasonable in our case, since the differences between diets are 3.2, 1.4 and 2.0 kg in the very overweight to moderately overweight blocks respectively. An alternative model is to assume that the difference between diets is constant as a percentage; for example, diet B may work out to be approximately 75% as effective as diet A in all blocks. This model could be fitted by logarithm transforming the data.

Confidence Interval for $\mu_1 - \mu_2$

To work out the likely range for $\mu_1 - \mu_2$ we need to calculate a confidence interval. From Table 12.5 we know that $Y.U_2 = \sqrt{3}(\overline{Y}_{2.} - \overline{Y}_{1.})/\sqrt{2}$ is distributed as $N\left[\sqrt{3}(\mu_2 - \mu_1)/\sqrt{2}, \sigma^2\right]$. Hence the random variable $W_1 = \sqrt{3}[(\overline{Y}_{1.} - \overline{Y}_{2.}) - (\mu_1 - \mu_2)]/\sqrt{2}$ is distributed as $N(0, \sigma^2)$, so the ratio

$$\frac{\sqrt{3}[(\overline{Y}_{1.} - \overline{Y}_{2.}) - (\mu_1 - \mu_2)]/\sqrt{2}}{\sqrt{[(Y.U_5)^2 + (Y.U_6)^2]/2}}$$

is a t_2 statistic. To obtain a 95% confidence interval for $\mu_1 - \mu_2$ we simply gamble that the realized value

$$t = \frac{\sqrt{3}[(\bar{y}_{1.} - \bar{y}_{2.}) - (\mu_1 - \mu_2)]/\sqrt{2}}{\sqrt{s^2}} = \frac{(\bar{y}_{1.} - \bar{y}_{2.}) - (\mu_1 - \mu_2)}{\sqrt{2s^2/3}}$$

lies between the 2.5 and 97.5 percentiles of the t_2 distribution, that is, between -4.303 and 4.303.

The resulting interval is

$$(\bar{y}_{1\cdot} - \bar{y}_{2\cdot}) - 4.303 \times \sqrt{2s^2/3} \leq \mu_1 - \mu_2 \leq (\bar{y}_{1\cdot} - \bar{y}_{2\cdot}) + 4.303 \times \sqrt{2s^2/3}$$
$$\text{that is,} \quad 2.2 - 2.28 \leq \mu_1 - \mu_2 \leq 2.2 + 2.28$$
$$\text{or,} \quad -.08 \leq \mu_1 - \mu_2 \leq 4.48$$

In summary we can state, with 95% confidence that we are correct, that the average advantage of diet A over diet B is between −.08 kg and 4.48 kg of weight loss over a six week period.

Report on Study

To report the results of the study we may well quote the 95% confidence interval for the advantage of diet A over diet B, as well as the average weight loss for diet A, 10.7 kg. Alternatively, we could present a summary of means, standard errors and LSD(5%), as shown in Table 12.6.

	Weight loss in kg during a 6 week period
Diet A	10.7
Diet B	8.5
SED	.53
LSD(5%)	2.3
Significance of difference	n.s.

Table 12.6: Summary of results for the diet study.

Minitab Analysis

To fit the model using Minitab we simply Regress the observation vector y onto the directions U_2, U_3, U_4 which span the model space together with U_1. Alternatively, we can use the Twoway command to calculate treatment, block and error sums of squares. The necessary commands for both possibilities are given in Table 12.7.

List of commands	Interpretation
name c1 = 'Wtloss' name c2 = 'Diets' name c3 = 'Blkcon1' name c4 = 'Blkcon2' name c5 = 'Trtments' name c6 = 'Blocks' name c7 = 'Errors' name c8 = 'Fits'	[Naming the columns
set c1 12.4 11.2 8.5 9.2 9.8 6.5	[Set observation vector into column one
set c2 −1 −1 −1 1 1 1 let c2 = c2/sqrt(6) set c3 −1 1 0 −1 1 0 let c3 =c3/sqrt(4) set c4 −1 −1 2 −1 −1 2 let c4 =c4/sqrt(12)	[Set U_2, U_3, U_4 into columns two, three, four
[regress c1 3 c2−c4	[Project y onto the model space
set c5 1 1 1 2 2 2	[Set treatment numbers into column five
set c6 1 2 3 1 2 3	[Set block numbers into column six
[twoway c1 c5 c6 c7 c8	[Do ANOVA
[print c1 c8 c7	[Print fitted model
[histogram c7	[Print histogram of errors
[plot c7 c8	[Plot errors versus fitted values
[stop	[End of job

Table 12.7: Minitab commands for diet study, randomized block design.

12.2   General Discussion

We now give a step-by-step summary of what happened when we analyzed the diet example in the previous section, placing things in a general setting.

1. With the randomized block design, the model space is expanded to include a block space as well as the usual overall mean and treatment spaces.

2. The overall mean space is, as before, spanned by the coordinate axis $U_1 = [1, \ldots, 1]^T$. The fitted overall mean vector is $\bar{y}_{..}$.

3. The treatment space is unchanged: it is $k - 1$ dimensional and the fitted treatment vector remains $\bar{y}_{i.} - \bar{y}_{..}$. Hypotheses of interest are specified in terms of contrasts among the treatment means, which in turn correspond to directions U_i within the treatment space. There is a one-to-one correspondence between a complete set of orthogonal contrasts and a coordinate system for the treatment space.

4. If there are n blocks then the block space is $n - 1$ dimensional, and the fitted block vector is $\bar{y}_{.j} - \bar{y}_{..}$. Each contrast among the block means corresponds to a direction in the block space. Therefore a coordinate system for the block space corresponds to a complete set of $n-1$ contrasts among the block means. The block space is in all ways analogous to the treatment space.

5. The fitted model vector is most symmetrically written as

$$\bar{y}_{..} + (\bar{y}_{i.} - \bar{y}_{..}) + (\bar{y}_{.j} - \bar{y}_{..})$$

This corresponds to a true model vector of $\mu + \alpha_i + \beta_j$, where $\alpha_i = \mu_i - \mu$ denotes a treatment effect and β_j denotes a block effect. In our development we have written the fitted model vector alternatively as

$$\bar{y}_{i.} + (\bar{y}_{.j} - \bar{y}_{..})$$

corresponding to a true model vector of $\mu_i + \beta_j$. This was done to avoid introducing a more complicated notation. However, the two forms are clearly equivalent.

6. The corrected form of the fitted model is

$$y - \bar{y}_{..} \;\; = \;\; (\bar{y}_{i.} - \bar{y}_{..}) \;+\; (\bar{y}_{.j} - \bar{y}_{..}) \;+\; \text{error vector}$$

where the error vector is obtained by subtraction. This was illustrated in Figure 12.1.

7. The usual ANOVA table is based on the Pythagorean decomposition

$$\|y - \bar{y}_{..}\|^2 \;\; = \;\; \|\bar{y}_{i.} - \bar{y}_{..}\|^2 \;+\; \|\bar{y}_{.j} - \bar{y}_{..}\|^2 \;+\; \|\text{error vector}\|^2$$

where the error sum of squares is obtained by subtraction. This decomposition can be obtained using the Twoway command of Minitab.

8. The error space is the complement of the model space, so has dimension

$$kn - [1 + (k-1) + (n-1)] = (k-1)(n-1)$$

If coordinate axes are desired, they can be obtained by "crossing" the $k-1$ treatment space axes with the $n-1$ block space axes. The error space is in fact the space of treatment by block interactions.

9. The estimate of σ^2 is

$$s^2 = \frac{\|\text{error vector}\|^2}{(k-1)(n-1)}$$

10. This estimate, s^2, is used as the baseline for hypothesis tests. For example, if U_c is the direction in the treatment space corresponding to the contrast $c = c_1\mu_1 + \cdots + c_k\mu_k$, the test statistic is $(y.U_c)^2/s^2$. Here $(y.U_c)^2$ can be calculated in Minitab by regressing y on U_c.

11. In our diet study we wrote down a complete set of orthogonal contrasts among the treatment means (rather trivially, $c_1 = \mu_2 - \mu_1$), and a complete set of orthogonal contrasts among the block means (we chose $d_1 = \beta_2 - \beta_1$ and $d_2 = \beta_3 - (\beta_1 + \beta_2)/2$). This enabled us to write down complete coordinate systems for the treatment space, U_2, and for the block space, U_3 and U_4. These were then used to fit the model as

$$
\underbrace{(y - \bar{y}..)}_{} = \underbrace{(y.U_2)U_2}_{\substack{\text{treatment} \\ \text{vector}}} + \underbrace{(y.U_3)U_3 + (y.U_4)U_4}_{\longleftarrow \text{ block vector } \longrightarrow} + \text{ error vector}
$$

This method can be used quite generally. In Minitab the method involves regressing y on the coordinate axes, here U_2, U_3 and U_4. When there are several independent hypotheses of interest this is often the tidiest computing method.

12. Confidence intervals are based on the same formulae as in preceding chapters, except that the number of error degrees of freedom used for the t value changes from $k(n-1)$ to $(k-1)(n-1)$.

In summary, the change in going from a completely randomized design to a randomized block design is that the $k(n-1)$ dimensional error space is partitioned into an $n-1$ dimensional block space and a $(k-1)(n-1)$ dimensional error space. If used intelligently, the blocking will account for a worthwhile percentage of the variation in the subject material. This is reflected in generally larger projection coefficients in the block space than in the new error space, so reducing the estimate of σ^2: in other words, the model hugs the data more closely. This leads to more powerful hypothesis tests and narrower confidence intervals.

In agricultural research, randomized block designs are the most common, with completely randomized designs being relatively uncommon. For this reason the designs of Examples A to D were simplified for use in Chapter 7. Of these, Examples A, B and D had a randomized block design, and are reanalyzed in Exercise 12.8. We now move on in §3 to analyze data from a randomized block experiment with a 2×3 factorial structure, before describing in general terms the "why" and "how" of blocking in §4.

12.3   A Realistic Case Study

To illustrate the partitioning of the treatment space and the setting up of the block space for a real example, we now consider a study of planting density for garlic. The trial layout and data were provided by Mike Malone and Jeremy Davison of the Marlborough Research Centre, Blenheim, New Zealand.

Objectives

Garlic is grown commercially in long, one metre wide beds with the outer rows of adjacent beds separated by 50 cm to allow for tractor movement. The objective of this experiment was to ascertain how many rows should be planted within each bed, and how close the garlic should be planted within each row.

Design

It was decided to use a 2×3 factorial treatment design, with factor A the spacing of plants within each row (5cm and 10cm), and factor B the number of rows per bed (3, 4 and 5). The trial was laid out as a randomized block design, as shown in Figure 12.2.

Data

At harvest time entire plots were harvested and the garlic weighed. The area occupied by each small bed (or plot) was taken as $4\text{m} \times 1.5\text{m} = 6\text{m}^2$, and the yields were scaled to units of kg per hectare. The resulting data is given in Table 12.8.

The corresponding observation vector is

$$y = [y_{11}, \ldots, y_{64}]^T = [5110, \ldots, 5875]^T$$

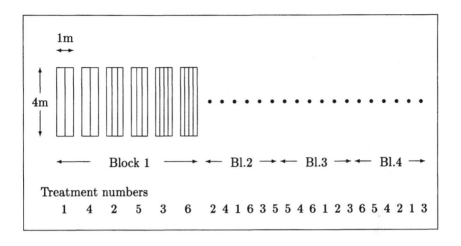

Figure 12.2: Field layout for garlic planting density trial.

Block:	1	2	3	4	Treatment mean
5cm spacing, 3 rows	5110	4040	8405	7045	6150
5cm spacing, 4 rows	6675	7000	8150	8145	7493
5cm spacing, 5 rows	7265	7230	6660	9259	7604
10cm spacing, 3 rows	3710	3785	4637	4635	4192
10cm spacing, 4 rows	4095	3730	4985	6540	4838
10cm spacing, 5 rows	6415	5355	5070	5875	5679
					Grand mean
Block mean:	5545	5190	6318	6917	5992

Table 12.8: Weight of garlic in kilograms per hectare, for each plot. Block, treatment and grand means are also given.

Model

The expected value of the theoretical population of garlic weights from the ith treatment and j^{th} block is assumed to be $\mu_i + \beta_j$. Here μ_i is the expected value of the ith treatment mean and β_j is the expected value of the difference between the jth block mean and the overall mean. The resulting model vector is

$$
\begin{bmatrix}
\mu_1 + \beta_1 \\
\mu_1 + \beta_2 \\
\mu_1 + \beta_3 \\
\mu_1 + \beta_4 \\
\cdot \\
\cdot \\
\cdot \\
\mu_6 + \beta_1 \\
\mu_6 + \beta_2 \\
\mu_6 + \beta_3 \\
\mu_6 + \beta_4
\end{bmatrix}
= \mu_1
\begin{bmatrix}
1 \\ 1 \\ 1 \\ 1 \\ \cdot \\ \cdot \\ \cdot \\ 0 \\ 0 \\ 0 \\ 0
\end{bmatrix}
+ \cdots + \mu_6
\begin{bmatrix}
0 \\ 0 \\ 0 \\ 0 \\ \cdot \\ \cdot \\ \cdot \\ 1 \\ 1 \\ 1 \\ 1
\end{bmatrix}
+ \beta_1
\begin{bmatrix}
1 \\ 0 \\ 0 \\ 0 \\ \cdot \\ \cdot \\ \cdot \\ 1 \\ 0 \\ 0 \\ 0
\end{bmatrix}
+ \cdots + \beta_4
\begin{bmatrix}
0 \\ 0 \\ 0 \\ 1 \\ \cdot \\ \cdot \\ \cdot \\ 0 \\ 0 \\ 0 \\ 1
\end{bmatrix}
$$

This defines the model space M.

An Appropriate Coordinate System

In order to produce an appropriate coordinate system for the treatment space we must first specify a complete set of five orthogonal contrasts among the treatment means. These are automatically generated by the appropriate sets of contrasts for factors A and B, as described in Chapter 10. For factor A, with only two levels of spacing, there is only one possible contrast; for factor B, with levels of 3, 4 and 5 rows per bed, it is most natural to specify linear and quadratic contrasts. The resulting set of five contrasts is shown in Table 12.9.

	Main effect contrasts			Interaction contrasts	
	A (spacing)	B (no. rows)		A × B	
		Linear	Quad	Spacing × lin	Spacing × quad
	c_1	c_2	c_3	c_4	c_5
μ_1	-1	-1	1	1	-1
μ_2	-1	0	-2	0	2
μ_3	-1	1	1	-1	-1
μ_4	1	-1	1	-1	1
μ_5	1	0	-2	0	-2
μ_6	1	1	1	1	1

Table 12.9: Factorial contrasts for garlic experiment.

The corresponding coordinate system for the treatment space is:

$$
U_2 = \begin{bmatrix} -1 \\ \cdot \\ -1 \\ \cdot \\ -1 \\ \cdot \\ 1 \\ \cdot \\ 1 \\ \cdot \\ 1 \\ \cdot \end{bmatrix}\frac{1}{\sqrt{24}}
\quad
U_3 = \begin{bmatrix} -1 \\ \cdot \\ 0 \\ \cdot \\ 1 \\ \cdot \\ -1 \\ \cdot \\ 0 \\ \cdot \\ 1 \\ \cdot \end{bmatrix}\frac{1}{\sqrt{16}}
\quad
U_4 = \begin{bmatrix} 1 \\ \cdot \\ -2 \\ \cdot \\ 1 \\ \cdot \\ 1 \\ \cdot \\ -2 \\ \cdot \\ 1 \\ \cdot \end{bmatrix}\frac{1}{\sqrt{48}}
\quad
U_5 = \begin{bmatrix} 1 \\ \cdot \\ 0 \\ \cdot \\ -1 \\ \cdot \\ -1 \\ \cdot \\ 0 \\ \cdot \\ 1 \\ \cdot \end{bmatrix}\frac{1}{\sqrt{16}}
\quad
U_6 = \begin{bmatrix} -1 \\ \cdot \\ 2 \\ \cdot \\ -1 \\ \cdot \\ 1 \\ \cdot \\ -2 \\ \cdot \\ 1 \\ \cdot \end{bmatrix}\frac{1}{\sqrt{48}}
$$

where each entry is repeated four times.

It remains for us to set up a coordinate system for the block space. For this purpose, any set of block contrasts will do the job. The most obvious is the set $d_1 = \beta_2 - \beta_1, d_2 = \beta_3 - (\beta_1 + \beta_2)/2$ and $d_3 = \beta_4 - (\beta_1 + \beta_2 + \beta_3)/3$. This set corresponds to a coordinate system for the block space of:

$$
U_7 = \frac{1}{\sqrt{12}}\begin{bmatrix} -1 \\ 1 \\ 0 \\ 0 \\ \vdots \\ -1 \\ 1 \\ 0 \\ 0 \end{bmatrix},
\quad
U_8 = \frac{1}{\sqrt{36}}\begin{bmatrix} -1 \\ -1 \\ 2 \\ 0 \\ \vdots \\ -1 \\ -1 \\ 2 \\ 0 \end{bmatrix},
\quad
U_9 = \frac{1}{\sqrt{72}}\begin{bmatrix} -1 \\ -1 \\ -1 \\ 3 \\ \vdots \\ -1 \\ -1 \\ -1 \\ 3 \end{bmatrix}
$$

where the sequence is repeated for each treatment.

Fitting the Model

To fit the model we project the observation vector $y = [5110, 4040, \ldots, 5875]^T$ onto the model space, $M = \mathrm{span}\{U_1, U_2, \ldots, U_9\}$. The resulting fitted model vector is

$$
(y.U_1)U_1 \;+\; \underbrace{(y.U_2)U_2 + \cdots + (y.U_6)U_6}_{\longleftarrow \text{ treatment vector } \longrightarrow} \;+\; \underbrace{(y.U_7)U_7 + \cdots + (y.U_9)U_9}_{\longleftarrow \text{ block vector } \longrightarrow}
$$

This can be fitted using Minitab by regressing y on $U_2, \cdots, U_9$.

An alternative form of the fitted model vector is

$$
\bar{y}_{..} \;+\; (\bar{y}_{i.} - \bar{y}_{..}) \;+\; (\bar{y}_{.j} - \bar{y}_{..})
$$

This can be fitted using the Twoway command of Minitab.

ANOVA Table

The appropriate Pythagorean breakups can be variously written as follows:

$$\|y - \bar{y}_{..}\|^2 = (y.U_2)^2 + \cdots + (y.U_6)^2 + (y.U_7)^2 + \cdots + (y.U_9)^2 + \|\text{error vector}\|^2$$
$$= (y.U_2)^2 + \cdots + (y.U_6)^2 + \quad \|\bar{y}_{.j} - \bar{y}_{..}\|^2 \quad + \quad \|\text{error vector}\|^2$$
$$= \quad \|\bar{y}_{i.} - \bar{y}_{..}\|^2 \quad + \quad \|\bar{y}_{.j} - \bar{y}_{..}\|^2 \quad + \quad \|\text{error vector}\|^2$$

The resulting ANOVA table is given in Table 12.10.

Source of Variation	df	SS	MS	F
Blocks	3	10823281	3607760	
Treatments	5	38181167		
Spacing	1	28496962	28496962	33.40(**)
Rows(lin)	1	8646540	8646540	10.14 (**)
Rows(quad)	1	357765	357765	.42
Spacing × Rows(lin)	1	1122	1122	.00
Spacing × Rows(quad)	1	678776	678776	.80
Error	15	12796692	853113	
Total	23	61801140		

Table 12.10: ANOVA table for garlic experiment.

From Table 12.10 we conclude that there is a significant difference in weight of garlic between the 5cm and 10cm spacings within the rows. We also conclude that there is a significant trend in weight of garlic with number of rows per bed, with no significant curvature about this trend. Before summarizing our results more formally, we shall check our assumptions.

Assumption Checking

The assumptions we need to check are those of independence, normality, constant variance and additivity of block effects and treatment means.

To check for exceptionally fertile or infertile patches in the field, which would affect our assumption of independence, we plot the errors according to position in the field (Figure 12.3). To show how the blocking has served to reduce the errors, we have plotted the errors about the general level of the observations in the block. The errors which would have resulted from an analysis of the data as a completely randomized design can then be easily seen as distances from the horizontal axis.

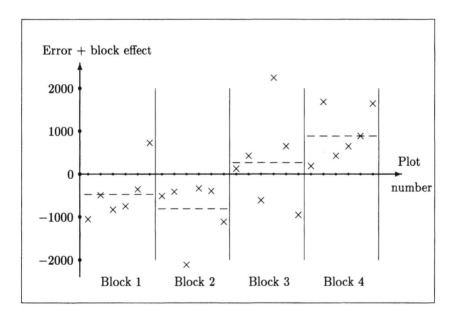

Figure 12.3: Error plus block effect (equal to observation minus treatment mean) plotted against position in field, garlic trial. The block effects are shown as dotted lines. The errors are the distances from the dotted lines.

As one would normally expect in a field experiment, there does appear to be a fertility gradient, with the first and second blocks being less fertile than the third and fourth blocks (Figure 12.3). However, the errors seem to be reasonably scattered about the block averages, so we shall accept the assumption of independence.

To check the assumption of normality we plot a histogram of the errors in Figure 12.4.

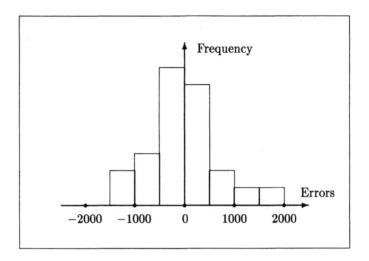

Figure 12.4: Histogram of the errors, garlic trial.

The histogram appears to be fairly normal. The largest standardized error is $1929/924 = 2.09$ which is not out of the ordinary.

To check the assumption of constant variance, we square the errors for each treatment and add. Since there are 15 d.f. for error and 6 treatments, we can divide the sum of squares within each treatment by $15/6 = 2.5$; the resulting variance estimates, given in Table 12.11, will then average out to the pooled variance, $s^2 = 853113$.

Treatment	Variance estimate
3 row, 5cm	2314032
4 row, 5cm	166820
5 row, 5cm	936377
3 row, 10cm	161317
4 row, 10cm	327094
5 row, 10cm	1213036
Average	853113

Table 12.11: Variance estimates for individual treatments, garlic trial.

In Table 12.11 the variance estimates follow no obvious pattern, so we accept the assumption of constant variance.

Finally we check the assumption of additivity of block and treatment effects. This we do by plotting the observations in each block against the treatment means, as illustrated in Figure 12.5. Lines can then be drawn by hand (or computer) through the values for each block. For the additivity assumption to hold these lines should be roughly parallel with a slope of one, as is the case in Figure 12.5.

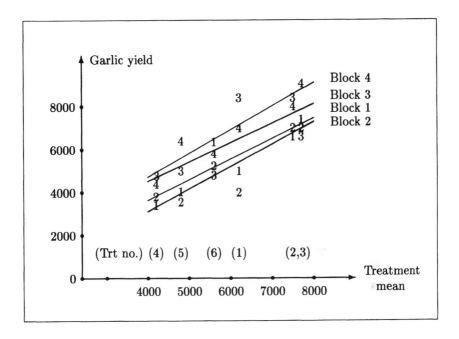

Figure 12.5: Observations, identified by block number (1, 2, 3, 4), plotted against treatment means. The block 1 line is drawn to fit the 1's, and so on.

Report on Study

One way of reporting the results is to present a list of treatment means, SED's and significance levels using either of the methods shown in Table 12.12. However, a much more attractive way of presenting the results is to graph the treatment means plus approximating polynomials as shown in Figure 12.6.

Method 1		Method 2	
	Garlic yield	**Main Effects**	**Garlic yield**
Treatment	**(kg/ha)**	**Spacing**	**(kg/ha)**
5 cm, 3 rows	6150	5 cm	7080
5 cm, 4 rows	7490	10 cm	4900
5 cm, 5 rows	7600	SED	377
10 cm, 3 rows	4190	Significance	**
10 cm, 4 rows	4840	**Number of rows**	
10 cm, 5 rows	5680	3 rows	5170
SED	653	4 rows	6170
Contrasts		5 rows	6640
Spacing	**	SED	462
Row(lin)	**	**Contrasts**	
Row(quad)	ns	Linear	**
Spacing × row (lin)	ns	Quadratic	ns
Spacing × row (quad)	ns	**Interaction contrasts**	
		Spacing × Row(lin)	ns
		Spacing × Row(quad)	ns

Table 12.12: Presentation of results, garlic study, by two methods.

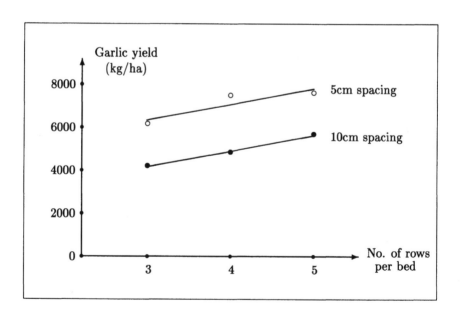

Figure 12.6: Plot of treatment means against number of rows per bed for 5cm (○) and 10cm (●) spacing. Approximating polynomials are parallel lines.

The only significant effects in our analysis of variance are the spacing effects and the "average" linear trend in the number of rows per bed. There is no evidence of any difference in the trends between the 5cm and 10cm spacings, nor of any curvature effects. This means that parallel straight lines can reasonably be used to approximate the data.

To obtain the fitted values on the lines at $x = 3, 4, 5$ shown in Figure 12.6, we can rerun our Minitab job with block effects and non-significant treatment effects excluded from the model; this involves regressing y onto just U_2 and U_3. If the equations of the lines are required, U_2 can be replaced by $[-1, -1, \ldots, 1, 1]^T$ and U_3 by the vector of x-values, $[3, 3, \ldots, 5, 5]^T$. In this case Minitab prints the fitted model as

$$c1 = 3052 - 1090 \ c2 + 735 \ c3$$

The equations of the two lines shown in Figure 12.6 are then

$$
\begin{aligned}
y &= 3052 - (-1090) + 735x \\
  &= \quad\quad 4142 \quad\quad + 735x \quad \text{(5cm spacing)} \\
y &= 3052 + (-1090) + 735x \\
  &= \quad\quad 1962 \quad\quad + 735x \quad \text{(10cm spacing)}
\end{aligned}
$$

12.4 Why and How to Block

The objective of blocking is to reduce the amount of unexplained variability in the most important variable in a study. In most studies several variables are measured; in our garlic study the number of bulbs and the average bulb weight were recorded as well as the total weight per plot. At the design stage we must settle on the most important variable, in this instance the total weight per plot. The blocking is then carried out in such a way as to trap as much as possible of the variation in this variable. Fortunately, the resulting increase in precision often carries through to the less important variables.

Using the garlic trial as an example, we must ask ourselves: what factors can cause variation in the total weight of garlic harvested from each plot? The most obvious answer is variations in fertility or other conditions in the field, such as variations along an irrigation spray line. A less obvious answer is variations between workers at the time of planting the garlic cloves or at the time of harvest; for example, one worker may be particular about which cloves are large enough for planting, or which bulbs are large enough to qualify as merchantable produce. Interruptions at time of planting or harvest can also introduce variability into the data; for example, field operations could be half completed when the remaining work is delayed for two weeks by storms, breakage of equipment or illness in the family. Intelligent blocking can simultaneously take account of, or insure against, any variability in our observations caused by the most important of these factors.

The aim of the experimenter should always be to ensure that variation occurs not within blocks, but between blocks. Then the bulk of the variability in our measurements will be reflected in the block space rather than the error space, giving rise to a more sensitive experiment. We now offer some examples of this in field and animal trials.

In field trials, plots should be blocked to cover the most important sources of variability known to the experimenter at the time of laying down. For example, blocks one and two could be put on a more fertile area and blocks three and four on a less fertile area, with blocks one and three being less effectively irrigated than blocks two and four. In general, it is wise to keep the plots within a block geographically contiguous, with blocks as compact as possible to insure against localized disasters such as one block being flooded, or eaten by cattle breaking through the fence. Field operations should be carried out one block at a time to insure against effects such as the gradual improvement in the skill of the workers, or changes in the condition of the machinery, as well as to insure against interruptions. If two workers or machines are being used, the best strategy is to have each work on half of each plot. If this is impractical, a single worker or machine should process an entire block, with a different worker or machine processing the next block. Following the harvest of the trial the produce may perhaps be placed in a coolstore to allow comparisons of storage life between the experimental treatments. If so, the produce should be kept in its original blocks with an eye to variations within the coolstore.

With trials involving animals   (including humans!)   the same basic principles apply. The important sources of variability here may be age, sex and weight. Suppose, for example, we wish to allocate 120 six-month-old lambs to six treatment groups, and that half the lambs are ewe lambs and half are ram lambs. The usual method would be to weigh the lambs, recording weights and ear tag numbers, then sort the 60 ewe lambs into order of decreasing weight, and similarly for the 60 ram lambs. The six heaviest ewe lambs would then form block one, within which one lamb would be allocated randomly to each treatment. The six next heaviest ewe lambs would form block two, and so on, until we end up with twenty blocks, ten blocks of ewe lambs and ten blocks of ram lambs. The twenty lambs allocated to each treatment in this manner would of course be half ewes and half rams, and would have an even spread of weights.

In certain types of experiment we need to be careful in defining the population of study to avoid violating the assumption of additivity of treatment and block effects. For example, in a study of the effect of phosphatic fertilizer on pasture production it would be unfortunate if two blocks were located on an extremely fertile area and two blocks on a very unfertile area. Hypothetical data from such a study is tabulated in Table 12.13, and the

Block	1	2	3	4	Treatment mean
Control (No P)	6000	6600	3000	4000	4900
20 kg P/ha	6200	6400	4000	4400	5250
40 kg P/ha	6200	6400	4800	5000	5600
60 kg P/ha	6000	6600	5400	5800	5950

Table 12.13: Example of treatment by block interaction. Blocks one and two are fertile and unresponsive, whereas blocks three and four are infertile and responsive. Data is pasture yield in kg/ha.

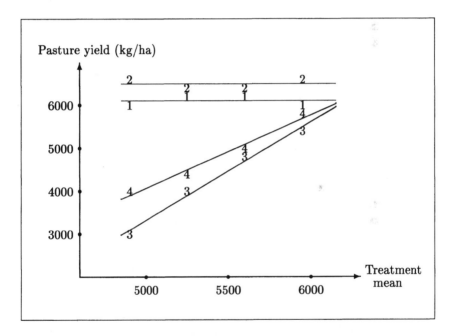

Figure 12.7: Example of violation of the additivity assumption. In the figure the observations in each block are numbered (1, 2, 3, 4).

check of additivity is graphed in Figure 12.7.

The moral is that while the method of blocking can cope with moderate variations in conditions, it should not be carried to the extreme. In the example just cited one can conceive of two populations of study: the population of fertile areas, and the population of infertile areas. In this case the experiment should perhaps have been concentrated on infertile areas if this was where the main interest lay. Alternatively, two experiments could have been conducted, one on infertile and one on fertile areas.

Similar problems arise in insect control experiments if some blocks are heavily infested with the target insect and other blocks have virtually no insects. Likewise in weed control experiments, it is ridiculous to locate some blocks on weedy areas and other blocks on weed free areas. In experiments on worm control in sheep, effects can be more subtle. If sheep are allocated to blocks on the basis of liveweight, it may be that the heavy sheep are the worm-resistant sheep while the lighter sheep are more susceptible to worms. Thus we may observe a response to worm control in the light liveweight blocks but not in the heavy liveweight blocks.

12.5 Summary

In a randomized block design the experimental units are allocated at random to treatments within blocks.   These blocks are arranged so that experimental units are as similar as possible within each block.   This allows us to isolate in the "block space" as much as possible of the variability caused by differences between experimental units, in a manner entirely analogous to the way in which the differences between treatments are isolated in the treatment space. The gains in precision due to blocking are often only modest; however, when natural or manmade disasters occur the gains can be substantial. For this reason blocking is often carried out routinely as a form of insurance.

The fitted model for this design is

$$y - \bar{y}_{..} = (\bar{y}_{i.} - \bar{y}_{..}) + (\bar{y}_{.j} - \bar{y}_{..}) + (y - \bar{y}_{i.} - \bar{y}_{.j} + \bar{y}_{..})$$

as illustrated in Figure 12.8.

The block space can be thought of in exactly the same way as the treatment space.   Contrasts between the block means correspond to directions in the block space in exactly the same way as contrasts between the treatment means correspond to directions in the treatment space. The block vector is $\bar{y}_{.j} - \bar{y}_{..}$ in the same way as the treatment vector is $\bar{y}_{i.} - \bar{y}_{..}$.

The error vector can be written as $(y - \bar{y}_{i.}) - (\bar{y}_{.j} - \bar{y}_{..})$, the completely randomized design error vector minus the vector of block effects. The error space now has dimension $(k - 1)(n - 1)$: the block space has stripped $n - 1$ dimensions from the $k(n - 1)$ dimensions of the error space in a completely randomized design.   The appropriate estimate of σ^2 is $s^2 = \|\text{error vector}\|^2/[(k - 1)(n - 1)]$.

This estimate of σ^2 is used for hypothesis testing and confidence interval estimation in an identical manner to that of Chapters 5–11. For example, the hypothesis $H_0 : c = c_1\mu_1 + \cdots + c_k\mu_k = 0$ is tested using the test statistic $(y.U_c)^2/s^2$, as previously.

For additional reading on the geometric aspects refer to Box, Hunter and Hunter (1978), pages 208–218. For more general reading refer to Cochran and Cox (1957), pages 106–110, or Little and Hills (1978), pages 53–60.

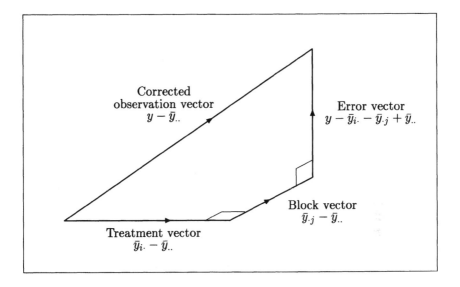

Figure 12.8: Orthogonal decomposition of the corrected observation vector, randomized block design.

Class Exercise

This class exercise has been designed to highlight the difference between a completely randomized and a randomized block design. A cautionary note: the exercise does tend to exaggerate the potential advantages of blocking.

Analyzing a Completely Randomized Design

Each class member begins by simulating an experiment with three treatments and three replicates laid down in a completely randomized design. This experiment was designed to assess the effects of lime and superphosphate on pasture production. Treatments applied in the spring were as follows:

1. Control

2. Superphosphate, at 250 kg/ha

3. Lime, at 2000 kg/ha

The data given in Table 12.14 is the amount of pasture growth during a one month period of the following autumn, for 64 hypothetical control plots.

| | Column | | | | | | | |
Row	1	2	3	4	5	6	7	8
1	510	570	720	570	570	360	630	690
2	480	510	750	720	570	510	690	840
3	630	480	450	480	630	480	720	750
4	360	450	690	630	570	630	570	720
5	510	480	630	690	720	570	720	870
6	690	720	750	630	480	840	690	840
7	720	570	720	750	720	570	930	870
8	570	840	690	720	570	630	750	930

Table 12.14: Control plot yields in kilograms per hectare from a production cut in the autumn.

(a) Each class member is first asked to use random numbers to select three plots for the control treatment (for example, the first plot chosen could be in row 3, column 5). The yields of these plots are the control treatment yields.

Second they randomly select three plots for treatment two. To simulate a true response of 200 kg/ha to superphosphate they add 200 to the plot yields to obtain the superphosphate treatment yields.

Third they randomly select three plots for treatment three, adding 100 to the corresponding yields to simulate a true response to lime of 100 kg/ha. Each class member should now have nine observations ready for analysis.

(b) Class members are now asked to calculate the completely randomized design vector decomposition $y = \bar{y}_{i\cdot} + $error vector, and to compute $s^2 = \|\text{error vector}\|^2/6$.

(c) They then compute the LSD(5%) and hence determine whether their superphosphate and lime responses were significant.

(d) The class instructor now draws a histogram of the s^2 values, a histogram of the responses to superphosphate, and a histogram of the responses to lime. Also, the class instructor is asked to count the number of nonsignificant and significant responses for each of superphosphate and lime. These results should be saved for comparison with the results to be obtained in (h).

Analyzing a Randomized Block Design

Suppose now that prior to laying down the trial the experimenter visually assessed the fertility of each plot on a 0 to 10 scale, where 0 corresponds to bare ground and 10 to very fertile ground. These assessments are given in Table 12.15.

	Column							
Row	1	2	3	4	5	6	7	8
1	5	6	7	6	5	4	6	7
2	5	5	8	7	6	5	7	8
3	6	5	4	5	6	5	7	8
4	4	4	7	6	6	6	6	7
5	5	5	6	7	7	6	7	9
6	7	7	8	6	5	8	7	8
7	7	6	7	8	7	6	9	9
8	6	8	7	7	6	6	8	9

Table 12.15: Scores for fertility assessed prior to the experiment in the spring.

(e) Each class member is now asked to choose three uniform blocks of three plots on the basis of the scores given in Table 12.15. Within each block they can then randomly assign the treatments, 1 to 3, to the plots. They can then read the corresponding plot yields from Table 12.14, again adding 200 to the superphosphate values and 100 to the lime values. They should now have a new observation vector ready for analysis.

(f) Each class member now calculates the randomized block design vector decomposition $y = \bar{y}_{i.} + (\bar{y}_{.j} - \bar{y}_{..})$ + error vector, and computes $s^2 = \|\text{error vector}\|^2/4$.

(g) They then calculate the LSD(5%) and determine whether their superphosphate and lime responses were significant.

(h) The class instructor now redraws the histograms from (d), using the same scales for ease of comparison, and again counts the number of students with non-significant and significant responses for each of superphosphate and lime. The new set of results is compared with those from the completely randomized design. Is the difference dramatic?

Exercises

(12.1) A grass grub research group was investigating biological methods of controlling grass grub numbers. In the laboratory they found that a particular type of nematode was a carrier for a disease which caused mortality in grass grubs. They therefore hypothesized that spraying nematodes onto grass grub infested pastures would be effective as a control measure.

As a first field study, they set up twelve experimental plots in a grass grub infested area. The experimental treatments were:

1. No nematodes

2. Nematodes at the rate of 1 million/ha

3. Nematodes at the rate of 2 million/ha

4. Nematodes at the rate of 4 million/ha

The plots were laid out in three blocks in a randomized block design.

(a) Write down an appropriate set of orthogonal unit vectors in the treatment space. Hint: Firstly contrast the no nematode treatment with the three nematode treatments.

(b) Write down a set of orthogonal unit vectors in the block space.

(c) Describe how you would calculate a set of orthogonal unit vectors in the error space. Calculate just one of these vectors to show that you know how to do it. How many such vectors are there?

(12.2) An animal research team (Andy Bray and John Munro, Ministry of Agriculture and Fisheries, New Zealand) was investigating ways of improving the feeding value of barley straw for sheep. It was known that the nitrogenous compound, ammonia, was effective for this purpose. However, the alternative nitrogenous compound, urea, was cheaper. An experiment was therefore set up to compare two rates of urea, a single rate of ammonia and a control treatment.

A randomized block design was used. Four large round bales of barley straw of a common cultivar were obtained from five different farms. The five sources of straw were treated as blocks and one round bale was allocated randomly to each of the four treatments. The treatments involved injecting the nitrogenous compounds, dissolved in water, into the bales, covering the bales with plastic sheeting, and leaving them for four weeks (the control treatment was injected with water). During this four week period twenty sheep were introduced to the idea of eating barley straw. The sheep were then allocated within liveweight blocks to the treatments, with the four heaviest sheep assigned to block one, and so on. The sheep were then housed in five blocks of four individual pens in the woolshed for the two week experimental period. The blocking was therefore designed to simultaneously account for three sources of variation: variation in the straw from farm to farm, variation from sheep to sheep associated with varying liveweight, and variation due to position within the woolshed.

The data in the table are the *in vivo* digestibilities of the straw fed to each animal. For example, the figure of 29.2% indicates that 29.2% of the straw fed to that sheep was digested.

Treatment	Block:	1	2	3	4	5
1. Control		29.2	42.6	38.9	42.6	33.6
2. Urea, 40 g/kg straw		42.2	41.9	6.1	48.3	45.7
3. Urea, 80 g/kg straw		34.6	49.0	34.6	50.8	42.8
4. Ammonia, 40g/kg straw		36.9	47.7	56.3	55.2	51.0

(a) Fit the randomized block model to these data using the Twoway command of Minitab.

(b) Use calculated numerical values to write out in full the vector decomposition

$$y = \bar{y}_{..} + (\bar{y}_{i.} - \bar{y}_{..}) + (\bar{y}_{.j} - \bar{y}_{..}) + \text{error vector}$$

Draw a sketch of this decomposition.

(c) Draw a histogram of the errors. Is there an "unusual value" in the data? Which value is it?

(d) Substitute a value of 44.8 in the place of this unusual value, and refit the model as in (a). Redraw your histogram of errors on the same scale as in (c). What change do you notice?

(e) Assuming we are interested in contrasting treatment 1 with treatments 2 to 4, treatments 2 and 3 with 4, and treatment 2 with 3, write down the corresponding orthogonal coordinate system for the treatment space. Also write down an orthogonal coordinate system for the block space.

(f) Use these coordinate systems to fit the model using the Regress command in Minitab. Summarize your sums of squares in an ANOVA table following the format of Table 12.10. Use the substituted value of 44.8 in these analyses.

(g) Write out a summary table with means, SED and significances of contrasts following the format of Table 12.12. Include a calculated LSD(5%) to enable comparison of treatments 2 and 4 which are at a common rate of 40 g/kg straw. Also summarize the results in a brief sentence or two.

(12.3) An experiment was carried out at a local sawmill to compare the cutting speeds, in seconds, of four blades. Each blade was tried out on three wood blocks of varying hardness (Pine, Kauri and Jarrah). The results are given in the following table.

	Pine	Kauri	Jarrah
Blade A	9	5	7
Blade B	11	6	10
Blade C	8	4	6
Blade D	12	6	9

(a) Rewrite the table with the addition of treatment means, block means and block effects as shown in Table 12.3. Note that treatments correspond to blades.

(b) Use these values to write out the vector decomposition

$$y = \bar{y}_{i.} + (\bar{y}_{.j} - \bar{y}_{..}) + \text{error vector}$$

(c) Calculate s^2 in the usual way. Use this to calculate an LSD(5%) and LSD(1%). Which blades are proven superior to which other blades?

(d) Present your results in a table of means, SED and LSD's.

(12.4) The following data was provided by George Tibbitts, of the University of California at Davis, who carried out an experiment comparing two varieties of rice at three seeding rates. The resulting $2 \times 3 = 6$ treatments were laid out in three replicate blocks. The harvest yields in tonnes per hectare are given in the table.

Variety		S201			M201	
Seeding rate (kg/ha)	90	150	210	90	150	210
Block 1	4.89	4.72	4.80	4.84	5.44	5.31
Block 2	3.46	3.79	3.68	5.11	4.55	4.72
Block 3	4.34	3.71	4.05	4.11	4.39	4.76

(a) Write down the five most appropriate orthogonal contrasts between the treatment means.

(b) Write down the five corresponding unit vectors in the treatment space, and two orthogonal unit vectors in the block space.

(c) Use the Regress command of Minitab to analyze the data. Summarize your results in an ANOVA table and a table of means, SED and contrasts in the usual way. Also plot a histogram of the errors.

(d) Plot the treatment means on a graph with rice yield up the y axis and seeding rate along the x axis. Use the symbols "×" for one variety and "•" for the other variety. Without doing any formal calculations draw a line through the means for each variety in turn.

(12.5) In the case of a randomized block design study with only two treatments, each block consists of a pair of experimental subjects. The randomized block analysis is then equivalent to the single population (paired samples) analysis described in Chapter 5. To illustrate this fact we shall reanalyze the illustrative example described in §5.1. For convenience, we now reproduce the relevant wheat grain yields, in tonnes per hectare, from Table 5.1.

	Farm 1	Farm 2	Farm 3
Control	5.0	4.3	5.9
Fungicide treated	6.1	5.7	7.0

(a) Work out the treatment means, block means and block effects.

(b) Use these to write down the vector decomposition in the form

$$y = \bar{y}_{..} + (\bar{y}_{i.} - \bar{y}_{..}) + (\bar{y}_{.j} - \bar{y}_{..}) + \text{error vector}$$

(c) Write out the resulting ANOVA table in the usual format. Is your F value the same as that given in Table 5.2?

(d) Write down coordinate axes for the overall mean space, U_1, the treatment space, U_2, and the block space, U_3 and U_4. Also write down coordinate axes for the error space, U_5 and U_6.

(e) Recalculate the F value as $\dfrac{(y.U_2)^2}{[(y.U_5)^2 + (y.U_6)^2]/2}$. Show your working.

(12.6) An entomologist and an agronomist, Dr Goldson and Dr Wynn-Williams, were researching ways of controlling sitona weevils in alfalfa stands. A third researcher, Dr Trought, had suggested using sheep at the very high stocking rate of 500 sheep/ha to trample the weevil eggs in mid winter, when the alfalfa is dormant and cannot be damaged by the trampling. An experiment was therefore laid out in a weevil infested alfalfa field in a randomized block design with four replicate blocks of four treatments (trampled for 0,1,2 or 4 weeks). The 16 plots, measuring 20m by 3m, were individually fenced and stocked with three sheep, mimicking 500 sheep/ha, for the appropriate length of time. The plots were laid out with the long sides adjacent in a single row. At a later date each plot was sampled to estimate the number of larvae which had hatched from the surviving eggs. The larvae numbers per square metre for each plot are given in the table.

	Block			
Treatment	1	2	3	4
Control	877	1062	730	1542
One week's trampling	646	1200	720	859
Two weeks' trampling	434	933	582	462
Four weeks' trampling	591	499	342	665

(a) Use the Twoway command of Minitab to fit a randomized block model.

(b) Draw a histogram of the errors.

(c) Plot the basic data on a graph with larvae numbers up the y axis and number of weeks of trampling along the x axis. Distinguish between blocks by using "1" for the data from block 1, "2" for the data from block 2, and so on. Hand draw a line through the data for each block. Do the responses seem roughly similar from block to block?

(d) Use all four rates (0, 1, 2 and 4) to work out the appropriate linear contrast. Write down the corresponding unit vector, U_2. Calculate the linear contrast sum of squares, $(y.U_2)^2$, either by hand or using the Regress command of Minitab.

(e) Summarize your results in an ANOVA table and a table of means, SED and contrast as usual. What are your conclusions?
Note: With counts, it is usual to transform the data using the square root transformation. The keen reader can see the effect of this by rerunning their Minitab job after inserting the command "let c1 = sqrt(c1)".

(12.7) An Animal Health Laboratory veterinarian was interested in evaluating a preventive drenching program for controlling nematode parasitism in kid goats. He therefore set up an experiment to compare a preventive drench program with both normal farmer practice and a suppression drench program. The three resulting treatments were:

1. **Preventive program**: drenched at weaning (19 Dec 1986), then after periods of 21, 21, 28 and 28 days (9 Jan, 30 Jan, 27 Feb and 27 March 1987).

2. **Normal farmer practice**: preventive program plus two more drenches (21 April, 19 June).

3. **Suppression program**: drenched every two weeks from weaning.

The animals available for the experiment were 30 doe kids and 75 wether kids. In order to allocate them to treatments, their weaning weights and ear tag numbers were recorded, then the doe and wether kids sorted separately into order of decreasing weight. Unhealthy and deformed kids were then

Block	Tag no.	Initial livewt.	Trt no.	Lwtgain
1	636	17.3	1	9.9
(Doe bl. 1)	625	17.0	2	13.8
	618	17.2	3	13.6
2	622	15.0	1	7.8
(Doe bl. 3)	606	14.7	2	11.3
	626	15.0	3	13.9
3	627	14.4	1	9.3
(Doe bl. 5)	602	14.5	2	8.4
	603	14.5	3	13.6
4	608	12.6	1	13.2
(Doe bl. 7)	605	12.3	2	11.0
	601	12.4	3	13.7
5	47	23.0	1	8.9
(Wether bl. 1)	12	22.5	2	15.2
	56	23.0	3	22.1
6	57	21.4	1	9.8
(Wether bl. 3)	34	21.0	2	10.8
	27	21.1	3	16.7
7	35	19.9	1	11.2
(Wether bl. 5)	11	20.6	2	10.8
	58	20.4	3	15.9
8	66	19.0	1	11.9
(Wether bl. 7)	38	19.3	2	14.1
	42	19.4	3	16.6
9	15	18.5	1	12.5
(Wether bl. 9)	8	18.9	2	11.2
	16	18.9	3	15.4
10	20	18.1	1	10.3
(Wether bl. 11)	22	18.1	2	13.5
	2	18.1	3	18.3
11	45	17.6	1	13.6
(Wether bl. 13)	71	17.3	2	12.5
	61	17.3	3	15.6
12	4	16.8	1	15.9
(Wether bl. 15)	69	16.8	2	14.1
	3	17.1	3	20.4
13	48	15.7	1	11.5
(Wether bl. 17)	1	15.8	2	14.6
	39	15.8	3	19.4
14	23	14.9	1	7.8
(Wether bl. 19)	17	14.7	2	14.9
	7	14.5	3	16.2

deleted from each list, along with some of the heaviest and lightest. This reduced the lists to 24 doe and 60 wether kids. These were then divided into 8 doe blocks and 20 wether blocks, with the first doe block containing the three heaviest doe kids, and so on. Within each block the three kids were then allocated randomly to the three treatments.

The three treatment groups were run together as a single mob throughout the experiment, which lasted from weaning on 19 Dec 1986 until 25 Nov 1987. The treatment to be applied to each kid was identified by a green, blue or red tag, so that on each drenching date it was a simple matter to pick out the kids to be drenched.

The most important variable measured was liveweight gain over the trial period. This data is given in the table, for every second block. Within each block the data have been sorted into a standard order: treatments 1, 2 and 3. The data is by courtesy of Alan Pearson, Veterinarian, Animal Health Laboratory, Ministry of Agriculture and Fisheries, Lincoln, New Zealand.

(a) Carry out an analysis of variance of the liveweight gains given in the table. Use the Twoway command of Minitab to calculate the sums of squares, and the Oneway command to calculate treatment means. Present a summary ANOVA table.

(b) Calculate the LSD(5%) and LSD(1%) to enable comparison of any pair of treatment means. Present a summary table of means and LSD's.

(c) Is there any evidence that the preventive drench program is inferior to normal farmer practice, or to the suppression drench program? How clearcut are these results?

(12.8) In Examples A,B and D of §7.2 the field layouts were simplified. These three experiments were really laid out in randomized blocks. We shall not give the actual randomizations here; however, in each of Tables 7.1, 7.2, and 7.4 the first column of data came from block 1, the second column of data from block 2, and so on.

(a) For Example A, write out a set of three orthogonal coordinate axes for the block space. Write down the block vector $\bar{y}_{.j} - \bar{y}_{..}$, calculate the block sum of squares, and revise Table 8.2. Was the blocking very effective in this instance?

(b) Repeat (a) for Example B, revising Table 9.3.

(c) Repeat (a) for Example D, revising Table 10.3.

(12.9) An experiment was carried out to investigate the effect of seeding rate on grain yield for two varieties of wheat, Kopara and Karamu. The treatments had a 2×4 factorial structure, and were numbered one to eight as shown in the table.

Variety	Seeding rate (kg/ha)			
	60	90	120	150
Kopara	μ_1	μ_2	μ_3	μ_4
Karamu	μ_5	μ_6	μ_7	μ_8

(a) Write down a complete set of orthogonal contrasts in the means $\mu_1, \cdots, \mu_8$ which is appropriate for this treatment design.

(b) The experiment was laid out with 40 plots arranged as 5 blocks of the 2 $\times$ 4 $=$ 8 treatments in a randomized block design. The mean grain yields in tonnes/ha are given in the table. The error mean square, s^2, was 0.072.

Variety	Seeding rate (kg/ha)			
	60	90	120	150
Kopara	5.4	5.6	5.9	5.9
Karamu	5.6	6.0	6.1	6.1

Calculate the sum of squares for the linear component of the main effect of seeding rate, and test the hypothesis H_0: linear component $= 0$. Show your working and state the tabular F value.

(12.10) There are sixteen hectares of foothill land available for a grazing trial designed to compare two grazing intensities; high intensity, meaning 80 animals per hectare, and low intensity, meaning 40 animals per hectare. The shape of the land is shown below:

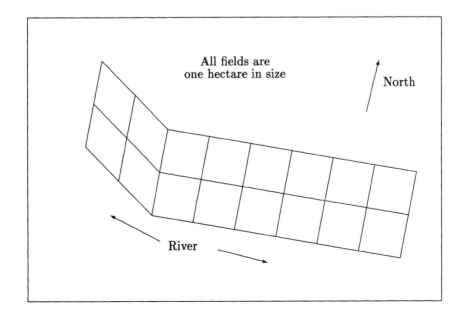

There is a river running along the south side of the land. How would you design the study so that the two treatments can be best compared? Assume there are at least 960 animals that can be used for the study, and that the animals are to be set stocked. Write out a skeleton of the ANOVA table for your design, treating fields/mobs as the basic experimental units.

(12.11) When a series of trials of identical design is to be analyzed as a single entity, it may at first glance appear sensible to fit a randomized block model, with trials as blocks. Such a case arose with the series of fertilizer beet trials described in Exercise 5.5. In the table we reproduce the data for the most important elements, which were N, P and Na. In (a) is shown the mean sugar yield for each treatment at each site, and in (b) the differences due to withholding each element from the fertilizer mix.

(a) Treatments	Sites					
	1	2	3	4	5	6
1. All	12.0	8.4	14.1	10.3	8.0	10.0
2. All − N	6.8	5.8	12.7	7.5	5.3	10.2
3. All − P	10.5	8.3	13.0	9.2	6.7	9.5
4. All − Na	8.9	7.6	12.1	8.8	6.9	9.4
(b) Differences						
N	5.2	2.6	1.4	2.8	2.7	−.2
P	1.5	.1	1.1	1.1	1.3	.5
Na	3.1	.8	2.0	1.5	1.1	.6

(a) Use the data in part (a) of the table to construct a randomized block ANOVA with four treatments and six blocks. Write out the ANOVA table.

(b) Calculate the LSD(5%) and LSD(1%). Use these to decide the significances of the depressions in sugar yield associated with withholding N,P and Na respectively.

   In part (b) of the table it is easy to see that the effect of withholding nitrogen is in general greater than the effect of withholding P or Na. This is because nitrogen is a very important element for plant growth, and one whose availability varies markedly from site to site. However, in the randomized block analysis we erroneously assumed that all treatments have a common variance; this is equivalent to assuming that all contrasts, such as those in part (b) of the table, have a common variance. To see the effect of this erroneous assumption, we shall analyze the effect of withholding N separately from that of P and Na.

(c) Analyze the data for the elements N, P and Na separately (three analyses). There are two equivalent methods you can use. The first is to use the single population methods of Chapter 5, mimicking the analysis of Exercise 5.5. The second method is to carry out three randomized block analyses, each with two treatments (treatments 1 and 2; 1 and 3; 1 and 4). For whichever method you use, write out the three ANOVA tables. How do the s^2 estimates vary between the three analyses?

(d) Write down the significances of the depressions in sugar yield due to withholding N,P and Na respectively. How do these compare with your results from (b)?

(12.12) Assume that we know a crop yield response is a quadratic form in the nitrogen fertilizer levels between 0 and 150 kg/ha. In a study to more precisely characterize the response curve, which of the following designs would you recommend?

> (a) 3 levels with 4 replications per level
> (b) 4 levels with 3 replications per level
> (c) 2 levels with 6 replications per level
> (d) 6 levels with 2 replications per level

Give reasons why you choose one particular design. What are the exact nitrogen levels you would apply? Hint: for simplicity, assume equal spacing.

(12.13) A $2 \times 4 \times 4 = 32$ treatment (pot) experiment was set up in a glasshouse to look at the effects of salinity, N status and P status on the growth of wheat. Three replicates of a randomized block design were set up. The mean plant heights, in cm, after 12 weeks' growth are shown in the table for each of the 32 treatments.

Salinity (ppm)	N rate (kg/acre)	P rate (P_2O_5/acre) 0	15	30	45	Mean
250	0	66.7	71.3	76.0	70.0	71.0
	30	84.3	86.3	87.7	84.0	85.6
	60	90.0	90.7	91.3	88.3	90.1
	90	90.0	90.3	93.7	90.3	91.1
	Mean	82.7	84.6	87.2	83.1	84.4
6000	0	63.3	66.7	70.3	65.0	66.3
	30	80.0	81.0	83.2	80.3	81.1
	60	82.0	85.7	88.3	83.7	84.9
	90	83.0	84.7	87.3	83.3	84.6
	Mean	77.1	79.5	82.3	78.1	79.2

(a) Take Factor A = salinity level, Factor B = rate of N, and Factor C = rate of P, and write down appropriate contrasts for each factor. For the first contrast for each of B and C, use "none" versus "some", $[-3, 1, 1, 1]^T$. For the second and third contrasts, specify linear and quadratic in the "some" class. N.B: Don't use the data — this is an easy question!

(b) Indicate briefly how you would generate the interaction contrasts.

(12.14) In Exercise 12.13, the $2 \times 4 \times 4$ pot trial, consider the plant height table of means. Suppose we are given $s^2 = 50$.

(a) Calculate the error degrees of freedom.

(b) Calculate the sums of squares for the three main effect of nitrogen contrasts. What are the F tests for these three contrasts, and what are your conclusions?

(c) Also calculate the F tests for the three contrasts for the nitrogen by salinity interaction, and state your conclusions.

(12.15) An experiment was set up to investigate the effect of "frequency of shift" (onto a new pasture) on the rate of growth of lambs. The three treatments were:

1. Lambs shifted every day

2. Lambs shifted every second day

3. Lambs shifted every seventh day

Nine one-hectare fields and nine balanced mobs of 30 lambs were used in the study. The mobs were allocated one to each field and three fields were allocated to each treatment in a randomized block design; for example, block 1 is shown in the figure. Each field was divided into 28, 14 or 4 areas, as appropriate, using electric fences, and the stock rotated around the areas, returning to the first area after 28 days. The shaded portion is a typical grazing area. The experiment lasted for three 28-day periods.

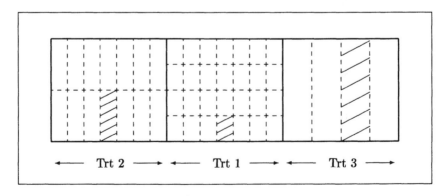

(a) Show that the linear and quadratic contrasts are

$$c_1 = -7\mu_1 - 4\mu_2 + 11\mu_3 \quad \text{and}$$
$$c_2 = 5\mu_1 - 6\mu_2 + \mu_3$$

assuming a spacing of 1, 2 and 7.

(b) Write down the corresponding treatment space unit vectors.

(c) Write down two block space unit vectors.

(d) Hence write down four error space unit vectors.

The experimental results were as shown in the table. Interestingly, with the 7-daily shift the lambs did relatively poorly. Field observations were that these lambs had a feast for the first 2–3 days after each shift, then suffered a famine for the next 4–5 days.

Treatment	Weight gains (g/lamb/day)
Daily shift	153
2-daily shift	112
7-daily shift	95
SED	17

(e) Calculate the linear and quadratic sums of squares.

(f) Compute s^2, the error degrees of freedom, and the linear and quadratic F tests. What are your conclusions?

(g) Calculate the 95% confidence interval for the true slope, β, where $\beta = (-7\mu_1 - 4\mu_2 + 11\mu_3)/62$. Is this interval consistent with the linear contrast F test?

Chapter 13

Latin Square Design

In this chapter we introduce the latin square design. This design is used to take account of two sources of variation between the experimental units, in an orthogonal fashion. In practice, the design is not widely used. However, we include it in this book since it illuminates the ideas of orthogonal blocking in a very elegant manner.

The plan for the chapter is to introduce the design in §1 by way of an example, then discuss the general case and possible uses of the design in §2. The chapter closes with a summary in §3.

13.1 Illustrative Example

In this section we use a fictional, but realistic study to illustrate the difference between a randomized block design and a latin square design.

Randomized Block Design

The study objective is to compare the drymatter yields of five pasture species under flood irrigation. Initially we assume that a randomized block design has been employed, with flood irrigation strips making up the blocks, as shown in Figure 13.1. The reasons for blocking in this way are

(a) there are variations in the depth of topsoil from one strip to the next, since some strips are cut by the earthmoving machinery to a lower level than other strips

(b) there are variations in the amount of water applied to the strips, since the irrigation "sills" are at slightly different heights

Both these effects can cause variations in productivity from strip to strip.

Block	1	2	3	4	5	

$\longrightarrow$ Incoming water (Headrace)

4	5	2	5	4
3	2	3	2	3
2	3	1	3	5
1	1	4	1	1
5	4	5	4	2

Overflow water (Tailrace) $\longrightarrow$

Figure 13.1: Randomized block layout for a pasture species comparison in a flood irrigated field. Treatment numbers are shown for each block.

The five treatments and an appropriate set of orthogonal contrasts are given in Table 13.1. Contrast d_1 compares ryegrass with the other species, d_2 compares the old ryegrass cultivar with the newer ryegrass cultivars, d_3 compares one of the new ryegrass cultivars with the other, and d_4 compares cocksfoot with fescue.

Treatments	d_1	d_2	d_3	d_4
1. Ariki ryegrass (old cultivar)	2	2	0	0
2. Ruanui ryegrass (new cultivar)	2	−1	1	0
3. Nui ryegrass (new cultivar)	2	−1	−1	0
4. Cocksfoot	−3	0	0	1
5. Fescue	−3	0	0	−1

Table 13.1: Treatment list and set of orthogonal contrasts.

Latin Square Design

The field layout shown in Figure 13.1 would in many cases be perfectly adequate. However, since the experiment is a long term study grazed by sheep, the researcher may be concerned to guard against possible trends in fertility down the strips. These trends occur because the sheep dung floats on the irrigation water and is transferred down the irrigation strips, resulting in higher fertility towards the bottom of the strip. We pose the question: is this effect large enough to worry about? In practice the answer is often "no". However, we answer "yes" so that we can proceed.

An appropriate field layout for this trial, which safeguards against possible trends down each strip, is a *latin square* design, as shown in Figure 13.2. With this design the rows (the positions down the strips) form blocks, as well as the columns (the irrigation strips). Treatments are then carefully arranged so that each occurs once in each row and once in each column. The method of allocation is described in §13.2.

Column	1	2	3	4	5
⟶ Incoming water (Headrace)					
Row 1	3	1	4	2	5
Row 2	4	2	3	5	1
Row 3	1	5	2	3	4
Row 4	5	3	1	4	2
Row 5	2	4	5	1	3
			Overflow water (Tailrace) ⟶		

Figure 13.2: Latin square layout for pasture species comparison. Treatment numbers are given for each plot.

Data

We now assume that the field layout followed the latter design, and introduce and analyze some appropriate data. These are dry matter yields in tonnes per hectare, totalled over all the cuts made in the first season, as shown in Table 13.2.

Distance	Irrigation strip number				
down strip	1	2	3	4	5
1	7.9 (3)	5.6 (1)	5.3 (4)	8.1 (2)	7.8 (5)
2	5.9 (4)	6.3 (2)	7.9 (3)	6.6 (5)	6.8 (1)
3	7.2 (1)	6.2 (5)	7.1 (2)	7.9 (3)	7.1 (4)
4	7.6 (5)	7.5 (3)	7.8 (1)	6.5 (4)	8.7 (2)
5	7.7 (2)	6.3 (4)	7.4 (5)	6.4 (1)	8.9 (3)

Table 13.2: Total drymatter yield, in tonnes per hectare, for the first season. Treatment numbers are given in brackets for ready reference.

We denote the drymatter yield from a plot, in tonnes per hectare, by y_{ijk}, where i refers to the treatment number, j the row number and k the column number. Hence, preserving the order of Table 13.2 and listing the data by rows (row 1 first, then row 2), our observation vector is

$$
y = \begin{bmatrix} y_{311} \\ y_{112} \\ y_{413} \\ y_{214} \\ y_{515} \\ \vdots \end{bmatrix} = \begin{bmatrix} 7.9 \\ 5.6 \\ 5.3 \\ 8.1 \\ 7.8 \\ \vdots \end{bmatrix}
$$

Notice that we have departed from our usual convention of sorting the data into treatment 1, treatment 2 and so on. In what follows this means that while the treatment vector is in an inconvenient order, the row and column vectors are in a standard order.

Model

The model must include terms accounting for differences between treatments, differences between positions down the strip (rows), and differences between strips (columns). The average or expected value of each observation is taken to be $\mu_i + r_j + c_k$, where μ_i is the usual treatment mean, r_j represents the difference in yield between the jth row and the average of the rows, and c_k represents the difference in yield between the kth column and the average of the columns. This is written out more fully in Table 13.3. Note that $r_1 + \cdots + r_5 = 0$ and $c_1 + \cdots + c_5 = 0$. Also note that the model assumes that these row and column effects are "additive" and do not vary between treatments.

	Column 1	Column 2	$\cdot$ $\cdot$ $\cdot$	Average
Row 1	$\mu_3 + r_1 + c_1$	$\mu_1 + r_1 + c_2$	$\cdot$ $\cdot$ $\cdot$	$\mu + r_1$
Row 2	$\mu_4 + r_2 + c_1$	$\mu_2 + r_2 + c_2$	$\cdot$ $\cdot$ $\cdot$	$\mu + r_2$
Row 3	$\mu_1 + r_3 + c_1$	$\mu_5 + r_3 + c_2$	$\cdot$ $\cdot$ $\cdot$	$\mu + r_3$
$\vdots$	$\vdots$	$\vdots$	$\vdots$ $\vdots$ $\vdots$	$\vdots$
Average	$\mu + c_1$	$\mu + c_2$	$\cdot$ $\cdot$ $\cdot$	μ

Table 13.3: Long term averages or expected values for each observation, each row and column mean, and the overall mean.

The resulting *model vector* is

$$
\begin{bmatrix}
\mu_3 + r_1 + c_1 \\
\mu_1 + r_1 + c_2 \\
\mu_4 + r_1 + c_3 \\
\mu_2 + r_1 + c_4 \\
\mu_5 + r_1 + c_5 \\
\vdots
\end{bmatrix}
=
\begin{bmatrix}
\mu_3 \\
\mu_1 \\
\mu_4 \\
\mu_2 \\
\mu_5 \\
\vdots
\end{bmatrix}
+
\begin{bmatrix}
r_1 \\
r_1 \\
r_1 \\
r_1 \\
r_1 \\
\vdots
\end{bmatrix}
+
\begin{bmatrix}
c_1 \\
c_2 \\
c_3 \\
c_4 \\
c_5 \\
\vdots
\end{bmatrix}
$$

Thanks to our layout, the treatment mean vector $[\mu_3, \mu_1, \ldots]^T$, the row vector $[r_1, r_1, \ldots]^T$, and the column vector $[c_1, c_2, \ldots]^T$ are mutually orthogonal, since $r_1 + \cdots + r_5 = 0$ and $c_1 + \cdots + c_5 = 0$. The treatment mean vector lies in a five dimensional subspace of 25-space whereas the row and column vectors each lie in four dimensional subspaces; this artificial difference arises because the first subspace is the sum of the one dimensional overall mean space and the four dimensional treatment space.

We see now the reason for our design: we have used our ability to control the allocation of treatments to plots to ensure that we can produce estimates of the μ_i, r_j and c_k values which are independent.

Coordinate axes for the overall mean space and the treatment space are

$$
\begin{array}{ccccc}
U_1 & U_2 & U_3 & U_4 & U_5 \\
\frac{1}{\sqrt{25}}\begin{bmatrix} 1 \\ 1 \\ 1 \\ 1 \\ 1 \\ \vdots \end{bmatrix} &
\frac{1}{\sqrt{150}}\begin{bmatrix} 2 \\ 2 \\ -3 \\ 2 \\ -3 \\ \vdots \end{bmatrix} &
\frac{1}{\sqrt{30}}\begin{bmatrix} -1 \\ 2 \\ 0 \\ -1 \\ 0 \\ \vdots \end{bmatrix} &
\frac{1}{\sqrt{10}}\begin{bmatrix} -1 \\ 0 \\ 0 \\ 1 \\ 0 \\ \vdots \end{bmatrix} &
\frac{1}{\sqrt{10}}\begin{bmatrix} 0 \\ 0 \\ 1 \\ 0 \\ -1 \\ \vdots \end{bmatrix}
\end{array}
$$

These follow from the contrasts listed in Table 13.1 and the design shown in Figure 13.2. For example, since U_2 corresponds to the contrast $d_1 = 2\mu_1 + 2\mu_2 + 2\mu_3 - 3\mu_4 - 3\mu_5$, we enter "2" for plots receiving treatments 1 to 3, and "−3" for plots receiving treatments 4 and 5. This is tedious, but unavoidable.

Coordinate axes for the row space are

$$
U_6 = \frac{1}{\sqrt{10}}\begin{bmatrix} 1 \\ \cdot \\ -1 \\ \cdot \\ 0 \\ \cdot \\ 0 \\ \cdot \\ 0 \\ \cdot \end{bmatrix},\;
U_7 = \frac{1}{\sqrt{30}}\begin{bmatrix} 1 \\ \cdot \\ 1 \\ \cdot \\ -2 \\ \cdot \\ 0 \\ \cdot \\ 0 \\ \cdot \end{bmatrix},\;
U_8 = \frac{1}{\sqrt{60}}\begin{bmatrix} 1 \\ \cdot \\ 1 \\ \cdot \\ 1 \\ \cdot \\ -3 \\ \cdot \\ 0 \\ \cdot \end{bmatrix},\;
U_9 = \frac{1}{\sqrt{100}}\begin{bmatrix} 1 \\ \cdot \\ 1 \\ \cdot \\ 1 \\ \cdot \\ 1 \\ \cdot \\ -4 \\ \cdot \end{bmatrix}
$$

where each entry occurs five times.

Finally, coordinate axes for the column space are

$$
U_{10} = \frac{1}{\sqrt{10}}\begin{bmatrix} 1 \\ -1 \\ 0 \\ 0 \\ 0 \\ \vdots \end{bmatrix},\;
U_{11} = \frac{1}{\sqrt{30}}\begin{bmatrix} 1 \\ 1 \\ -2 \\ 0 \\ 0 \\ \vdots \end{bmatrix},\;
U_{12} = \frac{1}{\sqrt{60}}\begin{bmatrix} 1 \\ 1 \\ 1 \\ -3 \\ 0 \\ \vdots \end{bmatrix},\;
U_{13} = \frac{1}{\sqrt{100}}\begin{bmatrix} 1 \\ 1 \\ 1 \\ 1 \\ -4 \\ \vdots \end{bmatrix}
$$

where each sequence of numbers occurs five times.

These 13 coordinate axes span the model space, M. Just as $U_2 \cdots U_5$ correspond to contrasts among the treatment means, so too do $U_6 \cdots U_9$ correspond to contrasts among the row effects, and $U_{10} \cdots U_{13}$ correspond to contrasts among the column effects. For example, U_6 corresponds to the contrast $r_1 - r_2$ and U_7 corresponds to the contrast $r_1 + r_2 - 2r_3$. In summary, then, the model space is a 13 dimensional subspace consisting of an overall mean space, a treatment space, a row space and a column space.

As usual we assume that each independently drawn observation y_{ijk} is normally distributed about its expected value, $\mu_i + r_j + c_k$, with variance σ^2.

Test Hypotheses

The test hypotheses are $H_0 : d_1 = 0$, $H_0 : d_2 = 0$, $H_0 : d_3 = 0$ and $H_0 : d_4 = 0$, where the contrasts d_i are as listed in Table 13.1. The corresponding directions are U_2, U_3, U_4 and U_5.

Our collection of raw materials is now complete. We have an observation vector $y = [7.9, 5.6, \ldots]^T$, a model space $M = span\{U_1, U_2, \cdots, U_{13}\}$ and directions $U_2, \cdots, U_5$ associated with our hypotheses of interest.

Fitting the Model

To fit the model we project the observation vector y onto the model space M. The resulting fitted model vector is $(y.U_1)U_1 + \cdots + (y.U_{13})U_{13}$. However, the reader should not be surprised to learn that this simplifies to

$$\bar{y}_{...} + (\bar{y}_{i..} - \bar{y}_{...}) + (\bar{y}_{.j.} - \bar{y}_{...}) + (\bar{y}_{..k} - \bar{y}_{...})$$

the overall mean vector plus the treatment vector, the row vector and the column vector. For our example this is

$$
\begin{bmatrix} \bar{y}_{...} \\ \cdot \\ \cdot \\ \cdot \\ \cdot \\ \vdots \end{bmatrix}
+
\begin{bmatrix} \bar{y}_{3..} - \bar{y}_{...} \\ \bar{y}_{1..} - \bar{y}_{...} \\ \bar{y}_{4..} - \bar{y}_{...} \\ \bar{y}_{2..} - \bar{y}_{...} \\ \bar{y}_{5..} - \bar{y}_{...} \\ \vdots \end{bmatrix}
+
\begin{bmatrix} \bar{y}_{.1.} - \bar{y}_{...} \\ \bar{y}_{.1.} - \bar{y}_{...} \\ \bar{y}_{.1.} - \bar{y}_{...} \\ \bar{y}_{.1.} - \bar{y}_{...} \\ \bar{y}_{.1.} - \bar{y}_{...} \\ \vdots \end{bmatrix}
+
\begin{bmatrix} \bar{y}_{..1} - \bar{y}_{...} \\ \bar{y}_{..2} - \bar{y}_{...} \\ \bar{y}_{..3} - \bar{y}_{...} \\ \bar{y}_{..4} - \bar{y}_{...} \\ \bar{y}_{..5} - \bar{y}_{...} \\ \vdots \end{bmatrix}
$$

where $\bar{y}_{...}$ denotes the overall mean, $\bar{y}_{3..}$ denotes the mean for treatment three, $\bar{y}_{.1.}$ denotes the mean for row one, $\bar{y}_{..1}$ denotes the mean for column one, and so on.

To obtain numerical values we must calculate the treatment means, row means and column means. These, together with the treatment effects, row effects and column effects, are given in Table 13.4.

	Treatment		Row		Column	
	Mean	Effect	Mean	Effect	Mean	Effect
1	6.76	−.38	6.94	−.20	7.26	.12
2	7.58	.44	6.70	−.44	6.38	−.76
3	8.02	.88	7.10	−.04	7.10	−.04
4	6.22	−.92	7.62	.48	7.10	−.04
5	7.12	−.02	7.34	.20	7.86	.72

Table 13.4: Treatment, row and column means and effects. Effects are obtained by subtracting the overall mean, 7.14, from the corresponding mean.

The resulting fitted model vector is

$$
\begin{bmatrix} 7.14 \\ 7.14 \\ 7.14 \\ 7.14 \\ 7.14 \\ \vdots \end{bmatrix}
+
\begin{bmatrix} .88 \\ -.38 \\ -.92 \\ .44 \\ -.02 \\ \vdots \end{bmatrix}
+
\begin{bmatrix} -.20 \\ -.20 \\ -.20 \\ -.20 \\ -.20 \\ \vdots \end{bmatrix}
+
\begin{bmatrix} .12 \\ -.76 \\ -.04 \\ -.04 \\ .72 \\ \vdots \end{bmatrix}
=
\begin{bmatrix} 7.94 \\ 5.80 \\ 5.98 \\ 7.34 \\ 7.64 \\ \vdots \end{bmatrix}
$$

$$
\begin{array}{ccccccccc}
\bar{y}_{\cdots} & & \bar{y}_{i\cdot\cdot} - \bar{y}_{\cdots} & & \bar{y}_{\cdot j\cdot} - \bar{y}_{\cdots} & & \bar{y}_{\cdot\cdot k} - \bar{y}_{\cdots} & & \\
\text{Overall} & + & \text{Treatment} & + & \text{Row} & + & \text{Column} & = & \text{Model} \\
\text{mean vector} & & \text{vector} & & \text{vector} & & \text{vector} & & \text{vector}
\end{array}
$$

In terms of the underlying model, shown in Table 13.3, we have now estimated all of the parameters. The estimates $\hat{\mu}_1 = 6.76, \cdots, \hat{\mu}_5 = 7.12$ are the estimated treatment means, $\hat{r}_1 = -.20, \cdots, \hat{r}_5 = .20$ are the estimated row effects, and $\hat{c}_1 = .12, \cdots, \hat{c}_5 = .72$ are the estimated column effects. However, only 13 of these estimates are independent, since we can write $\hat{r}_5 = -\hat{r}_1 - \hat{r}_2 - \hat{r}_3 - \hat{r}_4$ and $\hat{c}_5 = -\hat{c}_1 - \hat{c}_2 - \hat{c}_3 - \hat{c}_4$.

To complete the story, the fitted model is

$$
\begin{bmatrix} 7.9 \\ 5.6 \\ 5.3 \\ 8.1 \\ 7.8 \\ \vdots \end{bmatrix}
=
\begin{bmatrix} 7.94 \\ 5.80 \\ 5.98 \\ 7.34 \\ 7.64 \\ \vdots \end{bmatrix}
+
\begin{bmatrix} -.04 \\ -.20 \\ -.68 \\ .76 \\ .16 \\ \vdots \end{bmatrix}
$$

$$
\begin{array}{ccccc}
\text{Observation} & & \text{Model} & & \text{Error} \\
\text{vector} & & \text{vector} & & \text{vector}
\end{array}
$$

This can be written more fully, in corrected form, as

$$
y - \bar{y}_{\cdots} = (\bar{y}_{i\cdot\cdot} - \bar{y}_{\cdots}) + (\bar{y}_{\cdot j\cdot} - \bar{y}_{\cdots}) + (\bar{y}_{\cdot\cdot k} - \bar{y}_{\cdots}) + \text{error vector}
$$

$$
\begin{array}{ccccccccc}
\text{Corrected} & = & \text{Treatment} & + & \text{Row} & + & \text{Column} & + & \text{Error} \\
\text{obsn vector} & & \text{vector} & & \text{vector} & & \text{vector} & & \text{vector}
\end{array}
$$

Here the error vector lies in a 12-dimensional error space, since the model space is a 13-dimensional subspace of 25-space.

Geometrically this can be visualized as shown in Figure 13.3. Note that if we so desire we can think of the row space and column space as orthogonal subspaces within an eight dimensional *block space*.

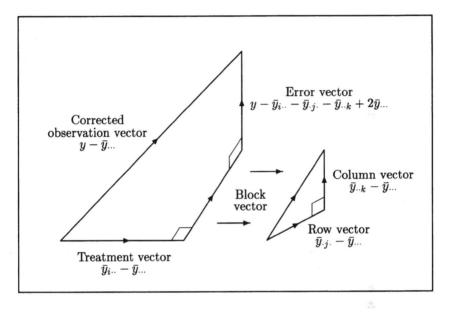

Figure 13.3: Orthogonal decomposition of corrected observation vector for latin square design.

Testing the Hypotheses

We are now in a position to test the hypotheses $H_0 : d_i = 0$. As usual, we compare appropriate squared distances, $(y.U_i)^2$. For example, in order to test the hypothesis $H_0 : d_1 = 0$, that ryegrass yields the same as other species, we calculate the test statistic

$$F = \frac{(y.U_2)^2}{\left[(y.U_{14})^2 + \cdots + (y.U_{25})^2\right]/12} = \frac{(y.U_2)^2}{\|\text{error vector}\|^2/12} = \frac{3.682}{.237} = 15.51$$

which comes from an $F_{1,12}$ distribution if H_0 is true.

The appropriate Pythagorean decomposition is

$$\|y - \bar{y}_{\cdots}\|^2 = \|\bar{y}_{i\cdots} - \bar{y}_{\cdots}\|^2 + \|\bar{y}_{\cdot j\cdot} - \bar{y}_{\cdots}\|^2 + \|\bar{y}_{\cdot\cdot k} - \bar{y}_{\cdots}\|^2 + \|\text{error vector}\|^2$$

with the treatment sum of squares, $\|\bar{y}_{i\cdots} - \bar{y}_{\cdots}\|^2$, further decomposed into contrast sums of squares $(y.U_2)^2 + \cdots + (y.U_5)^2$ in the usual manner. The resulting ANOVA table is given in Table 13.5.

Source of Variation	df	SS	MS	F
Row	4	2.528	.632	
Column	4	5.568	1.392	
Treatment	4	9.796		
$H_0 : d_1 = 0$	1	3.682	3.682	15.51 (**)
$H_0 : d_2 = 0$	1	3.605	3.605	15.19 (**)
$H_0 : d_3 = 0$	1	.484	.484	2.04 (ns)
$H_0 : d_4 = 0$	1	2.025	2.025	8.53 (*)
Error	12	2.848	.237	
Total	24	20.740		

Table 13.5: ANOVA table for latin square design.

Report on Study

To summarize the results of the study we can present a table such as that given in Table 13.6. This follows the usual format.

Treatments	Dry matter production (tonnes/ha) in first season
Ariki ryegrass (old)	6.8
Ruanui ryegrass (new)	7.6
Nui ryegrass (new)	8.0
Cocksfoot	6.2
Fescue	7.1
SED	.31
Contrasts	
Ryegrass vs. other species	**
Old vs. new ryegrasses	**
Ruanui vs. Nui ryegrass	ns
Cocksfoot vs. fescue	*

Table 13.6: Summary of results, pasture species comparison.

Minitab Analysis

To fit the model using Minitab the tidiest method is simply to Regress the observation vector y onto the directions $U_2, \ldots, U_{13}$ in the model space. The worst part of this method is setting up the unit vectors $U_2, \ldots, U_5$ in the treatment space, since these are not in a standard order. Rearranging the data does not help, since the row or column vectors would then be in non-standard order; such a reordering is also likely to lead to transcription errors.

A second method is to use the Oneway command three times, once for treatments, once for rows and once for columns. The outputs give us the treatment, row, and column sums of squares respectively; the error sum of squares can then be found by subtraction. In addition, the observation vector must be Regressed onto $U_2, \cdots, U_5$ to calculate the contrast sums of squares.

In practice, it is sensible to use a combination of the two methods, primarily using the first, but including the Oneway commands of the second method to serve as a check and to calculate treatment, row and column means.

Miscellaneous Comments

In analyzing this example we have skipped over certain aspects since they closely parallel the development of the illustrative example in the randomized block chapter (§12.1). In the "model" section we did not bother to write out the model vector in terms of individual directions $V_1 \cdots V_{15}$. In the "test hypotheses" section we did not check that the expected values, $E(Y.U_2)$ and so on, were as required, nor did we bother to spell out the distributions of the projection coefficients. Assumption checking was omitted since the data were contrived; the main change here would be in checking that treatment, row and column effects are additive. This could be achieved by plotting values for each row against the treatment means in one graph, and for each column in a second graph, using the format of §12.3. Confidence intervals were not mentioned; the method for producing these is the same as in previous chapters, except for a change in the error degrees of freedom.

13.2   General Discussion

We now give a brief summary of the general case for a latin square design.

1. With this design the number of experimental units (field plots, sheep, or runs in a factory) must be k^2, where k is the number of study populations. In other words, we are restricted to two replicates of two treatments, three replicates of three treatments, and so on.

2. The design is blocked in two mutually orthogonal ways, usually referred to as row blocking and column blocking. As a result we have a block space of dimension $2(k-1)$ which includes a row space of dimension $k-1$ and a column space of dimension $k-1$.

3. The treatment space is, as usual, of dimension $k-1$. This means the error space is of dimension $k^2 - [1 + (k-1) + 2(k-1)] = (k-1)(k-2)$.

4. The orthogonal decomposition is

$$y - \bar{y}_{...} = (\bar{y}_{i..} - \bar{y}_{...}) + (\bar{y}_{.j.} - \bar{y}_{...}) + (\bar{y}_{..k} - \bar{y}_{...}) + \text{error vector}$$

as illustrated in Figure 13.3.

5. The corresponding Pythagorean decomposition is summarized in Table 13.7.

Source of variation	df	SS
Rows	$k - 1$	$\|\bar{y}_{.j.} - \bar{y}_{...}\|^2$
Columns	$k - 1$	$\|\bar{y}_{..k} - \bar{y}_{...}\|^2$
Treatments	$k - 1$	$\|\bar{y}_{i..} - \bar{y}_{...}\|^2$
Error	$(k - 1)(k - 2)$	By subtraction
Total	$k^2 - 1$	$\|y - \bar{y}_{...}\|^2$

Table 13.7: Skeleton ANOVA table for a latin square design.

6. As usual we can specify $k - 1$ orthogonal contrasts among the treatment means and obtain $k - 1$ orthogonal unit vectors in the treatment space. These are used for hypothesis testing in the usual way.

7. Confidence interval formulae are the same as in preceding chapters except that the number of error degrees of freedom is now $(k - 1)(k - 2)$.

To summarize, in the latin square design the block space is of dimension $2(k - 1)$, double that for a randomized block design experiment of the same size. Correspondingly, the dimension of the error space is reduced by $k - 1$.

In practice the "5 × 5" latin square is the size which is most frequently used. The 2 × 2 square suffers from zero error degrees of freedom, while the 3 × 3 square has only two degrees of freedom for error. The 4 × 4 square still has only six degrees of freedom for error, generally regarded as too low. Larger squares such as the 6 × 6 tend to be too large, and it is uncommon to strike the required combination of 6 treatments, 6 rows and 6 columns.

Randomization

To obtain a randomized layout for a latin square design we need to use *standard squares*. These are latin squares with treatments assigned to the first row and first column in numerical order. For 2 × 2 and 3 × 3 layouts there is only one standard square. For a 4 × 4 layout there are four possible standard squares, listed in Table 13.8. For a 5 × 5 layout there are 56 standard squares, listed in Fisher and Yates (1963).

1	2	3	4
2	1	4	3
3	4	2	1
4	3	1	2

1	2	3	4
2	3	4	1
3	4	1	2
4	1	2	3

1	2	3	4
2	4	1	3
3	1	4	2
4	3	2	1

1	2	3	4
2	1	4	3
3	4	1	2
4	3	2	1

Table 13.8: Standard latin squares for the 4 × 4 layout.

To obtain a random layout, take the following steps:

1. Use random numbers to choose a standard square.

2. Reorder all the rows except the first of the standard square according to a sequence of random numbers; for example, 3 2 4 for a 4 × 4 square.

3. Reorder all the columns of the resulting square according to a new random sequence; for example 4 1 3 2 for a 4 × 4 square.

4. This final square gives you the treatment numbers in your random layout.

Field Trial Example

In a field trial, a researcher may suspect that there are fertility gradients running both along the fenceline and also at right angles to the fenceline, and therefore be tempted to use a latin square design as shown in Figure 13.4(a).

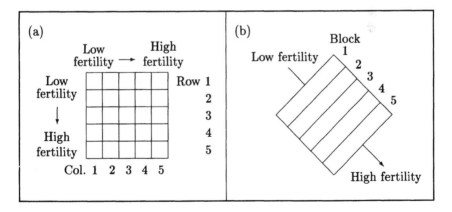

Figure 13.4: (a) Latin square design coping with fertility gradients in two directions; (b) Randomized block design with blocks at right angles to the resultant fertility gradient.

What the researcher may not realize is that the sum of the two fertility gradients is a single resultant fertility gradient running out into the field at some angle to the fenceline, as shown in Figure 13.4(b). Provided that this angle can be approximately determined, a better design is a randomized block design with low fertility blocks up to high fertility blocks, as shown in Figure 13.4(b). Within each block the fertility should now be roughly uniform.

In some cases, such as in Figure 13.1, we are unable to rotate the experimental plots, so this logic does not apply.

Industrial Example

In some industrial experiments the blocks may be operators (or perhaps human patients, or machines) who try out the treatments in a random order. In this case it is unfortunate if one treatment is on average tried much earlier than another treatment, since the operator may suffer from gradual fatigue, or conversely, slowly acquire more skill.

As an example, suppose that a manufacturer of calculators wishes to compare five prototype calculators for ease of use, using five staff members. Each staff member is to perform a given calculation using each of the five calculators, and the time taken is noted. A 5×5 latin square layout such as shown in Table 13.9 would be a sensible design. Here the treatments are the calculators, the rows are the staff members and the columns are the order in which each staff member tries out the various calculators. Note that we are assuming that the various staff members improve by similar amounts during the experiment, and differ between calculators by similar amounts.

Staff	Order of using the calculators				
members	1	2	3	4	5
1	D	E	A	C	B
2	B	A	C	D	E
3	E	B	D	A	C
4	A	C	B	E	D
5	C	D	E	B	A

Table 13.9: Latin square design specifying for each staff member the order in which they are to try out the calculators, labelled A,B,C,D and E.

Similar considerations may apply if several medical treatments are to be tried on a limited number of patients; the order in which the treatments are taken by a patient may well influence the response. Also, if several machines are being compared using several operators the order of testing may be important; for example, the machines may be warming up.

13.3   Summary

In a latin square design the experimental units are blocked to orthogonally take account of two sources of variation. The fitted model for this design is

$$y - \bar{y}_{...} = (\bar{y}_{i..} - \bar{y}_{...}) + (\bar{y}_{.j.} - \bar{y}_{...}) + (\bar{y}_{..k} - \bar{y}_{...}) + \text{error vector}$$

as illustrated in Fig 13.5.

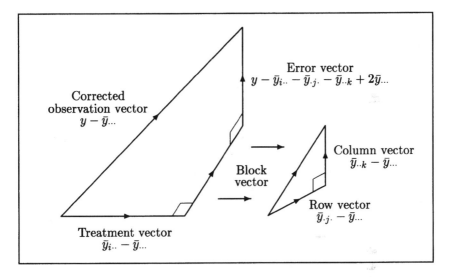

Figure 13.5: Orthogonal decomposition of corrected observation vector.

The model space consists of the one dimensional overall mean space and the $k - 1$ dimensional treatment, row and column spaces. The error space is of dimension $(k - 1)(k - 2)$, and the estimate of σ^2 is $s^2 = \|\text{error vector}\|^2 / [(k - 1)(k - 2)]$.

This estimate of σ^2 is used for hypothesis testing and confidence interval estimation in an identical manner to that of chapters 5 to 12.

For additional reading on the geometric aspects refer to Box, Hunter and Hunter (1978), pages 245–249. For more general reading refer to Cochran and Cox (1957), pages 117–124, and Little and Hills (1978), pages 77–85.

Exercise

(13.1) Reproduce the ANOVA table given in Table 13.5 by inputting the data in Table 13.2 and using Minitab to Regress the observation vector y onto the unit vectors $U_2, \cdots, U_{13}$ as given in the text. As a check also include three Oneway commands in your job, firstly inputting treatment numbers, then row numbers, and then column numbers.

Chapter 14

Split Plot Design

Here we introduce the simplest "hierarchical" design, the split plot design. This design has two error terms, corresponding to a subdivision of the error space into two orthogonal subspaces. Studies employing this design have (at least) two treatment factors; the effects of one factor, however, are estimated more accurately than the effects of the other factor. In practice this design is very commonly used.

The plan for the chapter is to introduce the new design in §1, then illustrate the method of analysis in §2. Further discussion and examples are contained in §3. The chapter closes with a summary in §4, and exercises.

14.1 Introduction

We introduce the split plot design using Example C. This example was initially presented in §7.2, and analyzed in §9.3 as a 2×3 factorial design. The aim of the experiment was to compare the establishment and survival of two turnip cultivars, and to investigate the influence of nitrogen fertilizer. Treatment factors were (A) cultivar, and (B) rate of nitrogen application. In this section we are able to reveal the true layout of this experiment.

The twenty four plots to be used in the experiment were positioned side by side in a single row, with their long sides adjacent. For the split plot design, the plots were first divided into four roughly uniform blocks. Each of these was then divided into two *mainplots* (mp). The two levels of factor A, Green globe and York globe cultivars of soft turnip, were then randomly assigned to the two mainplots within each block. Each mainplot was in turn divided into three *subplots* (sp), and the three levels of factor B, no N, 125N, and 250N, randomly assigned to the subplots. The resulting allocation of treatment numbers to subplots are shown in Figure 14.1. Note that if a randomized block layout had been employed, the $2 \times 3 = 6$ treatments would have been assigned purely at random within each block.

block 1						block 2					
mp1			mp2			mp3			mp4		
sp1	sp2	sp3	sp4	sp5	sp6	sp7	sp8	sp9	sp10	sp11	sp12
0N	125N	250N	125N	0N	250N	250N	125N	0N	0N	250N	125N
Trt1	Trt2	Trt3	Trt5	Trt4	Trt6	Trt6	Trt5	Trt4	Trt1	Trt3	Trt2
GG			YG			YG			GG		

block 3						block 4					
mp5			mp6			mp7			mp8		
sp13	sp14	sp15	sp16	sp17	sp18	sp19	sp20	sp21	sp22	sp23	sp24
250N	125N	0N	0N	250N	125N	250N	0N	125N	0N	125N	250N
Trt6	Trt5	Trt4	Trt1	Trt3	Trt2	Trt3	Trt1	Trt2	Trt4	Trt5	Trt6
YG			GG			GG			YG		

Figure 14.1: Correct field layout for Example C of §7.2.

Hierarchy of Variations

It is important now to be quite clear about what we mean by an experimental unit: these are the smallest units to which one treatment is applied. In the field layout shown in Figure 14.1 we distinguish between two different experimental units. There are the mainplots, the cultivar experimental units, and the subplots, the fertilizer experimental units.

The split plot design is the simplest design for which a hierarchy of experimental units, and the variations appropriate to them, must be discussed. For Example C the hierarchy of variations is as follows:

1. Variation between mainplots: since the mainplots are more distant from one another than the subplots, we would expect a larger variation of plant numbers between mainplots.

2. Variation between subplots: these units are close to one another, so we would expect the random variation between the subplots within a mainplot to be smaller than between mainplots.

Thus in a split plot design there are two levels of variation, σ_m^2 for mainplots and σ_s^2 for subplots. We anticipate that $\sigma_s^2 < \sigma_m^2$. By way of comparison, in our previous designs, such as the completely randomized or the randomized block design, there has been only one level of variation, σ^2.

In order to analyze the data from a split plot design experiment we need to separate the variation at the mainplot level from the variation at the subplot level. The way in which this is done is best illustrated by analyzing the data from Example C.

14.2   Analysis

For a split plot design the upshot of our analysis will be a separation of the space into two parts; one part will reflect the variations at the mainplot level, the other the variations at the subplot level. Within each space we operate in our usual fashion.

Data

The data presented in Table 7.3 were the soft turnip plant counts per 1.08m², sorted into a standard order. For convenience this is reproduced in Table 14.1 with the blocks and mainplots indicated.

Treatment Block:	1	2	3	4	Mean
Green globe soft turnips, no S/A	45	69	71	83	67.00
Green globe soft turnips, 125 kg/ha S/A	27	59	62	64	53.00
Green globe soft turnips, 250 kg/ha S/A	24	49	55	77	51.25
York globe soft turnips, no S/A	19	17	22	45	25.75
York globe soft turnips, 125 kg/ha S/A	29	13	27	21	22.50
York globe soft turnips, 250 kg/ha S/A	12	20	18	15	16.25

Table 14.1: Number of soft turnip plants per 1.8 by 0.6 square meters, Example C. The data for each of the eight mainplots is enclosed in a box.

In order to make our mainplot unit vectors straightforward to write down, we shall list the observations by mainplots. Hence we write our observation vector

$$y = [y_{111}, y_{121}, y_{131}, \ldots, y_{214}, y_{224}, y_{234}]^T$$
$$= [\ 45, \ \ 27, \ \ 24, \ldots, \ \ 45, \ \ 21, \ \ 15]^T$$

in the order: all of column 1 of Table 14.1, then column 2, and so on. Note that this departs from the order of previous chapters. Note also that we have changed to the "factorial notation" described in Appendix C, with y_{ijk} denoting the observation for the ith level of factor A, j^{th} level of factor B and k^{th} block.

Model

Here is the main idea: our observations are now assumed to be the sum of two independent normal random variables, the first with a large variance, the second with a small variance. How does this arise?

Considering just the eight mainplots, and ignoring the subplots within a mainplot, we have a randomized block design, as shown in Table 14.2. Then, as usual, the model vector has the form $\mu_{i\cdot} + \beta_k$.

Block:	1	2	3	4
Green globe	mp1	mp3	mp5	mp7
York globe	mp2	mp4	mp6	mp8

Table 14.2: The eight mainplots for Example C in standard order.

Overlaying this mainplot influence is an influence due to the subplot treatment which has been applied. The sum of these two influences results in a model vector of $\mu_{ij} + \beta_k$, which we now rewrite as the sum of the terms $(\mu_{i\cdot} + \beta_k)$ and $(\mu_{ij} - \mu_{i\cdot})$. That is, the subplot contribution to the model is the difference between the treatment mean, μ_{ij}, and the A main effect mean, $\mu_{i\cdot}$.

To describe the complete picture, we must recognize that the two parts of our observation have different variances. The first, the mainplot portion, has variance σ_m^2, while the second, the subplot portion, has a generally smaller variance, σ_s^2. In summary we can write the model in the form

$$Y_{ijk} = \left(\mu_{i\cdot} + \beta_k + \epsilon_{i\cdot k}^m\right) + \left(\mu_{ij} - \mu_{i\cdot} + \epsilon_{ijk}^s\right)$$

Here $\epsilon_{i\cdot k}^m \sim N\left[0, \sigma_m^2\right]$ is the mainplot error random variable for the mainplot receiving the ith level of factor A in the kth block, and $\epsilon_{ijk}^s \sim N\left[0, \sigma_s^2\right]$ is the subplot error random variable for the subplot receiving the $(i,j)^{th}$ treatment in the kth block. These sets of mainplot and subplot error random variables are assumed independent, both within and between sets. When written out in vector form our model is

$$\begin{bmatrix} Y_{111} \\ Y_{121} \\ Y_{131} \\ Y_{211} \\ Y_{221} \\ Y_{231} \\ \vdots \end{bmatrix} = \begin{bmatrix} \mu_{1\cdot} \\ \mu_{1\cdot} \\ \mu_{1\cdot} \\ \mu_{2\cdot} \\ \mu_{2\cdot} \\ \mu_{2\cdot} \\ \vdots \end{bmatrix} + \begin{bmatrix} \beta_1 \\ \beta_1 \\ \beta_1 \\ \beta_1 \\ \beta_1 \\ \beta_1 \\ \vdots \end{bmatrix} + \begin{bmatrix} \epsilon_{1\cdot1}^m \\ \epsilon_{1\cdot1}^m \\ \epsilon_{1\cdot1}^m \\ \epsilon_{2\cdot1}^m \\ \epsilon_{2\cdot1}^m \\ \epsilon_{2\cdot1}^m \\ \vdots \end{bmatrix} + \begin{bmatrix} \mu_{11} - \mu_{1\cdot} \\ \mu_{12} - \mu_{1\cdot} \\ \mu_{13} - \mu_{1\cdot} \\ \mu_{21} - \mu_{2\cdot} \\ \mu_{22} - \mu_{2\cdot} \\ \mu_{23} - \mu_{2\cdot} \\ \vdots \end{bmatrix} + \begin{bmatrix} \epsilon_{111}^s \\ \epsilon_{121}^s \\ \epsilon_{131}^s \\ \epsilon_{211}^s \\ \epsilon_{221}^s \\ \epsilon_{231}^s \\ \vdots \end{bmatrix}$$

Obsn vector	Mp trt vector	Block vector	Mp error vector	Sp trt vector	Sp error vector

$\longleftarrow$ Mainplot space $\longrightarrow$ $\longleftarrow$ Subplot space $\longrightarrow$

showing just the first replicate.

Mainplot Space

We focus now on the subspace which reflects the variations between main-plots, the mainplot space. This space is made up of three subspaces of interest: the mainplot treatment space, the block space and the mainplot error space. The first treatment contrast, involving York globe and Green globe cultivars, provides us with a ready made coordinate direction for the mainplot treatment space, namely

$$U_2 = \frac{1}{\sqrt{24}} \begin{bmatrix} -1 \\ 1 \\ -1 \\ 1 \\ -1 \\ 1 \\ -1 \\ 1 \end{bmatrix}$$

Here each entry occurs three times. The three block contrasts provide coordinate axes for the block space, as follows

$$U_3 = \frac{1}{\sqrt{12}} \begin{bmatrix} 1 \\ 1 \\ -1 \\ -1 \\ 0 \\ 0 \\ 0 \\ 0 \end{bmatrix}, \quad U_4 = \frac{1}{\sqrt{36}} \begin{bmatrix} 1 \\ 1 \\ 1 \\ 1 \\ -2 \\ -2 \\ 0 \\ 0 \end{bmatrix}, \quad U_5 = \frac{1}{\sqrt{72}} \begin{bmatrix} 1 \\ 1 \\ 1 \\ 1 \\ 1 \\ 1 \\ -3 \\ -3 \end{bmatrix}$$

Here again each entry occurs three times. We obtain coordinate axes for the mainplot error space by crossing U_2 with the block vectors, yielding

$$U_6 = \frac{1}{\sqrt{12}} \begin{bmatrix} -1 \\ 1 \\ 1 \\ -1 \\ 0 \\ 0 \\ 0 \\ 0 \end{bmatrix}, \quad U_7 = \frac{1}{\sqrt{36}} \begin{bmatrix} -1 \\ 1 \\ -1 \\ 1 \\ 2 \\ -2 \\ 0 \\ 0 \end{bmatrix}, \quad U_8 = \frac{1}{\sqrt{72}} \begin{bmatrix} -1 \\ 1 \\ -1 \\ 1 \\ -1 \\ 1 \\ 3 \\ -3 \end{bmatrix}$$

The corresponding projection coefficient random variables, $Y.U_i$, have one thing in common: their variances are identical, being $3\sigma_m^2 + \sigma_s^2$. Expected values and variances of these random variables are summarized in Table 14.3. The derivation of the result for $Y.U_2$ is given in the solution to the reader exercises at the end of the chapter.

	Mean	Variance
$Y.U_2$	$\sqrt{12}(\mu_{2\cdot} - \mu_{1\cdot})/\sqrt{2}$	$3\sigma_m^2 + \sigma_s^2$
$Y.U_3$	$\sqrt{6}(\beta_2 - \beta_1)/\sqrt{2}$	$3\sigma_m^2 + \sigma_s^2$
$Y.U_4$	$\sqrt{6}(2\beta_3 - \beta_1 - \beta_2)/\sqrt{6}$	$3\sigma_m^2 + \sigma_s^2$
$Y.U_5$	$\sqrt{6}(3\beta_4 - \beta_1 - \beta_2 - \beta_3)/\sqrt{12}$	$3\sigma_m^2 + \sigma_s^2$
$Y.U_6$	0	$3\sigma_m^2 + \sigma_s^2$
$Y.U_7$	0	$3\sigma_m^2 + \sigma_s^2$
$Y.U_8$	0	$3\sigma_m^2 + \sigma_s^2$

Table 14.3: Distributions of the mainplot projection coefficients.

Exercise for the reader

(14.1) (i) Show that $Y.U_2 = \left[-3\overline{Y}_{1\cdot 1} + 3\overline{Y}_{2\cdot 1} + \cdots \right]/\sqrt{24}$, where $\overline{Y}_{1\cdot 1} = (Y_{111} + Y_{121} + Y_{131})/3$ is the first mainplot random variable.

(ii) Show that $\overline{Y}_{1\cdot 1} \sim N\left[\mu_{1\cdot} + \beta_1,\ \sigma_m^2 + \sigma_s^2/3\right]$, using the distributions of Y_{111}, Y_{121} and Y_{131} given by the model.

(iii) Hence show that $Y.U_2 \sim N\left[\sqrt{12}(\mu_{2\cdot} - \mu_{1\cdot})/\sqrt{2},\ 3\sigma_m^2 + \sigma_s^2\right]$.

Subplot Space

We turn now to the subspace which reflects the variations between subplots, the subplot space. Our remaining four contrasts provide us with a ready made coordinate system for the subplot treatment space, namely

$$U_9 = \frac{1}{\sqrt{48}}\begin{bmatrix} -2 \\ 1 \\ 1 \\ -2 \\ 1 \\ 1 \\ \vdots \end{bmatrix}, \ U_{10} = \frac{1}{\sqrt{16}}\begin{bmatrix} 0 \\ -1 \\ 1 \\ 0 \\ -1 \\ 1 \\ \vdots \end{bmatrix}, \ U_{11} = \frac{1}{\sqrt{48}}\begin{bmatrix} 2 \\ -1 \\ -1 \\ -2 \\ 1 \\ 1 \\ \vdots \end{bmatrix}, \ U_{12} = \frac{1}{\sqrt{16}}\begin{bmatrix} 0 \\ 1 \\ -1 \\ 0 \\ -1 \\ 1 \\ \vdots \end{bmatrix}$$

Here each group of six numbers occurs four times.

The subplot error vectors are found by crossing the three block vectors with these four directions, giving

$$U_{13} = U_6 \times U_9, \quad \ldots, \quad U_{24} = U_8 \times U_{12}$$

For these coordinate axes the projection coefficient random variables, $Y.U_i$, also have one thing in common: their variances are σ_s^2. Expected

	Mean	Variance
$Y.U_9$	$\sqrt{8}(-2\mu_{.1} + \mu_{.2} + \mu_{.3})/\sqrt{6}$	σ_s^2
$Y.U_{10}$	$\sqrt{8}(\mu_{.3} - \mu_{.2})/\sqrt{2}$	σ_s^2
$Y.U_{11}$	$\sqrt{4}(2\mu_{11} - \mu_{12} - \mu_{13} - 2\mu_{21} + \mu_{22} + \mu_{23})/\sqrt{12}$	σ_s^2
$Y.U_{12}$	$\sqrt{4}(\mu_{12} - \mu_{13} - \mu_{22} + \mu_{23})/\sqrt{4}$	σ_s^2
$Y.U_{13}$	0	σ_s^2
$\vdots$	$\vdots$	$\vdots$
$Y.U_{24}$	0	σ_s^2

Table 14.4: Distributions of the subplot projection coefficients.

values and variances of these random variables are summarized in Table 14.4. The derivation of the result for $Y.U_9$ is given in the solution to the reader exercises at the end of the chapter.

Exercise for the reader

(14.2) (i) Show that $Y.U_9 = (V_{1.1} + V_{2.1} + V_{1.2} + \cdots)/\sqrt{48}$, where $V_{1.1} = -2Y_{111} + Y_{121} + Y_{131}$ and so on.

(ii) Show that $V_{1.1} \sim N\left[-2\mu_{11} + \mu_{12} + \mu_{13}, \ 6\sigma_s^2\right]$.

(iii) Hence show that $Y.U_9 \sim N\left[\sqrt{8}(-2\mu_{.1} + \mu_{.2} + \mu_{.3})/\sqrt{6}, \ \sigma_s^2\right]$, where $\mu_{.1}$, $\mu_{.2}$ and $\mu_{.3}$ are the factor B main effect means, in the notation of Appendix C.

To summarize, we have described a partitioning of the 24-space into a 1-dimensional overall mean space, a 7-dimensional mainplot space, and a 16-dimensional subplot space.

Model Space

Entries in a model vector are expected values of our observations, so the model space, the collection of all possible model vectors, is the span of

U_1	U_2		U_3	U_4	U_5		U_9	U_{10}	U_{11}	U_{12}
Mean	Mainplot			Block				Subplot		
axis	trt axis			axes				treatment axes		

That is, the model space is the sum of an overall mean space, a treatment space (mainplot and subplot), and a block space.

Note that the model space is built up from portions of both the mainplot and subplot spaces. The treatment space direction U_2, corresponding to the factor A contrast, and the block space directions, U_3, U_4 and U_5, lie in the mainplot space. The remaining treatment space directions, U_9, U_{10}, U_{11} and U_{12}, corresponding to the factor B and A$\times$B interaction contrasts, lie in the subplot space. Similarly, the error space is built from the mainplot error vectors, U_6 to U_8, and the subplot error vectors, U_{13} to U_{24}.

Test Hypotheses

The test hypotheses, previously listed in §7.2 and §9.3, take the form $H_0 : c_i = 0$. The contrasts c_i in our new notation are:

$$
\begin{aligned}
c_1 &= -\mu_{1\cdot} + \mu_{2\cdot} \\
c_2 &= -2\mu_{\cdot 1} + \mu_{\cdot 2} + \mu_{\cdot 3} \\
c_3 &= -\mu_{\cdot 2} + \mu_{\cdot 3} \\
c_4 &= 2\mu_{11} - \mu_{12} - \mu_{13} - 2\mu_{21} + \mu_{22} + \mu_{23} \\
c_5 &= \mu_{12} - \mu_{13} - \mu_{22} + \mu_{23}
\end{aligned}
$$

The associated directions are U_2 and U_9 to U_{12}.

In testing these hypotheses, we shall need to take special note of the level of variability of each projection coefficient, $Y.U_i$; a large realized value for one random variable may not be large for another.

We now have assembled all of the raw materials we need for our analysis: the observation vector y is $[45, 27, 24, \ldots, 45, 21, 15]^T$, the model space is $M = \text{span}\{U_1 - U_5, U_9 - U_{12}\}$, and the hypothesis directions are U_2, and U_9 to U_{12}.

Fitting the Model

Up to this point we have fitted the model by projecting the observation vector y onto the model space M. For a split plot design we extend this slightly by projecting also onto the mainplot error directions, U_6 to U_8. This allows us to separate the mainplot and subplot error terms. The resulting model is

$$
y - \bar{y}_{\cdots} = (\bar{y}_{i\cdots} - \bar{y}_{\cdots}) + (\bar{y}_{\cdot\cdot k} - \bar{y}_{\cdots}) + \text{Mp error} + (\bar{y}_{ij\cdot} - \bar{y}_{i\cdots}) + \text{Sp error}
$$

Corrected obsn vector	=	Mp trt vector	+	Block vector	+	Mp error vector	+	Sp trt vector	+	Sp error vector

$$
\begin{bmatrix} 5.7 \\ -12.3 \\ -15.3 \\ -20.3 \\ -10.3 \\ -27.3 \\ \vdots \end{bmatrix}
=
\begin{bmatrix} 17.8 \\ 17.8 \\ 17.8 \\ -17.8 \\ -17.8 \\ -17.8 \\ \vdots \end{bmatrix}
+
\begin{bmatrix} -13.3 \\ -13.3 \\ -13.3 \\ -13.3 \\ -13.3 \\ -13.3 \\ \vdots \end{bmatrix}
+
\begin{bmatrix} -11.8 \\ -11.8 \\ -11.8 \\ 11.8 \\ 11.8 \\ 11.8 \\ \vdots \end{bmatrix}
+
\begin{bmatrix} 9.9 \\ -4.1 \\ -5.8 \\ 4.3 \\ 1.0 \\ -5.3 \\ \vdots \end{bmatrix}
+
\begin{bmatrix} 3.1 \\ -0.9 \\ -2.2 \\ -5.3 \\ 8.0 \\ -2.8 \\ \vdots \end{bmatrix}
$$

where entries are given for just the first block. Note that the mainplot and subplot treatment vectors sum to the treatment vector used in previous chapters, $[27.7, 13.7, 12.0, -13.5, -16.8, -23.0]^T$. The mainplot treatment vector can also be written as the

$$\text{Factor A vector} \ = \ -87.2U_2 \ = \ \bar{y}_{i..} - \bar{y}_{...}$$

Similarly, the subplot treatment vector can be further subdivided into the sum of

Factor B vector	$=$	$-24.5U_9 - 8.0U_{10}$	$= \ \bar{y}_{.j.} - \bar{y}_{...}$
A×B Interaction vector	$=$	$9.8U_{11} - 4.5U_{12}$	$= \ \bar{y}_{ij.} - \bar{y}_{i..} - \bar{y}_{.j.} + \bar{y}_{...}$

as described in Appendix C. The resulting decomposition is illustrated in Figure 14.2.

Testing the Hypotheses

We are now able to test the hypotheses $H_0 : c_i = 0$. As usual we compare appropriate squared distances, $(y.U_i)^2$. What is new is that we must restrict ourselves to comparing projection coefficients $Y.U_i$ which have the same variances. In view of the results of Tables 14.3 and 14.4 this means that our test statistics must use either mainplot directions, or subplot directions.

To be specific, we first consider the hypothesis $H_0 : c_1 = 0$ corresponding to the factor A contrast. From Table 14.3, the relevant projection coefficient, $Y.U_2$, is distributed under H_0 as $N\left[0, 3\sigma_m^2 + \sigma_s^2\right]$; also $Y.U_6$, $Y.U_7$ and $Y.U_8$ follow this distribution. The appropriate test statistic is therefore

$$F = \frac{(y.U_2)^2}{\left[(y.U_6)^2 + (y.U_7)^2 + (y.U_8)^2\right]/3} = \frac{7597}{385} = 19.73$$

which comes from an $F_{1,3}$ distribution if H_0 is true. Note that all of U_2, U_6, U_7 and U_8 lie in the mainplot space.

Second, we consider the hypothesis $H_0 : c_2 = 0$ corresponding to the first factor B contrast. From Table 14.4, if H_0 holds then $Y.U_9$ is distributed as $N\left[0, \sigma_s^2\right]$; also $Y.U_{13}$ to $Y.U_{24}$ follow this distribution. The appropriate test statistic is therefore

$$F = \frac{(y.U_9)^2}{\left[(y.U_{13})^2 + \cdots + (y.U_{24})^2\right]/12} = \frac{602}{59} = 10.20$$

which comes from an $F_{1,12}$ distribution if H_0 is true. Here all of U_9 and U_{13} to U_{24} lie in the subplot space.

Similarly the directions U_{10} to U_{12} associated with the remaining hypotheses lie in the subplot space, so the appropriate test statistics have the same denominator as in the previous paragraph.

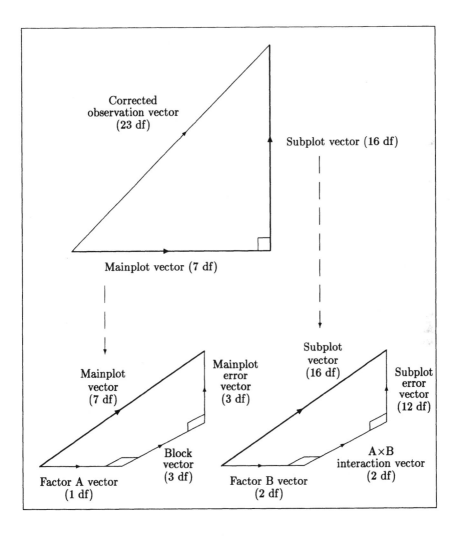

Figure 14.2: The corrected observation vector broken into mainplot and subplot vectors. In turn, these are broken into orthogonal vectors of interest.

ANOVA Table

The ANOVA table summarizing the calculation of the test statistics is given in Table 14.5. The corresponding Pythagorean breakup derives from the orthogonal decomposition pictured in Figure 14.2.

Source of Variation	df	SS	MS	F
Block	3	1934	645	
Cultivar	1	7597	7597	19.73 (*)
Mainplot error	3	1155	385	
(no N) v N	1	602	602	10.20 (**)
125 vs 250N	1	64	64	1.08
Cult. $\times$ (no N,N)	1	96	96	1.63
Cult. $\times$ (125,250)	1	20	20	.34
Subplot error	12	708	59	
Total	23	12177		

Table 14.5: ANOVA table for splitplot analysis, Example C, showing the mainplot and subplot sources of variation separately.

Minitab Analysis

The computations for Table 14.5 can be carried out in Minitab by regressing the observation vector y on the directions $U_2, \ldots, U_{12}$. The subplot error sum of squares is printed by Minitab since it is the residual sum of squares; the block and mainplot error sums of squares will need to be calculated by summing the appropriate $(y.U_i)^2$ terms.

Estimation of σ_m^2 and σ_s^2

For hypothesis testing and confidence interval estimation, the important variance estimates are the mainplot and subplot error mean squares, $s_1^2 = 385$ and $s_2^2 = 59$. The *variance components*, σ_m^2 and σ_s^2, are of interest only if the researcher wishes to estimate the relative importance of the two sources of random variation. At the subplot level, the subplot error mean square of $s_2^2 = 59$ is an unbiased estimate of the variation between subplots within a mainplot, σ_s^2, so we can write $s_s^2 = 59$. At the mainplot level, the mainplot error mean square of $s_1^2 = 385$ is an estimate of $3\sigma_m^2 + \sigma_s^2$.

Since

$$3s_m^2 + s_s^2 = 385 \quad \text{and} \quad s_s^2 = 59$$

we must have $s_m^2 = 109$. Obligingly $s_m^2 > s_s^2$ in support of our original assumption.

Assumption Checking

Assumptions should strictly be checked at both the mainplot and subplot levels of the hierarchy. However, since the number of mainplot error degrees of freedom is only three, we shall pass over the mainplot level and discuss only the subplot level.

The assumptions which require checking are those of independence and normality of the errors, constancy of the variance, and additivity of block and treatment effects. The assumption of independence can be checked by plotting the subplot errors, obtained from the subplot error vector, against the position in the field as given in Figure 14.1. This will not be done here, but the method is similar to that illustrated in Figure 12.3.

The second assumption, of normality, can be checked by plotting a histogram of the subplot errors, as given in Figure 14.3.

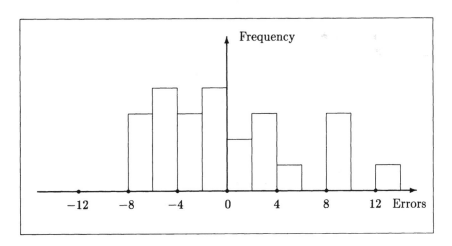

Figure 14.3: Histogram of subplot errors, Example C.

The histogram appears somewhat skewed, with a slightly longer tail on the right. This is what we expect when we are dealing with counts, since on the left of the distribution the counts are restricted to being greater than zero, whereas on the right of the distribution abnormally high counts can occur. In our case the problem is not very severe, so we shall take no action. One action we could take, however, would be to transform the data using the square root transformation; this tends to make the counts more normal in distribution.

To check the assumption of constant variance, we square the four subplot errors for each treatment and add. Since there are twelve degrees of freedom for subplot error and six treatments, each such sum of four squares can be considered to be the sum of $12/6 = 2$ independent squares, so we divide each sum of four squares by two. The resulting variance estimates, given in Table 14.6, will then average to $s_s^2 = 59$.

Treatment	Variance estimate
Green globe, no N	7
Green globe, 125N	36
Green globe, 250N	46
York globe, no N	126
York globe, 125N	74
York globe, 250N	64
Average	59

Table 14.6: Subplot variance estimates for separate treatments, Example C.

To our surprise the York globe cultivar, for which the mean plant counts are low, appears to produce treatments with a higher variance than the Green globe cultivar, for which the mean plant counts are higher. To formally check this apparent phenomenon we calculate the F test of the hypothesis $H_0 : (\sigma_1^2 + \sigma_2^2 + \sigma_3^2)/3 = (\sigma_4^2 + \sigma_5^2 + \sigma_6^2)/3$.

The appropriate test statistic is

$$F = \frac{(126 + 74 + 64)/3}{(7 + 36 + 46)/3} = 2.97$$

Since this is less than the 97.5 percentile of the $F_{6,6}$ distribution we conclude there is no strong evidence of a difference in variance between the two cultivars. We therefore accept the assumption of common variance.

To check the additivity assumption we mimic the procedure used to produce Figure 12.5. The resulting graph is given in Figure 14.4.

The figure is very informative. The value of 45 in treatment four (the fifth column from the left) is seen to be unusual when viewed in the context of the data for the York globe cultivar alone. More importantly, block effects are quite pronounced for the Green globe cultivar (the three treatments with large means) but almost nonexistent for the York globe cultivar. This violates the assumption of additivity of block and treatment effects, and casts serious doubt upon the validity of the analysis.

Since we do not wish to abandon our example in midstream, we shall complete the discussion of the example ignoring this problem. In practice, however, a statistician may well decide to abandon the split plot analysis at this stage, and analyze the data for Green globe and York globe cultivars separately. Each analysis would involve three nitrogen treatments and four blocks, and employ a randomized block model.

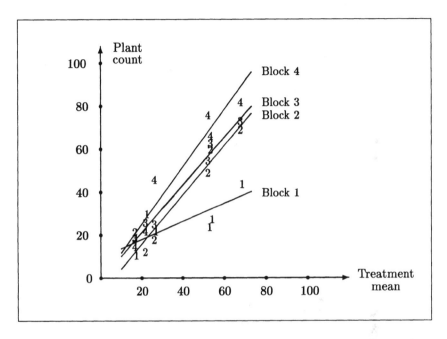

Figure 14.4: Checking the additivity assumption by plotting observations for each block, numbered 1 to 4, against the treatment means.

Report on Study

The usual method for presenting the results from a split plot analysis is given in Table 14.7. In some cases a worthwhile addition is the 2×3 table of treatment means.

The standard error of the difference, the SED, for the factor A main effect is calculated as $\sqrt{2s_1^2/nJ} = \sqrt{2(385)/12}$, using the mainplot error mean square and an effective replication of twelve. The SED for the factor B main effect is calculated as $\sqrt{2s_2^2/nI} = \sqrt{2(59)/8}$, using the subplot error mean square and an effective replication of eight. Here I is the number of levels of the mainplot factor and J the number of levels of the subplot factor. These formulae are derived in Exercise 14.8(c).

The evidence from Table 14.7 can be simply summarized. The Green globe cultivar of soft turnip had significantly higher plant numbers in early winter than the York globe cultivar. Also, the application of nitrogen significantly depressed plant numbers.

	Number of soft turnip plants per 1.08m^2
Mainplot factor (cultivar)	
Green globe	57.1
York globe	21.5
SED	8.0
Significance of difference	*
Subplot factor (nitrogen)	
no sulphate of ammonia	46.4
125 kg/ha S/A	37.8
250 kg/ha S/A	33.8
SED	3.8
Contrasts:	
(no N) v N	**
125 v 250 N	ns
Interaction contrasts	
Cultivar × (no N,N)	ns
Cultivar × (125,250)	ns

Table 14.7: Presentation of results for split plot design, Example C.

It is interesting to see how these results compare with those obtained in §9.3 using an incorrect model. The changes are in the significances of the cultivar and "no N versus N" contrasts. The cultivar contrast is now only 5% significant whereas before it was 1% significant; this is attributable to an increase in SED from 5.9 to 8.0 and a reduction in error degrees of freedom from eighteen to three. The "no N versus N" contrast is now 1% significant whereas before it was not significant; this is caused by the sizeable decrease in the SED, from 7.3 to 3.8, outweighing a not so sizeable decrease in error degrees of freedom from eighteen to twelve. In this latter case, the fitting of the mainplot averages before modelling the variations within each mainplot, means that the nitrogen effects are modelled much more tightly than in the previous, incorrect, model.

Confidence Intervals

The usual confidence interval formulae for contrasts apply in our present situation. The only proviso is that for mainplot contrasts (cultivar in our example) the mainplot error mean square and mainplot error degrees of freedom must be used, whereas for subplot contrasts (the other four in our example) the subplot error mean square and subplot error degrees of freedom must be used.

14.3 Discussion

In this section we discuss further aspects of the split plot design.

The Model

The analysis of the previous section has revealed for the first time how we can handle variability at more than one level. At the mainplot level the sources of variation corresponded to the factor A contrasts, the block contrasts, and the mainplot error contrasts. At the subplot level the sources of variation were the factor B contrasts, the A×B interaction contrasts, and the subplot error contrasts.

Fitting the model resulted in the decomposition

$$
\begin{aligned}
(y - \bar{y}_{...}) &= (\bar{y}_{i..} - \bar{y}_{...}) + (\bar{y}_{..k} - \bar{y}_{...}) + (\bar{y}_{i.k} - \bar{y}_{i..} - \bar{y}_{..k} + \bar{y}_{...}) \\
&+ (\bar{y}_{.j.} - \bar{y}_{...}) + (\bar{y}_{ij.} - \bar{y}_{i..} - \bar{y}_{.j.} + \bar{y}_{...}) + \text{subplot error vector}
\end{aligned}
$$

where the vectors on the right are the factor A vector, the block vector, the mainplot error vector, the factor B vector, the $A \times B$ interaction vector, and the subplot error vector respectively. This decomposition was illustrated in Figure 14.2.

Reasons for Usage

Practical considerations often lead to the use of a split plot design. Consider for example a study of the response of five wheat cultivars to flood irrigation with three levels: no irrigation, irrigated every three weeks, irrigated every six weeks. The smallest area conveniently flood irrigated may be a 12m wide × 200m long strip. To set up a randomized block design with only two replicates of the $5 \times 3 = 15$ treatments would therefore require 30 strips, occupying an area of 7.2 hectares, a prohibitively large area. However, four replicates of a split plot design with the five cultivars sown as subplots within each of the three irrigation mainplots, would occupy only twelve strips, a much more acceptable area. A consequence of this design would be greater precision for estimating differences between cultivars and cultivar by irrigation interactions, at the expense of lower precision for differences between irrigation levels. This is generally appropriate since differences due to irrigation are known to be large and variable between years, with little requirement for a high level of precision in any one experiment.

In industry, a split plot design often occurs naturally. For example, in an industrial process the study of one factor may involve large runs or batches with a machine which must be reset between runs; however, it may be possible to vary a second factor within each run.

A second reason for utilizing a split plot design is to deliberately allow greater precision for the estimation of the effects for one factor, compared to a second factor. This was discussed in the irrigation by cultivars example above. In general, the experiment would be set up so that the blocks would remove the maximum amount of variation between the experimental units, then mainplots would be arranged to take maximum account of the remaining variation. This would mean that the variation between subplots would be minimized. The factor requiring greater precision would become the subplot treatment factor. The interaction between the two factors would also enjoy this higher level of precision.

Arrangement of Mainplots

In Figure 14.1 the eight mainplots were arranged in a randomized block design. However, they could also have been arranged in a completely randomized design. For such a design the degrees of freedom are given in Table 14.8.

Source of variation	Degrees of freedom
Cultivar	1
Mainplot error	6
(no N) v N	1
125 v 250N	1
Cult. × (no N,N)	1
Cult. × (125,250)	1
Subplot error	12
Total	23

Table 14.8: Skeleton ANOVA table for splitplot design where mainplots are arranged in a completely randomized design.

In general, the mainplots can be arranged in a completely randomized design, a randomized block design, a latin square design, or many other designs which we haven't met. The design at the mainplot level has absolutely no effect on the breakdown at the subplot level, since it simply affects the modelling of the mainplot averages. Thus the number of subplot error degrees of freedom is always $(n-1)I(J-1)$, where n is the number of replicates, I is the number of levels of the mainplot factor and J is the number of levels of the subplot factor.

As a sidetrack, we note that randomized block designs can be converted to split plot designs if it becomes necessary to investigate another treatment factor. As an illustration, Example D of §7.2 was laid out in a randomized block design with six blocks and one treatment factor, the rates of seeding of the barley. Suppose, however, that the barley becomes infected with a previously unresearched disease during the course of the experiment. Also, the experimenter knows of a chemical which is reputed to have some effect on the disease, and wishes to measure its effectiveness in terms of grain yield at harvest-time. A sensible procedure in this case would be to apply the chemical to three of the blocks and leave the other three blocks untreated. This would convert the design to three replicates of a split plot design, by introducing a second treatment factor, chemical treatment, with two levels, none and some. The six old blocks would now be regarded as six mainplots. The treatment factors would be chemical treatment, factor A, and seeding rate, factor B.

Method of Randomization

In a split plot design the mainplots are randomized as appropriate to the design for the mainplots. The subplot treatments are then assigned in a completely random manner within each mainplot.

Method of Analysis

The general procedure is to firstly set up a complete set of $nI - 1$ coordinate axes spanning the mainplot space, so that all of the variation among the nI mainplots can be modelled. These axes will span the mainplot treatment space, the block space if blocking is present, and the mainplot error space. The estimate of the background variability of these vectors, $J\sigma_m^2 + \sigma_s^2$, is

$$s_1^2 = \|\text{mainplot error vector}\|^2/\text{mainplot error df}$$

which serves as the baseline for the mainplot treatment, or factor A, F tests.

To complete the split plot model, a further $(J - 1) + (I - 1)(J - 1)$ treatment axes must be specified, corresponding to the factor B main effects and the A $\times$ B interaction effects. After these terms have been added in to the model, the residual vector is the subplot error vector. This allows estimation of σ_s^2, the background variability of the subplot vectors, via

$$s_2^2 = \|\text{subplot error vector}\|^2/\text{subplot error df}$$

where the denominator is $(n-1)I(J-1)$. This serves as the baseline for the subplot treatment F tests, namely the factor B and the A$\times$B interaction.

The above model can be fitted using the Regress command in Minitab. Treatment means and main effect means for factors A and B can be calculated using the Oneway command.

Least Significant Differences

In Table 14.7 the LSD's for the main effects can be calculated by multiplying the appropriate SED by the appropriate t value. The cultivar

LSD(5%) = 8.0 × 3.182 = 25.5, using a t value with three degrees of freedom. The nitrogen LSD(5%) = 3.8 × 2.179 = 8.3, using a t value with twelve degrees of freedom. Notice that the greater precision at the subplot level is partly attributable to the lower t values associated with a higher number of degrees of freedom, as well as the lower variance estimates.

The LSD's for comparisons within the 3 × 2 interaction table are of two types. The first type allows for the more precise comparisons of levels of factor B at the same level of factor A, namely subplots within mainplots. The second type allows for the less precise remaining comparisons. The formulae for the SED's and corresponding t values are given in Cochran and Cox (1957).

More than Two Treatment Factors

For simplicity we have dealt with just two treatment factors, one assigned to mainplots and one to subplots. However, we can have any number of factors assigned to either mainplots or subplots. For example, suppose factors A and B were assigned to mainplots and factor C to subplots. Then the A,B and A×B contrasts would be mainplot contrasts and the C, A×C, B×C and A×B×C contrasts would be subplot contrasts. In general, the main effects and interactions among the mainplot factors are judged at the mainplot level, and all other main effects and interactions are judged at the subplot level.

Split Split Plot Design

When three or more treatment factors are under investigation it may be desirable, or necessary, to subdivide each subplot into subsubplots. This simply adds another level to the hierarchy of variations within the study. The details of the analysis can be deduced from the description given in Cochran and Cox (1957). The process can of course be continued ad infinitum, to split split split plot designs and so on.

Method of Estimation

For the record, we note that for a split plot design it can be shown that the least squares estimates of the model parameters no longer coincide with the maximum likelihood estimates discussed in §5.3.

14.4 Summary

In a split plot design the experimental units and the variation between them follow a hierarchy. The greatest variation is generally among the mainplots, with less variation among the subplots within each mainplot.

There are two main reasons for using this design: practical considerations, and a desire for greater precision for the subplot factor main effects or the mainplot × subplot interaction effects.

If the mainplots are laid out in a randomized block design the fitted model is

$$(y - \bar{y}_{...}) = (\bar{y}_{i..} - \bar{y}_{...}) + (\bar{y}_{..k} - \bar{y}_{...}) + \text{mainplot error vector}$$
$$+ (\bar{y}_{.j.} - \bar{y}_{...}) + (\bar{y}_{ij.} - \bar{y}_{i..} - \bar{y}_{.j.} + \bar{y}_{...}) + \text{subplot error vector}$$

where the mainplot error vector is $(\bar{y}_{i.k} - \bar{y}_{i..} - \bar{y}_{..k} + \bar{y}_{...})$. This decomposition was illustrated in Figure 14.2.

If the mainplots are laid out in a completely randomized design the first three terms on the right collapse to $(\bar{y}_{i..} - \bar{y}_{...}) + (\bar{y}_{i.k} - \bar{y}_{i..})$, the factor A vector plus the mainplot error vector. The other vectors are unchanged.

The A, B and A×B vectors are decomposed into contrast vectors in the usual manner. We judge the significance of A main effect contrasts using $s_1^2 = \|\text{mainplot error vector}\|^2 / (\text{mainplot error df})$ as the baseline. The significance of B main effect and A×B interaction contrasts are judged using $s_2^2 = \|\text{subplot error vector}\|^2 / (\text{subplot error df})$ as the baseline.

Confidence intervals for contrasts are constructed in the usual manner. However, care must be taken to use the appropriate mainplot or subplot error variance and degrees of freedom.

For background reading on the split plot design the reader is referred to Cochran and Cox (1957), pages 293–304, and Little and Hills (1978), pages 87–100.

Exercises

(14.1) (a) In Table 14.1 calculate the mean of the three subplot values within each mainplot.

(b) Analyze these eight mainplot averages as four replicates of two cultivar treatments using a randomized block model. Present your results in an ANOVA table.

(c) Can you spot the relationship between your sums of squares and those in the split plot analysis summarized in the ANOVA of Table 14.5?

(14.2)(a) Use the analysis of Example C as given in §14.2 to calculate the 95% confidence interval for the average difference in plant count between the two soft turnip cultivars.

(b) Similarly calculate the 95% confidence interval for the average difference in plant count between "no N" and "some N", that is, $(125N + 250N)/2 - noN$.

(14.3) An experiment was carried out to investigate the responsiveness of a barley crop to nitrogeneous fertilizer under three different levels of irrigation. Treatment factors and their levels were:

(A) Irrigation:

    1. No irrigation

    2. Irrigated at 15% soil moisture

    3. Irrigated at 20% soil moisture

(B) Nitrogen:

    1. No nitrogen

    2. 50 kg/ha N at time of seed drilling

    3. 50 kg/ha N at time of plant tillering

The resulting nine treatments were arranged in a split plot design with five replicates. Fifteen 80m × 6m portions of flood irrigation strips formed the mainplots; factor A was assigned to these mainplots in a randomized block design. Each mainplot was divided lengthwise into three 80m × 2m subplots, and factor B was randomly assigned to these subplots. The grain yields in kg/ha from each subplot are listed in the table.

Block:	1	2	3	4	5
Non irrigated					
No nitrogen	3055	3605	3315	2355	3220
50 N, drilling	4315	5210	4375	4165	4480
50 N, tillering	5210	5890	5005	5115	4750
Irrigation at 15% s.m.					
No nitrogen	3885	3595	3960	2670	3320
50 N, drilling	5515	4930	5360	4300	5005
50 N, tillering	5640	5005	5260	5215	5450
Irrigation at 20% s.m.					
No nitrogen	4625	3910	3710	3030	4315
50 N, drilling	5920	5600	5320	5590	5500
50 N, tillering	6500	5670	5920	5595	5625

(a) Draw a sketch of the field layout, roughly to scale, assuming the fifteen irrigation strips are in a single row with long sides adjacent.

(b) Specify appropriate contrasts for factors A and B.

(c) Use Minitab to statistically analyze the data. Summarize your output in an ANOVA table following the format of Table 14.5.

(d) Report the experimental results in a table such as that given in Table 14.7. Include a 3 × 3 table of treatment means. Plot these treatment means on a graph with yield up the y-axis and the levels of nitrogen equally spaced along the x-axis; use a different symbol for each level of irrigation. Summarize your results in a few brief sentences.

(14.4) In the two experiments following, specify all of the unit vectors except the subplot error vectors. Point out which are "mainplot treatment", "mainplot error", "block" if appropriate, and "subplot treatment" vectors. Also specify the order in which you would arrange the data.

(a) Mainplot treatments: Wheat, Barley (cereals) and Ryegrass (noncereal)
    Subplot treatments: 50, 100, 150 kg/ha N
    Arrangement of mainplots: Randomized block design
    Number of replicates: 2

(b) Mainplot treatment factors: Nitrogen: none and some
    　　　　　　　　　　　　　　　　　Phosphate: none and some
    Subplot treatments: Potassium: 100, 200, 400 kg/ha
    Arrangement of mainplots: Completely randomized design
    Number of replicates: 2

(14.5) Four field experiments were designed with a 2×2 factorial design, and the following treatments:

Treatment 1 : no sodium, no potassium
Treatment 2 : sodium, no potassium
Treatment 3 : no sodium, potassium
Treatment 4 : sodium, potassium

For each of the following field layouts, decide whether the experiment was "most probably" laid out as a completely randomized design, a randomized block design, or a split plot design.

Plot no.	1	2	3	4	5	6	7	8	9	10	11	12	13	14	15	16	17	18	19	20
Field layout 1																				
Treatment no.	4	2	1	3	4	2	3	1	3	2	4	1	1	4	2	3	2	4	3	1
Field layout 2																				
Treatment no.	3	1	4	2	2	4	1	3	1	3	2	4	3	1	4	2	1	3	2	4
Field layout 3																				
Treatment no.	1	4	3	2	2	3	4	2	1	3	4	3	2	1	1	4	2	4	1	3
Field layout 4																				
Treatment no.	4	3	1	2	3	4	2	1	2	1	4	3	1	2	4	3	4	3	1	2

(14.6) An experiment was carried out to investigate the effects of irrigation, nitrogen and seeding rate on grain quality in wheat. The trial design was a split plot design. Irrigation was the main plot treatment factor and nitrogen and seeding rate were subplot treatment factors. There were three levels of irrigation (none, 25 mm/week, 50 mm/week), two levels of nitrogen (none, 100 kg N/ha) and two seeding rates (60,120 kg/ha), making up a total of

$3 \times 2 \times 2 = 12$ treatments. There were five replicates and the main plots were laid out in a randomized block design.

The "main effect" means and ANOVA table for the quality score were as follows:

MAIN PLOT FACTOR   SUBPLOT FACTORS

Irrigation		**Nitrogen**		**Seeding rate**	
None	18.2	None	17.6	60 kg/ha	16.9
25 mm/week	16.1	100 kg/ha	15.6	120 kg/ha	16.3
50 mm/week	15.5	se (mean)	0.51	se (mean)	0.51
se (mean)	0.76				

Source of Variation	df	SS	MS	F
Blocks	4	74.43		
Irrigation				
None v some	1	76.80		
25 v 50	1	3.60		
Main plot error	8	92.42	11.552	
Nitrogen (N)	1	60.00		
Seeding rate (SR)	1	5.40		
N × SR	1	10.52		
Irrigation × N				
(None v some)×N	1	25.46		
(25 v 50)×N	1	9.43		
Irrigation × SR				
(None v some)×SR	1	6.40		
(25 v 50)×SR	1	8.75		
Irrigation × N × SR				
(None v some)×N×SR	1	3.17		
(25 v 50)×N×SR	1	14.91		
Subplot error	36	280.91	7.803	
Total	59	672.19		

(a) Calculate the 95% confidence interval for the irrigation main effect contrast $c_1 = \mu_{1..} - (\mu_{2..} + \mu_{3..})/2$ which compares "no irrigation" with "irrigation".

(b) Calculate the 95% confidence intervals for the nitrogen main effect contrast $c_2 = (\mu_{.1.} - \mu_{.2.})$ which compares "no N" with "N" and the seeding rate main effect contrast $c_3 = (\mu_{..1} - \mu_{..2})$ which compares "60 kg/ha" with "120 kg/ha".

(c) Use the ANOVA table shown to work out the F tests for the irrigation main effect contrast in (a) and the nitrogen and seeding rate contrasts in (b). Are your answers consistent with the results you obtained in parts (a) and (b)? Explain.

(14.7) The data below is rice yield in kg/plot for a split plot design with factors:

Main plots:	Depth of water	: 5, 10 and 15 cm
Subplots:	Seeding rate	: 90, 150 and 210 kg/ha
	Variety	: S201, M201

Three replicates were laid down. The nine main plots, one per row of data, were arranged in a completely randomized design. Each subplot occupied an area of 60m^2.

Water depth	Rep. No.	S201 Seeding rate kg/ha 90	150	210	M201 Seeding rate kg/ha 90	150	210
5cm	1	43.2	41.7	42.4	42.8	48.1	46.9
	2	30.6	33.5	32.5	45.2	40.2	41.7
	3	38.4	32.8	35.8	36.3	38.8	42.1
10cm	1	23.3	32.3	26.2	31.9	28.2	30.9
	2	26.2	37.9	30.7	43.0	43.3	38.9
	3	41.7	38.3	39.3	43.8	44.0	41.3
15cm	1	29.0	28.7	33.2	29.2	31.1	28.2
	2	38.3	39.3	46.7	46.5	43.5	44.4
	3	41.5	40.8	38.7	39.2	38.3	43.1

(a) For the main plot analysis write down the "main plot treatment" unit vectors, U_2 and U_3, corresponding to the linear and quadratic components of depth of water. For the data order, write down all of mainplot 1

= row 1, then mainplot 2 = row 2, and so on.

(b) Write down six more unit vectors corresponding to "mainplot error".

(c) Also write out the remaining (15) unit vectors for treatments.

(d) You now have the 23 unit vectors which span the model space. Use Minitab and Regress y onto $U_2 \cdots U_{24}$. If necessary, use several Regress commands, with subsets of the U_i vectors.

(e) Calculate the mainplot error mean square, s_1^2, and use it to test the hypotheses $H_0 : c_1 = 0$ and $H_0 : c_2 = 0$.

(f) Minitab will print the subplot error mean square, s_2^2. Use this to test the remaining hypotheses, $H_0 : c_i = 0$.

(g) Set out your results in an ANOVA table.

(h) In everyday language, what are your conclusions?

(14.8) (a) Use the model given in §14.2 to show that the mainplot mean random variables, $\overline{Y}_{i \cdot k}$, have the form

$$\overline{Y}_{i \cdot k} = \mu_{i \cdot} + \beta_k + \epsilon^m_{i \cdot k} + \bar{\epsilon}^s_{i \cdot k}$$

Hence show that $Var(\overline{Y}_{i \cdot k}) = \sigma^2_m + \sigma^2_s / J$.

(b) Show that the variance of the factor A main effect, $Var(\overline{Y}_{i \cdot \cdot})$, is $(J\sigma^2_m + \sigma^2_s)/nJ$. Hence show the SED for factor A main effects is $\sqrt{\dfrac{2s^2_1}{nJ}}$.

(c) Show that the variance of the factor B main effect, $Var(\overline{Y}_{\cdot j \cdot})$, is σ^2_2 / nI, and hence show that the SED for factor B main effects is $\sqrt{\dfrac{2s^2_2}{nI}}$.

Solutions to the Reader Exercises

(14.1) (i) $Y.U_2 = [-(Y_{111} + Y_{121} + Y_{131}) + (Y_{211} + Y_{221} + Y_{231}) - \cdots]/\sqrt{24}$

$$= \left[-3\overline{Y}_{1 \cdot 1} + 3\overline{Y}_{2 \cdot 1} - \cdots\right]/\sqrt{24}$$

(ii) $\overline{Y}_{1 \cdot 1} = [(\mu_{11} + \beta_1 + \epsilon^m_{1 \cdot 1} + \epsilon^s_{111}) + (\mu_{12} + \beta_1 + \epsilon^m_{1 \cdot 1} + \epsilon^s_{121})$

$$+ (\mu_{13} + \beta_1 + \epsilon^m_{1 \cdot 1} + \epsilon^s_{131})]/3$$

$$= \mu_{1 \cdot} + \beta_1 + \epsilon^m_{1 \cdot 1} + (\epsilon^s_{111} + \epsilon^s_{121} + \epsilon^s_{131})/3$$

so $\overline{Y}_{1 \cdot 1} \sim N\left[\mu_{1 \cdot} + \beta_1, \sigma^2_m + 3\sigma^2_s/9\right] = N\left[\mu_{1 \cdot} + \beta_1, \sigma^2_m + \sigma^2_s/3\right]$

(iii) $E(Y.U_2) = [-3(\mu_{1.} + \beta_1) + 3(\mu_{2.} + \beta_1) - \cdots]/\sqrt{24}$

$\qquad = 12(\mu_{2.} - \mu_{1.})/\sqrt{24} = \sqrt{12}(\mu_{2.} - \mu_{1.})/\sqrt{2}$

$Var(Y.U_2) = \left[9\,Var(\overline{Y}_{1.1}) + 9\,Var(\overline{Y}_{2.1}) + \cdots\right]/\sqrt{24}$

$\qquad = 8\left(9(\sigma_m^2 + \sigma_s^2/3)\right)/24 = 3\sigma_m^2 + \sigma_s^2$

(14.2) (i) $Y.U_3 = [-2Y_{111} + Y_{121} + Y_{131} - 2Y_{211} + Y_{221} + Y_{231} - \cdots]/\sqrt{48}$

$\qquad = [V_{1.1} + V_{2.1} + \cdots]/\sqrt{48}$

(ii) $V_{1.1} = -2(\mu_{11} + \beta_1 + \epsilon_{1.1}^m + \epsilon_{111}^s) + (\mu_{12} + \beta_1 + \epsilon_{1.1}^m + \epsilon_{121}^s)$

$\qquad\qquad\qquad\qquad\qquad\qquad + (\mu_{13} + \beta_1 + \epsilon_{1.1}^m + \epsilon_{131}^s)$

$\qquad = -2\mu_{11} + \mu_{12} + \mu_{13} - 2\epsilon_{111}^s + \epsilon_{121}^s + \epsilon_{131}^s$

so $V_{1.1} \sim N\left[-2\mu_{11} + \mu_{12} + \mu_{13}, 6\sigma_s^2\right]$

(iii) $E(Y.U_3) = 4(-2\mu_{11} + \mu_{12} + \mu_{13} - 2\mu_{21} + \mu_{22} + \mu_{23})/\sqrt{48}$

$\qquad = \sqrt{8}(-2\mu_{.1} + \mu_{.2} + \mu_{.3})/\sqrt{6}$

$Var(Y.U_3) = 8 \times 6\,\sigma_s^2/48 = \sigma_s^2$

Part V
Fundamentals of Regression

Earlier chapters have been primarily concerned with situations in which the mean of a random variable Y depends on another factor in a *qualitative* way. For example, in Chapter 8 we analyzed an example in which the mean of the random variable Y, namely greenfeed yield, depended on the mangel or fodder beet cultivar. We move on now to consider the situation in which the mean of the random variable depends on another factor in a *quantitative* way. For example, we shall consider how the seed yield of white clover, Y, depends on the density of the weed yarrow, x. We shall think of the mean of Y as a function of a variable x which can vary over a continuous range of values, although we will only have data for certain discrete values of x. Our primary interest is in the way in which $E(Y)$ depends on x. Until this point, contrasts have been used to compare treatments; now they are used to assess the form of the relationship.

For designed experiments this situation has already been introduced in Chapter 10. There the values of x were predetermined by the experimenter. More generally, however, the values of x are not fixed, but are obtained by a random sampling scheme. This means that the x values are usually all different from one another, so that there is no replication for any given x value; this makes assumption checking more difficult.

We shall cover the case of a straight line relationship between $E(Y)$ and x, so-called "simple regression", in Chapter 15. In Chapter 16 we look at the case of a polynomial relationship. Chapter 17 blends analysis of variance and regression techniques in analysis of covariance: in this case our relationships involve several parallel straight lines.

Chapter 15

Simple Regression

Over a century ago Sir Francis Galton investigated the manner in which a son's height depended on that of the father. He showed that the heights of tall fathers' sons were distributed about a value somewhat less than that of their father. The paper was entitled "Regression towards mediocrity in hereditary stature", and the word regression has stayed with the subject. However, the word has taken on a broader meaning in the world of statistics; it now refers to more general relationships between variables, not just the special linear relationship noted by Galton.

We begin by looking at the simplest relationship between two variables, the straight line relationship. Because this is just a continuous analogue of our earlier discrete analysis, it should not surprise the reader that our familiar methods for model fitting and hypothesis testing continue to suffice.

The plan for the chapter is to start with an illustrative example in §1, then give a summary of the general analysis in §2. This is followed in §3 by a discussion of confidence intervals, and in §4 by a discussion of the correlation coefficient. In §5 we discuss common pitfalls of regression analysis. The chapter closes with a summary in §6, a class exercise, and general exercises.

15.1 Illustrative Example

We introduce the linear regression technique by describing and analyzing the data from a study of the competition between two plants.

Objective

The purpose of the study was to estimate the effect of the agricultural weed yarrow, *Achillea millefolium*, on the yield of a white clover, *Trifolium repens*, seed crop. It was anticipated that the clover yield would decrease as the density of the yarrow increased.

Design

Three areas, shown as A, B and C in Figure 15.1, were selected in the clover seed crop. Ten $0.1m^2$ quadrats were positioned within each area using random coordinates on a 20×20 grid. In each quadrat the clover seed was harvested and weighed, and the density of yarrow assessed by counting the number of yarrow flower stems.

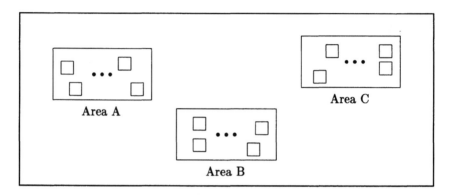

Figure 15.1: Random sampling scheme for competition study (not to scale).

Data

The data for area A, multiplied by ten to convert to a per square meter basis, are given in Table 15.1. These are the data we shall analyze in this section. The full data set will be analyzed as Exercise 17.2.

Number of yarrow flower stems per m^2	Clover seed yield in g/m^2
220	19.0
20	76.7
510	11.4
40	25.1
120	32.2
300	19.5
60	89.9
10	38.8
70	45.3
290	39.7

Table 15.1: Data from area A of the competition study.

The scattergram of clover seed yield y against yarrow density x is given in Figure 15.2. This reveals a tendency for clover seed yield to decrease as yarrow density increases.

To make our analysis easier to understand, we sort our quadrats into order of increasing yarrow density, as shown in Table 15.2. This is not an essential part of the analysis! We shall use this sorted data from now. Hence our observation vector is $y = [y_1, \ldots, y_{10}]^T = [38.8, \ldots, 11.4]^T$.

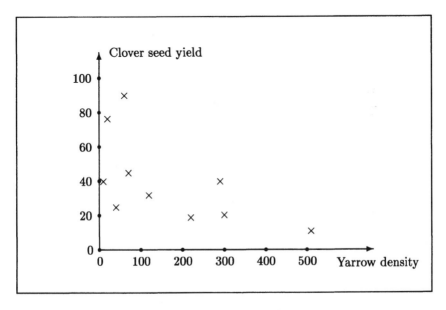

Figure 15.2: Scattergram of clover seed yield against yarrow density for the competition study.

Number of yarrow flower stems per m^2	Clover seed yield in g/m^2
10	38.8
20	76.7
40	25.1
60	89.9
70	45.3
120	32.2
220	19.0
290	39.7
300	19.5
510	11.4

Table 15.2: Competition study data in order of increasing yarrow density.

Model

When using the simple linear regression model we make four assumptions. The first three are:

1. The mean of an observation is assumed to depend on the x value with which it is associated, via the straight line relationship $E(Y) = \beta_0 + \beta_1(x - \bar{x})$. Here β_0 and β_1 are the unknown parameters of the line, and $\bar{x}$ the mean of all the x values in our sample.

2. For each x value, Y is assumed to be normally distributed about this mean.

3. For each x value, Y is assumed to have a common variance of σ^2.

Thus we can write the distribution of Y for a given x value as

$$Y \quad \sim \quad N\left[\beta_0 + \beta_1(x - \bar{x}),\ \sigma^2\right]$$

These first three model assumptions are conveniently summarized in Figure 15.3.

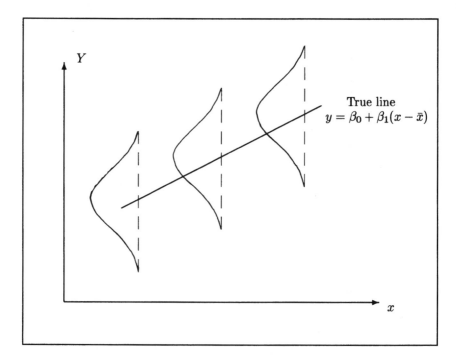

Figure 15.3: The assumptions of the linear regression model.

4. Finally, we assume that in sampling, our errors, the deviations of our observations from the line, are independent values from a $N[0, \sigma^2]$ distribution.

We always start out assuming the truth of the model and the independence of the errors. Whether these are reasonable assumptions must be checked at a later stage.

A distinction is sometimes drawn between the case where the x values are measured more or less without error, and the case where the x values arise from a random variable. In the experimental situation of Chapter 10 we were able to control the seeding rate of the malting barley, and hence gather repeated measurements for a given x value. In the observational situation of the example of this section both the x and y values are drawn from random variables. The regression model can be applied in either situation. In the latter case the straight line models the means of Y conditional upon the observed values of the random variable X.

The resulting model vector, the vector of expected values, is

$$\beta_0 \begin{bmatrix} 1 \\ \vdots \\ 1 \end{bmatrix} + \beta_1 \begin{bmatrix} x_1 - \bar{x} \\ \vdots \\ x_{10} - \bar{x} \end{bmatrix} = \beta_0 \begin{bmatrix} 1 \\ \vdots \\ 1 \end{bmatrix} + \beta_1 \begin{bmatrix} 10 - 164 \\ \vdots \\ 510 - 164 \end{bmatrix}$$

since $\bar{x} = 164$ flower stems per square meter. Hence the model space M is a two dimensional subspace of 10-space.

The appropriate orthogonal coordinate system for M is

$$U_1 = \frac{1}{\sqrt{10}} \begin{bmatrix} 1 \\ \vdots \\ 1 \end{bmatrix} \quad \text{and} \quad U_2 = \frac{1}{\sqrt{238640}} \begin{bmatrix} -154 \\ \vdots \\ 346 \end{bmatrix}$$

where $238640 = \|x - \bar{x}\|^2 = \sum_{i=1}^{N}(x_i - \bar{x})^2$.

Test Hypothesis

We are interested in whether there is evidence to suggest that clover seed yield is influenced by the density of the yarrow. We investigate this by testing the null hypothesis $H_0 : \beta_1 = 0$ against the alternative $H_1 : \beta_1 \neq 0$, where β_1 is the slope of the unknown true line.

Conveniently, the direction associated with this hypothesis is U_2. This is the case since the expected value of the projection coefficient random variable $Y.U_2$ is $\beta_1 \|x - \bar{x}\|$. Hence if $\beta_1 = 0$ the projection coefficient $y.U_2$ will be small whereas if $\beta_1 \neq 0$ it will be relatively large.

Exercise for the reader

(15.1) In order to show that $E(Y.U_2) = \beta_1 \|x - \bar{x}\|$, expand $y.U_2$ by substituting $y = [y_1, \ldots, y_N]^T$ and $U_2 = [x_1 - \bar{x}, \ldots, x_N - \bar{x}]^T / \|x - \bar{x}\|$, and take expected values. Note that the expected value of Y_1 is $\beta_0 + \beta_1(x_1 - \bar{x})$ and so on.

We now have all the raw materials for our analysis. We have an observation vector y, a model space M, and a direction U_2 associated with our hypothesis.

Fitting the Model

As usual, we fit the model by projecting the observation vector y onto the model space M. The resulting fitted model vector is

$$\hat{y} = (y.U_1)U_1 + (y.U_2)U_2 = 39.76 \begin{bmatrix} 1 \\ \vdots \\ 1 \end{bmatrix} - .0943 \begin{bmatrix} -154 \\ \vdots \\ 346 \end{bmatrix} = \begin{bmatrix} 54.3 \\ \vdots \\ 7.1 \end{bmatrix}$$

Hence $b_0 = 39.76$ and $b_1 = -.0943$ are the least squares estimates of the constant, β_0, and the slope, β_1. The algebraic formulae for b_0 and b_1 are:

$$b_0 = (y.U_1)/\sqrt{N} = \sum_{i=1}^{N} y_i / N = \bar{y} = 397.6/10 = 39.76$$

$$b_1 = \frac{y.U_2}{\|x - \bar{x}\|} = \frac{\sum_{i=1}^{N} y_i(x_i - \bar{x})}{\|x - \bar{x}\| \, \|x - \bar{x}\|} = \frac{\sum_{i=1}^{N} y_i(x_i - \bar{x})}{\sum_{i=1}^{N}(x_i - \bar{x})^2} = \frac{-22494.4}{238640}$$

$$= -.0943$$

The equation of the fitted line is

$$\begin{aligned} y &= \bar{y} &+& \quad b_1(x - \bar{x}) \\ \text{or,} \quad y &= 39.76 &-& \quad .0943(x - 164) \end{aligned}$$

This is shown in Figure 15.4, together with the observed and fitted y values.

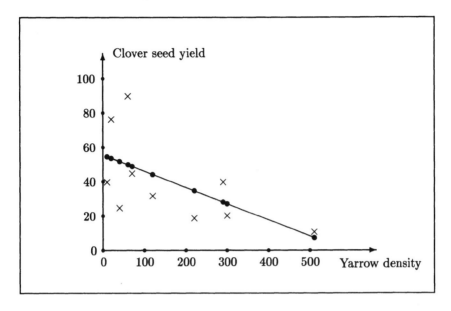

Figure 15.4: Least squares regression line, indicating the observed ($\times$) and fitted ($\bullet$) values.

The fitted model when written out in full is

$$
\begin{array}{ccccccc}
y & = & (y.U_1)U_1 & + & (y.U_2)U_2 & + & \text{error vector} \\[4pt]
\begin{bmatrix} 38.8 \\ \vdots \\ 11.4 \end{bmatrix} & = & \begin{bmatrix} 39.76 \\ \vdots \\ 39.76 \end{bmatrix} & -\ .0943 & \begin{bmatrix} -154 \\ \vdots \\ 346 \end{bmatrix} & + & \begin{bmatrix} -15.5 \\ \vdots \\ 4.3 \end{bmatrix} \\[4pt]
& = & \bar{y} & + & b_1(x - \bar{x}) & + & \text{error vector}
\end{array}
$$

where $\bar{y}$ and $x - \bar{x}$ here denote vectors. This fitted model is depicted in Figure 15.5. The error vector is obtained by subtracting the fitted model vector, $\hat{y}$, from the observation vector, y. For example, the error associated with the first observation is $y_1 - \hat{y}_1 = 38.8 - 54.3 = -15.5$. These values can be seen in Figure 15.4 as the vertical deviations of the observations about the fitted line.

The interplay between Figures 15.4 and 15.5 should be kept firmly in mind. A large constant b_0 corresponds to a large constant vector, while a steep slope b_1 corresponds to a large linear vector. Also, high variability s^2 about the line corresponds to a large error vector.

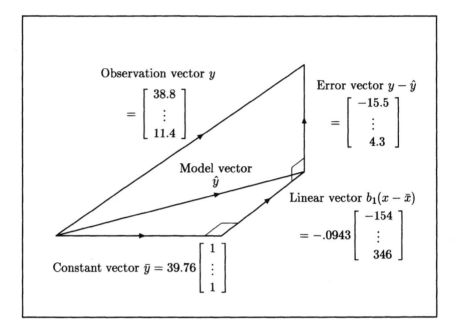

Figure 15.5: Orthogonal decomposition of the observation vector in the linear regression case. The symbol $\hat{y}$ is used for the vector of fitted values, $\hat{y} = \bar{y} + b_1(x - \bar{x})$.

Simplified Decomposition

Traditionally the orthogonal decomposition is written with the observation vector corrected for the mean. This yields the vector equation

$$
y - \bar{y} \quad = \quad b_1(x - \bar{x}) \quad + \quad \text{error vector}
$$

$$
\begin{bmatrix} -.96 \\ 36.94 \\ -14.66 \\ 50.14 \\ 5.54 \\ -7.56 \\ -20.76 \\ -.06 \\ -20.26 \\ -28.36 \end{bmatrix}
=
-.0943
\begin{bmatrix} -154 \\ -144 \\ -124 \\ -104 \\ -94 \\ -44 \\ 56 \\ 126 \\ 136 \\ 346 \end{bmatrix}
+
\begin{bmatrix} -15.5 \\ 23.4 \\ -26.3 \\ 40.3 \\ -3.3 \\ -11.7 \\ -15.5 \\ 11.8 \\ -7.4 \\ 4.3 \end{bmatrix}
$$

Geometrically, this move corresponds to focusing on the simpler 2-way

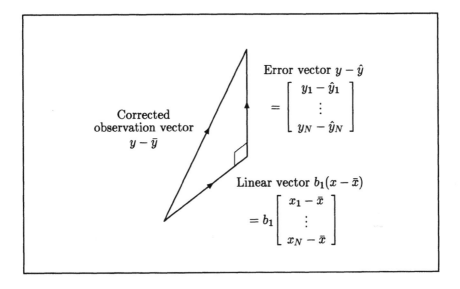

Figure 15.6: Simplified form of the orthogonal decomposition for a linear regression.

orthogonal decomposition of $y - \bar{y}$, rather than that of y itself. This decomposition is represented pictorially in Figure 15.6.

Testing the Hypothesis

To find out whether there is a relationship between yarrow density and clover seed yield we test whether β_1, the slope of the line, is zero. As usual, this is done by comparing the squared projection length in the direction of the hypothesis, $(y.U_2)^2$, with the average of the squared projection lengths in the error space. The resulting test statistic is

$$F = \frac{(y.U_2)^2}{\left[(y.U_3)^2 + \cdots + (y.U_{10})^2\right]/8} = \frac{(y.U_2)^2}{\|\text{error vector}\|^2/8} = \frac{2120.3}{463.5} = 4.57$$

where $U_3, \ldots, U_{10}$ are coordinate axis directions for the error space. The test statistic comes from an $F_{1,8}$ distribution if the hypothesis $H_0 : \beta_1 = 0$ is true. This distribution has a 95 percentile of 5.32, so we are unable to formally reject our hypothesis at the 5% level of significance.

ANOVA Table

The simplified orthogonal decomposition, shown in Figure 15.6, leads to the Pythagorean breakup

$$\|y - \bar{y}\|^2 \;=\; b_1^2 \|x - \bar{x}\|^2 \;+\; \|\text{error vector}\|^2$$

This in turn leads to the ANOVA table shown in Table 15.3, which provides the usual summary of the calculations leading to our test statistic.

Source of Variation	df	SS	MS	F
Regression	1	$b_1^2\|x - \bar{x}\|^2 = 2120.3$	2120.3	4.57
Error	8	$\|y - \hat{y}\|^2 = 3707.7$	463.5	
Total	9	$\|y - \bar{y}\|^2 = 5828.0$		

Table 15.3: ANOVA table for simple linear regression. Note that $(y.U_2)^2$ has been written in the alternative form, $b_1^2\|x - \bar{x}\|^2$.

Estimation of σ^2

In this example we have transformed the original independent set of observation random variables, $Y_i \sim N\left[\beta_0 + \beta_1(x_i - \bar{x}), \sigma^2\right]$, into a new independent set of normal random variables, $Y.U_i$, with means and variances as detailed in Table 15.4. With this new set, $Y.U_1$ and $Y.U_2$ are used to estimate the parameters β_0 and β_1, while the remaining $N - 2$ random variables are used to estimate the variance σ^2.

The directions $U_3, \ldots, U_{10}$ of the error space coordinate system will not be written down, since in general they are messy. Their existence is not in question, however, since we could generate $U_3, \ldots, U_{10}$ using the Gram-Schmidt method of orthogonalization described in Chapter 10.

	Average, or expected value		Variance
$Y.U_1$	$\sqrt{N}\beta_0 \quad = $	$\sqrt{10}\beta_0$	σ^2
$Y.U_2$	$|x - \bar{x}|\beta_1 \quad = $	$\sqrt{238640}\beta_1$	σ^2
$Y.U_3$	0		σ^2
$\vdots$	$\vdots$		$\vdots$
$Y.U_{10}$	0		σ^2

Table 15.4: Means and variances of the projection coefficients, $Y.U_i$.

Exercises for the reader

(15.2) Show that $E(Y.U_1) = \sqrt{10}\beta_0$, by substituting $y = [y_1, \ldots, y_N]^T$ and $U_1 = [1, \ldots, 1]^T/\sqrt{N}$ into the expression $y.U_1$, and taking expected values. Note that the expected value of each Y_i is $\beta_0 + \beta_1(x_i - \bar{x})$.

(15.3) Similarly, show that $E(Y.U) = 0$ for any direction $U = [a_1, \ldots, a_N]^T$ in the error space.

Minitab Analysis

The computations for Table 15.3 can be carried out in Minitab by Regressing the observation vector y onto the direction U_2. At the same time, a scattergram of the data can be drawn and assumptions checked, as detailed in Table 15.5.

Note the new form of the Regress command in Table 15.5. Minitab saves *standardized residuals* in column five, rather than ordinary residuals. When standardized, the residuals usually vary between -2 and $+2$. In standardizing, Minitab takes account of the fact that the regression line is more precisely determined at the center of the data than at the extremes. In this text, we shall deal only with ordinary residuals. These are saved into column seven in our sample job by using the "resids" subcommand.

Our straight line model $y = \beta_0 + \beta_1(x - \bar{x})$ led naturally to an *orthogonal* pair of vectors, $[1, \ldots, 1]^T$ and $[x_1 - \bar{x}, \ldots, x_N - \bar{x}]^T$, spanning the model space, and hence to our orthogonal coordinate system U_1, U_2. Note that Minitab is happy if we Regress y onto the vector $x - \bar{x}$ instead of U_2; this vector is stored in column three in the job listed in Table 15.5. This alternative has two advantages: it shortens the job and also allows direct computation of the estimates b_0 and b_1.

A second alternative arises if we respecify our model as $y = \alpha_0 + \alpha_1 x$. This leads naturally to a *nonorthogonal* pair of vectors, $[1, \ldots, 1]^T$ and $[x_1, \ldots, x_N]^T$, which span the model space. Since Minitab relies on matrix methods in its fitting procedures it is also happy to fit the model by Regressing y onto the vector x, which is stored in column two in the job listed in Table 15.5. This approach, though not especially recommended in this text, further shortens the job and provides the estimates a_0 and a_1 for the simplest form of fitted model, $y = a_0 + a_1 x$.

List of commands	Interpretation
⎡name c1 = 'ClovYld' ⎮name c2 = 'YarrDens' ⎮name c4 = 'HypoDirn' ⎮name c6 = 'Fitted' ⎣name c7 = 'Errors'	[Naming the columns
⎡set c1 ⎣38.8 76.7 ... 19.5 11.4	[Set y values into column one
⎡set c2 ⎣10 20 ... 300 510	[Set x values into column two
[plot c1 c2	[Draw scattergram of y on x
[let c3=c2−mean(c2)	[Calculate vector $x - \bar{x}$
⎡let k1=sqrt(sum(c3**2)) ⎣let c4=c3/k1	[Calculate unit vector U_2
⎡regress c1 1 c4 c5 c6; ⎣resids c7.	⎡Fit the model, saving ⎣errors and fitted values
[histogram c7	[Draw error histogram
[print c1 c6 c7	[Print vector decomposition
[plot c7 c6	⎡Plot errors against ⎣fitted values
[stop	[End of job

Table 15.5: Minitab commands for linear regression case, including plotting of scattergram and assumption checking.

Checking of Assumptions

The assumptions we must check are the linearity of the relationship, the normality and independence of the errors, and the constancy of the variance.

The linearity assumption can be checked by referring back to Figure 15.4. We visually check for any curvature about the line. In our example this reveals nothing unusual, so we accept the linearity assumption as reasonable.

To check whether the errors follow a normal distribution we plot a histogram as shown in Figure 15.7. This again reveals nothing unusual, so we also accept this assumption as reasonable.

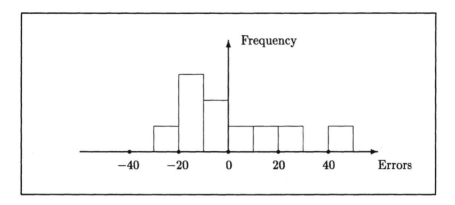

Figure 15.7: Histogram of errors for competition study.

Independence can be checked by plotting the errors against the position-ing of the quadrats in the field. Unfortunately in our case this information has been lost, so we are unable to carry out any such check.

Finally, to check the assumption of constant variance, we plot the errors against the fitted y values, as shown in Figure 15.8. Visually it appears that

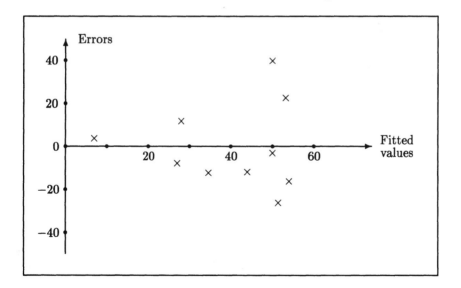

Figure 15.8: Checking the assumption of constant variance by plotting the errors against the fitted y values.

the errors are larger for large fitted y values than for small fitted y values. This tallies with our prior expectations, since in biological work the standard error of plant yields is usually proportional to the mean plant yield.

Refitting the Model

When the errors increase roughly linearly with the fitted y values, giving the funnel effect of Figure 15.8, a standard calculus argument, as in Scheffé, p.365, shows that if we use $\log y$ as the dependent variable then the variances will be made more uniform.

So we refit the model by taking logarithms of the seed yields in Table 15.2, giving a new observation vector, $y = [3.658, \ldots, 2.434]^T$, and repeating the fitting procedure. In our Minitab job we simply insert the command "let c1=log(c1)". The equation of the fitted line is now

$$\log y \quad = \quad 3.504 \quad - \quad .00290(x - 164)$$

This is shown in Figure 15.9, along with the new scattergram.

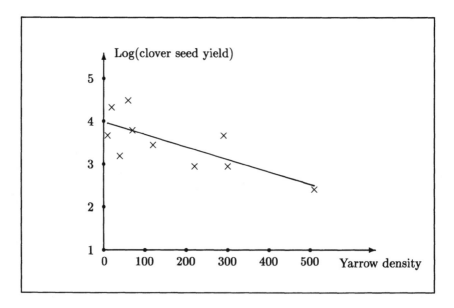

Figure 15.9: Relationship of log(clover seed yield) to yarrow density.

The revised ANOVA table is shown in Table 15.6. The F value has changed from 4.57 to 9.71, a considerable increase. This means that the slope β_1 is now proven different from zero at the 5% level of significance.

Source of Variation	df	SS	MS	F
Regression	1	2.0085	2.0085	9.71(*)
Error	8	1.6543	0.2068	
Total	9	3.6627		

Table 15.6: ANOVA table for regression of log (clover seed yield) on yarrow density.

We again need to check our assumptions. For the linearity check, we inspect Figure 15.9 and notice nothing untoward. The histogram of errors, shown in Figure 15.10, appears fairly normal. The plot of the errors against the fitted y values, shown in Figure 15.11, is somewhat improved compared to Figure 15.8, and no longer gives cause for concern. For these reasons, and because we know from experience that a log-model is in general more appropriate for modelling plant yields, we accept the new model as superior to the old.

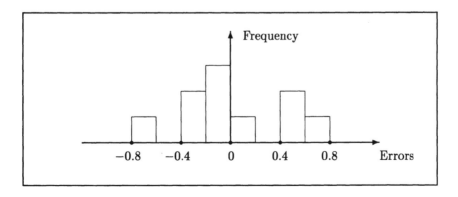

Figure 15.10: Histogram of errors for the regression of log(clover seed yield) on yarrow density.

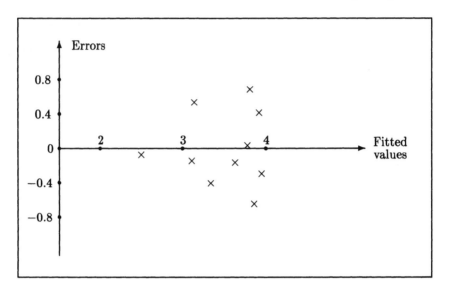

Figure 15.11: Plot of errors against fitted values for the transformed data.

The Transformed Curve

An important point must now be made. By transforming we have not only made our constant variance assumption more reasonable but we have also changed the assumed functional form of the curve. Originally we fitted a straight line $y = b_0 + b_1(x - \bar{x})$ using the raw data. By refitting the line as $\log y = b_0 + b_1(x - \bar{x})$ using the transformed y values, we have actually fitted the exponential curve

$$y = e^{b_0 + b_1(x - \bar{x})}$$

in the original (x, y) diagram.

Recall that the equation for the fitted line in the $(x, \log y)$ diagram is $\log y = 3.504 - .00290(x - 164)$. This back transforms to the curve $y = e^{[3.504 - .00290(x - 164)]}$ in the (x, y) diagram. The latter is shown along with the original scattergram in Figure 15.12.

Fitting the exponential curve as a straight line in the corresponding $(x, \log y)$ diagram, amounts to the same thing as fitting the exponential curve in the (x, y) diagram, using multiplicative rather than additive errors.

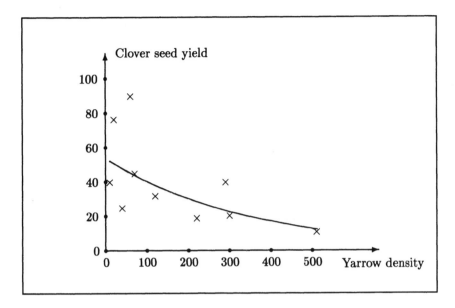

Figure 15.12: Exponential curve obtained by back-transforming the regression line of log(clover seed yield) on yarrow density.

Report on Study

A report on our study of the relationship between clover seed yield and yarrow density could be centered on a presentation of the scattergram and exponential curve as shown in Figure 15.12. Also included could be a statement that the downward trend was significant at the 5% level. The 95% confidence interval for the position of the true exponential curve could also be added to the figure. This will be derived in §2.

15.2   General Case

In this section we give a brief summary of the general case, then move on to express our results in terms of traditional algebraic formulae.

We begin with a general, step-by-step summary of what happened when we analyzed the competition study in the previous section. This summary is a simpler version of that given at the start of §10.2, so the reader may care to compare the two.

1. In the simple, or linear, regression case we assume a model of the form $y = \alpha_0 + \alpha_1 x$. This can also be rewritten in the form

$$y = \beta_0 + \beta_1(x - \bar{x})$$

Here the x values are treated as fixed values without error, while the corresponding y values are treated as realized values of random variables with mean $\beta_0 + \beta_1(x - \bar{x})$. The x variable is commonly known as the *independent* variable, while the y variable is known as the *dependent* variable.

2. In vector form our model is

$$
\begin{bmatrix} y_1 \\ \vdots \\ y_N \end{bmatrix} = \beta_0 \begin{bmatrix} 1 \\ \vdots \\ 1 \end{bmatrix} + \beta_1 \begin{bmatrix} x_1 - \bar{x} \\ \vdots \\ x_N - \bar{x} \end{bmatrix} + \text{error vector}
$$

This tells us that the model space M is of dimension two.

3. An orthogonal coordinate system for the model space is simply

$$
U_1 = \frac{1}{\sqrt{N}} \begin{bmatrix} 1 \\ \vdots \\ 1 \end{bmatrix}, \quad U_2 = \frac{1}{\|x - \bar{x}\|} \begin{bmatrix} x_1 - \bar{x} \\ \vdots \\ x_N - \bar{x} \end{bmatrix}
$$

4. The orientation of the model space is solely determined by the pattern of x-values. In the case of a designed experiment the x-values usually follow an orderly pattern, predetermined by the experimenter. The x-values may all be distinct or may constitute n replications of k distinct values, as in Chapter 10. In the more general case the x-values do not follow such an orderly pattern, and are usually distinct.

5. The fitted model is

$$
y = (y.U_1)U_1 + (y.U_2)U_2 + \text{error vector}
$$

6. This can be rewritten in the form

$$
y = b_0 \begin{bmatrix} 1 \\ \vdots \\ 1 \end{bmatrix} + b_1 \begin{bmatrix} x_1 - \bar{x} \\ \vdots \\ x_N - \bar{x} \end{bmatrix} + \text{error vector}
$$

This orthogonal decomposition is illustrated in Figure 15.5.

7. Here $b_0 = \bar{y}$ is the least squares estimate of β_0, and $b_1 = (y.U_2)/\|x - \bar{x}\|$ is the least squares estimate of β_1. These are also the maximum likelihood estimates.

8. The variance estimate is $s^2 = \|\text{error vector}\|^2/(N - 2)$.

9. The hypothesis $H_0 : \beta_1 = 0$ can be tested against the alternative $H_1 : \beta_1 \neq 0$ using the test statistic $(y.U_2)^2/s^2$, which comes from the $F_{1,N-2}$ distribution if H_0 is true. Usually this is the only hypothesis of interest. However, if desired, the hypothesis $H_0 : \beta_0 = 0$ can be similarly tested using the test statistic $(y.U_1)^2/s^2$.

10. The relevant calculations are conveniently summarized in an ANOVA table based on the orthogonal decomposition rewritten in the form

$$y - \bar{y} = b_1(x - \bar{x}) + \text{error vector}$$

for which the resulting Pythagorean breakup is

$$\|y - \bar{y}\|^2 = b_1^2 \|x - \bar{x}\|^2 + \|\text{error vector}\|^2$$

Here $b_1^2 \|x - \bar{x}\|^2 = (y.U_2)^2$ is called the *regression sum of squares*.

11. The fundamental assumption of linearity can be checked by a visual inspection of the scattergram of y on x with the regression line included. Independence can be checked by plotting error values against order or position of sampling, normality by inspecting a histogram of errors, and constancy of variance by plotting errors against fitted y values.

ANOVA Written Algebraically

If we wish to express our ANOVA table in terms of basic sums of squares and products, we need to examine the projection coefficient $y.U_2$. If we use the common abbreviation SS_x for the "x sum of squares", $\sum_{i=1}^{N}(x_i - \bar{x})^2$, and SP_{xy} for the "xy sum of products", $\sum_{i=1}^{N}(x_i - \bar{x})(y_i - \bar{y})$, then since

$$y = \begin{bmatrix} y_1 \\ \vdots \\ y_N \end{bmatrix} \quad \text{and} \quad U_2 = \frac{1}{\|x - \bar{x}\|} \begin{bmatrix} x_1 - \bar{x} \\ \vdots \\ x_N - \bar{x} \end{bmatrix}, \quad \text{it follows that}$$

$$y.U_2 = \frac{\sum_{i=1}^{N} y_i (x_i - \bar{x})}{\|x - \bar{x}\|} = \frac{\sum_{i=1}^{N}(y_i - \bar{y})(x_i - \bar{x})}{\sqrt{\sum_{i=1}^{N}(x_i - \bar{x})^2}} = \frac{SP_{xy}}{\sqrt{SS_x}}$$

Note that we have used the equality $\sum_{i=1}^{N} \bar{y}(x_i - \bar{x}) = 0$ to rearrange the numerator. The alternative form, SP_{xy}, is often preferred since it is symmetric in x and y.

It follows that the ANOVA table can be written in algebraic terms as shown in Table 15.7. Note that SS_y is the "y sum of squares", $\sum_{i=1}^{N}(y_i - \bar{y})^2$.

Source of Variation	df	SS
Regression	1	$(y.U_2)^2 = (SP_{xy})^2/SS_x$
Error	$N - 2$	$SS_y - (SP_{xy})^2/SS_x$
Total	$N - 1$	SS_y

Table 15.7: Simple regression ANOVA table using algebraic expressions.

For the fitted equation, the coefficients written algebraically are

$$b_0 = \bar{y} = (\sum_{i=1}^{N} y_i)/N \quad \text{and} \quad b_1 = \frac{(y.U_2)}{\|x - \bar{x}\|} = \frac{\sum_{i=1}^{N}(y_i - \bar{y})(x_i - \bar{x})}{\sum_{i=1}^{N}(x_i - \bar{x})^2} = \frac{SP_{xy}}{SS_x}$$

15.3 Confidence Intervals

In this section we derive confidence intervals for the true slope β_1, for the position of the true line at x_0, namely $\beta_0 + \beta_1(x_0 - \bar{x})$, and for the prediction of a new y value at x_0.

Confidence Interval for Slope, β_1

We derive an expression for a 95% confidence interval for the true slope, β_1. For this we recall from Table 15.4 that $y.U_2 = \|x - \bar{x}\|b_1$ comes from an $N[\|x - \bar{x}\|\beta_1, \sigma^2]$ distribution, whence $\|x - \bar{x}\|(b_1 - \beta_1)$ comes from an $N[0, \sigma^2]$ distribution and so $\|x - \bar{x}\|(b_1 - \beta_1)/\sqrt{s^2}$ comes from a t_{N-2} distribution. To obtain our confidence interval we gamble that the realized value lies within the limits

$$t_{N-2}(.025) \quad \le \quad \frac{\|x - \bar{x}\|(b_1 - \beta_1)}{s} \quad \le \quad t_{N-2}(.975)$$

which on rearrangement yields

$$b_1 - \frac{s}{\|x - \bar{x}\|}t_{N-2}(.975) \quad \le \quad \beta_1 \quad \le \quad b_1 + \frac{s}{\|x - \bar{x}\|}t_{N-2}(.975)$$

as the required 95% confidence interval.

For example, in our competition study of §1, the fitted slope in the log analysis was $b_1 = -.00290$. We had $\|x - \bar{x}\|^2 = 238640$ and $s^2 = 0.2068$, and a t value of $t_8(.975) = 2.306$. The 95% confidence interval for the true slope β_1 is therefore $-.00290 \pm .00215$, or

$$-.00505 \quad \le \quad \beta_1 \quad \le \quad -.00075$$

This confidence interval does not include the value $\beta_1 = 0$, as we would anticipate from the results of the hypothesis test in Table 15.6.

Standard Error of the Slope

Since $\|x - \bar{x}\|b_1$ comes from an $N[\|x - \bar{x}\|\beta_1, \sigma^2]$ distribution, it follows that b_1 comes from an $N[\beta_1, \sigma^2/\|x - \bar{x}\|^2]$ distribution. Hence the variance of the slope estimator, B_1, is $\sigma^2/\|x - \bar{x}\|^2$, which we estimate by $s^2/\|x - \bar{x}\|^2$. Hence the estimated standard error of the slope is

$$\text{se(slope)} = \frac{s}{\|x - \bar{x}\|}$$

For example, in the competition study this value is $\sqrt{.2068/238640} = .00093$. In these terms the 95% confidence interval can be written as

$$b_1 \quad \pm \quad se(slope) \times t\text{-value}$$

Here an important point needs to be made. The slope, β_1, in our competition study reflects the general growth of, and competition between, clover and yarrow in *the particular field and the particular season* in which the sampling was carried out. In a different field or season the slope β_1 is likely to be entirely different. For example, in another time or place the clover may yield ten times as much seed, and β_1 may be three times as large as in our study. If a researcher wishes to make a general statement about the slope of the relationship between the logarithm of clover seed yield and yarrow density, he or she needs to carry out a series of studies on a range of soil types, in several different seasons. The estimated standard error of the slope, .00093 in our study, provides no information on how the slope would vary in such a series of studies. It simply tells us about the accuracy of our estimate of β_1 for our one study. We stress this distinction since in the past some researchers have incorrectly used the resulting confidence interval to describe the variation within an agricultural district.

Confidence Interval for a Fitted Value

We move on now to assess the accuracy of the fitted line. This varies according to the position along the line, with the line being most accurately determined at the center of the range of x-values.

For a given value of x, say x_0, the fitted value lying on the line is $b_0 + b_1(x_0 - \bar{x})$. The variance of this estimate is

$$Var[B_0 + B_1(x_0 - \bar{x})] \quad = \quad Var(B_0) + (x_0 - \bar{x})^2 Var(B_1)$$
$$\text{since } B_0 \text{ and } B_1 \text{ are independent}$$

$$= \quad \frac{\sigma^2}{N} + (x_0 - \bar{x})^2 \frac{\sigma^2}{\|x - \bar{x}\|^2}$$

$$= \quad \sigma^2 \left[\frac{1}{N} + \frac{(x_0 - \bar{x})^2}{\|x - \bar{x}\|^2} \right]$$

This is estimated by substituting s^2 for σ^2. Hence the estimated standard error of the fitted value at $x = x_0$ is

$$se(\text{fitted value}) \quad = \quad \sqrt{s^2 \left[\frac{1}{N} + \frac{(x_0 - \bar{x})^2}{\|x - \bar{x}\|^2} \right]}$$

The resulting 95% confidence interval for the fitted value at $x = x_0$ is

$$\text{fitted value} \quad \pm \quad \text{se(fitted value)} \times t_{N-2}(.975), \quad \text{or,}$$

$$b_0 + b_1(x_0 - \bar{x}) \quad \pm \quad t_{N-2}(.975)\sqrt{s^2 \left[\frac{1}{N} + \frac{(x_0 - \bar{x})^2}{\|x - \bar{x}\|^2}\right]}$$

For example, in our competition study this interval is

$$3.504 - .00290(x_0 - 164) \quad \pm \quad 2.306\sqrt{.2068 \left[\frac{1}{10} + \frac{(x_0 - 164)^2}{238640}\right]}$$

As we would expect, this interval is centered on the fitted value on the line. However, its width depends on the value of x_0. The width is a minimum when $x_0 = \bar{x} = 164$, telling us that the position of the line is more accurately determined at the center of the range of x-values than elsewhere. The width increases as x_0 moves away from $\bar{x} = 164$, reflecting the increased uncertainty about the position of the true line. The lower and upper confidence limits are given in Table 15.8 for a range of x_0-values. The resulting 95% confidence band is graphed in Figure 15.13.

Value	Confidence Limits		Back transformed confidence limits	
of x_0	Lower	Upper	Lower	Upper
10	3.482	4.419	32.5	83.0
100	3.331	4.049	28.0	57.3
200	3.059	3.740	21.3	42.1
300	2.668	3.551	14.4	34.9
400	2.214	3.425	9.2	30.7
500	1.736	3.323	5.7	27.8

Table 15.8: The 95% confidence limits for the fitted values in our competition study for a range of x_0-values. Back transformed limits are also tabulated.

In Figure 15.13 the confidence limits are symmetric about the fitted line. If we decide, however, to back transform from the $(x, \log y)$ diagram to the (x, y) diagram, this symmetry is lost. The back transformation is easy; we simply use the exponential, or antilogarithm transformation to transform the lower and upper confidence limits back to the original units, as shown in Table 15.8. For example, $e^{3.482} = 32.5$.

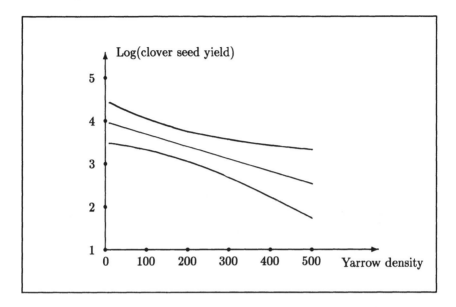

Figure 15.13: The 95% confidence limits for the fitted values, showing the uncertainty about the elevation and slope of the fitted line. Use your ruler to gauge the range of slopes which are plausible; this should match roughly with the confidence interval for the slope.

Confidence Interval for a Predicted Value

Lastly, we derive a confidence interval for a new, predicted value of y at x_0. For example, in our competition study, we may want to work out the likely range of clover seed yields which we would observe if we were to measure an eleventh quadrat which had a yarrow density of x_0.

To do this, we firstly work out the variance of Y_{pred}, defined as

$$Y_{\text{pred}} = b_0 + b_1(x_0 - \bar{x}) + \text{error}$$

that is, a best estimate plus an error term. This variance is

$$
\begin{aligned}
Var(Y_{\text{pred}}) &= Var(B_0) + (x_0 - \bar{x})^2 \, Var(B_1) + \sigma^2 \\
&\qquad \text{since the three estimators are mutually independent} \\
&= \frac{\sigma^2}{N} + (x_0 - \bar{x})^2 \frac{\sigma^2}{\|x - \bar{x}\|^2} + \sigma^2 \\
&= \sigma^2 \left[1 + \frac{1}{N} + \frac{(x_0 - \bar{x})^2}{\|x - \bar{x}\|^2} \right]
\end{aligned}
$$

This is estimated by substituting s^2 for σ^2. Hence the estimated standard error of the predicted value at x_0 is

$$\text{se(predicted value)} \;=\; \sqrt{ s^2 \left[1 + \frac{1}{N} + \frac{(x_0 - \bar{x})^2}{\|x - \bar{x}\|^2} \right] }$$

The resulting 95% confidence interval for the predicted value at x_0 is

$$\text{predicted value} \;\pm\; \text{se(predicted value)} \;\times\; t_{N-2}(.975), \quad \text{or,}$$

$$b_0 + b_1(x_0 - \bar{x}) \;\pm\; t_{N-2}(.975)\sqrt{ s^2 \left[1 + \frac{1}{N} + \frac{(x_0 - \bar{x})^2}{\|x - \bar{x}\|^2} \right] }$$

For example, in our competition study this interval is

$$3.504 - .00290(x_0 - 164) \;\pm\; 2.306\sqrt{ .2068 \left[1 + \frac{1}{10} + \frac{(x_0 - 164)^2}{238640} \right] }$$

The resulting 95% confidence band is shown in Figure 15.14.

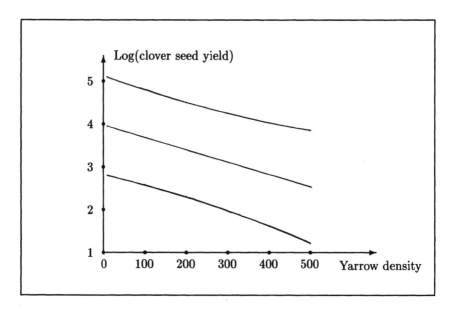

Figure 15.14: The 95% confidence limits for a predicted value, competition study.

The confidence band is much wider in Figure 15.14 than it was in Figure 15.13. Also, the band limits are more parallel than previously. This is a reflection of the dominant effect of the constant s^2 which has been added in to the variance estimate. Note that Figure 15.14 can be back transformed, as in the previous section.

We complete this subsection by presenting a more realistic example of the use of prediction. Consider the graph of the production of kiwifruit in New Zealand over the seven year period shown in Figure 15.15.

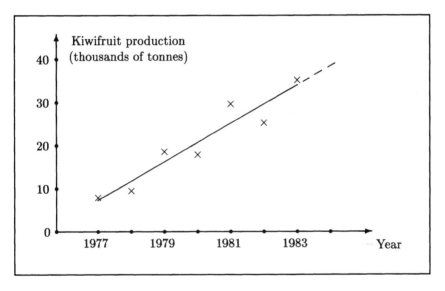

Figure 15.15: Use of linear regression to predict the likely level of New Zealand kiwifruit production for the year 1984.

As one would expect, there are ups and downs in production from year to year, owing to climatic and other variations. However, there is also a clear upward trend. For the kiwifruit marketing authority to plan coolstore and shipping capacity for the coming season, a prediction of the likely range of production in the coming season is of considerable interest. A simple way of obtaining such a prediction is to fit a linear regression to the existing data, and extrapolate to the coming season. The likely range of kiwifruit production would then be the 95% confidence interval for the predicted value for 1984. The width of this interval would reflect both the uncertainty in the estimation of the trend, and the seasonal variation about the overall trend. This is exactly the situation covered in the last subsection.

A more sophisticated approach would be to fit a quadratic equation, and to modify the prediction if the pattern of new plantings over the last few years suggested a sudden increase, or levelling off, of production in 1984.

Minitab to the Rescue

The reader will be relieved to realize that Minitab can easily be instructed to produce the 95% confidence bands tabulated and graphed in the last two subsections. For example, to produce Table 15.8 and Figure 15.13 we first set arbitrary values of x into a spare column, for example, values of $10, 50, 100, \ldots, 500$. Then the fitted values and their standard errors can be calculated and put into new columns. These columns can then be used to calculate confidence limits, which can be printed, plotted and back transformed. Table 15.9 lists the appropriate commands. As an aside, we note that the Predict subcommand of Regress will yield confidence limits for fitted and predicted values.

```
List of commands
name c1 = 'YarrDens'
name c2 = 'Fitted'
name c4 = 'LowerLim'
name c5 = 'UpperLim'
set c1
10 50 100 150 200 250 300 350 400 450 500
let c2 = 3.504 − .00290*(c1−164)
let c3 = 2.306 * sqrt(.2068*(1/10+(c1−164)**2/238640))
let c4 = c2 − c3
let c5 = c2 + c3
print c1 c4 c2 c5
mplot c2 c1, c4 c1, c5 c1
let c12 = exp(c2)
let c14 = exp(c4)
let c15 = exp(c5)
print c1 c14 c12 c15
mplot c12 c1, c14 c1, c15 c1
stop
```

Table 15.9: Minitab commands for calculating, printing, plotting and back transforming confidence bands, competition study.

15.4   Correlation Coefficient

We pause now to define the correlation coefficient, r, a term commonly used in statistics. This coefficient is a measure of how closely the points (x, y) cluster about the regression line, or the closeness of the "association" between the x and y variables.

Definition

How do we go about measuring how closely our scatter of points hugs a straight line? Consider Figure 15.16 which redisplays the simplified orthogonal decomposition in the competition study, and recall that the closer the hugging, the shorter the error vector. In turn, the shorter the error vector, the smaller the angle θ between the vectors $y - \bar{y}$ and $x - \bar{x}$.

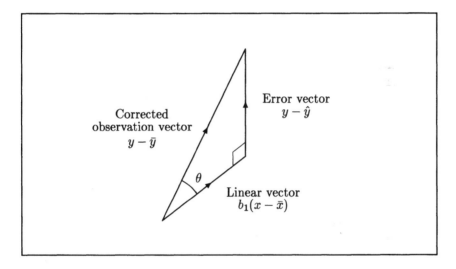

Figure 15.16: Simplified orthogonal decomposition for linear regression, showing the angle θ between the vectors $x - \bar{x}$ and $y - \bar{y}$.

The *correlation coefficient*, r, is defined as the cosine of the angle between the vectors $y - \bar{y}$ and $x - \bar{x}$. The closer this is to $+1$ or -1, the closer the scatterplot hugs a line. In the competition study, the correlation between log(clover seed yield) and yarrow density is

$$\cos \theta = \frac{x - \bar{x}}{\|x - \bar{x}\|} \cdot \frac{y - \bar{y}}{\|y - \bar{y}\|} = \frac{SP_{xy}}{\sqrt{SS_x}\sqrt{SS_y}} = \frac{-692.3}{\sqrt{238640}\sqrt{3.6627}} = -.740$$

This is not particularly close to $+1$ or -1, and corresponds to an angle of $\theta = 138^0$.

We think of r as an estimate of the true correlation coefficient ρ, where

$$\rho = \frac{E\left[(X - \mu_X)(Y - \mu_Y)\right]}{\sqrt{E\left[(X - \mu_X)^2\right] E\left[(Y - \mu_Y)^2\right]}} = \frac{Cov(X, Y)}{\sigma_X \sigma_Y}$$

Range of Values

Since the correlation coefficient is the cosine of an angle, it must lie in the interval from -1 to $+1$. We illustrate the various cases which can arise in Figure 15.17.

If the vectors $x - \bar{x}$ and $y - \bar{y}$ differ in magnitude but not in direction, that is, $y_i - \bar{y} = b_1(x_i - \bar{x})$ where $b_1 > 0$, for each $i = 1, 2, \ldots, N$, then $\theta = 0^0$ so the correlation coefficient is 1. This occurs when there is an exact linear relationship between x and y, as in a change of scale from degrees Celsius to degrees Fahrenheit, or pounds to kilograms. In the (x, y) plane the (x_i, y_i) pairs lie on a straight line with positive slope. This case is summarized in Figure 15.17(a).

If the vectors $x - \bar{x}$ and $y - \bar{y}$ are in precisely opposite directions, that is, $y_i - \bar{y} = b_1(x_i - \bar{x})$ where $b_1 < 0$, then $\theta = 180^0$ so the correlation coefficient is -1, as shown in Figure 15.17(e). This occurs when there is an exact linear relationship between x and y with negative slope. Such would be the case if data on the percentage of weeds surviving, x, after chemical treatment, were transformed to data on the percentage control of weeds, $y = 100 - x$.

Figure 15.17(b) and (d) illustrate the cases which occur most commonly in practice. In these cases the value of x is of some help in estimating the corresponding value of y, via the predictor $y = \bar{y} + b_1(x - \bar{x})$, but the value of x does not determine y exactly.

Notice that the closer θ is to 0^0 or 180^0, the less is the scatter about the line in the (x, y) plane, and the more use is the x value as a predictor of y. Conversely, the closer θ is to 90^0, the greater is the scatter about the line in the (x, y) plane, and the less use is the x value as a predictor of y. The extreme situation is shown in Figure 15.17(c), where $\theta = 90^0$ and the x value provides absolutely no information about the value of y.

Coefficient of Determination

The square of the correlation coefficient, r^2, is known as the *coefficient of determination*. Figure 15.16 provides the clue to this naming. The regression sum of squares, the squared length of the linear vector, can be written as

$$\|y - \bar{y}\|^2 \cos^2 \theta \quad = \quad \|y - \bar{y}\|^2 r^2$$

so that r^2 is the "proportion of the total corrected sum of squares explained by the regression". In our competition study, for example, $r^2 = (-.740)^2 = 0.548$. This value is routinely computed by many regression computer programs; for example, Minitab prints it as a percentage, labelled "R-sq", or 54.8% in our example. The range of values of r^2 is from 0 to 1, with low values indicating low correlation, and high values high correlation, either positive or negative.

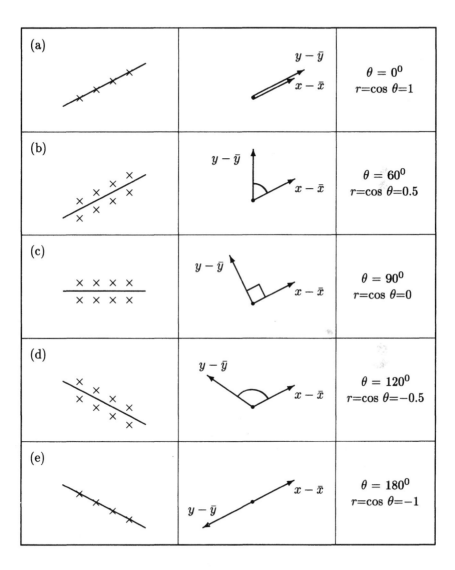

Figure 15.17: Behavior of the correlation coefficient, $r = \cos\theta$. In the left column the data values are shown in the (x, y) plane, while in the middle column the corresponding vectors are sketched in N-space.

Link Between F and r

By rewriting the regression sum of squares as $r^2\|y - \bar{y}\|^2$, it follows that we can rewrite our Pythagorean breakup as

$$\|y - \bar{y}\|^2 \quad = \quad r^2\|y - \bar{y}\|^2 \quad + \quad (1 - r^2)\|y - \bar{y}\|^2$$

Therefore the test statistic F, used for testing the hypothesis $H_0 : \beta_1 = 0$, can be rewritten in terms of r^2 as follows:

$$F = \frac{\text{Regression SS}}{(\text{Error SS})/(N - 2)} = \frac{r^2\|y - \bar{y}\|^2}{(1 - r^2)\|y - \bar{y}\|^2/(N - 2)} = \frac{(N - 2)r^2}{1 - r^2}$$

Conversely, a rearrangement of this gives us $r^2 = F/(N - 2 + F)$. The upshot is that the test statistic F and correlation coefficient r are functionally related; hence given one of F or r, we can easily calculate the other (to within a $+$ or $-$ sign).

Test for Correlation

One often hears the question "Are x and y correlated?" In other words, is the correlation coefficient zero, or is there sufficient evidence for us to reject the hypothesis $H_0 : \rho = 0$?

The hypothesis $H_0 : \rho = 0$ is true if and only if the hypothesis $H_0 : \beta_1 = 0$ is true, where β_1 is the slope of the true regression line. Under the null hypothesis $H_0 : \rho = 0$, the distribution of the estimated correlation coefficient, $r = \pm\sqrt{F/(N - 2 + F)}$, is a direct transformation of the reference distribution $F_{1,N-2}$, which is the distribution followed by the realized F values under the null hypothesis $H_0 : \beta_1 = 0$. Hence the percentiles of one distribution can be obtained from the other. For example, in the competition study, the reference distribution $F_{1,8}$ has 95% of realized values lying in the range from 0 to 5.32; therefore the corresponding reference distribution of correlation coefficients, under $H_0 : \rho = 0$, has 95% of realized values, r, lying in the range $\pm\sqrt{5.32/(10 - 2 + 5.32)} = \pm 0.632$, where -0.632 is the 2.5 percentile and 0.632 is the 97.5 percentile.

The 97.5 and 99.5 percentiles of the distribution of correlation coefficients, r, under the null hypothesis $H_0 : \rho = 0$, are given in Table T.3. In order to test whether the estimated correlation coefficient, such as the value of $-.740$ in our competition study, is 5% or 1% significantly different from zero, we compare its numerical value with these tabulated values, such as 0.632 and 0.765 in our example. Note that the number of degrees of freedom is again listed as $N - 2$, or in our example, 8. Thus our correlation coefficient is 5% significant, so we can formally reject the hypothesis $H_0 : \rho = 0$ at the 5% level of significance, and conclude that log(clover seed yield) and yarrow density are correlated.

The test for significant correlation, r, is equivalent to the test for significant slope, b_1, so there is no point in carrying out both tests. Why have two topics, regression and correlation, when one would suffice? One answer is that for historical and other reasons, some users of statistics find it more convenient to work with correlation coefficients than regression line slopes. For example, some users may be solely interested in the closeness of association between two variables, and have no interest in the nature of the relationship; for these users the correlation coefficient is a convenient summary statistic.

Goodness of Fit

Users of regression and correlation often make the mistake of thinking that a high correlation coefficient implies that a straight line provides a good fit to the data. This is not necessarily so. For example, Figure 15.18 displays a data set with an r value of 0.99, where the x, y relationship is clearly not linear, but curvilinear. The high r value simply means that a high proportion of the variability in the y values is explained by the straight line model; an even higher proportion, however, may be explained by a quadratic polynomial or other curvilinear model.

Checking the linearity assumption is best carried out by an examination of the scattergram of y on x. Alternatively, the plot of errors against fitted values can be examined.

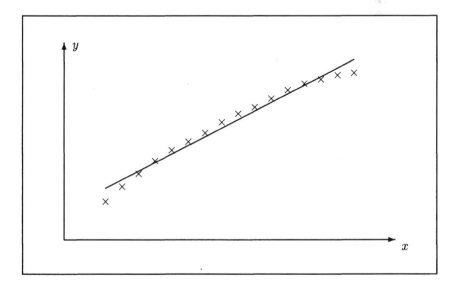

Figure 15.18: Sample data set showing that a high r value does not necessarily imply a linear relationship.

15.5   Pitfalls for the Unwary

In this section we first sound some words of warning about the interpretation of a significant regression, then discuss the dangers of extrapolation. We move on to discuss the effect of unusual values, and finish by stressing the value of graphs.

Cause and Effect

A high correlation between two variables does not imply a *causal* relationship between them. That is, changes in the x variable do not necessarily cause changes in the y variable. They are merely "associated", perhaps through some common intermediary.

As an example, Figure 15.19 shows that from 1970 to 1980, the incidence of cancer in the U.S.A. rose, as did the consumption of kiwifruit. Therefore these two variables are correlated. Nevertheless, it would be a mistake (we hope!) to suggest that the increase in kiwifruit consumption caused the increase in the incidence of cancer. The two variables are simply linked through a common intermediary, "time".

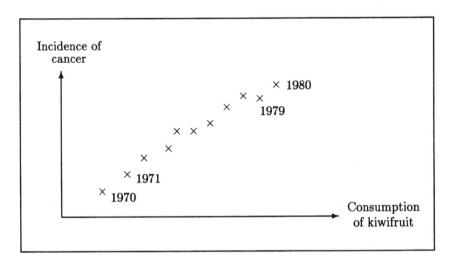

Figure 15.19: The relationship between incidence of cancer and consumption of kiwifruit per capita, in the U.S.A., for the period 1970-1980. The data is fictitious.

Similarly, in a recent survey of the density of the weed "wild oats" in wheat crops in Canterbury, New Zealand, it was found that on average, fields which had been sprayed with a chemical for the purpose of controlling the wild oats had a higher wild oat density at harvest time than fields which had not been sprayed. Thus, wild oat density and spraying were positively correlated. Does this suggest that the chemical promoted the growth of the wild oats? No! This "spurious" correlation is explained by two facts. Firstly, farmers only spray fields with high wild oat densities, since the chemicals are expensive. Secondly, the chemicals are not 100% effective. The result is that the infested fields which were sprayed still had higher wild oat densities at harvest time than the relatively clean fields which were not sprayed.

In a controlled experiment the researcher may believe that the problem of spurious correlation does not exist. It is true that an experimenter is in a much better position to make inferences than a survey worker, since as many as possible of the extraneous factors are controlled in an experiment. Nevertheless, unknown or uncontrollable factors can still make a nonsense of conclusions drawn from an experiment. For example, suppose that an experiment has been carried out in an alfalfa crop to investigate the effect of boron fertilizer, at rates of 0, 2, 4 and 6 kg/ha, on the growth of the crop. If a response is observed, one might reasonably assume that the boron response was due to the correction of a boron deficiency in the alfalfa. Hypothetically, however, the response could alternatively be due to the removal of an unsuspected predator of alfalfa, such as a weevil to which boron is toxic. In practice, experimenters are often in the position of having to reinterpret the work of their forebears in the light of such new evidence.

The Dangers of Extrapolation

While a straight line may be a reasonable approximation to the underlying functional relationship for the range of the data, it may be quite inadequate when extrapolated outside that range. For example, Figure 15.20 shows a graph presented in the "Christchurch Press" on December 12, 1987.

The data gives the carbon dioxide content in atmospheric gases trapped in Antarctic ice over the last two hundred years, and has been used to predict future levels of carbon dioxide in the atmosphere. The curve fitted was not linear, but the same principle applies; the conclusions in the article heavily depend upon the assumed functional form of the relationship between x = age and y = carbon dioxide.

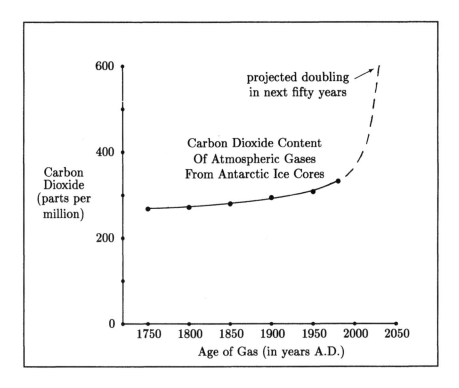

Figure 15.20: The graph shows the increase in the carbon dioxide content of the atmosphere over the last 200 years. The rise since 1950 suggests that the carbon dioxide content will double in the next 50 years, leading to a temperature increase of 5 to 10 degrees Celsius in the polar regions. Reproduced with the kind permission of "The Press", Christchurch.

Outliers

Unusual values, or outliers, can have a large influence on the position of the regression line. This occurs since the deviations from the line, $y_i - \hat{y}_i$, are squared in the calculation of s^2. For example, a single careless mistake in data entry can prejudice the entire statistical analysis by inflating s^2 and reducing the results to nonsignificance. Also, genuinely unusual values can have a large influence on the fitted line. The effect of a single outlier is illustrated for two cases in Figure 15.21. The variance estimate, s^2, is increased in both cases. Also the regression slope, b_1, is changed in case (b), when the outlier is not centrally located along the x-axis, that is, its x-value is not close to $\bar{x}$.

A measure commonly used to flag an outlier is Cook's distance. It serves to alert the investigator to a value which is strongly influencing the fitted line. Minitab allows the user to produce these distances via the COOKD subcommand of Regress.

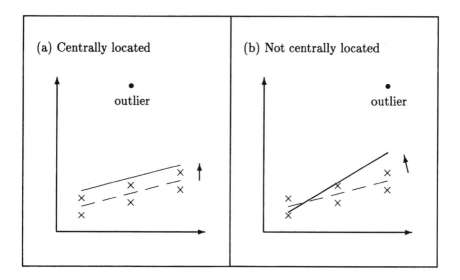

Figure 15.21: The effect of an outlier on the fitted regression line when the x-value for the outlier is (a) centrally located and (b) not centrally located. In each case the solid line is fitted using all the data, and the dashed line is fitted excluding the outlier.

The Value of Graphs

The saying "one picture is worth a thousand words" is very true. In our context, a scattergram is much more revealing than an r^2 value, an F value, or any other statistic or set of statistics. This is amply demonstrated by Anscombe's quartet, reproduced in Figure 15.22. In these four data sets, all of the statistics $(N, \bar{x}, \bar{y}, b_0, b_1, se(b_1), t, SS_x, SS_y, SP_{xy}, s, F, r)$ are identical, yet obviously the interpretation of the graphs differs immensely.

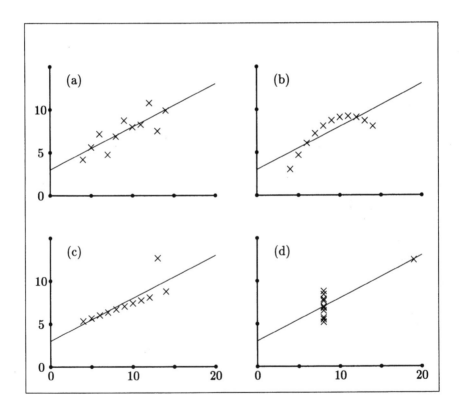

Figure 15.22: Anscombe's quartet, reproduced from F.J.Anscombe "Graphs in Statistical Analysis" *The American Statistician* 27 (1973) 17-21, with kind permission from *The American Statistician*.

15.6 Summary

The simple linear regression model, in orthogonal form, is

$$y \;=\; \beta_0 + \beta_1(x - \bar{x}) + \epsilon$$

The corresponding model space M has coordinate axis directions

$$U_1 = \frac{1}{\sqrt{N}} \begin{bmatrix} 1 \\ \vdots \\ 1 \end{bmatrix}, \quad U_2 = \frac{1}{\|x - \bar{x}\|} \begin{bmatrix} x_1 - \bar{x} \\ \vdots \\ x_N - \bar{x} \end{bmatrix}$$

The fitted model is

$$\begin{bmatrix} y_1 \\ \vdots \\ y_N \end{bmatrix} = b_0 \begin{bmatrix} 1 \\ \vdots \\ 1 \end{bmatrix} + b_1 \begin{bmatrix} x_1 - \bar{x} \\ \vdots \\ x_N - \bar{x} \end{bmatrix} + \begin{bmatrix} y_1 - \hat{y}_1 \\ \vdots \\ y_N - \hat{y}_N \end{bmatrix}$$

where $b_0 = \bar{y}$, $b_1 = SP_{xy}/SS_x$ and $\hat{y}_i = b_0 + b_1(x_i - \bar{x})$ for each i. This is illustrated in Figure 15.23.

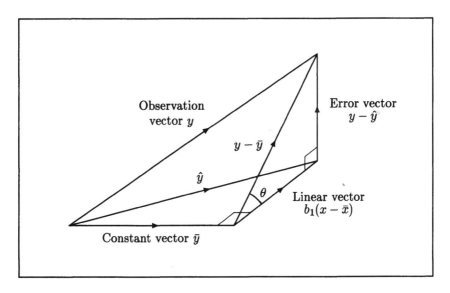

Figure 15.23: Orthogonal decomposition of the observation vector into constant, linear and error vectors. The decomposition $y - \bar{y} = b_1(x - \bar{x}) + (y - \hat{y})$ is also displayed.

This decomposition is redisplayed in Figure 15.24 for a single data point in the (x, y) diagram, as

$$y_i \;=\; \bar{y} + b_1(x_i - \bar{x}) + (y_i - \hat{y}_i)$$

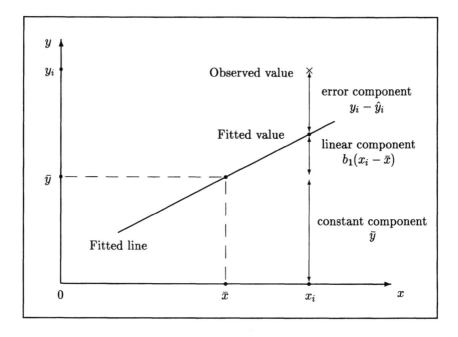

Figure 15.24: The orthogonal decomposition redrawn in the (x, y) plane for a single data point (x_i, y_i).

The duality between the two figures is important. The constant vector is large when the line is far from the x-axis in the region of the given x-values. The linear vector reflects the size of the slope, β_1. The error vector is large when the points in the scattergram are far from the fitted line: the smaller the variance about the line, the smaller the error vector.

The hypothesis of interest, $H_0 : \beta_1 = 0$, is tested using the statistic $(y.U_2)^2/s^2$, which follows the $F_{1,N-2}$ distribution if H_0 is true. Here $(y.U_2)^2$, the regression sum of squares, can be rewritten as $b_1^2\|x - \bar{x}\|^2 = SP_{xy}^2/SS_x$. The relevant calculations are usually summarized in an ANOVA table which is based on the Pythagorean breakup

$$\|y - \bar{y}\|^2 \;\; = \;\; b_1^2\|x - \bar{x}\|^2 \;\; + \;\; \|\text{error vector}\|^2$$

The correlation coefficient, r, is the cosine of the angle between the corrected observation vector, $y - \bar{y}$, and the vector $x - \bar{x}$. It follows that r^2 is the proportion of the total sum of squares which is explained by the regression.

The 95% confidence interval for the slope β_1 is

$$b_1 \;\; \pm \;\; \text{se}(B_1) \times t_{N-2}(.975)$$

where $\text{se}(B_1) = s/\|x - \bar{x}\|$.

The 95% confidence interval for the position of the true line at x_0, given by $\beta_0 + \beta_1(x_0 - \bar{x})$, is

$$y_0 \pm \text{se}(Y_0) \times t_{N-2}(.975)$$

where $y_0 = \bar{y} + b_1(x_0 - \bar{x})$ and $\text{se}(Y_0) = \sqrt{s^2 \left[\frac{1}{N} + \frac{(x_0 - \bar{x})^2}{\|x - \bar{x}\|^2} \right]}$.

The 95% confidence interval for a new, predicted y value is

$$y_0 \pm \text{se}(\text{prediction}) \times t_{N-2}(.975)$$

where $\text{se}(\text{prediction}) = \sqrt{s^2 \left[1 + \frac{1}{N} + \frac{(x_0 - \bar{x})^2}{\|x - \bar{x}\|^2} \right]}$.

For a discussion of the special case of regression through the origin, $y = \beta x$, see Appendix D.

Class Exercise

In this class exercise we shall simulate an experiment which examines the influence of boron fertilizer on the seed yield of alfalfa. Experimental treatments will consist of borax applications at rates of 2, 4, 6, 8 and 10 kg borax/ha, assigned at random to five experimental plots. We shall generate our data by randomly selecting values from normal distributions centered on the straight line

$$y \ = \ 80 \ + \ 10x$$

where y = alfalfa seed yield in kg/ha, and
x = rate of borax in kg/ha,

as illustrated in Figure 15.25. We shall then analyze our data using the linear regression model, and see how well we estimate the true line.

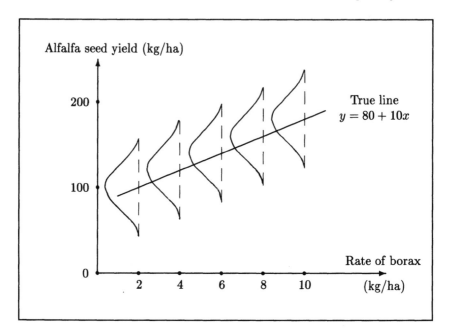

Figure 15.25: The line $y = 80+10x$ for use in the class exercise, with y-values for a given x normally distributed about the associated point on the line with variance $\sigma^2 = 400$.

(a) The first step is for each student to generate a set of five y-values in the following manner. Pick a random number in the range 1 to 20 from Table T.1, then go to the column for $x = 2$ kg borax/ha in Table 15.10 and find the corresponding y-value. This will be your alfalfa seed yield, in kg/ha, for the plot receiving boron at a rate of 2 kg borax/ha. Repeat this process for $x=4$, 6, 8 and 10 kg borax/ha using the other four columns of Table 15.10. Don't worry if you strike the same random number more than once.

Write down the resulting observation vector, y, and the vector of corresponding x-values, x.

(b) Using graph paper, draw a scattergram of your data.

(c) Model space coordinate axis directions are U_1 and $U_2 = (x - \bar{x})/\|x - \bar{x}\|$. Write down these two vectors.

(d) Fit the model in the form

$$y - \bar{y} \quad = \quad (y.U_2)U_2 \quad + \quad \text{error vector}$$

writing out each vector in 5-space in full. Sketch this orthogonal decomposition as a vector diagram, labelling each vector appropriately.

(e) Rewrite the linear vector, $(y.U_2)U_2$, in the form $b_1(x - \bar{x})$. What is your least squares estimate b_1 of the slope β_1?

(f) Add your regression line, $y = \bar{y} + b_1(x - \bar{x})$, to your scattergram. Does it look correct? Mark in the fitted values and error values.

At this point the class instructor is asked to draw up two histograms of class results. The first is the histogram of calculated $\bar{y}$ values and the second is the histogram of calculated slopes, b_1.

Since the true line written in orthogonal form is $y = 140 + 10(x - 6)$, we know that $\bar{y}$ will in the long run average to $\beta_0 = 140$. Also, from Table 15.4 we know that the estimator $\bar{Y}$ has a standard error of $\sigma/\sqrt{N} = 20/\sqrt{5} = 8.9$. Does the class histogram of $\bar{y}$ values support these statements?

Similarly the slopes b_1 should average to $\beta_1 = 10$ in the long run. Also, from Table 15.4 we know that the estimator of the slope, B_1, has a standard error of $\sigma/\|x - \bar{x}\| = 20/\sqrt{40} = 3.2$. Does the class histogram of b_1 values support these statements?

| Random | Alfalfa seed yields, y, for x values of: | | | | |
number	2 kg/ha	4 kg/ha	6 kg/ha	8 kg/ha	10 kg/ha
1	60	78	102	120	140
2	72	90	108	132	152
3	78	96	114	138	158
4	84	102	120	144	164
5	84	102	126	144	164
6	84	108	126	150	170
7	90	108	132	150	170
8	96	114	138	150	170
9	96	114	138	156	176
10	96	120	138	156	176
11	102	120	144	162	182
12	102	126	144	162	182
13	102	126	150	168	182
14	108	132	150	168	188
15	114	132	150	168	194
16	114	138	156	174	194
17	120	138	156	180	200
18	126	144	162	186	206
19	132	150	168	192	212
20	138	162	180	198	218

Table 15.10: Samples of twenty y values normally distributed with mean $E(Y) = 80 + 10x$ and standard deviation $\sigma = 20$ kg/ha, for five rates of applied borax, $x = 2, 4, 6, 8, 10$ kg/ha.

(g) Each class member is now asked to use their results from (d) to draw up an ANOVA table based on the Pythagorean decomposition

$$\|y - \bar{y}\|^2 \quad = \quad (y.U_2)^2 \quad + \quad \|\text{error vector}\|^2$$

(h) Is the calculated F value significant? How strong is the evidence to support the idea that increasing the rate of boron will increase the alfalfa seed yield?

At this point the class instructor is asked to draw up two more class histograms. The first is the histogram of s^2 values; these estimates of σ^2 will in the long run average to $\sigma^2 = 400$, though individual estimates may not be very close to this value. The second is the histogram of the calculated test statistics, F; these will follow a noncentral $F_{1,3}$ distribution since we know that in reality $\beta_1 \neq 0$. In what percentage of simulated experiments was $H_0 : \beta_1 = 0$ rejected at the 5% level of significance? This percentage, p, is an approximation to the power of the 5% level test for our assumed values, $\beta_1 = 10$ and $\sigma^2 = 400$, and our experimental design.

(i) Each class member is now asked to calculate their own estimate of the accuracy of the slope estimator, B_1, using the formula

$$\text{se}(\text{slope}) = s/\|x - \bar{x}\|$$

This can then be used to calculate a 95% confidence interval for the true slope β_1 using the formula $b_1 \pm \text{se}(\text{slope}) \times t_{N-2}(.975)$.

The class instructor can now determine what percentage of the class members' intervals contain the true value, $\beta_1 = 10$. In the long run, the percentage will be 95%. Also, the class instructor can determine what percentage of the intervals contain the hypothesized value $\beta_1 = 0$. This percentage should be the same as the percentage p in part (h) above, since $\beta_1 = 0$ is included in the 95% confidence interval if and only if the hypothesis $H_0 : \beta_1 = 0$ is rejected at the 5% level of significance.

(j) If time permits, each class member can go on to calculate and plot the 95% confidence band for the position of the true line. The method is to take a selection of x values, for example $x_0 = 1, 3, 5, 6, 7, 9$ and 11, then for each x value, x_0, calculate

$$\text{se}(\text{fitted value}) = \sqrt{s^2 \left[\frac{1}{N} + \frac{(x_0 - \bar{x})^2}{\|x - \bar{x}\|^2}\right]}$$

and hence the 95% confidence interval

$$[\bar{y} + b_1(x_0 - \bar{x})] \quad \pm \quad \text{se}(\text{fitted value}) \times t_{N-2}(.975)$$

After the band has been plotted, the "true line" can be drawn in to see if it is included in the band.

(k) Lastly, the correlation coefficient, r, can be calculated by each class member using the formula

$$r \;=\; \cos\theta \;=\; \frac{(x - \bar{x})}{\|x - \bar{x}\|} \cdot \frac{(y - \bar{y})}{\|y - \bar{y}\|}$$

The hypothesis $H_0 : \rho = 0$, that the true correlation is zero, can then be tested using the critical values of the correlation coefficient as given in Table T.3. The significance level of the result should agree with that from the test of $H_0 : \beta_1 = 0$, since the two tests are mathematically equivalent.

Exercises

(15.1) Our cholesterol-conscious world is demanding leaner lambmeat. In order to study the relationship between fatness and size in ram lambs, a sample of twelve ram lambs were slaughtered and a standard fatness measurement recorded along with the carcase weight for each lamb. These data are given in the table.

Fatness	7	8	7	9	12	14	16	15	16	20	15	17
Carcase weight	14.4	14.8	15.2	16.1	16.7	17.2	17.5	17.8	18.6	18.8	19.6	19.7

(a) Draw a scattergram of fatness, y, against carcase weight, x. Use Minitab for this purpose if you so desire.

(b) Fit the regression equation in orthogonal form $y = b_0 + b_1(x - \bar{x})$. Use Minitab if you wish.

(c) Write out as vectors in 12-space the orthogonal decomposition

$$y - \bar{y} \;=\; b_1(x - \bar{x}) \;+\; \text{error vector}$$

Minitab again can assist in this. Draw a sketch of the decomposition, labelling vectors appropriately.

(d) Add the regression line to your scattergram. Is the assumption of linearity reasonable?

(e) Check the assumptions of normality and constancy of variance by doing a histogram of the errors and plotting the errors against the fitted values. Again, use Minitab if you wish. Are the assumptions reasonable?

(f) Write out the ANOVA table corresponding to the decomposition in (c) above. Test the hypothesis $H_0 : \beta_1 = 0$ by calculating the appropriate test statistic. What is your conclusion?

(g) Write down the estimated correlation coefficient, r. Test the hypothesis that the true correlation is zero, $H_0 : \rho = 0$, by comparing your calculated value with the values in Table T.3. Did you obtain a result similar to that in (f) above?

(15.2) An experiment was carried out at the Winchmore Irrigation Research Station in Canterbury, New Zealand, to determine the performance of "gummy" ewes wintered over at varying levels of pasture allowance. Gummy ewes are old ewes with teeth so poor that they have to use their gums to harvest their food. For the experiment 150 ewes were weighed and randomly allocated, within liveweight blocks, to five treatment groups. These five groups of 30 ewes were fed pasture allowances of 1.05, 1.20, 1.35, 1.50 and 2.00 kg dry matter/ewe/day. These allowances were fed out on a weekly basis by moving the mobs each Wednesday morning onto fresh breaks of feed with areas proportional to the stated allowances. The resulting average liveweight gains of the ewes in the five mobs are given in the table.

Allowance (kg DM/ewe/day)	1.05	1.20	1.35	1.50	2.00
Liveweight gain (g/ewe/day)	−65	−42	−19	−8	20

(a) Draw a scattergram of these data, treating liveweight gain as the dependent variable, y, and allowance as the independent variable, x.

(b) Write down the unit vector $U_2 = (x - \bar{x})/\|x - \bar{x}\|$.

(c) Calculate the projection coefficient $y.U_2$.

(d) Write down the fitted model in the form

$$y - \bar{y} \quad = \quad (y.U_2)U_2 \quad + \quad \text{error vector}$$

as vectors in 5−space.

(e) Rewrite this in the form

$$y - \bar{y} \quad = \quad b_1(x - \bar{x}) \quad + \quad \text{error vector}$$

to enable you to estimate the slope, β_1.

(f) Write down the equation of the line and draw the line on your scattergram. Is the straight line model appropriate?

(g) Disregarding any problems noticed in (f), calculate the terms in the Pythagorean breakup

$$\|y - \bar{y}\|^2 \quad = \quad (y.U_2)^2 \quad + \quad \|\text{error vector}\|^2$$

and summarize your results in an ANOVA table.

(h) Test the hypothesis $H_0 : \beta_1 = 0$.

(i) Use your regression line to estimate the pasture allowance required for the gummy ewes to maintain their weight over the winter period, under the pasture quality and other conditions of this experiment.

(15.3) An experiment was carried out to investigate the bee-pollination requirements of kiwifruit orchards. Before flowering, a large orchard was divided into six equally sized and environmentally similar areas. These were allocated, two per treatment, in a completely random manner to the three levels of beehive density listed in the table. After the fruit had "set" on the vines the number of fruit were counted in twenty one-metre lengths of row, randomly positioned within each area of the kiwifruit orchard. The average number of fruit per metre in each area is given in the table.

Treatment	Number of kiwifruit set/metre	
8 hives per hectare	33.4	30.6
12 hives per hectare	61.8	64.2
16 hives per hectare	63.7	61.3

(a) Draw a scattergram of these data.

(b) Write down the unit vector $U_2 = (x - \bar{x})/\|x - \bar{x}\|$, and reduce it to its simplest terms.

(c) Fit the linear regression model. Write down the equation of the line and add it to the scattergram. Summarize your Pythagorean breakup in an ANOVA table. How large is your estimate of σ^2? Are the assumptions of the model reasonable?

(d) For the linear regression model in our simple exercise, a coordinate system for the error space is given by the directions

$$U_3 = \frac{1}{\sqrt{12}} \begin{bmatrix} 1 \\ 1 \\ -2 \\ -2 \\ 1 \\ 1 \end{bmatrix}, \ U_4 = \frac{1}{\sqrt{2}} \begin{bmatrix} -1 \\ 1 \\ 0 \\ 0 \\ 0 \\ 0 \end{bmatrix}, \ U_5 = \frac{1}{\sqrt{2}} \begin{bmatrix} 0 \\ 0 \\ -1 \\ 1 \\ 0 \\ 0 \end{bmatrix}, \ U_6 = \frac{1}{\sqrt{2}} \begin{bmatrix} 0 \\ 0 \\ 0 \\ 0 \\ -1 \\ 1 \end{bmatrix}$$

assuming the data order is $y = [33.4, 30.6, 61.8, \ldots]^T$. Calculate the projection coefficients $y.U_3$, $y.U_4$, $y.U_5$ and $y.U_6$. Which of these is big relative to the others? Show that the average of the squares of these coefficients is s^2, the estimate of σ^2 found in (c).

(e) Reanalyze the data as two replicates of three treatments, with linear and quadratic contrasts specified. Summarize your analysis in an ANOVA table following the format established in Table 10.4.

(f) Add the new fitted y values, which are just the treatment means, to your scattergram and sketch a curve through these points.

(g) Compare the sums of squares calculated in (c) and (d) with those calculated in (e). You should spot two equalities, apart from the total sums of squares being equal. What are these? Also, how does your new estimate of σ^2 compare with the old estimate? Which model is more appropriate? (The first assumes a straight line relationship and the second assumes a quadratic curve.)

(15.4) One of the authors acquired a kitten in late November, 1986. The kitten was born on October 12, 1986 and was given the name "Paw Paw". Since Paw Paw was the first kitten ever owned by the author, he decided to plot a graph of its increasing weight. Paw Paw was initially weighed on the kitchen scales, but when she became too heavy the bathroom scales were used. The resulting weights are given in the table.

Age (days):	55	65	73	86	109	153	171	210
Weight (kg):	.9	1.1	1.4	1.6	2.1	3.2	3.2	3.9

Age (days):	230	248	270	293	313	393	459
Weight (kg) :	4.3	4.5	4.5	4.3	4.5	5.0	5.0

(a) Draw a scattergram of these data, treating Paw Paw's weight, in kilograms, as the dependent variable y and age, in days, as the independent variable x. Paw Paw was spayed when she was 155 days old, causing a check in her growth. Also, her weight reached a plateau when she was about eight months old.

(b) Fit a regression line to the portion of the growth curve prior to the plateau (include the first ten values, covering the period 55-248 days of age inclusive). Write down the equation of the line and add the line to the relevant portion of the scattergram. Also write out the appropriate ANOVA table, and plot the error values against the fitted values as a check on the linearity assumption.

(c) Work out the 95% confidence interval for the true slope, and convert it to units of grams/day. This will give confidence limits for Paw Paw's average daily growth rate over the period 55-248 days of age.

(15.5) An experiment is conducted on ten cars to investigate the relationship between the amount of a petrol additive and the reduction in the quantity of nitrogen oxides in the exhaust. The data is shown in the table.

(a) On graph paper, draw a scattergram of y against x.

(b) Use Minitab to fit a straight line model to the data. Write down the regression equation and add the fitted line to the scattergram.

Amount of additive (x)	1	1	2	3	4	4	5	6	6	7
Reduction in nitrogen oxides (y)	2.1	2.5	3.1	3.0	3.8	3.2	4.3	3.9	4.4	4.8

(c) Present the ANOVA table and use it to test $H_0 : \beta_1 = 0$ against $H_1 : \beta_1 \neq 0$. Produce a 95% confidence interval for β_1. Sum up in a sentence your conclusions from the experiment.

(d) Check the residuals by producing
   (i) a histogram of the residuals,
   (ii) a plot of the residuals against the x-values.
   Is there any evidence that the model is unsatisfactory?

(e) Produce a 95% confidence interval for the mean reduction in nitrogen oxides when the amount of additive, x, is 4. Check that your answer is reasonable.

(15.6) In the table we present the New Zealand kiwifruit production, in thousands of tonnes, for the years 1977 to 1983. This dataset was discussed in §15.2 and graphed in Figure 15.15.

Year	1977	1978	1979	1980	1981	1982	1983
Production (thousands of tonnes)	7.97	9.58	18.65	17.97	29.79	25.35	35.30

(a) Fit a simple regression line to the data, write down its equation, and draw it onto a scattergram of the data.

(b) Write out the appropriate ANOVA table, and plot the residuals against the fitted values.

(c) Imagine you are the marketing manager for the New Zealand Kiwifruit Authority. Use the regression line to predict the production for 1984, and calculate the 95% confidence interval for this prediction. Do you have any comments on the dangers of such an extrapolation?

(15.7) Show that the estimate of the slope, b_1, and the estimate of the correlation, r, are related by the formula

$$r = \frac{\|x - \bar{x}\| b_1}{\|y - \bar{y}\|}$$

For this reason the distribution of r is quite complicated. The term $\|x - \bar{x}\|$ is a constant, b_1 is a linear combination of y_i values, and $\|y - \bar{y}\|$ is the square root of a quadratic function of y_i values.

Solutions to the Reader Exercises

(15.1) Now $\quad y.U_2 = \dfrac{y_1(x_1 - \bar{x}) + \cdots + y_N(x_N - \bar{x})}{\|x - \bar{x}\|}$

so the associated random variable is

$$Y.U_2 = \frac{Y_1(x_1 - \bar{x}) + \cdots + Y_N(x_N - \bar{x})}{\|x - \bar{x}\|}$$

where the x_i and the $\bar{x}$ values are constants. The observation random variables Y_i have expected values $\beta_0 + \beta_1(x_i - \bar{x})$, so it follows that $E(Y.U_2)$ is given by

$$\frac{[\beta_0 + \beta_1(x_1 - \bar{x})](x_1 - \bar{x}) + \cdots + [\beta_0 + \beta_1(x_N - \bar{x})](x_N - \bar{x})}{\|x - \bar{x}\|}$$

$$= \frac{\beta_0[(x_1 - \bar{x}) + \cdots + (x_N - \bar{x})] + \beta_1\left[(x_1 - \bar{x})^2 + \cdots + (x_N - \bar{x})^2\right]}{\|x - \bar{x}\|}$$

$$= \quad \beta_1 \frac{\|x - \bar{x}\|^2}{\|x - \bar{x}\|} \quad = \quad \beta_1\|x - \bar{x}\|$$

(15.2) Now $\quad y.U_1 = (y_1 + \cdots + y_N)/\sqrt{N}$

$$\begin{aligned}
\text{so} \quad E(Y.U_1) &= [E(Y_1) + \cdots + E(Y_N)]/\sqrt{N} \\
&= [\beta_0 + \beta_1(x_1 - \bar{x}) + \cdots + \beta_0 + \beta_1(x_N - \bar{x})]/\sqrt{N} \\
&= [N\beta_0 + \beta_1(x_1 - \bar{x} + \cdots + x_N - \bar{x})]/\sqrt{N} \\
&= \sqrt{N}\beta_0
\end{aligned}$$

(15.3) Now $\quad y.U = a_1 y_1 + \cdots + a_N y_N$

$$\begin{aligned}
\text{so} \quad E(Y.U) &= a_1 E(Y_1) + \cdots + a_N E(Y_N) \\
&= a_1[\beta_0 + \beta_1(x_1 - \bar{x})] + \cdots + a_N[\beta_0 + \beta_1(x_N - \bar{x})] \\
&= (a_1 + \cdots + a_N)\beta_0 + [a_1(x - \bar{x}) + \cdots + a_N(x_N - \bar{x})]\beta_1
\end{aligned}$$

Here the two bracketted expressions are both zero, since U is orthogonal to both U_1 and U_2, implying that $a_1 + \cdots + a_N = 0$ and $a_1(x_1 - \bar{x}) + \cdots + a_N(x_N - \bar{x}) = 0$.

Chapter 16

Polynomial Regression

In Chapter 15 we made a strong assumption about the form of the relationship between the x and y variables: we assumed that y depended *linearly* on x. In some of the exercises in that chapter such an assumption was clearly unreasonable. In this chapter we study the fitting of quadratic, cubic and higher order polynomials to sets of data, and look at the problem of how to choose the best fitting curve.

Much of the hard work for this chapter has already been done under the heading of "polynomial contrasts" in Chapter 10, where we dealt with the fitting of polynomials for designed experiments. In that case we had several replications of the y variable for each value of the x variable, making assessment of the error, σ^2, easy. In the more general case no two x-values are the same, so we have to think more carefully about how to estimate the error term.

The plan for this chapter is to firstly analyze a dataset in §1 in which no two x-values are the same; this means there is no "pure error" term. Then in §2 we analyze a second dataset in which there are replications for some x-values but not others. The chapter closes with a summary in §3, followed by exercises.

16.1   No Pure Error Term

We illustrate the procedure to follow in the case when no two x-values are the same by showing how to fit a polynomial of appropriate order to the data described in Exercise 15.2.

Data

The data consist of the liveweight gains, in grams per ewe per day, of five mobs of gummy (old and toothless) ewes fed at varying levels of pasture allowance, measured in kilograms of dry matter (DM) per ewe per day. These are reproduced in Table 16.1, and graphed in Figure 16.1.

Here the observation vector is $y = [-65, -42, -19, -8, 20]^T$, while the vector of x-values is $[1.05, 1.20, 1.35, 1.50, 2.00]^T$.

Allowance (kg DM/ewe/day)	1.05	1.20	1.35	1.50	2.00
Liveweight gain (g/ewe/day)	−65	−42	−19	−8	20

Table 16.1: Liveweight gains of gummy ewes fed at different levels of pasture allowance.

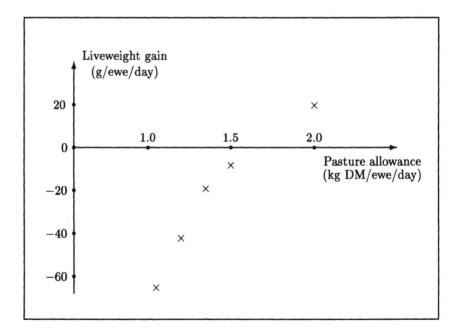

Figure 16.1: Scattergram of liveweight gain against pasture allowance.

Models

Since there are five distinct x-values, we can fit a polynomial of up to order four, one less than the number of distinct x-values. The sequence of possible polynomial models, from a constant (order 0) to a quartic (order 4) is as follows:

Constant: $y = \alpha_0$
Linear: $y = \alpha_0 + \alpha_1(x - \bar{x})$
Quadratic: $y = \alpha_0 + \alpha_1(x - \bar{x}) + \alpha_2(x - \bar{x})^2$
Cubic: $y = \alpha_0 + \alpha_1(x - \bar{x}) + \alpha_2(x - \bar{x})^2 + \alpha_3(x - \bar{x})^3$
Quartic $y = \alpha_0 + \alpha_1(x - \bar{x}) + \alpha_2(x - \bar{x})^2 + \alpha_3(x - \bar{x})^3 + \alpha_4(x - \bar{x})^4$

where the α_i values vary from line to line.

The choice of an "appropriate" order of polynomial is to some extent arbitrary. Different statisticians advocate different procedures for making this choice, and we shall advocate our favorite procedure. In addition, researchers sometimes modify the formal choice using their knowledge of the shape of the underlying relationship, generally opting for a low-order polynomial in preference to a more complicated higher-order polynomial.

In reality few relationships are truly polynomial, so we always think in terms of polynomial approximations. Formally, however, we assume that some order of polynomial provides an exact fit to the underlying relationship for our set of x-values. For this order of polynomial we make the usual assumptions that the y values are normally distributed with constant variance about a true mean on the curve. These model assumptions are illustrated for the case of a quadratic polynomial in Figure 16.2. In addition we assume that our sampling errors are independent.

Orthogonalization

As already discussed in Chapter 10 in some detail, it is advantageous to convert our sequence to an orthogonal sequence of polynomials:

Constant: $y = \beta_0$ (Order 0)
Linear: $y = \beta_0 + \beta_1 p_1(x)$ (Order 1)
Quadratic: $y = \beta_0 + \beta_1 p_1(x) + \beta_2 p_2(x)$ (Order 2)
Cubic: $y = \beta_0 + \beta_1 p_1(x) + \beta_2 p_2(x) + \beta_3 p_3(x)$ (Order 3)
Quartic: $y = \beta_0 + \beta_1 p_1(x) + \beta_2 p_2(x) + \beta_3 p_3(x) + \beta_4 p_4(x)$ (Order 4)

Here $p_0(x) = 1$, $p_1(x)$, $p_2(x)$, $p_3(x)$ and $p_4(x)$ will be the constant, linear, quadratic, cubic and quartic components, which are predetermined by the pattern of the x-values. The advantage of using the orthogonal sequence is that the estimated coefficients, b_0, b_1, b_2, b_3 and b_4, do not change from one order of polynomial to the next. Hence it is sufficient to fit the full quartic model, then simply drop unimportant components from the model.

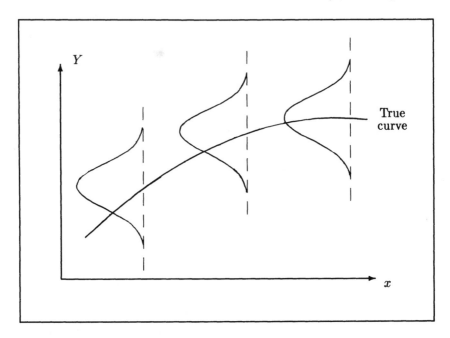

Figure 16.2: Illustration of assumptions for the quadratic regression model.

Vectors

When written out in vector form, in terms of the vectors $X_1 = 1$, $X_2 = (x - \bar{x})$, $X_3 = (x - \bar{x})^2$, $X_4 = (x - \bar{x})^3$ and $X_5 = (x - \bar{x})^4$, the full nonorthogonal model is as follows:

$$
\begin{bmatrix} Y_1 \\ Y_2 \\ Y_3 \\ Y_4 \\ Y_5 \end{bmatrix} = \alpha_0 \begin{bmatrix} 1 \\ 1 \\ 1 \\ 1 \\ 1 \end{bmatrix} + \alpha_1 \begin{bmatrix} -.37 \\ -.22 \\ -.07 \\ .08 \\ .58 \end{bmatrix} + \alpha_2 \begin{bmatrix} .137 \\ .048 \\ .005 \\ .006 \\ .336 \end{bmatrix} + \alpha_3 \begin{bmatrix} -.0507 \\ -.0106 \\ -.0003 \\ .0005 \\ .1951 \end{bmatrix} + \alpha_4 \begin{bmatrix} .01874 \\ .00234 \\ .00002 \\ .00004 \\ .11316 \end{bmatrix}
$$

Here X_2 and X_4 are quite highly correlated, as are X_3 and X_5. This can cause problems for some computer programs.

To orthogonalize this set of vectors, we must use the Gram-Schmidt method as described in §10.1. This method simply involves making each vector orthogonal to its predecessors by subtracting from the vector its projection onto the space spanned by its predecessors.

We shall call the orthogonalized vectors $T_1, \ldots, T_5$ and the corresponding unit vectors $U_1, \ldots, U_5$. Since X_1 and X_2 are already orthogonal, the first two vectors are easy to write down:

$$
T_1 = \begin{bmatrix} 1 \\ 1 \\ 1 \\ 1 \\ 1 \end{bmatrix}, \quad
T_2 = \begin{bmatrix} -.37 \\ -.22 \\ -.07 \\ .08 \\ .58 \end{bmatrix}
\quad \text{and} \quad
U_1 = \frac{1}{\sqrt{5}} \begin{bmatrix} 1 \\ 1 \\ 1 \\ 1 \\ 1 \end{bmatrix}, \quad
U_2 = \frac{1}{\sqrt{.533}} \begin{bmatrix} -.37 \\ -.22 \\ -.07 \\ .08 \\ .58 \end{bmatrix}
$$

Next we orthogonalize X_3 by calculating

$$
\begin{aligned}
T_3 &= \quad X_3 \quad - (X_3.U_1)U_1 - \quad (X_3.U_2)U_2 \\
&= \begin{bmatrix} .137 \\ .048 \\ .005 \\ .006 \\ .336 \end{bmatrix} - .107 \begin{bmatrix} 1 \\ 1 \\ 1 \\ 1 \\ 1 \end{bmatrix} - .251 \begin{bmatrix} -.37 \\ -.22 \\ -.07 \\ .08 \\ .58 \end{bmatrix} = \begin{bmatrix} .123 \\ -.003 \\ -.084 \\ -.120 \\ .084 \end{bmatrix}
\end{aligned}
$$

Hence $U_3 = [.123, -.003, -.084, -.120, .084]^T / \sqrt{.0438}$.

Instead of proceeding further by hand, we shall use the Minitab computer program given in Table 10.9 to calculate T_4, U_4, T_5 and U_5. The full orthogonalized model is found to be

$$
\begin{bmatrix} Y_1 \\ Y_2 \\ Y_3 \\ Y_4 \\ Y_5 \end{bmatrix} = \beta_0 \begin{bmatrix} 1 \\ 1 \\ 1 \\ 1 \\ 1 \end{bmatrix} + \beta_1 \begin{bmatrix} -.37 \\ -.22 \\ -.07 \\ .08 \\ .58 \end{bmatrix} + \beta_2 \begin{bmatrix} .123 \\ -.003 \\ -.084 \\ -.120 \\ .084 \end{bmatrix} + \beta_3 \begin{bmatrix} -.0128 \\ .0187 \\ .0100 \\ -.0186 \\ .0027 \end{bmatrix} + \beta_4 \begin{bmatrix} .00053 \\ -.00188 \\ .00231 \\ -.00100 \\ .00004 \end{bmatrix}
$$

Hence the two remaining unit vectors are:

$$
\begin{aligned}
U_4 &= [-.0128, .0187, .0100, -.0186, .0027]^T / \sqrt{.000967} \\
U_5 &= [.00053, -.00188, .00231, -.00100, .00004]^T / \sqrt{.0000102}
\end{aligned}
$$

In the orthogonalized model the vector $T_1 = [1, 1, 1, 1, 1]^T$ is the vector of values for the constant polynomial component $p_0(x) = 1$, while $T_2 = [-.37, \cdots, .58]^T$ is the vector of values for the linear polynomial component $p_1(x) = (x - 1.42)$. The vector $T_3 = [.123, \ldots, .084]^T$ is the vector of values for the quadratic polynomial component

$$
p_2(x) = (x - 1.42)^2 - .107 - .251(x - 1.42)
$$

as can be seen by inspecting the derivation of T_3 above. These three components are shown pictorially in Figure 16.3.

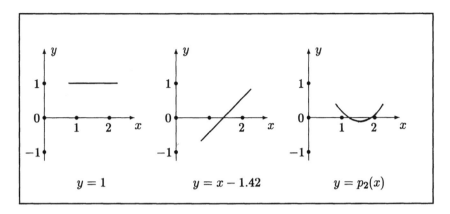

Figure 16.3: The constant, linear and quadratic "building blocks" for our polynomial approximation.

In these terms the orthogonal polynomial model is

$$Y = \beta_0 + \beta_1(x - 1.42) + \beta_2[(x - 1.42)^2 - .251(x - 1.42) - .107]$$
$$+ \beta_3 p_3(x) + \beta_4 p_4(x)$$

Here we have not given the equations for the cubic and quartic components, $p_3(x)$ and $p_4(x)$. If required, these can be obtained by following the pattern described in Chapter 10.

This completes the assembly of the raw materials. We have an observation vector $y = [-65, \ldots, 20]^T$, a model space $M = \text{span}\{U_1, \ldots, U_5\}$, and directions $U_1, \ldots, U_5$ corresponding to orthogonal polynomial components of order 0 to 4.

Fitting the Full Model

For the full quartic model, the model space fills the entire 5-space, and the appropriate orthogonal coordinate system is $U_1, \ldots, U_5$ as follows:

$$\frac{1}{\sqrt{5}}\begin{bmatrix} 1 \\ 1 \\ 1 \\ 1 \\ 1 \end{bmatrix}, \begin{bmatrix} -.507 \\ -.301 \\ -.096 \\ .110 \\ .794 \end{bmatrix}, \begin{bmatrix} .589 \\ -.014 \\ -.402 \\ -.575 \\ .401 \end{bmatrix}, \begin{bmatrix} -.411 \\ .601 \\ .321 \\ -.599 \\ .087 \end{bmatrix}, \begin{bmatrix} .165 \\ -.589 \\ .725 \\ -.314 \\ .013 \end{bmatrix}$$

To fit the model we project the observation vector y onto the model space $M = \text{span}\{U_1, \ldots, U_5\}$. The resulting orthogonal decomposition is

$$y = -22.8 + 62.4U_2 - 17.4U_3 + 1.9U_4 + 3.0U_5$$

This is depicted in Figure 16.4.

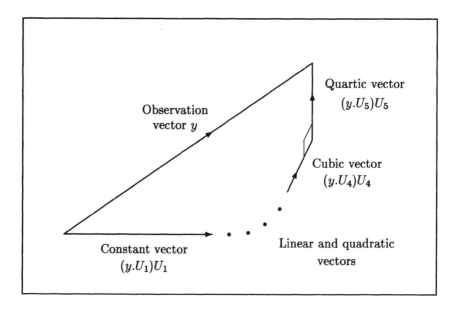

Figure 16.4: Orthogonal decomposition of the observation vector into the constant, linear, quadratic, cubic and quartic vectors.

In its corrected form this decomposition is

$$y - (-22.8) \quad = \quad 62.4U_2 \; - \; 17.4U_3 \; + \; 1.9U_4 \; + \; 3.0U_5$$

The corresponding ANOVA table is given in Table 16.2.

Source of variation	df	SS=MS
Linear	1	3897.8
Quadratic	1	304.5
Cubic	1	3.6
Quartic	1	8.9
Total	4	4214.8

Table 16.2: Pythagorean breakup of the total sum of squares in terms of polynomial component sums of squares.

What Order of Polynomial?

In biology and many other fields, the true relationship between two variables is commonly a smoothly changing one, with the y variable often rising to an asymptote within the range of interest. In these cases it is normal to observe the phenomenon exhibited in Table 16.2; the component sums of squares decrease rapidly with increasing order. The problem is "how do we decide upon a cut-off point, beyond which the component projection coefficients can reasonably be regarded as purely random?"

In our example the cut-off point is fairly obviously a polynomial of degree two, with the cubic and quartic coefficients being reasonably regarded as purely random. In general, however, the decision is not always so clearcut. The decision-making procedure which we advocate is as follows:

1. Start with a constant polynomial and increase the order of polynomial until the next polynomial component is considered "unimportant".

2. A component is judged to be "unimportant" only when it is nonsignificant when tested both against

(a) a variance estimate, s^2, obtained by pooling the sum of squares for all higher order components, and

(b) a variance estimate obtained by pooling all higher order components except the next highest order of component.

Condition (b) is included to safeguard against the occasions on which estimate (a) is inflated by a large component of order one more than the component under test. To see how this works in the case of our present example, consider Table 16.3.

Component under test	SS	Variance est.(a)	df	F value	Variance est. (b)	df	F value
Linear	3897.8	105.7	3	36.9(**)	6.25	2	623.6(**)
Quadratic	304.5	6.25	2	48.7(*)	8.9	1	34.2(ns)
Cubic	3.6	8.9	1	0.4(ns)	–	–	–
Quartic	8.9	–	–	–	–	–	–

Table 16.3: Calculations for deciding order of polynomial approximation.

When the linear component is tested the variance estimate (a) is inflated to 105.7 by the inclusion of the large quadratic component in the error term. In our case, however, the linear component sum of squares of 3897.8 is sufficiently large for the $F_{1,3}$ value of 36.9 to still attain significance. Variance estimate (b) is not similarly inflated, so yields a much higher $F_{1,2}$ value. When the quadratic component is tested, the variance estimate (a) is no longer inflated, and provides a more accurate test than that based on estimate (b) because of the greater degrees of freedom. In our example, the suggested procedure returns the intuitively correct decision of a quadratic. Note that it would have returned this decision even without requirement (b). However, if the linear and quadratic sums of squares had been similar in size, requirement (b) would have been essential.

Occasions will arise in which this procedure will produce ridiculous answers, as with any general procedure. The sums of squares in Table 16.2 should always be looked at intelligently, and the assumptions of the model should be checked after the order of polynomial has been decided. We shall now complete our analysis assuming a quadratic model.

The Quadratic Equation

Recall that our orthogonal decomposition was

$$y = -22.8 + 62.4U_2 - 17.4U_3 + 1.9U_4 + 3.0U_5$$

This can be rewritten in an earlier form by substituting $U_2 = T_2/\|T_2\|$, and so on. This gives us

$$y = -22.8 + \frac{62.4}{\|T_2\|}T_2 - \frac{17.4}{\|T_3\|}T_3 + \frac{1.9}{\|T_4\|}T_4 + \frac{3.0}{\|T_5\|}T_5$$

Here T_2 is the vector of $p_1(x) = x - \bar{x}$ values, T_3 is the vector of $p_2(x) = (x - 1.42)^2 - .251(x - 1.42) - .107$ values, and so on. Hence the equation of the fitted full quartic polynomial is

$$y = -22.8 + \frac{62.4}{\|T_2\|}p_1(x) - \frac{17.4}{\|T_3\|}p_2(x) + \frac{1.9}{\|T_4\|}p_3(x) + \frac{3.0}{\|T_5\|}p_4(x)$$

Thus the estimated parameters for our quadratic polynomial are

$$b_0 = \frac{y.U_1}{\|T_1\|} = -22.8 = \bar{y} \quad \text{(as usual)}$$

$$b_1 = \frac{y.U_2}{\|T_2\|} = \frac{62.4}{\sqrt{.533}} = 85.5$$

$$b_2 = \frac{y.U_3}{\|T_3\|} = \frac{-17.4}{\sqrt{.0438}} = -83.4 \quad \text{(using more accuracy)}$$

Therefore the equation for our fitted quadratic is

$$y = -22.8 + 85.5(x - 1.42) - 83.4[(x - 1.42)^2 - .251(x - 1.42) - .107]$$

In Figure 16.5 we have drawn this fitted curve onto our original scattergram.

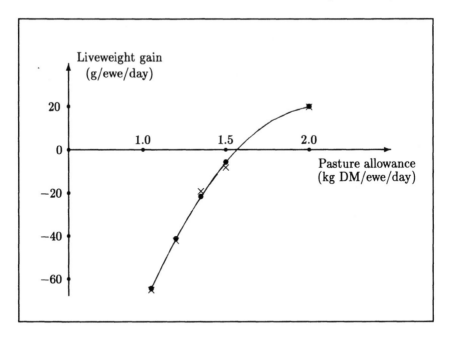

Figure 16.5: Quadratic approximation to the relationship between liveweight gain and pasture allowance, showing observed ($\times$) and fitted ($\bullet$) values.

Back to the Geometry

In deciding to fit a quadratic, we have settled upon a model space of dimension three and an error space of dimension two. Our fitted model is

$$y = (y.U_1)U_1 + (y.U_2)U_2 + (y.U_3)U_3 + \text{error vector}$$

or

$$y = \hat{y} + (y - \hat{y})$$

or

$$
\begin{bmatrix} -65 \\ -42 \\ -19 \\ -8 \\ 20 \end{bmatrix}
=
\begin{bmatrix} -64.7 \\ -41.4 \\ -21.8 \\ -5.9 \\ 19.8 \end{bmatrix}
+
\begin{bmatrix} -.3 \\ -.6 \\ 2.8 \\ -2.1 \\ .2 \end{bmatrix}
$$

where the fitted values, $\hat{y}_1, \ldots, \hat{y}_5$, are the values of the quadratic for $x = 1.05, \ldots, 2.00$, as shown in Figure 16.5. Note that the corresponding error values, $y_i - \hat{y}_i$, can also be seen in Figure 16.5 as the vertical deviations of the observations from the fitted curve.

Corrected for the mean, and expanded in terms of the linear and quadratic vectors, the fitted model is

$$
\begin{array}{ccccccc}
y - \bar{y} & = & (y.U_2)U_2 & + & (y.U_3)U_3 & + & \text{error vector}
\end{array}
$$

or
$$
\begin{array}{ccccccc}
y - \bar{y} & = & b_1 T_2 & + & b_2 T_3 & + & \text{error vector}
\end{array}
$$

$$
=
\begin{bmatrix}
-42.2 \\
-19.2 \\
3.8 \\
14.8 \\
42.8
\end{bmatrix}
=
\begin{bmatrix}
-31.6 \\
-18.8 \\
-6.0 \\
6.8 \\
49.6
\end{bmatrix}
+
\begin{bmatrix}
-10.3 \\
.3 \\
7.0 \\
10.0 \\
-7.0
\end{bmatrix}
+
\begin{bmatrix}
-.3 \\
-.6 \\
2.8 \\
-2.1 \\
.2
\end{bmatrix}
$$

| Corrected obsn vector | Linear vector | Quadratic vector | Error vector |

Here the linear vector is the vector of values of the fitted linear component, $b_1 p_1(x_i)$, and the quadratic vector is the vector of values of the fitted quadratic component, $b_2 p_2(x_i)$. This decomposition is illustrated in Figure 16.6. The corresponding Pythagorean decomposition is summarized in Table 16.4.

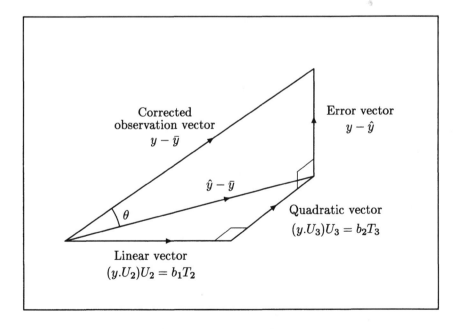

Figure 16.6: Orthogonal decomposition for a quadratic model.

Source of Variation	df	SS	MS	F
Linear component	1	3897.8	3897.8	623.6(**)
Quadratic component	1	304.5	304.5	48.7(*)
Error	2	12.5	6.25	
Total	4	4214.8		

Table 16.4: Traditional form of ANOVA table for a quadratic model.

Assumption Checking

In assuming that a quadratic model is appropriate we have assumed that our observations are independently and normally distributed, with constant variance, about a quadratic curve. The independence assumption can be justified since the five mobs were rotationally grazed on separate pastures, with new breaks of pasture being allocated weekly to the mobs in a purely random manner. We take the normality assumption to be reasonable, without bothering to plot a histogram, since we have only five error values.

To check the remaining assumptions we plot the errors against the fitted values in Figure 16.7. This reveals no evidence of any trend remaining in the data, or non-constancy of the variance, so we accept the quadratic model as reasonable. Note that we could have simply examined Figure 16.5 as an alternative to plotting Figure 16.7.

Report on Study

A report on our study of the relationship between liveweight gain of gummy ewes and their level of pasture allowance can be based on a presentation of the scattergram and quadratic curve as shown in Figure 16.5. The significance level of the linear and quadratic coefficients, $P < .01$ and $P < .05$, could be mentioned. The 95% confidence band for the position of the true quadratic curve could also be added to the figure; the relevant formula for this band will be given shortly.

Note that in our analysis we are in the position of *describing* the trends in the data, rather than testing particular hypotheses. The main reason for fitting a curve through the data is to help the eye follow the general trends, without the distraction of the noise in the data. As an aside, we must remember that the underlying quadratic relationship is probably quite specific to the conditions under which the experiment was conducted. While the overall pattern in the results may apply more generally, the particular parameters, β_0, β_1 and β_2, are likely to vary from one set of experimental conditions to another, with our estimated standard errors providing no clue as to the magnitude of this variation.

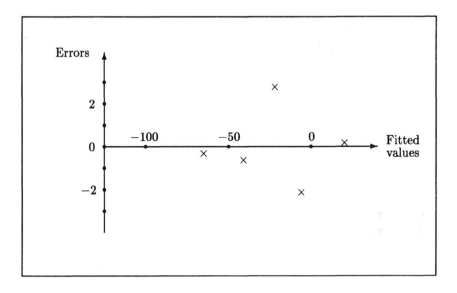

Figure 16.7: Checking for any violations of the assumptions of a quadratic polynomial and constant variance.

Minitab Analysis

For analysis using Minitab, some or all of the commands given in Table 10.9 can be used. The simplest approach is to use just the first 14 lines of the job, fitting a sequence of polynomials in nonorthogonal form; however, problems of "collinearity" may necessitate fitting the sequence in orthogonal form.

Confidence Intervals for β_0, β_1, and β_2

The distributions we need for deriving confidence intervals are given in Table 16.5, for our assumed quadratic model.

	Average, or expected value			Variance
$Y.U_1$	$\beta_0\|T_1\|$	=	$\sqrt{5}\beta_0$	σ^2
$Y.U_2$	$\beta_1\|T_1\|$	=	$\sqrt{.533}\beta_1$	σ^2
$Y.U_3$	$\beta_2\|T_2\|$	=	$\sqrt{.0438}\beta_2$	σ^2
$Y.U_4$		0		σ^2
$Y.U_5$		0		σ^2

Table 16.5: Distribution of the projection coefficients, $Y.U_i$, for our set of x-values, assuming a quadratic model.

For β_0, note that $y.U_1 = \sqrt{5}b_0$ is from an $N(\sqrt{5}\beta_0, \sigma^2)$ distribution. Hence $\sqrt{5}(b_0 - \beta_0)$ comes from an $N(0, \sigma^2)$ distribution, and $\sqrt{5}(b_0 - \beta_0)/s$ comes from a t_2 distribution. Our 95% level gamble, then, is that $t_2(.025) \leq \sqrt{5}(b_0 - \beta_0)/s \leq t_2(.975)$, leading to a confidence interval for β_0 of

$$b_0 \ \pm \ t_2(.975)\frac{s}{\sqrt{5}}$$

Substituting $b_0 = -22.8$, $t_2(.975) = 4.303$ and $s = \sqrt{6.25} = 2.5$ leads to the 95% confidence interval $-27.6 \leq \beta_0 \leq -18.0$.

Similarly, the 95% confidence intervals for β_1 and β_2 are

$$b_1 \ \pm \ t_2(.975)s/\|T_2\| \quad \text{and} \quad b_2 \ \pm \ t_2(.975)s/\|T_3\|$$

respectively. Substituting $b_1 = 85.5$, $b_2 = -83.4$, $\|T_2\| = \sqrt{.533}$ and $\|T_3\| = \sqrt{.0438}$ leads us to the intervals $70.8 \leq \beta_1 \leq 100.2$ and $-134.8 \leq \beta_2 \leq -32.0$.

Note that in each of these cases the form of the confidence interval is

$$\text{estimate} \quad \pm \quad t\text{-value} \times \text{se(estimator)}$$

where $\text{se(estimator)} = \text{se}(B_i) = s/\|T_{i+1}\|$. In our example, $\text{se}(B_0) = s/\sqrt{5}$, $\text{se}(B_1) = s/\sqrt{.533}$ and $\text{se}(B_2) = s/\sqrt{.0438}$.

Confidence Interval for a Fitted Value

We move on now to derive a 95% confidence interval for the true position of the quadratic for a pasture allowance of say $x_0 = 1.6$ kg DM/ewe/day. That is, we want a confidence interval for $\beta_0 + \beta_1 p_1(x_0) + \beta_2 p_2(x_0) = \beta_0 + \beta_1 p_1(1.6) + \beta_2 p_2(1.6)$, where $p_1(1.6) = 1.6 - 1.42 = 0.18$ and $p_2(1.6) = (1.6 - 1.42)^2 - .251(1.6 - 1.42) - .107 = .120$. The corresponding estimate is

$$\begin{aligned}
\hat{y} &= b_0 + b_1 p_1(x_0) + b_2 p_2(x_0) \\
&= b_0 + .18b_1 - .120b_2 = 2.56
\end{aligned}$$

and the variance of the estimator is

$$\text{Var}(\hat{Y}) = \text{Var}(B_0) + p_1(x_0)^2\,\text{Var}(B_1) + p_2(x_0)^2\,\text{Var}(B_2)$$

since B_0, B_1 and B_2 are independently and normally distributed. Hence the estimated standard error is

$$\begin{aligned}
\text{se}(\hat{Y}) &= \sqrt{s^2/N + p_1(x_0)^2 \times s^2/\|T_2\|^2 + p_2(x_0)^2 \times s^2/\|T_3\|^2} \\
&= s\sqrt{1/5 + .18^2/.533 + .120^2/.0438} \\
&= 1.92
\end{aligned}$$

The resulting 95% confidence interval is

$$\hat{y} \quad \pm \quad t\text{-value} \times se(\hat{Y})$$

$$\text{or} \quad 2.56 \quad \pm \quad 4.303 \times 1.92,$$

$$\text{or} \quad -5.7 \quad \text{to} \quad 10.8 \ \text{grams/ewe/day}$$

Note here that the general formula is

$$\text{fitted } y\text{-value at } x_0 \quad \pm \quad t\text{-value} \times \sqrt{s^2 \left[\frac{1}{N} + \frac{p_1(x_0)^2}{\|T_2\|^2} + \frac{p_2(x_0)^2}{\|T_3\|^2} \right]}$$

where $p_1(x_0) = x_0 - \bar{x}$ and $p_2(x_0)$ is the quadratic component at x_0. This is identical to the simple regression formula except that the term $p_2(x_0)^2/\|T_3\|^2$ has been added.

Confidence Interval for a Predicted Value

To obtain a predicted value, say for a sixth mob of sheep fed an allowance of $x_0 = 1.6$ kg DM/ewe/day, we simply add in the term "1", so that we have "$1 + \frac{1}{N} + \cdots$" in the above expression. Explicitly,

$$y_{\text{pred}} \quad = \quad b_0 + b_1 p_1(x_0) + b_2 p_2(x_0) = 2.56, \quad \text{as before, and}$$

$$Var(Y_{\text{pred}}) \quad = \quad Var(B_0) + p_1(x_0)^2 \, Var(B_1) + p_2(x_0)^2 \, Var(B_2) + \sigma^2$$

Note that this expression includes a term for the variation of the predicted value about the true value, σ^2. This leads to an estimated standard error of

$$se(Y_{\text{pred}}) \quad = \quad \sqrt{s^2/N + p_1(x_0)^2 \times s^2/\|T_2\|^2 + p_2(x_0)^2 \times s^2/\|T_3\|^2 + s^2}$$

$$= \quad \sqrt{s^2 \left[1 + 1/5 + .18^2/.533 + .120^2/.0438 \right]} = 3.15$$

Hence the resulting 95% confidence interval is -11.0 to 16.1 grams/ewe/day. Here the general formula is

$$\text{predicted } y\text{-value at } x_0 \quad \pm \quad t\text{-value} \times \sqrt{s^2 \left[1 + \frac{1}{N} + \frac{p_1(x_0)^2}{\|T_2\|^2} + \frac{p_2(x_0)^2}{\|T_3\|^2} \right]}$$

Again this is identical to the simple regression formula except that the term $p_2(x_0)^2/\|T_3\|^2$ has been added.

These formulae can easily be expanded to the case of a cubic polynomial model by adding in the term $p_3(x_0)^2/\|T_4\|^2$; for a quartic polynomial two terms are added, and so on.

As in the simple regression case, Minitab can be used to calculate the appropriate confidence limits for a suitable set of x_0-values, for example, $x_0 = 1.05, 1.10, 1.15, \ldots, 2.00$.

Multiple Correlation Coefficient, R

Both this and the following section are "optional extras", so can be skipped without losing the thread of the book.

In simple regression, the correlation coefficient, r, is the cosine of the angle between the corrected observation vector $y - \bar{y}$ and the vector $x - \bar{x}$. Its values range from -1 to $+1$. In numerical value it is the same as the cosine of the angle between $y - \bar{y}$ and the corrected model vector

$$\hat{y} - \bar{y} = (y.U_2)U_2 = b_1 T_2 = b_1(x - \bar{x})$$

For positive slopes, however, it has a positive value, and for negative slopes it has a negative value.

In quadratic regression, the corresponding quantity is the *multiple correlation coefficient*, R, defined as the cosine of the angle between the corrected observation vector, $y - \bar{y}$, and the corrected model vector,

$$\hat{y} - \bar{y} = (y.U_2)U_2 + (y.U_3)U_3 = b_1 T_2 + b_2 T_3$$

No meaning can be attached to the direction of a multiple regression when there is more than one independent variable, so R is always defined to be non-negative. The appropriate angle, θ, is shown in Figure 16.6. By definition, θ is restricted to the range from 0^0 to 90^0, and R to the range from 0 to $+1$.

In our example, the relevant vectors are

$$y - \bar{y} = \begin{bmatrix} -42.2 \\ -19.2 \\ 3.8 \\ 14.8 \\ 42.8 \end{bmatrix} \quad \text{and} \quad \hat{y} - \bar{y} = \begin{bmatrix} -41.9 \\ -18.6 \\ 1.0 \\ 16.9 \\ 42.6 \end{bmatrix}$$

Hence $R = \cos\theta = 4202.5/(64.92 \times 64.83) = 0.9985$, a very high correlation coefficient. Note that this corresponds to an angle of $\theta = \cos^{-1}(.9985) = 3^0$.

Coefficient of Determination, R^2

The *coefficient of determination*, R^2, is the square of the multiple correlation coefficient. In our case $R^2 = .9985^2 = .9970$.

From Figure 16.8 we see that it is possible to rewrite our Pythagorean breakup as

$$\|y - \bar{y}\|^2 = R^2\|y - \bar{y}\|^2 + (1 - R^2)\|y - \bar{y}\|^2$$

in which the regression sum of squares, $\|\hat{y} - \bar{y}\|^2$, has been rewritten as $\|y - \bar{y}\|^2 \cos^2\theta = R^2\|y - \bar{y}\|^2$. Hence R^2 is the "proportion of the total sum of squares accounted for by the regression". This value is printed by Minitab as a percentage, labelled "R-sq"; this is 99.7% in our example.

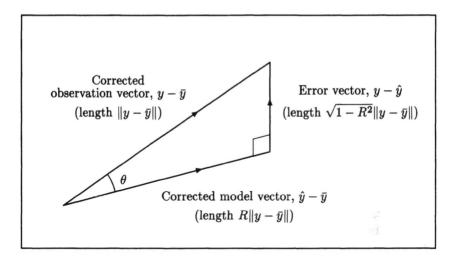

Figure 16.8: Writing the lengths of $\hat{y} - \bar{y}$ and $y - \hat{y}$ in terms of the length of $y - \bar{y}$ and R, the cosine of θ.

Link Between F and R

Percentage points, or critical values, of the distribution of R^2 under the composite null hypothesis $H_0 : \beta_1 = \beta_2 = 0$ are tabulated in some textbooks. The equivalent F test statistic is

$$
F = \frac{\text{Regression mean square}}{\text{Error mean square}} = \frac{[(y.U_2)^2 + (y.U_3)^2]/2}{s^2}
$$

$$
= \frac{\|\hat{y} - \bar{y}\|^2/2}{\|y - \hat{y}\|^2/(N - 3)} = \frac{R^2\|y - \bar{y}\|^2/2}{(1 - R^2)\|y - \bar{y}\|^2/(N - 3)}
$$

from Figure 16.8. Hence

$$
F = \frac{(N - 3)R^2}{2(1 - R^2)}
$$

Here the F value comes from the $F_{2,N-3}$ distribution if the null hypothesis is true. In practice this overall F test, and the equivalent R^2 test, is of less use than the individual F tests of the separate hypotheses, $H_0 : \beta_1 = 0$ and $H_0 : \beta_2 = 0$.

Percentage of Variance Accounted for by the Regression

The coefficient of determination, R^2, is sometimes used as an indication of the success of the model. It suffers, however, from the drawback that it always increases when another term is added to the model. In the context of our example, the R^2 values increase from .9248 to 1.0000 as we increase the order of polynomial from one to four, shown in Table 16.6.

Order of polynomial	Regression SS	R^2	Variance est. (s^2)	% Variance accounted for
1 (line)	3897.8	.9248	105.7	90.0%
2 (quadratic)	4202.3	.9970	6.25	99.4%
3 (cubic)	4205.9	.9979	8.9	99.2%
4 (quartic)	4214.8	1.0000	0	100.0%
Total SS = 4214.8			1053.7 = Total MS	

Table 16.6: The behaviour of R^2, and an alternative measure, as the order of polynomial model is increased.

A preferred alternative to R^2, the proportion of the SS accounted for by the regression model, is "the percentage of the variance accounted for by the regression model". This is calculated as the difference between the total variance (the total MS) and the variance of the fitted model (the error MS), expressed as a percentage of the total variance. In our example, this increases from 90.0% to 99.4% when we add the quadratic term to the model, then drops back to 99.2% when we add the cubic term. In general, if purely random effects are added to the model, the percentage variance accounted for by the regression will hover about a fixed value. Both measures go to 100% when zero degrees of freedom remain for error. Minitab prints the alternative measure under the label "R-sq(adj)".

16.2   Pure Error Term

In Chapter 10 we presented examples of how to fit polynomial approximations when there are several replications available for each of the x-values. We now elaborate on this, introducing the *lack of fit* and *pure error* sums of squares, plus a general procedure for deciding upon the order of an approximating polynomial. For variety, we shall use an example in which the number of replications varies from one x-value to the next.

The Study

Our example is based on an experiment conducted to determine the growth curves of the agricultural weed yarrow, under four levels of shading, namely 100%, 46.8%, 23.7% and 6.4% daylight. The underlying idea was that if yarrow proved sensitive to shading, then there would be some hope of suppressing the species by growing an aggressive crop, such as barley, in yarrow infested fields. The real-life experiment consisted of 24 mainplots with six replicates of the four shading treatments; each mainplot consisted of twelve seedlings, with two being harvested on each of six harvest dates, as described by Bourdôt et al (1984).

For our example we shall consider just one mainplot, consisting of a shade-house providing 46.8% daylight and containing twelve yarrow seedlings. For variety, we assume that different numbers of seedlings were harvested on the six harvest dates. The objective of our analysis is to fit a polynomial which adequately approximates the growth curve of yarrow under 46.8% daylight.

Data

On each harvest date, randomly selected yarrow seedlings were taken into the laboratory, washed, dried and weighed. The total dry weights of the plants are presented in Table 16.7.

	Days from start of experiment					
	7	13	17	21	25	29
Total dry	1.26	2.42	3.34	5.75	6.75	7.93
weights (g)	1.41		3.67	4.97		9.82
	1.07					8.47
log(weights)	.231	.884	1.206	1.749	1.910	2.071
	.344		1.300	1.603		2.284
	.068					2.137

Table 16.7: Total dry weights, in grams, and their natural logarithms, for the yarrow plants within a 46.8% daylight shade-house, arranged by harvest date. The numbers of plants harvested on each occasion were 3, 1, 2, 2, 1 and 3, respectively.

For plant weights we expect the standard error to increase in proportion to the mean, so we automatically apply the logarithm transformation. We therefore take as our observation vector $y = [.231, .344, \ldots, 2.284, 2.137]^T$ contenting ourselves with the accuracy of three decimal places. The corresponding vector of x values is $x = [7, 7, \ldots, 29, 29]^T$. The appropriate scattergram is shown in Figure 16.9.

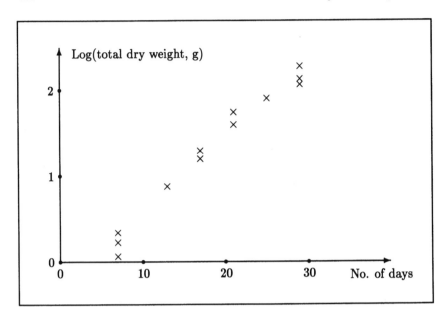

Figure 16.9: Scattergram for the yarrow shading experiment.

Models

Since there are six distinct x-values, we can fit a polynomial of up to order five, a quintic polynomial. In nonorthogonal form, the full model is

$$y = \alpha_0 + \alpha_1(x - \bar{x}) + \alpha_2(x - \bar{x})^2 + \alpha_3(x - \bar{x})^3 + \alpha_4(x - \bar{x})^4 + \alpha_5(x - \bar{x})^5$$

where $\bar{x} = 18.5$ days. In orthogonal form, the full model is

$$y = \beta_0 + \beta_1 p_1(x) + \beta_2 p_2(x) + \beta_3 p_3(x) + \beta_4 p_4(x) + \beta_5 p_5(x)$$

where we do not have explicit expressions for the polynomial components $p_i(x)$ at this stage.

Fitting the Full Model

Using Minitab we can fit the full model in nonorthogonal form as in Table 10.9, assuming we encounter no collinearity problems. These occur when one vector is almost a linear combination of the other vectors; Minitab issues a warning and calculations stop. Alternatively, we can fit the model in orthogonal form, again as in Table 10.9. In the latter case Minitab calculates U_2, U_3, U_4, U_5 and U_6, the coordinate axes corresponding to linear, quadratic, cubic, quartic and quintic components, and projects y onto the model space spanned by $U_1, U_2, \ldots, U_6$. The resulting ANOVA table (minus the mean square column) is presented in the left half of Table 16.8.

Source of variation	df	SS	$F_{1,6}$	Order of polynomial	df	Lack of fit MS	F
Linear	1	6.5129	505 (**)	Constant	5	1.3210	102 (**)
Quadratic	1	0.0852	6.60(*)	Linear	4	0.0230	1.78(ns)
Cubic	1	0.0020	0.16(ns)	Quadratic	3	0.0023	0.18(ns)
Quartic	1	0.0001	0.01(ns)	Cubic	2	0.0024	0.19(ns)
Quintic	1	0.0047	0.36(ns)	Quartic	1	0.0047	0.36(ns)
Error	6	0.0774			6	0.0129	
Total	11	6.6823					

Table 16.8: Pythagorean breakup in terms of individual polynomial component sums of squares and a "pure error" term. Also presented are the lack of fit mean squares for approximating polynomials of increasing order.

Notice that in our present example we have a pure error term which involves no assumptions concerning the correct order of polynomial. We shall use the resulting variance estimate, $s^2 = .0129$, as the baseline for our F tests when we look for an appropriate order of polynomial with which to approximate the growth curve of yarrow. Incidentally, this variance estimate could also have been calculated by fitting six treatment means, corresponding to the six dates, and using the Pythagorean breakup

$$\|y - \bar{y}_{..}\|^2 = \|\bar{y}_{i.} - \bar{y}_{..}\|^2 + \|y - \bar{y}_{i.}\|^2$$

$$6.6823 = 6.6049 + .0774$$

for k unequally replicated populations, as discussed in Appendix B.

What Order of Polynomial?

When a pure error term is available, the general procedure advocated by the authors is as follows:

> Start with a constant polynomial. Increase the order of approximating polynomial until both the next polynomial component and the "lack of fit" are nonsignificant when tested against the pure error mean square. (The lack of fit mean square for a particular order of polynomial is the average of the $(y.U_i)^2$ terms for all higher order components.)

The calculations are summarized in Table 16.8 for our example. In our case a quadratic polynomial is indicated since significant F values occur in the first two rows of the table.

The Equation

Our Minitab output, using the job of Table 10.9, gives us the quadratic equation in nonorthogonal form as

$$\log y \;=\; 1.422 + .0875(x - 18.5) - .00156(x - 18.5)^2$$

In orthogonal form, Minitab gives us the equation as

$$\log y = 1.316 + .0893 p_1(x) - .00156 p_2(x)$$

Here we know that the linear component is $p_1(x) = (x - 18.5)$, but we do not yet know the quadratic component, $p_2(x)$. To obtain this we use the formula which we gave towards the end of §10.1, namely

$$
\begin{aligned}
p_2(x) \;&=\; (x - \bar{x})^2 - [(X_3.U_1)/\|T_1\|]\,p_0(x) - [(X_3.U_2)/\|T_2\|]\,p_1(x) \\
&=\; (x - 18.5)^2 - 68.0833 p_0(x) + 1.1714 p_1(x) \\
&=\; (x - 18.5)^2 - 68.0833 + 1.1714(x - 18.5)
\end{aligned}
$$

Hence the equation in orthogonal form is

$$
\begin{aligned}
\log y \;=\; 1.316 \;+\; &.0893(x - 18.5) \\
- \;&.00156[(x - 18.5)^2 + 1.1714(x - 18.5) - 68.0833]
\end{aligned}
$$

This has been drawn onto the original scattergram in Figure 16.10.

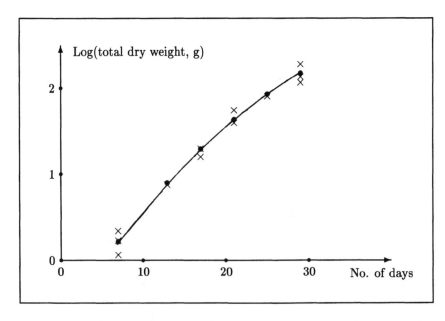

Figure 16.10: Quadratic approximation to the growth curve of yarrow under 46.8% daylight, showing observed ($\times$) and fitted ($\bullet$) values.

The Geometry

In deciding upon a quadratic model, we have settled upon a model space of dimension three, and an error space of dimension nine. Our fitted model is

$$y = (y.U_1)U_1 + (y.U_2)U_2 + (y.U_3)U_3 + \text{error vector}$$

$$\text{or} \quad y = \hat{y} + (y - \hat{y})$$

$$\text{or} \quad \begin{bmatrix} .231 \\ \vdots \\ 2.137 \end{bmatrix} = \begin{bmatrix} .2097 \\ \vdots \\ 2.1681 \end{bmatrix} + \begin{bmatrix} .0213 \\ \vdots \\ -.0311 \end{bmatrix}$$

where the fitted values are the values of the quadratic for $x = 7, 7, \ldots, 29$. These fitted values are shown in Figure 16.10, where the error values, $y_i - \hat{y}_i$, can also be seen as the vertical deviations of the observations from the fitted curve.

In Figure 16.11 we present the geometric picture. There we see that the

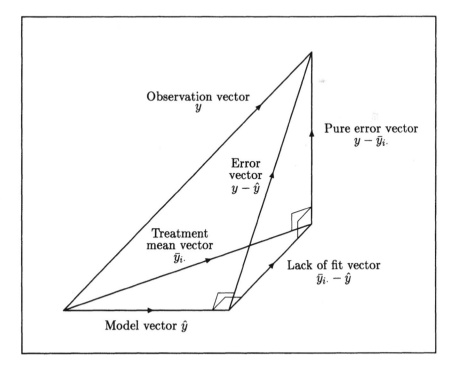

Figure 16.11: The orthogonal decompositions for the quadratic model $y = \hat{y} + (y - \hat{y})$, and the full model, $y = \bar{y}_{i.} + (y - \bar{y}_{i.})$, showing how the error vector can be split into a pure error vector plus a lack of fit vector.

error vector is the sum of the "pure error" vector, consisting of deviations of observations from the appropriate treatment (or date) means, and the "lack of fit" vector, consisting of deviations of treatment means from the appropriate values on the fitted curve. That is,

$$
\begin{bmatrix}
.0213 \\
.1343 \\
-.1417 \\
-.0096 \\
-.0811 \\
.0129 \\
.1183 \\
-.0277 \\
-.0144 \\
-.0971 \\
.1159 \\
-.0311
\end{bmatrix}
=
\begin{bmatrix}
.0046 \\
.0046 \\
.0046 \\
-.0096 \\
-.0341 \\
-.0341 \\
.0453 \\
.0453 \\
-.0144 \\
-.0041 \\
-.0041 \\
-.0041
\end{bmatrix}
+
\begin{bmatrix}
.0167 \\
.1297 \\
-.1463 \\
0 \\
-.0470 \\
.0470 \\
.0730 \\
-.0730 \\
0 \\
-.0930 \\
.1200 \\
-.0270
\end{bmatrix}
$$

$$
\begin{array}{ccc}
\text{Error} & \text{Lack of fit} & \text{Pure error} \\
\text{vector} & \text{vector} & \text{vector} \\
y - \hat{y} & \bar{y}_{i\cdot} - \hat{y} & y - \bar{y}_{i\cdot}
\end{array}
$$

Our model can also be written in the form

$$
\begin{array}{rcccl}
y - \bar{y} & = & (y.U_2)U_2 & + & (y.U_3)U_3 & + \text{ error vector} \\
\text{or} \quad y - \bar{y} & = & b_1 T_2 & + & b_2 T_3 & + \text{ error vector}
\end{array}
$$

This is the form which corresponds to the usual ANOVA table for a quadratic model, shown here as Table 16.9, in which

$$
\begin{array}{ccccc}
\text{Error SS} & = & \text{Lack of fit SS} & + & \text{Pure error SS} \\
\text{or} \quad .0842 & = & .0068 & + & .0774
\end{array}
$$

Source of Variation	df	SS	MS	F
Linear component	1	6.5129	6.5129	696 (**)
Quadratic component	1	.0852	.0852	9.11(**)
Error	9	.0842	.0094	
Total	11	6.6823		

Table 16.9: Traditional form of ANOVA table for a quadratic model.

Behavior of R^2

It may be of interest to see how the coefficient of determination, R^2, and the percentage of variance accounted for by the regression, behave as the order of polynomial is increased from one to five. The relevant figures are given in Table 16.10. The R^2 value increases to .988, while the percentage of variance accounted for by the regression declines when the order of polynomial is increased beyond two.

Order of polynomial	Regression SS	R^2	Error MS (s^2)	Percent variance explained
1 (line)	6.5129	.975	.0169	97.2
2 (quadratic)	6.5981	.987	.0094	98.5
3 (cubic)	6.6001	.988	.0103	98.3
4 (quartic)	6.6002	.988	.0117	98.1
5 (quintic)	6.6049	.988	.0129	97.9
Total SS =	6.6823		.6075 =	Total MS

Table 16.10: The behavior of R^2, and the percentage variance accounted for by the regression, as the order of polynomial model is increased.

Assumption Checking

The assumption of independence appeared reasonable with our data. In the field, individual plants were sufficiently far apart to prevent any competition or shading of neighbouring plants; also, there was adequate ventilation within the shade-house provided by a 5cm gap between the bottom of the polyshade cloth and the ground, so temperature gradients were minimized. The assumption of normality would usually be checked by plotting a histogram of the errors, but we shall not present such a figure, since nothing untoward would be seen. The appropriateness of the quadratic model and the constancy of variance can be checked by looking at the pattern of the errors in Figure 16.10, or by plotting the errors against the fitted values; in our case the model appears perfectly adequate and the variance appears constant.

Report on Study

The study results would be well summarized by presenting Figure 16.10, with the possible addition of a 95% confidence band for the position of the true quadratic curve.

16.3 Summary

In this chapter we have reworked the theory covered in Chapter 10, using "regression" terminology instead of "analysis of variance" terminology.

In order to fit polynomial models of increasing order, we orthogonalized the sequence of vectors,

$$X_1 = 1, \quad X_2 = (x - \bar{x}), \quad X_3 = (x - \bar{x})^2, \quad X_4 = (x - \bar{x})^3, \quad \ldots$$

using the Gram-Schmidt method, to yield

$$T_1 = 1, \quad T_2 = x - \bar{x}, \quad T_3, \quad T_4, \quad \ldots$$

These vectors comprise the values of the polynomial components

$$p_0(x) = 1, \quad p_1(x) = (x - \bar{x}), \quad p_2(x), \quad p_3(x), \quad \ldots$$

The corresponding sequence of unit vectors is

$$U_1 = \frac{T_1}{\|T_1\|}, \quad U_2 = \frac{T_2}{\|T_2\|}, \quad U_3, \quad U_4, \quad \ldots$$

This sequence of unit vectors can be used to obtain fitted values for any order of polynomial.

When deciding upon an appropriate order of polynomial with which to approximate the unknown true curve, two main cases occur. Either all of the x-values are distinct, in which case there is no pure error space, or some of the x-values are identical, in which case there is a pure error space. In the former case we need to assume that the polynomial coefficients, β_i, are zero for i greater than some value, so that we have some directions to make up our error space.

When all of the x-values are distinct the suggested procedure is as follows:

> Start with a constant polynomial and increase the order of polynomial until the next order component is judged to be "unimportant". See Table 16.3 for details.

When some of the x-values are identical the procedure is as follows:

> Start with a constant polynomial and increase the order of polynomial until both the next order component and the "lack of fit" are nonsignificant when tested against the pure error mean square. See Table 16.8 for details.

In both cases the adequacy of the assumed model should be checked by plotting the errors against the fitted values.

The formulae for the 95% confidence intervals for the coefficients β_0, β_1, β_2, ..., the position of the true curve for a given value of x, and the prediction of a new y-value for a given value of x, are detailed in §16.1.

The methods used in this chapter are identical to those used in Part II, in which the experiments and surveys have a completely random layout. If the layout is blocked, as in Part III, the block unit vectors must be included as directions in the model space. The resulting model then consists of parallel polynomials, with the polynomial for block one being parallel to the polynomial for block two, and so on.

Exercises

(16.1) The trays used for packing New Zealand peaches for export are of a standard size. To accommodate fruit of varying size grades, plastic liners are made for the fruit to nestle in, with varying sizes of pocket. Liners which are commercially available will accommodate 40 small fruit, 32 slightly larger fruit, or 28, 25, 23 or 20 fruit of increasing size.

A horticultural researcher was interested in estimating the relationship between the average weight of the peaches and the number in each tray. He therefore went to a packing shed and weighed the peaches in eleven trays. The resulting data, giving the average peach weight in grams for each tray, is shown in the table below. The data is by courtesy of John Eiseman, Lincoln College, Canterbury, New Zealand.

Weight(g)	211.6	165.6	192.0	182.7	154.8	143.9
Number	20	25	20	23	25	28

Weight(g)	133.2	124.8	119.8	100.0	97.0
Number	28	32	32	40	40

(a) Plot a scattergram of these data, with dependent variable the average weight of a peach, and independent variable the number of fruit per tray.

(b) Calculate the pure error estimate, s^2, by regarding the six liner sizes as six populations. The relevant formula is $s^2 = \|y - \bar{y}_i\|^2/5$ (see Appendix B).

(c) Use the procedure given in this chapter to decide upon an appropriate order of polynomial. Use Minitab or an alternative statistical package to calculate the component sums of squares to as high an order as necessary, and summarize your calculations as shown in Table 16.8.

(d) Write down the equation of the polynomial approximation in nonorthogonal or orthogonal form, whichever you prefer. Draw the curve onto the scattergram.

(e) Check the model by plotting the errors against the fitted values. Does the variance appear to be constant?

(f) Repeat (a) to (e) using logarithm-transformed weights. This will be tedious unless you have used Minitab to assist you.

(16.2) In this exercise we fit an appropriate polynomial approximation to the growth curve of the kitten Paw Paw, over the *entire period* for which data is available. The relevant data is given in Exercise 15.4.

(a) Draw a scattergram of the 15 data points.

(b) Use the procedure given in this chapter to decide upon an appropriate order of polynomial. Use Minitab or an alternative means to calculate the component sums of squares to as high an order as necessary, summarizing your calculations as shown in Table 16.3.

(c) Write down the equation of the approximating polynomial, and draw the curve onto the scattergram.

(d) Check the model by plotting the errors against the fitted values. What are your comments?

(16.3) The kiwifruit experiment described in Exercise 15.3 was harvested in the following autumn, with the exception of one area which was flooded by Cyclone Bola. The resulting harvest yields, in tonnes per hectare, are given below.

Treatment	Kiwifruit yield (tonnes/ha)	
8 hives per hectare	3.3	3.0
12 hives per hectare	6.0	
16 hives per hectare	6.2	6.0

(a) Draw a scattergram of these data.

(b) Work out the linear and quadratic component unit vectors using the Gram-Schmidt method of orthogonalization.

(c) Write down the equations of the linear and quadratic components, $p_1(x)$ and $p_2(x)$.

(d) By hand, fit the quadratic model. Show your working. Write out the fitted model in terms of the 5-space vectors

$$y = \hat{y} + (y - \hat{y})$$

where $\hat{y}$ is the fitted model vector.

(e) Write out the corresponding ANOVA table. Are the linear and quadratic coefficients significant?

(f) What is the equation of the polynomial approximation in orthogonal form? Draw the curve onto your scattergram.

(g) Predict the kiwifruit yield which would have been obtained with 10 hives per hectare. Predict also the yield for 40 hives per hectare. Is it sensible to make these two predictions?

(16.4) For the real-life experiment on the shading of yarrow as described in §16.2, we consider data from just four of the 24 mainplots, comprising two blocks of two levels of shading (100% and 6.4% daylight). The average total dry weights, in grams, of the two plants harvested on each occasion are shown in the table, by courtesy of Dr.G.W.Bourdôt, MAFTech, Lincoln, New Zealand.

Harvest date	Block 1		Block 2	
(in days)	100%	6.4%	100%	6.4%
7	2.13	.82	1.61	.95
13	3.11	1.05	3.59	1.00
17	4.58	.93	4.81	1.34
21	6.76	1.41	5.85	1.52
25	8.83	1.33	7.37	2.05
29	14.34	2.28	10.46	1.76

We shall now analyze the *logarithms* of these total dry weights in a split plot analysis of variance as described in Chapter 14, and decide upon an appropriate order of approximating polynomial.

(a) Draw a scattergram of the log data, identifying the data for each main-plot separately.

(b) Decide upon a data order for your observation vector, and write out the unit vector in the block space, U_2, the main plot factor A space, U_3, and the main plot error space, $U_4 = U_2 \times U_3$.

(c) Work out the unit vector, U_5, for the linear component of the harvest date factor B. Cross this with U_3 to find the linear component of the shading by date interaction, A×B. Call this U_6.

(d) Project y onto U_2 to U_6 using Minitab or an alternative package. Write out the resulting sums of squares in an ANOVA table, distinguishing main-plot and subplot sums of squares. At this stage the fitted model corresponds to independent straight lines for the two levels of shading.

(e) To add in the quadratic components in a nonorthogonal manner, square the $x - \bar{x}$ values and work out the $(x - \bar{x})^2$ direction. Call the result X_7. Then cross X_7 with U_3, calling the result X_8. Regress y onto U_2 to U_6, X_7 and X_8 and add the additional terms in to your ANOVA table. Independent quadratics have now been fitted.

(f) To add in the cubic components, cube the $x - \bar{x}$ values and obtain the $(x - \bar{x})^3$ direction. Call the result X_9. Then cross X_9 with U_3, calling the result X_{10}. Regress y onto $U_2 - U_6$, $X_7 - X_{10}$ and add the cubic terms in to your ANOVA table. These terms should be quite small relative to their predecessors.

This process could be continued until the quartic and quintic components of the B main effect and the A × B interaction have been added in to the model. At this stage the residual error term would consist of pure subplot error. However, we shall not persist to this point.

(g) Use the subplot error mean square from the independent cubics model in (f) to decide upon the appropriate order for the polynomial approximations. Is there any good reason for fitting independent rather than parallel polynomials?

(h) Calculate the harvest date log means for each level of shading and add them to your scattergram. Assuming straight lines are appropriate, draw independent lines through these points by hand.

Chapter 17

Analysis of Covariance

The analysis of covariance, or ANCOVA, technique is an amalgam of the ANOVA and regression techniques. In analysis of covariance we fit parallel straight lines to approximate the relationship between two variables, such as fatness and weight, for several groups, such as male and female sheep. This is usually done for one of two reasons. Firstly, a researcher may simply wish to describe the relationships. Alternatively, a researcher may be trying to increase the precision of comparisons between groups by explaining some of the variation in the y variable, say fatness, using a related x variable, say weight, which *covaries* with y. To rephrase this, we may be primarily interested in accounting for some of the extraneous variation which may affect the precision of a study, or bias the estimates of population means.

In this chapter we work with fatness and carcase weight data for five ewe lambs and five ram lambs. In §1 we use the analysis of covariance technique to fit parallel straight lines to the data. In §2 we check the "parallel" assumption by fitting independent lines, and testing the difference between the two slopes. Other examples of the use of ANCOVA are given in §3, and a summary in §4. The chapter closes with general exercises.

17.1 Illustrative Example

We now introduce our illustrative example, specify an appropriate model, and fit parallel lines to the data.

Study Objectives

Our primary objectives are:

(a) To describe the relationships between fatness and carcase weight for both ewe and ram lambs

(b) To ascertain whether ewes are in general fatter than rams at a given carcase weight

Data

The ten lambs in the study were all slaughtered on the same day. The resulting fatness and carcase weight data are given in Table 17.1, and graphed in Figure 17.1. Fatness was measured as the total depth of tissue, in mm, over the twelfth rib, 11cm from the dorsal midline.

							Means
Ewes	Fatness (y)	4	8	11	15	15	10.6
	Carcase wt (x)	10.3	11.9	15.0	17.1	18.7	14.6
Rams	Fatness (y)	5	4	10	8	14	8.2
	Carcase wt (x)	12.4	14.3	16.6	18.8	21.9	16.8

Table 17.1: Fatness and carcase weight, in kilograms, for five ewe lambs and five ram lambs.

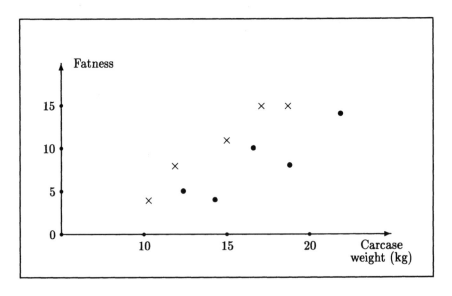

Figure 17.1: The scattergrams of fatness versus carcase weight for ewe ($\times$) and ram ($\bullet$) lambs.

The resulting observation vector is

$$y = [y_{11}, y_{12}, y_{13}, y_{14}, y_{15}, y_{21}, y_{22}, y_{23}, y_{24}, y_{25}]^T$$
$$= [\ 4,\ \ 8,\ \ 11,\ \ 15,\ \ 15,\ \ \ 5,\ \ \ 4,\ \ 10,\ \ \ 8,\ \ 14]^T$$

The corresponding vector of x-values is

$$
\begin{aligned}
x &= [\ x_{11},\ x_{12}, \ldots,\ x_{24},\ x_{25}]^T \\
&= [10.3, 11.9, \ldots, 18.8, 21.9]^T
\end{aligned}
$$

Model

The analysis of covariance model assumes that the underlying true curves are the parallel lines

$$
\begin{aligned}
y &= \beta_{01} + \beta_1(x - \bar{x}_{1\cdot}) \quad \text{for ewes} \\
\text{and} \quad y &= \beta_{02} + \beta_1(x - \bar{x}_{2\cdot}) \quad \text{for rams}
\end{aligned}
$$

That is, we have separate constant terms, β_{01} and β_{02}, but a common slope, β_1. As usual, we assume our observations are normally distributed around these mean values, with constant variance, and that the errors are independent. This is illustrated in Figure 17.2.

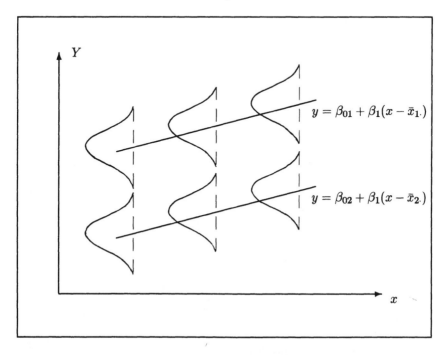

Figure 17.2: Illustration of the assumptions of the analysis of covariance model. A common slope β_1 and variance σ^2 are assumed.

The resulting model vector is

$$
\beta_{01}
\begin{bmatrix}
1 \\ 1 \\ 1 \\ 1 \\ 1 \\ 0 \\ 0 \\ 0 \\ 0 \\ 0
\end{bmatrix}
+ \beta_{02}
\begin{bmatrix}
0 \\ 0 \\ 0 \\ 0 \\ 0 \\ 1 \\ 1 \\ 1 \\ 1 \\ 1
\end{bmatrix}
+ \beta_1
\begin{bmatrix}
x_{11} - \bar{x}_{1.} \\
x_{12} - \bar{x}_{1.} \\
x_{13} - \bar{x}_{1.} \\
x_{14} - \bar{x}_{1.} \\
x_{15} - \bar{x}_{1.} \\
x_{21} - \bar{x}_{2.} \\
x_{22} - \bar{x}_{2.} \\
x_{23} - \bar{x}_{2.} \\
x_{24} - \bar{x}_{2.} \\
x_{25} - \bar{x}_{2.}
\end{bmatrix}
$$

These three vectors are mutually orthogonal, so the model space M is a three dimensional subspace of 10-space, with a coordinate system, U_1, U_2, and U_3 obtained by dividing each vector by its length.

An alternative form for the model vector is

$$
\beta_0
\begin{bmatrix}
1 \\ \vdots \\ \vdots \\ 1
\end{bmatrix}
+ \frac{\beta_{02} - \beta_{01}}{2}
\begin{bmatrix}
-1 \\ \vdots \\ 1 \\ \vdots
\end{bmatrix}
+ \beta_1
\begin{bmatrix}
x_{11} - \bar{x}_{1.} \\ \vdots \\ x_{21} - \bar{x}_{2.} \\ \vdots
\end{bmatrix}
$$

This leads to an alternative coordinate system, U_1, U_2, U_3, for the model space.

Hypotheses of Interest

Questions of interest are "Is the common slope, β_1, nonzero?" and "Is one line at a higher elevation than the other?"

The direction associated with the first question is $U_3 = (x - \bar{x}_{i.})/\|x - \bar{x}_{i.}\|$, the third axis in both of the above coordinate systems for the model space.

The second question is not associated with any of the directions listed to date, since the difference in elevation is not simply $\beta_{01} - \beta_{02}$, but is instead given, at $x = x_0$, by the difference in the y values

$$
[\beta_{01} + \beta_1(x_0 - \bar{x}_{1.})] - [\beta_{02} + \beta_1(x_0 - \bar{x}_{2.})] = (\beta_{01} - \beta_{02}) - \beta_1(\bar{x}_{1.} - \bar{x}_{2.})
$$

which is independent of x_0, since the lines are parallel. It eventuates that the associated direction is a linear combination of $U_2 = [-1, \ldots, 1]^T/\sqrt{10}$ and U_3. We delay further discussion of this, however, until we have fitted the model.

The raw materials for our analysis are now assembled. We have the observation vector y, a model space M, and a direction U_3 associated with one of our hypotheses of interest, $H_0 : \beta_1 = 0$.

Fitting the Model

To fit the model we project y onto the coordinate axes for the model space, U_1, U_2 and U_3, using Minitab or any other multiple regression program. To fit the first form of the model as given above, we regress y onto U_1 to U_3 using the Noconstant subcommand to suppress the fitting of the overall mean. In Table 17.2 this is done by regressing c1 onto c11, c12 and c13. To fit the second form of the model we regress y onto the new U_2 and U_3 in the alternative system, with U_1 being automatically fitted. This is shown in Table 17.2 as the regression of c1 onto c3 and c4.

When fitting the model in its first form in Table 17.2, we chose to not divide c11 and c12 by $\sqrt{5}$. This meant that Minitab directly printed the fitted equations

$$y \;=\; 10.6 + 1.1275(x - 14.6) \quad \text{for ewes, and}$$
$$y \;=\; 8.2 + 1.1275(x - 16.8) \quad \text{for rams}$$

These equations are both of the form $y = \bar{y}_{i\cdot} + b_1(x - \bar{x}_{i\cdot})$. Figure 17.3 shows these lines plotted on the original scattergrams.

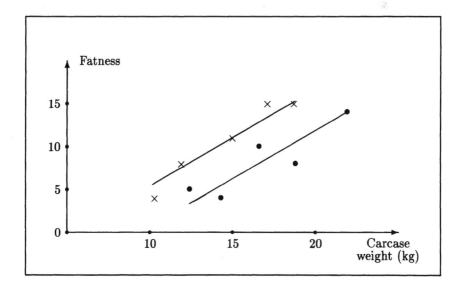

Figure 17.3: Parallel straight lines fitted by the "analysis of covariance" technique.

List of commands	Interpretation
name c1 = 'Fatness' name c2 = 'CarcWt' name c11 = 'Const1' name c12 = 'Const2' name c13 = 'Avslope' name c3 = 'NewU2' name c4 = 'Av-slope' name c6 = 'Fitted' name c7 = 'Errors'	[Naming the columns
set c1 4 ... 14 set c2 10.3 ... 21.9 plot c1 c2	[Data entry and scattergram
set c11 (1,0)5 set c12 (0,1)5	[Directions of U_1 and U_2
let k1=sum(c2*c11)/5 let k2=sum(c2*c12)/5 print k1−k2	[Calculating $\bar{x}_{1\cdot}$ and $\bar{x}_{2\cdot}$
let c13=(c2−k1)*c11 +(c2−k2)*c12	[Direction of $U_3 = (x - \bar{x}_{i\cdot})/\|x - \bar{x}_{i\cdot}\|$
regress c1 3 c11−c13; noconstant.	[Fitting model in first form
set c3 (−1,1)5 let c4=c13	[Directions of new U_2 and U_3
regress c1 2 c3−c4 c5 c6; resids c7.	[Fitting model in second form saving errors and fitted values
histogram c7 print c1 c6 c7 plot c7 c6	[Checking of assumptions
stop	[End of job

Table 17.2: Minitab commands for fitting parallel straight lines (ANCOVA) in two equivalent forms, plus assumption checking.

The Geometry

The orthogonal decomposition corresponding to the second form of the fitted model is

$$
\begin{aligned}
y &= (y.U_1)U_1 + (y.U_2)U_2 + (y.U_3)U_3 + \text{error vector} \\
&= \bar{y}_{..} + (\bar{y}_{i.} - \bar{y}_{..}) + b_1(x - \bar{x}_{i.}) + \text{error vector}
\end{aligned}
$$

When written in corrected form this becomes

$$
y - \bar{y}_{..} = (\bar{y}_{i.} - \bar{y}_{..}) + b_1(x - \bar{x}_{i.}) + \text{error vector}
$$

Corrected observation vector = Treatment vector + Linear vector + Error vector

$$
\begin{bmatrix} -5.4 \\ -1.4 \\ 1.6 \\ 5.6 \\ 5.6 \\ -4.4 \\ -5.4 \\ 0.6 \\ -1.4 \\ 4.6 \end{bmatrix}
=
\begin{bmatrix} 1.2 \\ 1.2 \\ 1.2 \\ 1.2 \\ 1.2 \\ -1.2 \\ -1.2 \\ -1.2 \\ -1.2 \\ -1.2 \end{bmatrix}
+ 1.1275
\begin{bmatrix} -4.3 \\ -2.7 \\ 0.4 \\ 2.5 \\ 4.1 \\ -4.4 \\ -2.5 \\ -0.2 \\ 2.0 \\ 5.1 \end{bmatrix}
+
\begin{bmatrix} -1.75 \\ 0.44 \\ -0.05 \\ 1.58 \\ -0.22 \\ 1.76 \\ -1.38 \\ 2.03 \\ -2.45 \\ 0.05 \end{bmatrix}
$$

This decomposition is illustrated in Figure 17.4.

The corresponding Pythagorean breakup is

$$
\|y - \bar{y}_{..}\|^2 = \|\bar{y}_{i.} - \bar{y}_{..}\|^2 + b_1^2\|x - \bar{x}_{i.}\|^2 + \|\text{error vector}\|^2
$$
$$
168.40 = 14.40 + 133.04 + 20.96
$$

as summarized in Table 17.3. The resulting pooled variance estimate is $s^2 = 2.994$.

Source of Variation	df	SS	MS	F
$H_0 : \beta_{01} = \beta_{02}$	1	14.40	14.40	
$H_0 : \beta_1 = 0$	1	133.04	133.04	44.44(**)
Error	7	20.96	2.994	
Total	9	168.40		

Table 17.3: ANOVA table from the analysis of covariance.

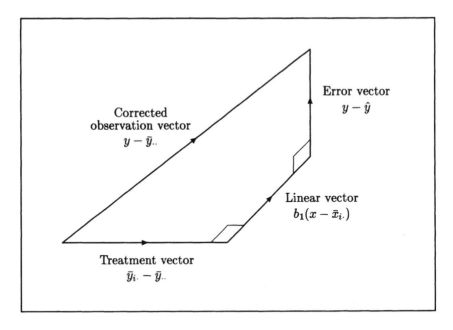

Figure 17.4: Orthogonal decomposition of corrected observation vector for analysis of covariance.

Testing the Hypotheses

We are now in a position to test our first hypothesis, that the slope, β_1, is zero. The relevant test statistic is

$$F = \frac{(y.U_3)^2}{\left[(y.U_4)^2 + \cdots + (y.U_{10})^2\right]/7} = \frac{b_1^2\|x - \bar{x}_i.\|^2}{\|\text{error vector}\|^2/7} = \frac{133.04}{20.96/7} = 44.44$$

If the hypothesis is true, then this comes from an $F_{1,7}$ distribution, for which the 99-percentile is 12.25. We therefore reject the hypothesis at the 1% level of significance, concluding that fatness is related to carcase weight, within groups of ewe and ram lambs.

We move on now to test whether there is a difference in elevation between the two lines. This difference, at any point x_0, is

$$[\beta_{01} + \beta_1(x_0 - \bar{x}_1.)] - [\beta_{02} + \beta_1(x_0 - \bar{x}_2.)] = (\beta_{01} - \beta_{02}) - \beta_1(\bar{x}_1. - \bar{x}_2.)$$

which is estimated as

$$[b_{01} + b_1(x_0 - \bar{x}_1.)] - [b_{02} + b_1(x_0 - \bar{x}_2.)] = (b_{01} - b_{02}) - b_1(\bar{x}_1. - \bar{x}_2.)$$
$$= (10.6 - 8.2) - 1.1275(14.6 - 16.8) = 4.88$$

What is $Var\left[(B_{01} - B_{02}) - B_1(\bar{x}_{1\cdot} - \bar{x}_{2\cdot})\right]$, the variance of the corresponding estimator? To estimate this variance we need the formulae for the variances of the estimators $(B_{01} - B_{02})$ and B_1. These can be obtained from the distributions of the projection coefficients, summarized in Table 17.4.

Realized value	Expected value	Variance
$y.U_1 = \sqrt{10}\,\bar{y}_{\cdot\cdot}$	$\sqrt{10}(\beta_{01} + \beta_{02})/2$	σ^2
$y.U_2 = \sqrt{5}(b_{02} - b_{01})/\sqrt{2}$	$\sqrt{5}(\beta_{02} - \beta_{01})/\sqrt{2}$	σ^2
$y.U_3 = b_1\|x - \bar{x}_{i\cdot}\|$	$\beta_1\|x - \bar{x}_{i\cdot}\|$	σ^2
$y.U_4$	0	σ^2
$\vdots$	$\vdots$	$\vdots$
$y.U_{10}$	0	σ^2

Table 17.4: Distribution of projection coefficients, $Y.U_i$, for the analysis of covariance, using the second coordinate system.

From Table 17.4 the estimator $(B_{01} - B_{02})$ has variance $2\sigma^2/5$, while the estimator B_1 has variance $\sigma^2/\|x - \bar{x}_{i\cdot}\|^2$. Therefore

$$Var[(B_{01} - B_{02}) - B_1(\bar{x}_{1\cdot} - \bar{x}_{2\cdot})] = Var(B_{01} - B_{02}) + (\bar{x}_{1\cdot} - \bar{x}_{2\cdot})^2 Var B_1$$

$$= \frac{2\sigma^2}{5} + \frac{(\bar{x}_{1\cdot} - \bar{x}_{2\cdot})^2}{\|x - \bar{x}_{i\cdot}\|^2}\sigma^2$$

Hence the estimated standard error of the difference in elevation is

$$\sqrt{s^2\left[\frac{2}{5} + \frac{(14.6 - 16.8)^2}{104.66}\right]} = 1.156$$

The resulting 95% confidence interval for the difference in fatness is 4.88 ± 2.73, using a t_7 critical value of 2.365. Similarly, the 99% interval is 4.88 ± 4.04. Since zero is not included in either interval we conclude that the ewe lambs were significantly fatter, with $P = 0.01$, than the ram lambs, for common carcase weights.

This conclusion could also have been reached by calculating the t_7 value of $4.88/1.156 = 4.22$, which is 1% significant, or by calculating the $F_{1,7}$ value of $4.88^2/1.156^2 = 17.82$, which is again 1% significant.

Exercises for the reader

(17.1) To show that $E(Y.U_1) = \sqrt{10}(\beta_{01} + \beta_{02})/2$, expand the random variable $Y.U_1$ in terms of the observation random variables Y_{11}, Y_{12} and so on, take expected values, and substitute $E(Y_{11}) = \beta_{01} + \beta_1(x_{11} - \bar{x}_{1\cdot})$, and so on.

(17.2) Similarly, show that $E(Y.U_3) = \beta_1\|x - \bar{x}_{i\cdot}\|$.

Adjusted Means

Ewe lambs had a mean fatness of 10.6 while the corresponding figure for ram lambs was 8.2. Hence the unadjusted difference in fatness was 2.4 units. However, the ewe lambs were on average lighter than the ram lambs, so this difference grew to 4.88 when ewes and rams *of the same carcase weight* were compared. This adjusted difference in the y-values is often expressed by quoting *adjusted means*, which are defined to be the fitted y-values on each line for the overall mean x-value of $\bar{x}_{..} = 15.7$kg. These are

$$
\begin{aligned}
\bar{y}_{1.} \text{ (adjusted)} &= b_{01} + b_1(\bar{x}_{..} - \bar{x}_{1.}) \\
&= \bar{y}_{1.} + b_1(\bar{x}_{..} - \bar{x}_{1.}) \\
&= 10.6 + 1.1275(15.7 - 14.6) = 11.84, \text{ and} \\
\bar{y}_{2.} \text{ (adjusted)} &= b_{02} + b_1(\bar{x}_{..} - \bar{x}_{2.}) \\
&= \bar{y}_{2.} + b_1(\bar{x}_{..} - \bar{x}_{2.}) \\
&= 8.2 + 1.1275(15.7 - 16.8) = 6.96
\end{aligned}
$$

The situation is illustrated in Figure 17.5 where the mean value for ewes of $\bar{y}_{1.} = 10.6$ has been moved up the ewe line to the adjusted mean value of 11.84, which represents the average fatness of a 15.7 kg ewe lamb carcase. Similarly, the mean value for rams of $\bar{y}_{2.} = 8.2$ has been moved down the

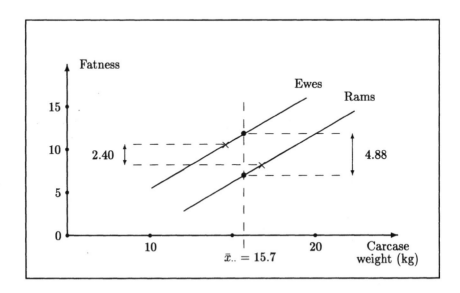

Figure 17.5: The relationship between unadjusted means ($\times$) and adjusted means ($\bullet$) in analysis of covariance. The unadjusted difference between the means is 2.40, whereas the adjusted difference is 4.88.

ram line to the adjusted mean value of 6.96, which represents the average fatness of a 15.7 kg ram lamb carcase. Note that the adjusted difference in fatness of $11.84 - 6.96 = 4.88$ has a standard error of 1.156, as shown in the last section.

Assumption Checking

The assumptions of independence, normality of errors, linearity and constancy of variance need to be checked as usual. To check the last two assumptions, Figure 17.3 can be examined, or errors can be plotted against fitted y values. Since these checks are straightforward and reveal nothing untoward, we shall not occupy space with them here.

The additional assumption of equality of slopes, $\beta_{11} = \beta_{12}$, must also be checked. This check can either be carried out informally by inspecting Figure 17.3, or formally by examining the appropriate test statistic which we shall derive in §17.2.

Report on Study

The most illuminating method of summarizing the analysis would be to present Figure 17.3. The equations of the lines could also be presented along with the standard errors of the estimators of the slope, β_1, and the difference in elevation.

If the main interest in the study, however, is the difference in fatness, with carcase weight being regarded solely as a nuisance factor, then a second method of summarizing the results may be appropriate. This simply involves presenting the adjusted means and their SED in the same way as we would present the results of an ANOVA involving two treatments. This is shown in Table 17.5. The disadvantage of this method is that the readers of the report do not have the same check on the validity of the method as they have when Figure 17.3 is displayed.

	Fatness (adjusted means)
Ewes	11.84
Rams	6.96
SED	1.156
Significance of difference	**

Table 17.5: An alternative, less preferred method of presenting the results from an ANCOVA.

Precision and Bias

In our example the comparison of fatness between ewe and ram lambs has increased in precision by carrying out an ANCOVA instead of a simple ANOVA of the fatness values. This is because the variations in carcase weight have been very useful in explaining the variations in fatness within each group. This is evidenced by a reduction in variance estimate from $s^2 = 19.3$ with ANOVA to $s^2 = 2.994$ with ANCOVA.

Another point to note is that the comparison by ANOVA of the fatness of ewe and ram lambs is an unfair comparison, since lighter ewes are being compared with heavier rams, and fatness increases with weight. The method of ANCOVA adjusts for this bias, and allows a fairer comparison.

The ANOVA results are summarized in Table 17.6 so that the reader can easily see the difference in result between the two analyses.

	Fatness
Ewes	10.6
Rams	8.2
SED	2.78
Significance of difference	n.s.

Table 17.6: ANOVA of fatness values for comparison with Table 17.5.

Confidence Interval for β_1

We now derive the 95% confidence interval for β_1, the common slope. From Table 17.4 it follows that the estimate, b_1, of the common slope, β_1, comes from an $N(\beta_1, \sigma^2/\|x - \bar{x}_i.\|^2)$ distribution. Hence the estimator B_1 has a standard error of $\sigma/\|x - \bar{x}_i.\|$, which we estimate as $s/\|x - \bar{x}_i.\|$. In our example this yields $\text{se}(B_1) = \sqrt{2.994}/\sqrt{104.66} = 0.169$. The resulting 95% confidence interval for the true common slope, β_1, is therefore

$$
\begin{aligned}
b_1 &\pm t_7(.975) \times \text{se}(B_1), & \text{or} \\
1.1275 &\pm 2.365 \times 0.169, & \text{or} \\
1.13 &\pm 0.40 &
\end{aligned}
$$

The corresponding confidence interval for the difference in elevation has been calculated earlier in this section.

Algebraic Formulae

Using the methods of earlier chapters it is easy to show that the first two coefficients are $b_{01} = \bar{y}_1.$ and $b_{02} = \bar{y}_2.$, the first and second treatment means. To find the formula for the common slope, b_1, we evaluate the projection vector $(y.U_3)U_3$ and equate coefficients with those in $b_1(x - \bar{x}_i.)$.

Now $U_3 = (x - \bar{x}_i.)/\|x - \bar{x}_i.\| = (x - \bar{x}_i.)/\sqrt{SS_{x1} + SS_{x2}}$, so

$$y.U_3 = \frac{\sum_{j=1}^n (x_{1j} - \bar{x}_1.)y_{1j} + \sum_{j=1}^n (x_{2j} - \bar{x}_2.)y_{2j}}{\|x - \bar{x}_i.\|} = \frac{SP_{xy1} + SP_{xy2}}{\sqrt{SS_{x1} + SS_{x2}}}$$

Hence $(y.U_3)U_3 = \dfrac{SP_{xy1} + SP_{xy2}}{SS_{x1} + SS_{x2}}(x - \bar{x}_i.)$, so by equating coefficients we see that the common slope is

$$b_1 = \frac{SP_{xy1} + SP_{xy2}}{SS_{x1} + SS_{x2}} = \frac{\sum_{i=1}^2 \sum_{j=1}^n (x_{ij} - \bar{x}_i.)(y_{ij} - \bar{y}_i.)}{\sum_{i=1}^2 \sum_{j=1}^n (x_{ij} - \bar{x}_i.)^2}$$

This formula generalizes to more than two groups.

The reader may be interested to see how this common slope, b_1, could be calculated from the independently fitted slopes, $b_{11} = SP_{xy1}/SS_{x1}$ and $b_{12} = SP_{xy2}/SS_{x2}$. For this purpose we substitute $SP_{xy1} = SS_{x1}b_{11}$ and $SP_{xy2} = SS_{x2}b_{12}$ into the formula for b_1, thus obtaining

$$b_1 = \frac{SS_{x1}b_{11} + SS_{x2}b_{12}}{SS_{x1} + SS_{x2}}$$

Hence b_1 is a weighted average of the independently fitted slopes, reducing to a simple average if $SS_{x1} = SS_{x2}$.

General Case

The ANCOVA method generalizes easily to more than two groups. For example, with three groups we fit three constants, b_{01}, b_{02} and b_{03}, and a common slope, b_1, by projecting the observation vector y onto three unit vectors involving 0's and 1's, plus a fourth unit vector in the direction $x - \bar{x}_i.$. The algebraic formulae given above also extend naturally. For example, the common slope becomes

$$b_1 = (SP_{xy1} + SP_{xy2} + SP_{xy3})/(SS_{x1} + SS_{x2} + SS_{x3})$$

17.2 Independent Lines

We shall now show how to fit independent lines, assuming a common variance, σ^2, and derive a test for the assumption of a common slope, $H_0 : \beta_{11} = \beta_{12}$.

Model

In this section we assume that the underlying true curves are the independent lines

$$y = \beta_{01} + \beta_{11}(x - \bar{x}_{1.}) \quad \text{for ewes,}$$
$$\text{and} \quad y = \beta_{02} + \beta_{12}(x - \bar{x}_{2.}) \quad \text{for rams}$$

Here β_{01} and β_{02} denote the constant terms, and β_{11} and β_{12} denote the slopes. As usual, we assume that our observations are normally distributed about these mean values, with constant variance and independence of errors, as illustrated in Figure 17.6.

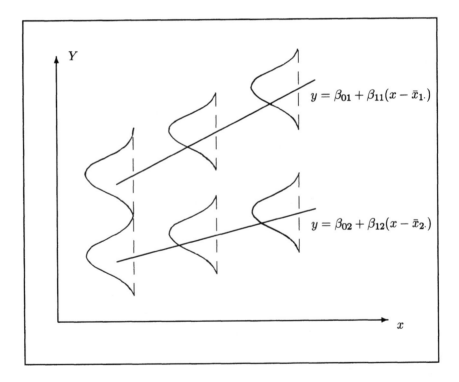

Figure 17.6: Illustration of the assumptions of the model for independent lines with a common variance, σ^2.

The resulting model vector is

$$
\beta_{01}
\begin{bmatrix} 1 \\ 1 \\ 1 \\ 1 \\ 1 \\ 0 \\ 0 \\ 0 \\ 0 \\ 0 \end{bmatrix}
+ \beta_{02}
\begin{bmatrix} 0 \\ 0 \\ 0 \\ 0 \\ 0 \\ 1 \\ 1 \\ 1 \\ 1 \\ 1 \end{bmatrix}
+ \beta_{11}
\begin{bmatrix} x_{11} - \bar{x}_{1\cdot} \\ x_{12} - \bar{x}_{1\cdot} \\ x_{13} - \bar{x}_{1\cdot} \\ x_{14} - \bar{x}_{1\cdot} \\ x_{15} - \bar{x}_{1\cdot} \\ 0 \\ 0 \\ 0 \\ 0 \\ 0 \end{bmatrix}
+ \beta_{12}
\begin{bmatrix} 0 \\ 0 \\ 0 \\ 0 \\ 0 \\ x_{21} - \bar{x}_{2\cdot} \\ x_{22} - \bar{x}_{2\cdot} \\ x_{23} - \bar{x}_{2\cdot} \\ x_{24} - \bar{x}_{2\cdot} \\ x_{25} - \bar{x}_{2\cdot} \end{bmatrix}
$$

where $x_{11} - \bar{x}_{1\cdot} = 10.3 - 14.6 = -4.3$, and so on. The above four vectors are mutually orthogonal, so the model space M is a four dimensional subspace of 10-space. A coordinate system, U_1, U_2, U_3, U_4, for the model space can be obtained simply by dividing each vector by its length.

The model vector can also be written in the form

$$
\beta_0
\begin{bmatrix} 1 \\ \vdots \\ 1 \\ \vdots \end{bmatrix}
+ \frac{\beta_{02} - \beta_{01}}{2}
\begin{bmatrix} -1 \\ \vdots \\ 1 \\ \vdots \end{bmatrix}
+ \beta_1
\begin{bmatrix} x_{11} - \bar{x}_{1\cdot} \\ \vdots \\ x_{21} - \bar{x}_{2\cdot} \\ \vdots \end{bmatrix}
+ \frac{\beta_{12} - \beta_{11}}{SS_{x1} + SS_{x2}}
\begin{bmatrix} -SS_{x2}(x_{11} - \bar{x}_{1\cdot}) \\ \vdots \\ SS_{x1}(x_{21} - \bar{x}_{2\cdot}) \\ \vdots \end{bmatrix}
$$

where $SS_{x1} = 49$ and $SS_{x2} = 55.66$. This may look horrendous, but in words it is not so bad. The first two terms represent the average, β_0, and the difference of the two constants, and the last two terms represent the average, β_1, and the difference of the two slopes. The average slope β_1 is actually a weighted average, $(SS_{x1}\beta_{11} + SS_{x2}\beta_{12})/(SS_{x1} + SS_{x2})$, which helps to explain the peculiar form of the fourth term. In its new form, the model vector gives us four new orthogonal directions which can serve as an alternative coordinate system, $U_1, \ldots, U_4$, for the model space.

Hypotheses of Interest

Our main purpose in this section is to check whether the line relating fatness and weight for ewes is parallel to that for rams. Put succinctly, we would like to test the hypothesis $H_0 : \beta_{11} = \beta_{12}$. The corresponding direction in 10-space, written in its simplest form, is

$$
U_4 = \frac{1}{\sqrt{1/SS_{x1} + 1/SS_{x2}}}
\begin{bmatrix} -(x_{11} - \bar{x}_{1\cdot})/SS_{x1} \\ \vdots \\ (x_{21} - \bar{x}_{2\cdot})/SS_{x2} \\ \vdots \end{bmatrix}
$$

the fourth coordinate axis for the alternative form of the model.

The raw materials for our analysis are now assembled. We have an observation vector $y = [4, \ldots, 5, \ldots]^T$, a four-dimensional model space M, and a direction U_4 associated with our hypothesis of major interest, $H_0 : \beta_{11} = \beta_{12}$.

Fitting the Model

We initially fit the model by projecting y onto the original set of coordinate axes, U_1, U_2, U_3 and U_4. To achieve this in Minitab we regress y onto U_1 to U_4, using the Noconstant subcommand to suppress the fitting of the overall mean vector, as described in the first part of Table 17.7. The resulting fitted equations are

$$y \ = \ 10.6 \ + \ 1.318(x - 14.6) \quad \text{for ewes, and}$$
$$y \ = \ \ \ 8.2 \ + \ 0.959(x - 16.8) \quad \text{for rams.}$$

These are of the form $y = b_{0i} + b_{1i}(x - \bar{x}_{i.})$. Figure 17.7 exhibits these lines on the original scattergrams.

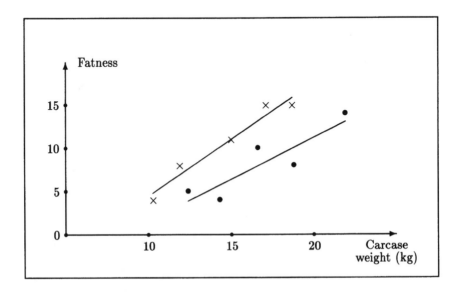

Figure 17.7: Independently fitted straight lines.

List of commands	Interpretation
name c1 = 'Fatness' name c2 = 'CarcWt' name c11 = 'Const1' name c12 = 'Const2' name c13 = 'Slope1' name c14 = 'Slope2' name c3 ='NewU2' name c4 = 'Avslope' name c5 = 'DiffSlop' name c6 = 'Fitted' name c8 = 'Errors'	[Naming the columns
set c1 4 ... 14 set c2 10.3 ... 21.9 plot c1 c2	[Data entry and scattergram
set c11 (1,0)5 set c12 (0,1)5	[Directions of U_1 and U_2
let k1=sum(c2*c11)/5 let k2=sum(c2*c12)/5 print k1−k2	[Calculating and printing $\bar{x}_1$ and $\bar{x}_2$
let c13=(c2−k1)*c11 let c14=(c2−k2)*c12	[Directions of U_3 and U_4
regress c1 4 c11−c14; noconstant.	[Fitting model in first form
set c3 (−1,1)5 let c4=c13+c14	[Directions of new U_2 and U_3
let k3=sum(c13**2) let k4=sum(c14**2) print k3−k4	[Calculating and printing SS_{x1} and SS_{x2}
let c5=(−k4*c13+k3*c14) /(k3+k4)	[Direction of new U_4
regress c1 3 c3−c5 c6 c7; resids c8.	[Fitting the model in second form
histogram c8 print c1 c7 c8 plot c8 c7	[Checking of assumptions

Table 17.7: Minitab commands for fitting the model in the original and revised forms, for the case of two independent straight lines.

The Geometry

In corrected form, the orthogonal decomposition for the second form of the model is

$$y - (y.U_1)U_1 \;=\; (y.U_2)U_2 + (y.U_3)U_3 + (y.U_4)U_4 + \text{error vector, or}$$
$$y - \bar{y}_{..} \;=\; (\bar{y}_{i.} - \bar{y}_{..}) + b_1(x - \bar{x}_{i.}) + (y.U_4)U_4 + \text{error vector}$$

This model is fitted in Minitab by regressing y onto the coordinate axes U_2 to U_4, as described in the second part of Table 17.7.

The resulting Pythagorean decomposition is

$$\|y - \bar{y}_{..}\|^2 \;=\; \|\bar{y}_{i.} - \bar{y}_{..}\|^2 + b_1^2\|x - \bar{x}_{i.}\|^2 + (y.U_4)^2 + \|\text{error vector}\|^2$$
$$168.40 \;=\; 14.40 \;+\; 133.04 \;+\; 3.36 \;+\; 17.60$$

as summarized in Table 17.8.

Source of Variation	df	SS	MS	F
$H_0 : \beta_{01} = \beta_{02}$	1	14.40	14.40	
$H_0 : \beta_1 = 0$	1	133.04	133.04	45.35(**)
$H_0 : \beta_{11} = \beta_{12}$	1	3.36	3.36	1.14(ns)
Error	6	17.60	2.934	
Total	9	168.40		

Table 17.8: ANOVA table for comparing two regression slopes.

Note that the 7-dimensional error space from the analysis of covariance in §17.1 has now been split into two subspaces, of dimension one and six respectively. The old error sum of squares of 20.96 has been broken into the sum of $(y.U_4)^2 = 3.36$ and the new error sum of squares of 17.60. The new pooled variance estimate is $s^2 = 2.934$.

Testing the Hypothesis

In order to test the hypothesis $H_0 : \beta_{11} = \beta_{12}$, we calculate the test statistic

$$F = \frac{(y.U_4)^2}{\left[(y.U_5)^2 + \cdots + (y.U_{10})^2\right]/6} = \frac{(y.U_4)^2}{\|\text{error vector}\|^2/6} = \frac{3.36}{17.60/6} = 1.14$$

If the hypothesis is true, then this comes from an $F_{1,6}$ distribution, which has a 95-percentile of 5.99. Our test is nonsignificant, so we conclude that there is no reason to suspect that the slopes differ. This confirms the validity of the analysis of covariance carried out in §17.1.

Algebraic Formulae

For the interested reader we now digress to give the algebraic formulae for the estimated constants, b_{01} and b_{02}, and the estimated slopes, b_{11} and b_{12}. For this purpose the first form of the model is the easier to use. The coordinate system, U_1, U_2, U_3 and U_4, is

$$
\frac{1}{\sqrt{n}}\begin{bmatrix} 1 \\ \vdots \\ 0 \\ \vdots \end{bmatrix}, \quad \frac{1}{\sqrt{n}}\begin{bmatrix} 0 \\ \vdots \\ 1 \\ \vdots \end{bmatrix}, \quad \frac{1}{\sqrt{SS_{x1}}}\begin{bmatrix} x_{11} - \bar{x}_{1\cdot} \\ \vdots \\ 0 \\ \vdots \end{bmatrix}, \quad \frac{1}{\sqrt{SS_{x2}}}\begin{bmatrix} 0 \\ \vdots \\ x_{21} - \bar{x}_{2\cdot} \\ \vdots \end{bmatrix}
$$

where $n = 5$ is the sample size for each group. Hence the fitted model vector is $(y.U_1)U_1 + (y.U_2)U_2 + (y.U_3)U_3 + (y.U_4)U_4$, or

$$
\begin{bmatrix} \bar{y}_{1\cdot} \\ \vdots \\ 0 \\ \vdots \end{bmatrix} + \begin{bmatrix} 0 \\ \vdots \\ \bar{y}_{2\cdot} \\ \vdots \end{bmatrix} + \frac{SP_{xy1}}{SS_{x1}}\begin{bmatrix} x_{11} - \bar{x}_{1\cdot} \\ \vdots \\ 0 \\ \vdots \end{bmatrix} + \frac{SP_{xy2}}{SS_{x2}}\begin{bmatrix} 0 \\ \vdots \\ x_{21} - \bar{x}_{2\cdot} \\ \vdots \end{bmatrix}
$$

By comparing this fitted model vector with the true model vector given earlier in this section we can see that the estimates of β_{01}, β_{02}, β_{11} and β_{12} are

$$
b_{01} = \bar{y}_{1\cdot}\,, \quad b_{02} = \bar{y}_{2\cdot}\,, \quad b_{11} = SP_{xy1}/SS_{x1}, \quad b_{12} = SP_{xy2}/SS_{x2}
$$

These are the same as the formulae derived in Chapter 15.

General Case

For our example we have chosen to deal with the simplest case, involving only two regression lines. When three or more lines are being fitted, the first method of fitting can easily be generalized. The second method, involving unit vectors corresponding to contrasts among the slopes, β_{1i}, can also be generalized, as we show in the next two pages; however, the unit vectors will not necessarily be orthogonal unless $SS_{x1} = SS_{x2} = \ldots$.

17.3 Use of ANCOVA

In this section we shall discuss the use of ANCOVA in relation to three practical examples. The first example will extend our familiar ewe and ram example to include a third group of wether (castrated ram) lambs. The second example will illustrate the usage of ANCOVA for increasing the precision of treatment comparisons in a study of the effect of drenching on weight gain in goats. The third example will illustrate the usage of ANCOVA for establishing an average relationship between the number of weevils and the age of an alfalfa stand, after the removal of farm effects.

Ewes, Wethers and Rams

In Table 17.9 we present some real life data on the leanness of twelve ewe, twelve wether and twelve ram lambs which were run together in a single mob, grazing the same pasture and being similarly treated in all ways.

Ewes		Wethers		Rams	
Fatness	Carcase Wt	Fatness	Carcase Wt	Fatness	Carcase Wt
14	13.8	11	13.6	7	14.4
8	13.2	7	13.9	8	14.8
14	13.7	14	14.0	7	15.2
14	14.8	11	14.5	9	16.1
17	14.6	12	15.4	12	16.7
19	15.4	15	15.7	14	17.2
19	16.2	13	16.2	16	17.5
21	16.4	19	17.3	14	17.8
17	16.5	16	17.9	16	18.8
23	17.1	22	18.4	20	18.8
24	17.8	19	18.6	15	19.6
22	18.7	23	19.6	17	19.7

Table 17.9: Fatness and carcase weight (kg) data for ewe, wether and ram lambs. Data by courtesy of Dr. Andy Bray, Winchmore Irrigation Research Station, Ashburton, New Zealand.

Here the observation vector, y, consists of the 36 fatness values, while the x vector consists of the 36 carcase weights. To fit three parallel lines (ANCOVA) we simply project y onto the model space spanned by the vectors

$$\begin{bmatrix} 1 \\ \vdots \\ 0 \\ \vdots \\ 0 \\ \vdots \end{bmatrix}, \begin{bmatrix} 0 \\ \vdots \\ 1 \\ \vdots \\ 0 \\ \vdots \end{bmatrix}, \begin{bmatrix} 0 \\ \vdots \\ 0 \\ \vdots \\ 1 \\ \vdots \end{bmatrix}, \begin{bmatrix} x_{11} - \bar{x}_{1\cdot} \\ \vdots \\ x_{21} - \bar{x}_{2\cdot} \\ \vdots \\ x_{31} - \bar{x}_{3\cdot} \\ \vdots \end{bmatrix}$$

Since there are three groups in the study we may specify two orthogonal contrasts among the groups. Wethers are intermediate between ewes and rams in many respects, so we shall make the simple assumption that they are *midway* between ewes and rams. This means we have equally spaced rates, on a scale of maleness for example, so that the linear and quadratic contrasts are $(-1, 0, 1)$ and $(1, -2, 1)$ respectively. The corresponding comparisons among the elevations of the lines are

Linear:

$$[\beta_{03} + \beta_1(\bar{x}_{..} - \bar{x}_{3.})] - [\beta_{01} + \beta_1(\bar{x}_{..} - \bar{x}_{1.})] = (\beta_{03} - \beta_{01}) - \beta_1(\bar{x}_{3.} - \bar{x}_{1.})$$

Quadratic:

$$[\beta_{01} + \beta_1(\bar{x}_{..} - \bar{x}_{1.})] \quad - \quad 2[\beta_{02} + \beta_1(\bar{x}_{..} - \bar{x}_{2.})] \quad + \quad [\beta_{03} + \beta_1(\bar{x}_{..} - \bar{x}_{3.})]$$
$$= \quad (\beta_{01} - 2\beta_{02} + \beta_{03}) \quad - \quad \beta_1(\bar{x}_{1.} - 2\bar{x}_{2.} + \bar{x}_{3.})$$

The estimates of these differences among the elevations are obtained by replacing the β's by b's. The standard errors can be obtained by mimicking the methods used in §17.1, and t or F tests derived as in that section.

If the assumption $\beta_{11} = \beta_{12} = \beta_{13}$ needs checking, the methods of §17.2 can be mimicked to fit three lines independently. The standard errors of B_{11}, B_{12} and B_{13} can then be used to test the linear and quadratic contrasts among the slopes β_{11}, β_{12} and β_{13}. Alternatively, the numerators $(y.U_i)^2$ for the linear and quadratic test statistics, $(y.U_i)^2/s^2$, can be calculated directly by projecting y onto the directions

$$\begin{bmatrix} -(x_{11} - \bar{x}_{1.})/SS_{x1} \\ \vdots \\ 0 \\ \vdots \\ (x_{31} - \bar{x}_{3.})/SS_{x3} \\ \vdots \end{bmatrix} \quad \text{and} \quad \begin{bmatrix} (x_{11} - \bar{x}_{1.})/SS_{x1} \\ \vdots \\ -2(x_{21} - \bar{x}_{2.})/SS_{x2} \\ \vdots \\ (x_{31} - \bar{x}_{3.})/SS_{x3} \\ \vdots \end{bmatrix}$$

Notice that these two directions are not orthogonal since $SS_{x1} \neq SS_{x3}$, so two different regress commands must be used in Minitab.

Drenching of Goats

Experiments were carried out on six commercial goat farms to determine whether the standard worm drenching program was adequate. Forty goats were used in each experiment. Twenty of these, chosen completely at random, were drenched according to the standard program, while the remaining twenty were drenched more frequently. The goats were individually tagged, and weighed at the start and end of the year-long study. For the first farm in the study the resulting liveweight gains are given in Table 17.10, along with the initial liveweights.

In each experiment the main interest was in the comparison of the liveweight gains between the two treatments. This comparison could be made using an ANOVA as detailed in Chapter 6. However, a commonly

Standard drench program										
Weightgain	5	3	8	7	6	4	8	6	7	5
Initial wt	21	24	21	22	23	26	22	23	24	20
Weightgain	5	2	4	5	10	3	8	6	6	3
Initial wt	27	28	28	30	18	27	19	20	19	22
More intensive program										
Weightgain	9	8	8	8	11	8	7	6	5	7
Initial wt	18	18	19	19	19	21	20	21	23	23
Weightgain	6	9	3	7	5	7	7	5	3	8
Initial wt	25	25	26	24	24	25	26	27	30	29

Table 17.10: Liveweight gains and initial liveweights, in kg, for each of the treatment groups on farm one. This example is based on a study conducted by Mr Alan Pearson, Veterinarian, Animal Health Laboratory, Lincoln, New Zealand.

observed biological phenomenon allows us to increase the precision of the analysis. This is the phenomenon of "regression to the norm", which in our instance means that the lighter animals gain more weight than the heavier animals (Figure 17.8). Since this can be assumed to occur within both

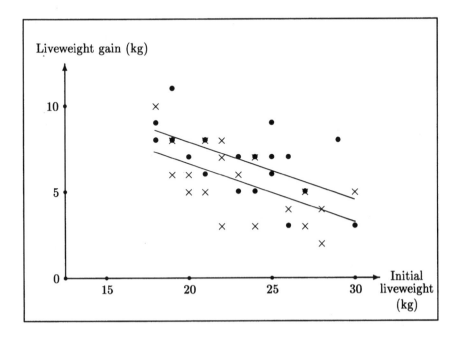

Figure 17.8: Scattergrams of liveweight gains versus initial liveweights for treatments 1 (×) and 2 (•). Parallel lines fitted by ANCOVA are also shown.

treatment groups, it is appropriate to use ANCOVA to fit two parallel lines to the data, as shown in Figure 17.8. The corresponding ANCOVA comparison of adjusted mean liveweight gains will in general be more sensitive than the ANOVA comparison of unadjusted mean liveweight gains.

Weevils on Alfalfa

Our last example comes from a survey of sitona weevils on alfalfa stands of varying ages. The survey covered six farmers' properties, with varying numbers of stands per farm. The numbers of sitona weevil larvae per square metre are given for each stand in Table 17.11.

Farmer's name	Age of alfalfa stand (in years)	Sitona weevil larval numbers (per m^2)
Band	2	1315
Band	4	725
Band	6	90
Fechney	1	520
Fechney	3	285
Fechney	9	30
Mulholland	2	725
Mulholland	6	20
Adams	2	150
Adams	3	225
Forrester	1	455
Forrester	3	75
Bilborough	2	850
Bilborough	3	650

Table 17.11: Sitona weevil larval numbers per m^2 with age of alfalfa stand for each farm. Data by courtesy of Dr Steve Goldson, MAFTech, Lincoln, New Zealand.

The survey objective was to confirm a previously observed phenomenon, that weevil populations decline as the alfalfa stand gets older. In Table 17.11 this trend occurs within five out of the six farms. However, a simple regression model is inadequate to model the data (see Figure 17.9) since there are differences in larval numbers between the farms, presumably due to differences in management or other characteristics. This suggests parallel lines are required, one per farm.

Other problems for the analysis are the likely greater variability of the higher larval numbers, and the constraint that larval numbers cannot be negative. These problems are both solved by logarithm transforming the larval numbers. The resulting decision was to use ANCOVA to fit parallel lines in the (age, $\log_e$(larval nos)) picture. This corresponds to fitting exponential curves in the (age, larval nos) picture.

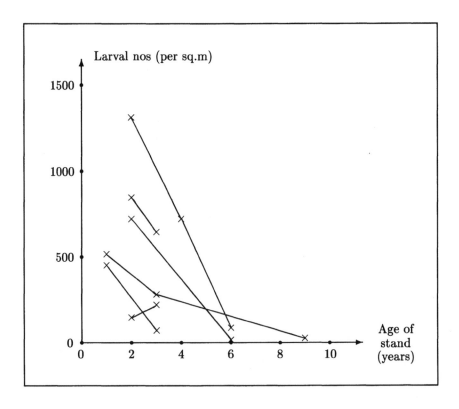

Figure 17.9: Relationship between alfalfa stand age and larval density, on six farms.

For the parallel line fitting, the required directions in the model space are as follows:

$$
\begin{bmatrix} 1 \\ 1 \\ 1 \\ 0 \\ 0 \\ 0 \\ 0 \\ 0 \\ 0 \\ 0 \\ 0 \\ 0 \\ 0 \\ 0 \end{bmatrix}
\begin{bmatrix} 0 \\ 0 \\ 0 \\ 1 \\ 1 \\ 1 \\ 0 \\ 0 \\ 0 \\ 0 \\ 0 \\ 0 \\ 0 \\ 0 \end{bmatrix}
\begin{bmatrix} 0 \\ 0 \\ 0 \\ 0 \\ 0 \\ 0 \\ 1 \\ 1 \\ 0 \\ 0 \\ 0 \\ 0 \\ 0 \\ 0 \end{bmatrix}
\begin{bmatrix} 0 \\ 0 \\ 0 \\ 0 \\ 0 \\ 0 \\ 0 \\ 0 \\ 1 \\ 1 \\ 0 \\ 0 \\ 0 \\ 0 \end{bmatrix}
\begin{bmatrix} 0 \\ 0 \\ 0 \\ 0 \\ 0 \\ 0 \\ 0 \\ 0 \\ 0 \\ 0 \\ 1 \\ 1 \\ 0 \\ 0 \end{bmatrix}
\begin{bmatrix} 0 \\ 0 \\ 0 \\ 0 \\ 0 \\ 0 \\ 0 \\ 0 \\ 0 \\ 0 \\ 0 \\ 0 \\ 1 \\ 1 \end{bmatrix}
\begin{bmatrix} x_{11} - \bar{x}_{1.} \\ x_{12} - \bar{x}_{1.} \\ x_{13} - \bar{x}_{1.} \\ x_{21} - \bar{x}_{2.} \\ x_{22} - \bar{x}_{2.} \\ x_{23} - \bar{x}_{2.} \\ x_{31} - \bar{x}_{3.} \\ x_{32} - \bar{x}_{3.} \\ x_{41} - \bar{x}_{4.} \\ x_{42} - \bar{x}_{4.} \\ x_{51} - \bar{x}_{5.} \\ x_{52} - \bar{x}_{5.} \\ x_{61} - \bar{x}_{6.} \\ x_{62} - \bar{x}_{6.} \end{bmatrix}
$$

The resulting fitted equations are of the form

$$\log_e y = b_{0i} + b_1(x - \bar{x}_{i.})$$

These can be back transformed to the exponential curves

$$y = \exp[b_{0i} + b_1(x - \bar{x}_{i.})]$$

The "average" fitted equation in this example turned out to be

$$
\begin{aligned}
\log_e y &= \bar{y}_{..} + b_1(x - \bar{x}_{..}) \\
&= 5.517 - .499(x - 3.357) \\
&= 7.19 - .499x
\end{aligned}
$$

This back transforms to the exponential curve

$$
\begin{aligned}
y &= e^{7.19} \times e^{-.499x} = 1326e^{-.499x} \\
&= 1326(.61)^x
\end{aligned}
$$

which demonstrates a mean decline of 39% in larval density from one year to the next.

17.4   Summary

The analysis of covariance model for k independent populations specifies k parallel lines

$$y = \beta_{0i} + \beta_1(x - \bar{x}_{i.}), \quad \text{for } i = 1, \dots, k$$

The resulting model is

$$
\begin{bmatrix} Y_{11} \\ \vdots \\ Y_{21} \\ \vdots \\ Y_{k1} \\ \vdots \end{bmatrix}
= \beta_{01} \begin{bmatrix} 1 \\ \vdots \\ 0 \\ \vdots \\ 0 \\ \vdots \end{bmatrix}
+ \cdots + \beta_{0k} \begin{bmatrix} 0 \\ \vdots \\ 0 \\ \vdots \\ 1 \\ \vdots \end{bmatrix}
+ \beta_1 \begin{bmatrix} x_{11} - \bar{x}_{1\cdot} \\ \vdots \\ x_{21} - \bar{x}_{2\cdot} \\ \vdots \\ x_{k1} - \bar{x}_{k\cdot} \\ \vdots \end{bmatrix}
+ \begin{bmatrix} \epsilon_{11} \\ \vdots \\ \epsilon_{21} \\ \vdots \\ \epsilon_{k1} \\ \vdots \end{bmatrix}
$$

As usual independence, normality of errors, and constancy of variance are assumed.

The corresponding coordinate system for the model space is

$$
U_1 = \frac{1}{\sqrt{n_1}} \begin{bmatrix} 1 \\ \vdots \\ 0 \\ \vdots \\ 0 \\ \vdots \end{bmatrix}, \quad \cdots \quad ,
\quad U_k = \frac{1}{\sqrt{n_k}} \begin{bmatrix} 0 \\ \vdots \\ 0 \\ \vdots \\ 1 \\ \vdots \end{bmatrix}, \quad
U_{k+1} = \frac{x - \bar{x}_{i\cdot}}{\|x - \bar{x}_{i\cdot}\|}
$$

where $n_1, \ldots, n_k$ are the sample sizes. As usual the model can be fitted by projecting the observation vector y onto the coordinate axes $U_1, \ldots, U_{k+1}$. This can be done in Minitab using the Regress command in conjunction with the Noconstant subcommand.

The resulting fitted lines are

$$
y = \bar{y}_{i\cdot} + b_1(x - \bar{x}_{i\cdot})
$$

where for each population the line goes through the centroid, $(\bar{x}_{i\cdot}, \bar{y}_{i\cdot})$, of the data. The constant terms are estimated by $b_{0i} = \bar{y}_{i\cdot}$, and the common slope is estimated by

$$
b_1 = \frac{y \cdot (x - \bar{x}_{i\cdot})}{\|x - \bar{x}_{i\cdot}\|^2} = \frac{(y - \bar{y}_{i\cdot}) \cdot (x - \bar{x}_{i\cdot})}{\|x - \bar{x}_{i\cdot}\|^2}
$$

$$
= \frac{\displaystyle\sum_{i=1}^{k}\sum_{j=1}^{n_i}(y_{ij} - \bar{y}_{i\cdot})(x_{ij} - \bar{x}_{i\cdot})}{\displaystyle\sum_{i=1}^{k}\sum_{j=1}^{n_i}(x_{ij} - \bar{x}_{i\cdot})^2}
$$

The alternative form of the model arises by specifying $k - 1$ orthogonal contrasts among the populations, as in earlier chapters. This corresponds to breaking the model space into an overall mean space, a treatment space, and a linear space. These have axis directions $U_1 = [1, \ldots, 1]^T / \sqrt{N}$ where $N = n_1 + \cdots + n_k$ is the total number of observations, $U_2, \ldots, U_k$ corresponding to $k-1$ treatment contrasts, and $U_{k+1} = (x - \bar{x}_{i.}) / \|x - \bar{x}_{i.}\|$ corresponding to the linear term. In this form the model can be fitted in Minitab by regressing y onto $U_2, \ldots, U_{k+1}$ in the usual way. The resulting orthogonal decomposition is illustrated in Figure 17.10.

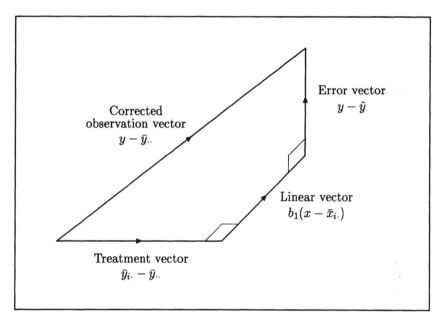

Figure 17.10: Orthogonal decomposition of corrected observation vector for analysis of covariance.

The hypothesis $H_0 : \beta_1 = 0$ can be tested by calculating the test statistic $(y.U_{k+1})^2 / s^2$. This statistic follows the $F_{1, N-k-1}$ distribution if $\beta_1 = 0$.

The elevations of the parallel lines are usually calculated at a common x value of $\bar{x}_{..}$, the overall mean x value. These elevations are referred to as adjusted means, and are given by the formula

$$\bar{y}_{i.}(\text{adjusted}) = \bar{y}_{i.} + b_1(\bar{x}_{..} - \bar{x}_{i.})$$

Here the underlying true elevations are $\beta_{0i} + \beta_1(\bar{x}_{..} - \bar{x}_{i.})$. Contrasts among these elevations can be specified in the usual manner, with significance tests obtained by calculating the variance of each contrast in terms of the variance of the estimators B_{0i} and B_1.

The common slope assumption can be checked by fitting the k lines independently, then calculating and testing important contrasts among the slopes, β_{1i}. Other assumptions can be checked in the usual manner.

Exercises

(17.1) In this exercise we analyze the ewe, wether, ram data given in Table 17.9.

(a) Using the Mplot command of Minitab or otherwise, draw a scattergram of the data with ewes, wethers and rams distinguished.

(b) Use Minitab to fit parallel lines to the data, using model space directions in the first of the two forms given in §17.1. Write down the equations of the lines.

(c) Add these lines to the scattergram. Does the assumption that the lines are parallel seem reasonable?

(d) Calculate the adjusted means and present them in a table following the format of Table 17.5, but omitting the SED and significance.

(e) Calculate the linear contrast in the elevations of the lines, the standard error of this contrast, and the corresponding 95% confidence interval. Is there evidence of any difference in fatness among the sexes? Hint: See the text of §17.3, which tells us that the linear contrast is the difference between the ewe and ram line elevations.

(17.2) (a) In Exercise 17.1, refit the three lines without requiring them to be parallel. What are the three slope estimates b_{11}, b_{12}, and b_{13}?

(b) The linear contrast among these three slopes is $\beta_{13} - \beta_{11}$, the difference in slope between ewes and rams. What is your estimate of this difference?

(c) The corresponding estimator, $B_{13} - B_{11}$, has a variance of

$$Var(B_{13}) + Var(B_{11}) = \frac{\sigma^2}{SS_{x3}} + \frac{\sigma^2}{SS_{x1}}$$

Estimate this variance, and calculate the corresponding standard error.

(d) Hence test the hypothesis $H_0 : \beta_{11} = \beta_{13}$, showing your working. Was the parallel lines assumption reasonable in Exercise 7.1?

(17.3) In this exercise we analyze the goat drenching data given in Table 17.10.

(a) Use Minitab to carry out a simple ANOVA to compare the liveweight gains of the two treatment groups. Present your results in a table of means, with SED and significance of difference.

(b) Reanalyze the data using analysis of covariance, presenting your results in a table of adjusted means, with SED and significance of difference.

(c) What fraction of the ANOVA error sum of squares has been explained by the common regression slope? To rephrase the question, by what fraction has the error sum of squares been reduced between ANOVA and ANCOVA? This fraction is the analog of the r^2 value mentioned in the simple regression chapter.

(d) What is your slope estimate, b_1, and the associated standard error, $se(B_1)$? Calculate the test statistic for the hypothesis $H_0 : \beta_1 = 0$ using the formula $F = [b_1/se(B_1)]^2$. Does your answer equal $(y.U_4)^2/s^2$?

(**17.4**) In Exercise 12.7 we analyzed a goat drenching experiment similar to that analyzed in exercise 17.3. The main difference was that its design was a randomized block.

(a) Reanalyze the liveweight gains in Exercise 12.7 as an ANCOVA, using initial liveweight as a covariate and ignoring blocks. Calculate the adjusted means and the standard error of the difference (SED) between the adjusted means for treatments 1 and 2.

(b) The randomized block ANOVA yielded a variance estimate of $s^2 = 3.904$ and an SED of 0.75. How did these two values compare with those from (a)?

The moral of the story is that it is better to set up blocks at the design stage than to rely upon ANCOVA at the analysis stage. In essence, the reason for this is that the randomized block ANOVA corresponds to general parallel curves rather than the parallel lines used in ANCOVA.

(**17.5**) In this exercise we analyze the sitona weevil data given in Table 17.11. (a) Logarithm transform the larval numbers and draw a scattergram of $\log_e$(larval numbers) against age of alfalfa stand, with farms distinguished. This can be accomplished in Minitab using the Mplot command.

(b) Fit parallel lines to the $(x, \log_e y)$ data by projecting the observation vector on to the directions given in the text.

(c) Write down the fitted equations for the six farms, and draw them onto the scattergram. Do the assumptions of the model appear reasonable?

(d) Write out the summary ANCOVA table.

(e) Calculate the 95% confidence interval for the true slope, β_1, in the $(x, \log_e y)$ picture.

(f) In the text we back transformed the estimated slope, $b_1 = -.499$, to obtain the figure of $e^{b_1} = e^{-.499} = .61$, which led to the estimated mean decline rate of 39% per annum. Back-transform the lower and upper confidence limits for β_1 in the same way. What is the resulting 95% confidence interval for the mean annual decline rate?

(g) Draw the average curve, $y = 1326(.61)^x$, onto the (x, y) scattergram.

(17.6) In the yarrow-clover competition study described in §15.1 we fitted a straight line through the scattergram of log(clover seed yield) against yarrow density, for area A only. In this exercise we shall use ANCOVA to fit parallel lines through the data for all three areas, A,B and C, as given in the table. These data were provided by courtesy of Dr. Graeme Bourdôt, Ministry of Agriculture and Fisheries, Lincoln, New Zealand.

Clover seed yield (in g/m^2)									
Area A:	19.0	76.7	11.4	25.1	32.2	19.5	89.9	38.8	45.3	39.7
Area B:	16.5	1.8	82.4	54.2	27.4	25.8	69.3	28.7	52.6	34.5
Area C:	49.7	23.3	38.9	79.4	53.2	30.1	4.0	20.7	29.8	68.5
Number of yarrow flower stems (per m^2)									
Area A:	220	20	510	40	120	300	60	10	70	290
Area B:	460	320	0	80	0	450	30	250	20	100
Area C:	0	220	160	0	120	150	450	240	250	0

(a) Using the Mplot command of Minitab, or otherwise, draw a scattergram of log(yield) against yarrow density. Are there any one or two values which appear unusual?

(b) Fit parallel straight lines to the data by extending the Minitab job given in Table 17.2. Do not reject any data values at this stage. Use the first form of the model for simplicity.

(c) Add these lines to the scattergram.

(d) Use Minitab to plot a histogram of the errors, and to plot the errors against the fitted values. What peculiarities can you detect?

(e) Obtain a second copy of the scattergram drawn in (a) and circle the two values which are most unusual on the basis of (d).

(f) Reject these two values and rerun your Minitab job with *'s inserted for the rejected x and y values. Don't forget to adjust the calculation of $\bar{x}_{1.}$, $\bar{x}_{2.}$ and $\bar{x}_{3.}$ as necessary.

(g) How does the variance estimate, s^2, change between (b) and (f)?

(h) Add your new lines to the scattergram in (e).

(i) Replot the histogram of errors and replot the errors against the fitted values. Is the situation improved?

(j) Calculate a 95% confidence interval for the true slope, β_1. How does this interval compare with the interval derived in §15.2 using just the data from area A?

(17.7) A study was carried out to show that the giant buttercups on two different dairy farms had different tolerances to the chemical "MCPA" (presumably due to the fact that on one farm the chemical had been seldom used whereas on the other farm it had been intensively used). For the study, seed from the buttercups on the two farms were collected and sown under standard laboratory conditions at the Lincoln research area. A split plot design experiment was set up, with the two buttercup biotypes on main plots and nine rates of chemical (0, 0.198, 0.296, 0.444, 0.667, 1, 1.5, 2.25 and 3.375 kg of active ingredient per ha) on subplots. Main plots were arranged as four replicates of a randomized block design.

At the completion of the study each plot was scored for the amount of living tissue as a percentage of the control plot in the same mainplot. For each biotype the average of these scores over the four replicates is given in the table, by courtesy of Dr Graeme Bourdôt, Ministry of Agriculture and Fisheries, Lincoln, New Zealand.

Rate of MCPA (kg active ingredient/ha)							
.198	.296	.444	.667	1.0	1.5	2.25	3.375
Mean scores for Silcock's farm							
62.5	47.5	40	37.5	30	17.5	11.25	7.75
Mean scores for Jones' farm							
26.25	30	18.75	15	8.75	8.75	2.5	1.625

(a) Each rate of chemical is 1.5 times the preceding rate, so that on a log scale the rates are equally spaced. Draw a scattergram of the scores versus the $\log_e$(rates) with the buttercup biotypes from the two farms separately identified.

(b) Use the Minitab program listed in Table 17.2 to fit two independent straight lines. Add these lines to the scattergram.

(c) Draw up a summary ANOVA table corresponding to the second form of the model. Test the hypothesis that the two slopes are equal, $H_0 : \beta_{11} = \beta_{12}$. Would it be reasonable to use the ANCOVA model in this instance?

Aside: The real-life analysis of the data from this experiment was as a splitplot ANOVA using polynomial contrasts for the log rates. This analysis confirmed that the trends were in fact linear, rather than curved.

Solutions to the Reader Exercises

(17.1) Now $Y.U_1 = (Y_{11} + \cdots + Y_{21} + \cdots)/\sqrt{10} = \sqrt{10}\bar{Y}_{..}$,
so that

$$
\begin{aligned}
E(Y.U_1) &= [\beta_{01} + \beta_1(x_{11} - \bar{x}_{1.}) + \cdots + \beta_{02} + \beta_1(x_{21} - \bar{x}_{2.}) + \cdots]/\sqrt{10} \\
&= [5\beta_{01} + \beta_1 \sum_{j=1}^{5}(x_{1j} - \bar{x}_{1.}) + 5\beta_{02} + \beta_1 \sum_{j=1}^{5}(x_{2j} - \bar{x}_{2.})]/\sqrt{10} \\
&= 5(\beta_{01} + \beta_{02})/\sqrt{10} \quad = \quad \sqrt{10}(\beta_{01} + \beta_{02})/2
\end{aligned}
$$

(17.2) Now $Y.U_3 = [Y_{11}(x_{11} - \bar{x}_{1.}) + \cdots + Y_{21}(x_{21} - \bar{x}_{2.}) + \cdots]/\|x - \bar{x}_{i.}\|$,
so that

$$
\begin{aligned}
E(Y.U_3) &= \{ \ [\beta_{01} + \beta_1(x_{11} - \bar{x}_{1.})](x_{11} - \bar{x}_{1.}) + \cdots \\
&\quad + \ [\beta_{02} + \beta_1(x_{21} - \bar{x}_{2.})](x_{21} - \bar{x}_{2.}) + \cdots \ \} / \|x - \bar{x}_{i.}\| \\
&= \{\beta_{01} \sum_{j=1}^{5}(x_{1j} - \bar{x}_{1.}) + \beta_1 \sum_{j=1}^{5}(x_{1j} - \bar{x}_{1.})^2 \\
&\quad + \beta_{02} \sum_{j=1}^{5}(x_{2j} - \bar{x}_{2.}) + \beta_1 \sum_{j=1}^{5}(x_{2j} - \bar{x}_{2.})^2\} / \|x - \bar{x}_{i.}\| \\
&= \beta_1 \|x - \bar{x}_{i.}\|^2 / \|x - \bar{x}_{i.}\| \\
&= \beta_1 \|x - \bar{x}_{i.}\|
\end{aligned}
$$

Chapter 18

General Summary

In §1 of this chapter we review the method that has been used throughout this textbook, and in §2 we discuss its extension to more complicated problems using matrix methods.

18.1 Review

In this book we have shown that the natural mathematical setting for a discussion of ANOVA, regression and ANCOVA is provided by the theory of vectors in N-dimensional space. In a specific situation the design of the study determines a p dimensional model space, plus a residual $N - p$ dimensional error space, as illustrated in Figure 18.1. Questions of interest correspond to directions in the model space, and are tested by comparing the squared projection length in that direction, $(y.U_c)^2$, with the average of the squared projection lengths in the error space directions, s^2.

For a proper statistical treatment of the topic we need to regard the observations $y_1, \cdots, y_N$ as realized values of random variables $Y_1, \cdots, Y_N$. We assume these random variables have a common variance σ^2, with expected values in general nonzero and different from one another. For example, observations 1 and 2 may come from populations 1 and 2, with expected values μ_1 and μ_2. When we project the observation vector $y = [y_1, \ldots y_N]^T$ onto the orthogonal directions $U_1, \ldots, U_N$ we create a new set of realized values, $y.U_1, \ldots, y.U_N$, associated with a new set of random variables $Y.U_1, \ldots, Y.U_N$. That is, we transform a set of independent normally distributed random variables, $Y_1, \ldots, Y_N$, with variance σ^2, into a new set of independent normally distributed random variables, $Y.U_1, \ldots, Y.U_N$, also with variance σ^2, and with all except the first p having zero mean. This second set is very useful. The first p of these new random variables are used to estimate the p parameters of the model, while the last $N - p$ random variables, with zero mean, are available for estimating the variance σ^2.

In estimating σ^2, we note that $(y.U_{p+1})^2$ is an estimate of σ^2 since $y.U_{p+1}$ comes from an $N(0, \sigma^2)$ distribution. Similarly $(y.U_{p+2})^2, \ldots, (y.U_N)^2$ are estimates of σ^2. The best estimate, s^2, is obtained by averaging these independent estimates, yielding

$$s^2 = \frac{(y.U_{p+1})^2 + \cdots + (y.U_N)^2}{N - p}$$

This was discussed in more detail in Chapter 3.

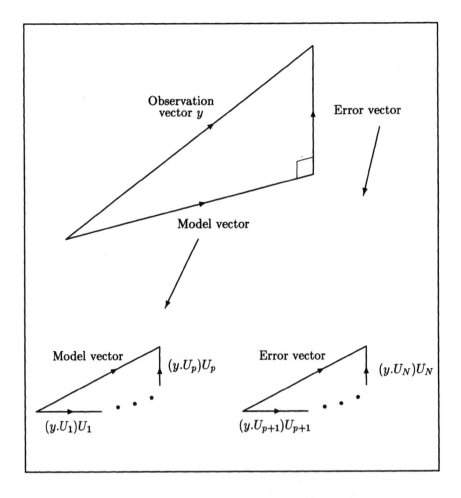

Figure 18.1:   The N-dimensional observation vector broken into a p-dimensional model vector and an $N - p$ dimensional error vector. These in turn are broken into sums of projections onto one dimensional subspaces.

Throughout this book many different models have been specified, leading to many different model spaces and coordinate systems. We shall now attempt to summarize the various types of model within each of the categories of ANOVA, regression and ANCOVA.

ANOVA Models

When we deal with questions about k populations, the model space consists of the overall mean space, the treatment space and the block space, if blocks are present. This is illustrated in Figure 18.2.

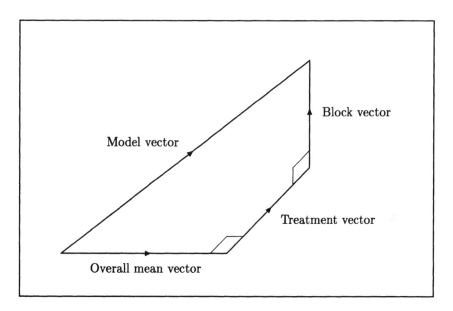

Figure 18.2: Decomposition of the model vector into overall mean, treatment and block vectors.

The *overall mean space* is a one dimensional subspace with coordinate axis direction $U_1 = [1, \cdots, 1]^T / \sqrt{N}$, corresponding to the hypothesis $H_0 : \mu = 0$. The projection onto this subspace is $(y.U_1)U_1 = \bar{y}_{..}$, the overall mean vector. Only in the single population case is the hypothesis $H_0 : \mu = 0$ of any interest. The corresponding test statistic is $F = (y.U_1)^2 / s^2$, which comes from the $F_{1,N-p}$ distribution if H_0 is true.

The *treatment space* is a $k-1$ dimensional subspace with axis directions $U_2, \ldots, U_k$. The projection onto this subspace is

$$(y.U_2)U_2 + \cdots + (y.U_k)U_k = \bar{y}_{i.} - \bar{y}_{..}$$

known as the treatment vector. The orthogonal directions $U_2, \ldots, U_k$ may be arbitrarily chosen, or may correspond to contrasts of interest. These contrasts are of the form $c = c_1\mu_1 + \cdots + c_k\mu_k$, and can be categorized into the following types:

- Class comparisons

- Factorial contrasts

- Polynomial, or rates, contrasts

- Pairwise comparisons

Contrasts, c, are tested using the associated directions, U_c. The hypothesis $H_0 : c_1\mu_1 + \cdots + c_k\mu_k = 0$ corresponds to the direction $U_c = [c_1, \ldots, c_2, \ldots, c_k, \ldots]^T / \sqrt{n \sum_{i=1}^{k} c_i^2}$, assuming equal replications. The corresponding test statistic is $F = (y.U_c)^2/s^2$, which follows the $F_{1,N-p}$ distribution if H_0 is true.

The *block space* is the last component of the model space. The purpose of blocking is to remove some of the variation in the experimental material, so that the residual variance, σ^2, is reduced. The block space has dimension $n-1$ for the randomized block design, and dimension $2(n-1)$ for the $n \times n$ Latin square design.

The subspace which completes the N-space is the *error space*. The error space has dimension $k(n-1)$ for the completely randomized design, dimension $(k-1)(n-1)$ for the randomized block design, and dimension $(n-1)(n-2)$ for the $n \times n$ Latin square design.

In the case of the splitplot design the treatment and error spaces are each further decomposed into mainplot and subplot spaces, reflecting two levels of variation, that between mainplots and that between subplots within each mainplot. When hypotheses are tested care must be taken to ensure the appropriate error term is used.

Regression Models

In the case of a simple, or *linear* regression the model space is of dimension two, with axis directions $U_1 = [1, \ldots, 1]^T / \sqrt{N}$ and $U_2 = (x - \bar{x})/\|x - \bar{x}\|$. The model space is the sum of these two one-dimensional subspaces, the overall mean space and the linear space. The projection onto the model space is

$$(y.U_1)U_1 + (y.U_2)U_2 = \bar{y} + b_1(x - \bar{x})$$

where $b_1 = SP_{xy}/SS_x$ is the estimated slope. The significance of the relationship between x and y is ascertained by testing the hypothesis $H_0 : \beta_1 = 0$, which corresponds to the direction U_2. The corresponding test statistic is $F = (y.U_2)^2/s^2$, which comes from the $F_{1,N-2}$ distribution if H_0 is true.

For a *quadratic regression* the model space is of dimension three. The direction of the third axis U_3 is found by orthogonalizing the set of vectors 1, $(x - \bar{x})$ and $(x - \bar{x})^2$; this involves subtracting from $X_3 = (x - \bar{x})^2$ its projection onto U_1 and U_2, yielding

$$U_3 = \frac{X_3 - (X_3.U_1)U_1 - (X_3.U_2)U_2}{\|X_3 - (X_3.U_1)U_1 - (X_3.U_2)U_2\|}$$

The model space is now the sum of the overall mean space, the linear space and the quadratic space. The projection onto the model space is

$$(y.U_1)U_1 + (y.U_2)U_2 + (y.U_3)U_3 = \bar{y} + b_1(x - \bar{x}) + b_2 p_2(x)$$

where b_2 is the estimated quadratic coefficient and $p_2(x)$ is the quadratic polynomial component. The hypothesis $H_0 : \beta_2 = 0$, where β_2 is the quadratic coefficient, can be tested using the test statistic $F = (y.U_3)^2/s^2$, which comes from the $F_{1,N-3}$ distribution if H_0 is true.

Cubic, quartic and higher order polynomial regressions can be fitted by adding higher order terms in a similar manner. In the simplest case of k equally replicated, equally spaced x-values the coefficients for the unit vectors $U_2, \ldots, U_k$ can be obtained from standard tables such as Table 10.10. This situation occurs in designed experiments in which the k treatments correspond to k rates of some commodity.

For choosing an appropriate order of polynomial with which to approximate an unknown relationship, the obvious method is to increase the order of polynomial until the next polynomial coefficient is nonsignificant. In some circumstances, however, this is a poor method; refinements are therefore advocated in §16.1 and §16.2.

ANCOVA Model

In the case of analysis of covariance, which involves the fitting of parallel lines to approximate the relationship between x and y for k treatment groups, the model space is of dimension $k + 1$. Its subspaces are the usual overall mean and treatment spaces of dimension one and $k - 1$ respectively, plus a linear space of dimension one. The latter has a single coordinate axis direction of $U_{k+1} = (x - \bar{x}_{i.})/\|x - \bar{x}_{i.}\|$. The fitted equations resulting from the projection of y onto the model space are

$$y = \bar{y}_{i.} + b_1(x - \bar{x}_{i.})$$

where b_1 is the estimated common slope. The significance of the common slope, β_1, is ascertained by testing the hypothesis $H_0 : \beta_1 = 0$ using the test statistic $(y.U_{k+1})^2/s^2$, which comes from the $F_{1,N-k-1}$ distribution if H_0 is true. Treatment contrasts between the elevations of the lines, or the adjusted y means, cannot be immediately tested using the contrast sums of squares $(y.U_2)^2, \ldots, (y.U_k)^2$, since the latter are calculated using unadjusted y means.

Comparison of Models

Figure 18.3 illustrates the essential differences between the ANOVA, regression and ANCOVA models. The ANOVA model emphasizes the comparison of mean y values between treatments. The regression model emphasizes the relationship between two variables, y and x. The ANCOVA model includes both aspects, allowing for differences between treatments as well as the relationship between x and y within each treatment.

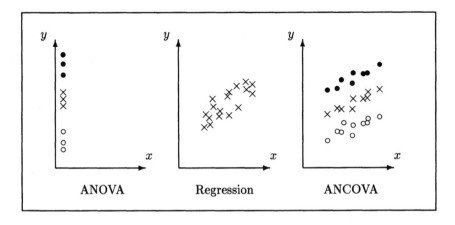

Figure 18.3: Illustration of the essential differences between the ANOVA, regression and ANCOVA models.

While it is helpful to make these basic distinctions, it should by now be evident to the reader that the distinctions are to some extent artificial. All three models involve the projection of the observation vector onto an appropriate model space, and in Minitab we use the regression command Regress to fit all three models. In many situations more than one label can be attached to the analysis. For example, an experiment with a completely randomized design and treatments comprising rates of some commodity, x, can be analyzed equally well as an ANOVA with polynomial contrasts, or as a polynomial regression. The same experiment with a randomized block design can be analyzed as an ANOVA with polynomial contrasts and block effects, or as an ANCOVA with the blocks being treated as treatment groups and covariates the polynomial components $p_i(x)$. In fact, if the two models produce different s^2 values, the questions to ask are "what have we done wrong?" or "what are the differences in the assumptions?" For example, we may have forgotten to include the block effects in the regression model, or assumed a high order polynomial in the ANOVA model but only a straight line in the regression model.

18.2 Where to from Here?

In this textbook we have based our presention on *orthogonal* coordinate systems. When the natural vectors spanning the model space have not been orthogonal, such as in polynomial regression, we have removed the problem by orthogonalizing these vectors. This solution can also be applied in cases not covered in this book. Examples are multiple regression, with say three correlated x-variables x_1, x_2, x_3, and its ANCOVA counterpart, with say three correlated covariates x_1, x_2, x_3. In these two cases the resulting uncorrelated x variables would be x_1, plus x_2 adjusted for its correlation with x_1, plus x_3 adjusted for its correlation with x_1 and x_2. If there is a sensible ordering for x_1, x_2 and x_3 this could be an ideal method. If not, the method favours x_1 over the other variables, and to be fair one would need to try all possible orders (x_1, x_2, x_3), (x_1, x_3, x_2), (x_2, x_3, x_1), (x_2, x_1, x_3), (x_3, x_1, x_2) and (x_3, x_2, x_1). For the sophisticated reader, this would amount to examining all simple and partial regression coefficients β_1, β_2, β_3, $\beta_{1.2}$, $\beta_{1.3}$, $\beta_{2.3}$, $\beta_{2.1}$, $\beta_{3.1}$, $\beta_{3.2}$, $\beta_{1.23}$, $\beta_{2.13}$ and $\beta_{3.12}$, which could be a very good idea!

In other cases, the geometric method becomes difficult to apply, as with more complicated experimental designs such as incomplete block designs. In these cases it is sensible to use *matrix* methods to project the observation vector y onto a model space spanned by a nonorthogonal coordinate system. In this text we have deliberately chosen not to discuss matrix methods since we firmly believe that the fundamental ideas are more easily understood in a vector geometric setting. However, the answer to our question "Where to from here?" is "To matrix methods." To ease the transition to such methods for the student who is advancing beyond this book, we close by presenting a simple example of the use of matrix methods.

Matrix Methods

The example we shall use is a simplified experiment laid out in a completely randomized design with two replicates of three treatments, and natural contrasts of the class comparison type. The treatments, observations and contrasts are detailed in Table 18.1.

Treatments	Yields (t/ha)		Contrasts	
Lupins (legume)	4.3	4.5	1	1
Peas (legume)	4.8	5.0	1	-1
Mustard (nonlegume)	3.6	4.2	-2	0

Table 18.1: A simplified green manure experiment. The first contrast compares legumes with nonlegumes, while the second contrast compares one legume with the other legume.

The model in vector form is

$$
\begin{bmatrix} y_{11} \\ y_{12} \\ y_{21} \\ y_{22} \\ y_{31} \\ y_{32} \end{bmatrix} = \mu_1 \begin{bmatrix} 1 \\ 1 \\ 0 \\ 0 \\ 0 \\ 0 \end{bmatrix} + \mu_2 \begin{bmatrix} 0 \\ 0 \\ 1 \\ 1 \\ 0 \\ 0 \end{bmatrix} + \mu_3 \begin{bmatrix} 0 \\ 0 \\ 0 \\ 0 \\ 1 \\ 1 \end{bmatrix} + \begin{bmatrix} e_{11} \\ e_{12} \\ e_{21} \\ e_{22} \\ e_{31} \\ e_{32} \end{bmatrix}
$$

In matrix form this is

$$
\begin{bmatrix} y_{11} \\ y_{12} \\ y_{21} \\ y_{22} \\ y_{31} \\ y_{32} \end{bmatrix} = \begin{bmatrix} 1 & 0 & 0 \\ 1 & 0 & 0 \\ 0 & 1 & 0 \\ 0 & 1 & 0 \\ 0 & 0 & 1 \\ 0 & 0 & 1 \end{bmatrix} \begin{bmatrix} \mu_1 \\ \mu_2 \\ \mu_3 \end{bmatrix} + \begin{bmatrix} e_{11} \\ e_{12} \\ e_{21} \\ e_{22} \\ e_{31} \\ e_{32} \end{bmatrix}
$$

Here the vectors spanning the model space have become the columns of the *design matrix*, which is denoted by the symbol X. The vector of parameters is usually denoted by β, and the vector of errors by e. In general the model is written in matrix form as

$$
y = X\beta + e
$$

The model space can be thought of as the set of all vectors $X\beta$, with the parameters μ_1, μ_2 and μ_3 in $\beta = [\mu_1, \mu_2, \mu_3]^T$ being allowed to assume any real values. For a given observation vector y, the least squares estimates, $\hat{\mu}_1, \hat{\mu}_2$ and $\hat{\mu}_3$ of the parameters are those values for which $X\hat{\beta}$ is at minimum distance from y; this condition is satisfied when $X\hat{\beta}$ is the projection of y onto the model space. Matrix theory then tells us that the parameter estimates are given by

$$
\hat{\beta} = (X^T X)^{-1}(X^T y)
$$

Substituting the X and y in our example gives us

$$
X^T X = \begin{bmatrix} 1 & 1 & 0 & 0 & 0 & 0 \\ 0 & 0 & 1 & 1 & 0 & 0 \\ 0 & 0 & 0 & 0 & 1 & 1 \end{bmatrix} \begin{bmatrix} 1 & 0 & 0 \\ 1 & 0 & 0 \\ 0 & 1 & 0 \\ 0 & 1 & 0 \\ 0 & 0 & 1 \\ 0 & 0 & 1 \end{bmatrix} = \begin{bmatrix} 2 & 0 & 0 \\ 0 & 2 & 0 \\ 0 & 0 & 2 \end{bmatrix}, \text{ so}
$$

$$
(X^T X)^{-1} = \begin{bmatrix} .5 & 0 & 0 \\ 0 & .5 & 0 \\ 0 & 0 & .5 \end{bmatrix}
$$

Also
$$X^T y = \begin{bmatrix} 1 & 1 & 0 & 0 & 0 & 0 \\ 0 & 0 & 1 & 1 & 0 & 0 \\ 0 & 0 & 0 & 0 & 1 & 1 \end{bmatrix} \begin{bmatrix} 4.3 \\ 4.5 \\ 4.8 \\ 5.0 \\ 3.6 \\ 4.2 \end{bmatrix} = \begin{bmatrix} 8.8 \\ 9.8 \\ 7.8 \end{bmatrix}$$

Hence
$$\hat{\beta} = \begin{bmatrix} \hat{\mu}_1 \\ \hat{\mu}_2 \\ \hat{\mu}_3 \end{bmatrix} = \begin{bmatrix} .5 & 0 & 0 \\ 0 & .5 & 0 \\ 0 & 0 & .5 \end{bmatrix} \begin{bmatrix} 8.8 \\ 9.8 \\ 7.8 \end{bmatrix} = \begin{bmatrix} 4.4 \\ 4.9 \\ 3.9 \end{bmatrix}$$

the vector of treatment means.

The projection vector $X\hat{\beta}$ gives us the vector of fitted values:

$$X\hat{\beta} = \begin{bmatrix} 1 & 0 & 0 \\ 1 & 0 & 0 \\ 0 & 1 & 0 \\ 0 & 1 & 0 \\ 0 & 0 & 1 \\ 0 & 0 & 1 \end{bmatrix} \begin{bmatrix} 4.4 \\ 4.9 \\ 3.9 \end{bmatrix} = \begin{bmatrix} 4.4 \\ 4.4 \\ 4.9 \\ 4.9 \\ 3.9 \\ 3.9 \end{bmatrix}$$

The error vector is then obtained by differencing the observed and fitted values:

$$y - X\hat{\beta} = \begin{bmatrix} 4.3 \\ 4.5 \\ 4.8 \\ 5.0 \\ 3.6 \\ 4.2 \end{bmatrix} - \begin{bmatrix} 4.4 \\ 4.4 \\ 4.9 \\ 4.9 \\ 3.9 \\ 3.9 \end{bmatrix} = \begin{bmatrix} -.1 \\ .1 \\ -.1 \\ .1 \\ -.3 \\ .3 \end{bmatrix}$$

leading to a variance estimate of

$$s^2 = \frac{\|y - X\hat{\beta}\|^2}{3} = \frac{.22}{3} = .0733$$

The corresponding orthogonal decomposition is

$$y = X\hat{\beta} + (y - X\hat{\beta})$$

with Pythagorean breakup

$$\|y\|^2 = \|X\hat{\beta}\|^2 + \|y - X\hat{\beta}\|^2$$

In our example this is $117.38 = 117.16 + .22$.

For hypothesis testing purposes an alternative form of the model is more suitable. This is the form

$$
\begin{bmatrix} y_{11} \\ y_{12} \\ y_{21} \\ y_{22} \\ y_{31} \\ y_{32} \end{bmatrix}
= \mu \begin{bmatrix} 1 \\ 1 \\ 1 \\ 1 \\ 1 \\ 1 \end{bmatrix}
+ \frac{\mu_1 + \mu_2 - 2\mu_3}{6} \begin{bmatrix} 1 \\ 1 \\ 1 \\ 1 \\ -2 \\ -2 \end{bmatrix}
+ \frac{\mu_1 - \mu_2}{2} \begin{bmatrix} 1 \\ 1 \\ -1 \\ -1 \\ 0 \\ 0 \end{bmatrix}
+ \begin{bmatrix} e_{11} \\ e_{12} \\ e_{21} \\ e_{22} \\ e_{31} \\ e_{32} \end{bmatrix}
$$

$$
= \begin{bmatrix} 1 & 1 & 1 \\ 1 & 1 & 1 \\ 1 & 1 & -1 \\ 1 & 1 & -1 \\ 1 & -2 & 0 \\ 1 & -2 & 0 \end{bmatrix}
\begin{bmatrix} \mu \\ c_1 \\ c_2 \end{bmatrix}
+ \begin{bmatrix} e_{11} \\ e_{12} \\ e_{21} \\ e_{22} \\ e_{31} \\ e_{32} \end{bmatrix}
$$

where $\mu = (\mu_1 + \mu_2 + \mu_3)/3$ is the average of the three population means, and $c_1 = (\mu_1 + \mu_2 - 2\mu_3)/6$ and $c_2 = (\mu_1 - \mu_2)/2$ are the two contrasts of interest.

Recalculating our solutions yields $\hat{\mu} = 4.4$, $\hat{c}_1 = .25$ and $\hat{c}_2 = -.25$, plus the same vectors of fitted and error values, and the same overall Pythagorean breakup. However, we can also rewrite our fitted model in matrix form as

$$
\begin{bmatrix} y_{11} \\ y_{12} \\ y_{21} \\ y_{22} \\ y_{31} \\ y_{32} \end{bmatrix}
= \begin{bmatrix} 1 \\ 1 \\ 1 \\ 1 \\ 1 \\ 1 \end{bmatrix} [\hat{\mu}] +
\begin{bmatrix} 1 \\ 1 \\ 1 \\ 1 \\ -2 \\ -2 \end{bmatrix} [\hat{c}_1] +
\begin{bmatrix} 1 \\ 1 \\ -1 \\ -1 \\ 0 \\ 0 \end{bmatrix} [\hat{c}_2] +
\begin{bmatrix} -.1 \\ .1 \\ -.1 \\ .1 \\ -.3 \\ .3 \end{bmatrix}
$$

and obtain the more detailed Pythagorean breakup

$$117.38 \quad = \quad 116.16 + .75 + .25 + .22$$

enabling the hypotheses $H_0 : c_1 = 0$ and $H_0 : c_2 = 0$ to be tested in the usual way.

An Aside

We note that the Regress command of Minitab uses matrix methods for fitting the model. This explains why we have been able to use axes which are not of length one, or nonorthogonal coordinate axes, when it has been more convenient for our purposes.

18.3 Summary

In this chapter we have reviewed this book's use of vector geometric methods for the analysis of data using the ANOVA, regression and ANCOVA techniques. We have also supplied a brief introduction to matrix methods for advancing students.

Appendix A

Unequal Replications: Two Populations

In the main body of the text we have usually assumed that our samples are of the same size, or in the language of the subject, that the treatments are equally replicated. This is a desirable design feature. However, it will sometimes be the case that the samples differ in size, or the treatments are *unequally replicated*. For example, this occurs when nature or human fallibility intervenes in an equally replicated experiment, and certain results are totally lost.

In this appendix we deal with the simplest unequal replications case, involving just two populations. In §1 we analyze an illustrative example, then in §2 we summarize the general case. We conclude with some general exercises.

A.1   Illustrative Example

Suppose that in the bulkometer experiment of Chapter 6, subsamples 1−5 were mistakenly washed with an extra powerful detergent, "X45", and had to be discarded. The remaining data is retabulated in Table A.1. By a freak of the randomization, our layout has resulted in treatment one having subsamples 6−7 and treatment two having subsamples 8−10. The resulting observation vector is $y = [29.53, \cdots, 30.80]^T$.

Subsample nos	Subsample means
Treatment 1 (two day conditioning)	
6	29.53
7	29.90
Treatment 2 (three day conditioning)	
8	30.37
9	29.10
10	30.80

Table A.1: The remaining bulkometer readings.

Model

The model is

$$
\begin{bmatrix} 29.53 \\ 29.90 \\ 30.37 \\ 29.10 \\ 30.80 \end{bmatrix} = \mu_1 \begin{bmatrix} 1 \\ 1 \\ 0 \\ 0 \\ 0 \end{bmatrix} + \mu_2 \begin{bmatrix} 0 \\ 0 \\ 1 \\ 1 \\ 1 \end{bmatrix} + \begin{bmatrix} e_{11} \\ e_{12} \\ e_{21} \\ e_{22} \\ e_{23} \end{bmatrix}
$$

where the $e_{ij} = y_{ij} - \mu_i$ are independent values from an $N[0, \sigma^2]$ distribution.

The model space M is a 2-dimensional subspace of 5-space, with natural coordinate axes $U_1 = [1, 1, 0, 0, 0]^T / \sqrt{2}$ and $U_2 = [0, 0, 1, 1, 1]^T / \sqrt{3}$.

Hypothesis

As in the equal replications case, we wish to test the null hypothesis $H_0 : \mu_1 = \mu_2$ against the alternative hypothesis $H_1 : \mu_1 \neq \mu_2$. To find the direction associated with the hypothesis $H_0 : \mu_1 - \mu_2 = 0$, we look for a unit vector U for which the projection coefficient $y.U$ has an expected value of $k(\mu_1 - \mu_2)$. For this we require $y.U = k(\bar{y}_{1.} - \bar{y}_{2.})$.

The only suitable direction is given by the unit vector

$$
U = \frac{1}{\sqrt{\frac{1}{2} + \frac{1}{3}}} \begin{bmatrix} 1/2 \\ 1/2 \\ -1/3 \\ -1/3 \\ -1/3 \end{bmatrix}
$$

To check that U is correct, we calculate the projection coefficient

$$
y.U = \begin{bmatrix} y_{11} \\ y_{12} \\ y_{21} \\ y_{22} \\ y_{23} \end{bmatrix} \cdot \frac{1}{\sqrt{\frac{1}{2} + \frac{1}{3}}} \begin{bmatrix} 1/2 \\ 1/2 \\ -1/3 \\ -1/3 \\ -1/3 \end{bmatrix} = \frac{\bar{y}_{1.} - \bar{y}_{2.}}{\sqrt{\frac{1}{2} + \frac{1}{3}}}
$$

which is of the required form.

This suggests an alternative coordinate system for the model space:

$$U_1 = \frac{1}{\sqrt{5}}\begin{bmatrix} 1 \\ 1 \\ 1 \\ 1 \\ 1 \end{bmatrix} \quad \text{and} \quad U_2 = \frac{1}{\sqrt{\frac{1}{2}+\frac{1}{3}}}\begin{bmatrix} 1/2 \\ 1/2 \\ -1/3 \\ -1/3 \\ -1/3 \end{bmatrix} = \frac{1}{\sqrt{30}}\begin{bmatrix} 3 \\ 3 \\ -2 \\ -2 \\ -2 \end{bmatrix}$$

where we have also written U_2 in an alternative form with integer components. These two axis directions correspond to the null hypotheses $H_0 : \mu = (2\mu_1 + 3\mu_2)/5 = 0$ and $H_0 : \mu_1 - \mu_2 = 0$.

We now have all the raw materials required for our analysis. We have an observation vector $y = [29.53, \cdots, 30.80]^T$, a model space $M = \text{span}\{U_1, U_2\}$, and a direction U_2 associated with the hypothesis $H_0 : \mu_1 = \mu_2$.

Fitting the Model

As usual, we fit the model by projecting y onto the model space, $M = \text{span}\{U_1, U_2\}$. Using the original coordinate system for M, this yields the fitted model vector

$$(y.U_1)U_1 + (y.U_2)U_2 = \begin{bmatrix} 29.715 \\ 29.715 \\ 0 \\ 0 \\ 0 \end{bmatrix} + \begin{bmatrix} 0 \\ 0 \\ 30.090 \\ 30.090 \\ 30.090 \end{bmatrix} = \begin{bmatrix} 29.715 \\ 29.715 \\ 30.090 \\ 30.090 \\ 30.090 \end{bmatrix} = \bar{y}_i.$$

as in the equal replications case. Thus $\bar{y}_1. = 29.715$ is the least squares estimate of μ_1, and $\bar{y}_2. = 30.090$ is the least squares estimate of μ_2 .

Refitting the model using the alternative coordinate system yields the alternative fitted model vector

$$(y.U_1)U_1 + (y.U_2)U_2 = \begin{bmatrix} 29.940 \\ 29.940 \\ 29.940 \\ 29.940 \\ 29.940 \end{bmatrix} + \begin{bmatrix} -.225 \\ -.225 \\ .150 \\ .150 \\ .150 \end{bmatrix} = \bar{y}.. + (\bar{y}_i. - \bar{y}..)$$

as in the equal replications case. Here $\bar{y}.. = 29.940$ is the simple average of the observations, and can be rewritten as $(2\bar{y}_1. + 3\bar{y}_2.)/5$, a weighted average of the treatment means.

The resulting fitted model is

$$
\begin{bmatrix} 29.53 \\ 29.90 \\ 30.37 \\ 29.10 \\ 30.80 \end{bmatrix}
=
\begin{bmatrix} 29.940 \\ 29.940 \\ 29.940 \\ 29.940 \\ 29.940 \end{bmatrix}
+
\begin{bmatrix} -.225 \\ -.225 \\ .150 \\ .150 \\ .150 \end{bmatrix}
+
\begin{bmatrix} -.185 \\ .185 \\ .280 \\ -.990 \\ .710 \end{bmatrix}
$$

Observation vector	Overall mean vector	Treatment vector	Fitted error vector
y	$\bar{y}_{..}$	$(\bar{y}_{i\cdot} - \bar{y}_{..})$	$(y - \bar{y}_{i\cdot})$

$$
y \quad = \quad \bar{y}_{..} \quad + \quad (\bar{y}_{i\cdot} - \bar{y}_{..}) \quad + \quad (y - \bar{y}_{i\cdot})
$$

as in the equal replications case.

Testing the Hypothesis

The hypothesis $H_0 : \mu_1 = \mu_2$ can now be tested by calculating the test statistic

$$
F = \frac{(y.U_2)^2}{\left[(y.U_3)^2 + (y.U_4)^2 + (y.U_5)^2\right]/3} = \frac{\|\bar{y}_{i\cdot} - \bar{y}_{..}\|^2}{\|y - \bar{y}_{i\cdot}\|^2/3}
$$

where U_3, U_4 and U_5 are coordinate axes for the error space, such as

$$
U_3 = \frac{1}{\sqrt{2}}\begin{bmatrix} 1 \\ -1 \\ 0 \\ 0 \\ 0 \end{bmatrix}, \quad
U_4 = \frac{1}{\sqrt{2}}\begin{bmatrix} 0 \\ 0 \\ 1 \\ -1 \\ 0 \end{bmatrix} \quad \text{and } U_5 = \frac{1}{\sqrt{6}}\begin{bmatrix} 0 \\ 0 \\ 1 \\ 1 \\ -2 \end{bmatrix}
$$

The appropriate Pythagorean breakup is

$$
\|y\|^2 = \|\bar{y}_{..}\|^2 + \|\bar{y}_{i\cdot} - \bar{y}_{..}\|^2 + \|y - \bar{y}_{i\cdot}\|^2
$$

$$
\text{or,} \quad 4483.8178 = 4482.0180 + .1688 + 1.6310
$$

Our test statistic is therefore

$$
F = \frac{0.1688}{1.6310/3} = \frac{0.1688}{0.5437} = 0.31
$$

which is not significant when compared with the percentiles of the $F_{1,3}$ distribution.

ANOVA Table

The simplified form of the orthogonal decomposition is

$$
y - \bar{y}_{..} = (\bar{y}_{i\cdot} - \bar{y}_{..}) + (y - \bar{y}_{i\cdot})
$$

$$
\text{or,} \quad
\begin{bmatrix} -.41 \\ -.04 \\ .43 \\ -.84 \\ .86 \end{bmatrix}
=
\begin{bmatrix} -.225 \\ -.225 \\ .150 \\ .150 \\ .150 \end{bmatrix}
+
\begin{bmatrix} -.185 \\ .185 \\ .280 \\ -.990 \\ .710 \end{bmatrix}
$$

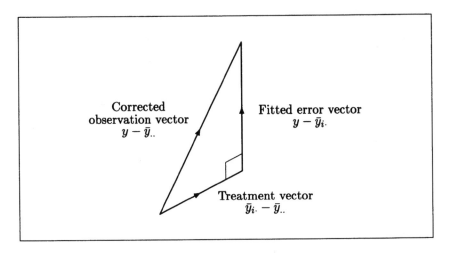

Figure A.1: Simplified decomposition, unequal replications example.

as illustrated in Figure A.1.

The corresponding Pythagorean breakup is

$$\|y - \bar{y}_{..}\|^2 = \|\bar{y}_{i.} - \bar{y}_{..}\|^2 + \|y - \bar{y}_{i.}\|^2$$

$$\text{or,} \quad 1.7998 = 0.1688 + 1.6310$$

This is summarized in the traditional ANOVA table, given in Table A.2.

Source of Variation	df	SS	MS	F
Treatments ($H_0 : \mu_1 = \mu_2$)	1	0.1688	.1688	0.31 (ns)
Error	3	1.6310	.5437	
Total (corrected)	4	1.7998		

Table A.2: ANOVA table for our unequal replications example.

Estimation of σ^2

In our example we have transformed our set of independent random variables, $Y_{11}, Y_{12} \sim N[\mu_1, \sigma^2]$ and $Y_{21}, Y_{22}, Y_{23} \sim N[\mu_2, \sigma^2]$, into a new set of independent random variables

$$Y.U_1 \sim N\left[\sqrt{5}\mu, \sigma^2\right], \qquad Y.U_2 \sim N\left[(\mu_1 - \mu_2)/\sqrt{\tfrac{1}{2} + \tfrac{1}{3}}, \sigma^2\right],$$

$$Y.U_3, \ldots, Y.U_5 \sim N\left[0, \sigma^2\right]$$

The first and second of these were used to estimate $\mu = (2\mu_1 + 3\mu_2)/5$ and $\mu_1 - \mu_2$ respectively, while the last three were used to estimate σ^2 via

$$s^2 = \frac{(y.U_3)^2 + (y.U_4)^2 + (y.U_5)^2}{3} = \frac{\|y - \bar{y}_{i\cdot}\|^2}{3} = \frac{1.6311}{3} = .5437$$

Confidence Interval for $\mu_1 - \mu_2$

The best estimate for the mean difference in bulkometer reading between samples conditioned for two and three days, $\mu_1 - \mu_2$, is given by $\bar{y}_{1\cdot} - \bar{y}_{2\cdot} = 29.715 - 30.090 = -0.375$. To derive the 95% confidence interval for the difference $\mu_1 - \mu_2$, we examine the distribution of the associated random variable $Y.U_2$.

Now the random variable $Y.U_2 = (\overline{Y}_{1\cdot} - \overline{Y}_{2\cdot})/\sqrt{\frac{1}{2} + \frac{1}{3}}$ has a $N\left[(\mu_1 - \mu_2)/\sqrt{\frac{1}{2} + \frac{1}{3}}, \sigma^2\right]$ distribution. It follows that the random variable

$$\frac{\left[(\overline{Y}_{1\cdot} - \overline{Y}_{2\cdot}) - (\mu_1 - \mu_2)\right]/\sqrt{\frac{1}{2} + \frac{1}{3}}}{\sqrt{\left[(Y.U_3)^2 + (Y.U_4)^2 + (Y.U_5)^2\right]/3}}$$

is a t_3-statistic. To obtain a 95% confidence interval for the difference between the two means we gamble that the realized value

$$t = \frac{\left[(\bar{y}_{1\cdot} - \bar{y}_{2\cdot}) - (\mu_1 - \mu_2)\right]/\sqrt{\frac{1}{2} + \frac{1}{3}}}{\sqrt{\left[(y.U_3)^2 + (y.U_4)^2 + (y.U_5)^2\right]/3}} = \frac{(\bar{y}_{1\cdot} - \bar{y}_{2\cdot}) - (\mu_1 - \mu_2)}{\sqrt{s^2(\frac{1}{2} + \frac{1}{3})}}$$

lies between the 2.5 and 97.5 percentiles of the t_3 distribution. This leads to a 95% confidence interval for $\mu_1 - \mu_2$ of

$$(\bar{y}_{1\cdot} - \bar{y}_{2\cdot}) - 3.182\sqrt{s^2(\tfrac{1}{2} + \tfrac{1}{3})} \le \mu_1 - \mu_2 \le (\bar{y}_{1\cdot} - \bar{y}_{2\cdot}) + 3.182\sqrt{s^2(\tfrac{1}{2} + \tfrac{1}{3})}$$

or, $\qquad 0.38 \pm 2.14$

Minitab Analysis

The commands for an analysis using Minitab are virtually identical to those listed in Tables 6.7 and 6.9. In Table 6.7 the unit vector is now $U_2 = [3, 3, -2, -2, -2]^T/\sqrt{30}$, and in Table 6.9 the treatment numbers are now 1, 1, 2, 2, 2.

Report on Study

Our results can be summarized in the same manner as in the equal replications case, as shown in Table A.3. Here the SED is calculated as $\sqrt{s^2(\frac{1}{2} + \frac{1}{3})} = \sqrt{.5437(\frac{1}{2} + \frac{1}{3})} = 0.67$.

Mean bulkometer reading, in cm^3/g	
Treatment	
Conditioned for 2 days	29.7
Conditioned for 3 days	30.1
SED	0.67
Significance of difference	ns

Table A.3: Presentation of results for unequal replications example.

A.2   General Case

In this section we summarize the general case, then comment upon the optimal allocation of experimental subjects to two treatment groups.

Summary

In the general two population case, we assume the sample sizes to be unequal, with n_1 being the sample size for population one, and n_2 the sample size for population two. The raw materials for the analysis are:

(1) The observation vector is $y = [y_{11}, \cdots, y_{1n_1}, y_{21}, \cdots, y_{2n_2}]^T$.

(2) Model space coordinate axes, U_1 and U_2, are

$$\frac{1}{\sqrt{n_1 + n_2}} \begin{bmatrix} 1 \\ \vdots \\ 1 \\ \vdots \end{bmatrix} \quad \text{and} \quad \frac{1}{\sqrt{\frac{1}{n_1} + \frac{1}{n_2}}} \begin{bmatrix} 1/n_1 \\ \vdots \\ -1/n_2 \\ \vdots \end{bmatrix} = \frac{1}{\sqrt{n_1 n_2(n_1 + n_2)}} \begin{bmatrix} n_2 \\ \vdots \\ -n_1 \\ \vdots \end{bmatrix}$$

The model space is of dimension two, while the error space is of dimension $(n_1 + n_2 - 2)$.

(3) The direction associated with the hypothesis $H_0 : \mu_1 = \mu_2$ is the coordinate axis U_2.

The fitted model is $y = \bar{y}_{i\cdot} + (y - \bar{y}_{i\cdot})$, as in the equal replications case. This can be rewritten in corrected form as

$$y - \bar{y}_{\cdot\cdot} = (\bar{y}_{i\cdot} - \bar{y}_{\cdot\cdot}) + (y - \bar{y}_{i\cdot})$$

as shown in Figure A.2.

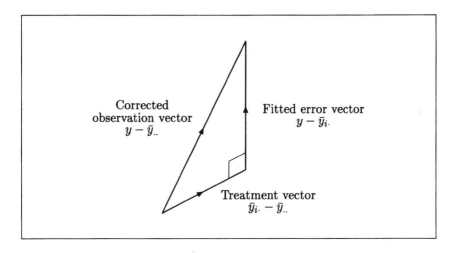

Figure A.2: Orthogonal decomposition for unequal replications case.

The corresponding Pythagorean breakup is given by

$$\|y - \bar{y}_{\cdot\cdot}\|^2 = \|\bar{y}_{i\cdot} - \bar{y}_{\cdot\cdot}\|^2 + \|y - \bar{y}_{i\cdot}\|^2$$

This can also be written algebraically as

$$\sum_{i=1}^{2}\sum_{j=1}^{n_i}(y_{ij} - \bar{y}_{\cdot\cdot})^2 = \sum_{i=1}^{2} n_i(\bar{y}_{i\cdot} - \bar{y}_{\cdot\cdot})^2 + \sum_{i=1}^{2}\sum_{j=1}^{n_i}(y_{ij} - \bar{y}_{i\cdot})^2$$

The hypothesis $H_0 : \mu_1 = \mu_2$ is tested by calculating the test statistic

$$F = \frac{(y.U_2)^2}{s^2} = \frac{\|\bar{y}_{i\cdot} - \bar{y}_{\cdot\cdot}\|^2}{\|y - \bar{y}_{i\cdot}\|^2/(n_1 + n_2 - 2)}$$

This statistic follows the F_{1,n_1+n_2-2} distribution if H_0 is true.

The 95% confidence interval for the difference $\mu_1 - \mu_2$ is

$$(\bar{y}_{1\cdot} - \bar{y}_{2\cdot}) \pm s\sqrt{\frac{1}{n_1} + \frac{1}{n_2}}\, t_{n_1+n_2-2}(.975)$$

Optimal Design

We now digress to consider an interesting design question: if the total number of available plots (N) is fixed, how should we choose the sample sizes n_1 and n_2 in order to make the confidence interval for $\mu_1 - \mu_2$ as narrow as possible?

The answer is that we must make $\sqrt{\frac{1}{n_1} + \frac{1}{n_2}}$, and so $\frac{1}{n_1} + \frac{1}{n_2}$, as small as possible. But

$$\frac{1}{n_1} + \frac{1}{n_2} = \frac{n_1 + n_2}{n_1 n_2}$$

and since $n_1 + n_2$ is fixed we must make the product $n_1 n_2$ as large as possible. Now $n_1 n_2 = n_1(N - n_1)$ whose plot is shown in Figure A.3 for $n_1 = 0, \ldots, N$. It is greatest when n_1 is as close as possible to the middle of the interval from 0 to N. For example, if $N = 6$ then $n_1 = n_2 = 3$ is best, whereas if $N = 7$ then $n_1 = 3, n_2 = 4$ or $n_1 = 4, n_2 = 3$ would be equally good.

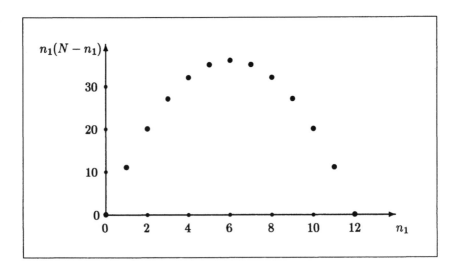

Figure A.3: Plot of $n_1(N - n_1)$ against n_1, for $N = 12$.

To summarize, the best design is to allocate half of the plots to one treatment and half to the second treatment. This conclusion is of course dependent on the assumption that the two populations have a common variance σ^2.

Exercises

(A.1) Suppose that in exercise 6.1, two of the seedling plots were drowned out. The remaining data were:

1. Control	80, 68 cm
2. Oxadiazon	62, 56, 68, 66 cm

Consider the following vectors:

y	U_1	U_2	U_3	U_4	U_5	U_6
80	1	-2	1	0	0	0
68	1	-2	-1	0	0	0
62	1	1	0	1	1	1
56	1	1	0	-1	1	1
68	1	1	0	0	-2	1
66	1	1	0	0	0	-3
	$\sqrt{6}$	$\sqrt{12}$	$\sqrt{2}$	$\sqrt{2}$	$\sqrt{6}$	$\sqrt{12}$

(a) Calculate the squared length of the projection of y onto each of these directions.

(b) Assume the data are independently drawn from normal populations with true means μ_1 and μ_2 and a common variance σ^2. Compute an F test of the hypothesis $H_0 : \mu_1 = \mu_2$ versus $H_1 : \mu_1 \neq \mu_2$ by dividing $(y.U_2)^2$ by the average of $(y.U_3)^2$, $(y.U_4)^2$ and $(y.U_6)^2$.

(c) Write out the vector decomposition

$$y = \bar{y}_{..} + (\bar{y}_{i.} - \bar{y}_{..}) + (y - \bar{y}_{i.})$$

and recalculate F. Is there enough evidence to reject your hypothesis?

(d) Use Minitab to obtain the F value by regressing y on U_2.

(e) Present your results in a table using the format given in Table A.3.

(f) Calculate the 95% confidence interval for the true difference $(\mu_1 - \mu_2)$.

(A.2) For a completely randomized design experiment having two treatments with 3 and 5 plots for treatments 1 and 2 respectively, write down the coordinate axes U_1 and U_2 corresponding to tests of the hypotheses $H_0 : \mu = 0$ and $H_0 : \mu_1 = \mu_2$.

(A.3) Suppose that in Exercise 6.3 a herd of hungry cows accidentally strayed into the field with the six experimental plots. Before the farmer could reverse this disaster they had devoured the bulk of one plot. The remaining data is as follows:

1. Control   9.6   7.4
2. Phosphate   11.3   10.1   12.2

(a) Recalculate the orthogonal decomposition $y - \bar{y}_{..} = (\bar{y}_{i.} - \bar{y}_{..}) + (y - \bar{y}_{i.})$

(b) Recalculate $\|y - \bar{y}_{..}\|^2 = \|\bar{y}_{i.} - \bar{y}_{..}\|^2 + \|y - \bar{y}_{i.}\|^2$ and the F ratio, and rewrite the ANOVA table.

(c) Test the hypothesis $H_0 : \mu_1 = \mu_2$. What are your conclusions?

(d) Use Minitab to check your calculations.

(A.4) The heights and sexes of students in a statistics course were recorded as follows:

Sex (M or F):	F	M	F	M	F	M	F	F	M
Height (cm):	140	165	168	184	170	177	156	157	170
Sex:	M	M	M	M	M	M	F	M	
Height:	190	178	187	183	178	175	156	191	

(a) Draw two histograms of the heights, one for females and one for males. Use the same class intervals for both histograms, and position one histogram immediately above the other for ease of comparison.

(b) Draw up an observation vector, with the female heights first and the male heights second.

To enable us to test the hypothesis "females and males are on average equal in height", we shall pretend that these two samples of female and male heights are random samples from the populations of all adult female heights and all adult male heights. In practice this assumption would require serious examination!! We shall also assume that these two populations are normally distributed with means μ_1 and μ_2 and a common variance σ^2.

(c) Write down the orthogonal decomposition in the form

$$y = \bar{y}_{..} + (\bar{y}_{i.} - \bar{y}_{..}) + (y - \bar{y}_{i.})$$

and the Pythagorean breakup in the form

$$\|y - \bar{y}_{..}\|^2 = \|\bar{y}_{i.} - \bar{y}_{..}\|^2 + \|y - \bar{y}_{i.}\|^2$$

(d) Write out the corresponding ANOVA table, and test the hypothesis $H_0 : \mu_1 = \mu_2$. What are your conclusions?

(e) Check your calculations using Minitab (rounding errors may mean your hand calculated values do not agree precisely with those given by Minitab).

(f) Calculate the 95% confidence interval for $(\mu_2 - \mu_1)$, the average difference in height between males and females.

Appendix B

Unequal Replications: Several Populations

In this appendix we describe how to cope with unequal replications in the general k populations case. In §1 we discuss the implications for class comparisons, in §2 we deal with factorial contrasts, and in §3 we cover the remaining cases. Each section concludes with a summary of the general case; the appendix as a whole concludes with a summary in §4, and exercises.

B.1 Class Comparisons

In this section we assume all contrasts are of the class comparison type. We first present an illustrative example, then summarize the general case.

The Study

An experiment was carried out to determine the relative merits of peas, lupins and mustard as green manure crops. The first two species, peas and lupins, are legumes, while mustard is a nonlegume. The experimenter was especially interested in whether the two legumes produced more or less dry matter than the nonlegume. She was also interested in whether one legume produced more or less than the other legume.

The experiment was laid down in a completely randomized design, with four replicates. Unfortunately, five of the experimental plots were flooded, and data were obtained from only seven out of the twelve plots.

Data

The dry matter yields, expressed in tonnes per hectare, from the seven plots
are given in Table B.1.   The resulting observation vector is
$y = [3.0, 3.6, 3.4, 3.8, 3.3, 4.4, 4.0]^T$.

Treatment	Yields (t/ha of dry matter)		
Peas	3.0	3.6	
Lupins	3.4	3.8	3.3
Mustard	4.4	4.0	

Table B.1: Dry matter yield for each plot.

Model

The model is

$$
\begin{bmatrix} 3.0 \\ 3.6 \\ 3.4 \\ 3.8 \\ 3.3 \\ 4.4 \\ 4.0 \end{bmatrix} = \mu_1 \begin{bmatrix} 1 \\ 1 \\ 0 \\ 0 \\ 0 \\ 0 \\ 0 \end{bmatrix} + \mu_2 \begin{bmatrix} 0 \\ 0 \\ 1 \\ 1 \\ 1 \\ 0 \\ 0 \end{bmatrix} + \mu_3 \begin{bmatrix} 0 \\ 0 \\ 0 \\ 0 \\ 0 \\ 1 \\ 1 \end{bmatrix} + \begin{bmatrix} e_{11} \\ e_{12} \\ e_{21} \\ e_{22} \\ e_{23} \\ e_{31} \\ e_{32} \end{bmatrix}
$$

where the $e_{ij} = y_{ij} - \mu_i$ are assumed to be independent values from an
$N[0, \sigma^2]$ distribution.   The model space M is a three dimensional subspace of 7-space, with natural coordinate axes $U_1 = [1, 1, 0, 0, 0, 0, 0]^T/\sqrt{2}$,
$U_2 = [0, 0, 1, 1, 1, 0, 0]^T/\sqrt{3}$ and $U_3 = [0, 0, 0, 0, 0, 1, 1]^T/\sqrt{2}$.

Test Hypotheses

The questions of interest are:

1. Do the legumes differ from the nonlegume?

2. Do the lupins differ from the peas?

In the unequally replicated case, there are two approaches to hypothesis
formulation. With the first approach, the hypotheses are expressed in terms
of comparisons of simple averages of the population means, and have a simple practical interpretation. However in general the hypothesis tests are not
statistically independent since the associated directions are not orthogonal.
With the second approach, suitable orthogonal directions are chosen to correspond to interesting comparisons of simple averages of the observations.

The corresponding contrasts are then derived. In general, these are comparisons of weighted averages of the population means, with the weights depending on the numbers of replicates. This makes interpretation more difficult. However, with this approach the hypothesis tests are statistically independent. To illustrate the two approaches, we now set up our hypotheses in these alternative ways.

First Approach

With the first approach, we set up the two test hypotheses as, firstly, a comparison of the simple average of the leguminous species means with that of the nonleguminous species means, and, secondly, a comparison of one legume with the other legume. This leads to the test hypotheses:

1. $H_0 : \frac{\mu_1 + \mu_2}{2} = \mu_3$ versus $H_1 : \frac{\mu_1 + \mu_2}{2} \neq \mu_3$

2. $H_0 : \mu_1 = \mu_2$ versus $H_1 : \mu_1 \neq \mu_2$

The corresponding contrasts are $c_1 = \mu_1 + \mu_2 - 2\mu_3$ and $c_2 = \mu_1 - \mu_2$, with associated directions

$$U_2 = \frac{1}{\sqrt{\frac{1}{2} + \frac{1}{3} + 2}} \begin{bmatrix} 1/2 \\ 1/2 \\ 1/3 \\ 1/3 \\ 1/3 \\ -2/2 \\ -2/2 \end{bmatrix} \quad \text{and} \quad U_3 = \frac{1}{\sqrt{\frac{1}{2} + \frac{1}{3}}} \begin{bmatrix} 1/2 \\ 1/2 \\ -1/3 \\ -1/3 \\ -1/3 \\ 0 \\ 0 \end{bmatrix}$$

Here $y.U_2 = (\bar{y}_{1\cdot} + \bar{y}_{2\cdot} - 2\bar{y}_{3\cdot})/\sqrt{\frac{1}{2} + \frac{1}{3} + 2}$ and $y.U_3 = (\bar{y}_{1\cdot} - \bar{y}_{2\cdot})/\sqrt{\frac{1}{2} + \frac{1}{3}}$, with expected values of the form $k(\mu_1 + \mu_2 - 2\mu_3)$ and $k(\mu_1 - \mu_2)$, as required. Note that U_2 and U_3 are not orthogonal, although they span the treatment space.

Second Approach

With the second approach, we write down the orthogonal directions which correspond to simple comparisons of, firstly, the legume plots with the non-legume plots, and, secondly, the plots of one legume with those of the other legume. These directions are

$$U_2 = \frac{1}{\sqrt{\frac{1}{5} + \frac{1}{2}}} \begin{bmatrix} 1/5 \\ 1/5 \\ 1/5 \\ 1/5 \\ 1/5 \\ -1/2 \\ -1/2 \end{bmatrix} \quad \text{and} \quad U_3 = \frac{1}{\sqrt{\frac{1}{2} + \frac{1}{3}}} \begin{bmatrix} 1/2 \\ 1/2 \\ -1/3 \\ -1/3 \\ -1/3 \\ 0 \\ 0 \end{bmatrix}$$

Here $y.U_2 = (2\bar{y}_{1.} + 3\bar{y}_{2.} - 5\bar{y}_{3.})/\sqrt{\frac{1}{5} + \frac{1}{2}}$ and $y.U_3 = (\bar{y}_{1.} - \bar{y}_{2.})/\sqrt{\frac{1}{2} + \frac{1}{3}}$, with expected values of the form $k(2\mu_1 + 3\mu_2 - 5\mu_3)$ and $k(\mu_1 - \mu_2)$. Hence the contrasts associated with U_2 and U_3 are $c_1 = 2\mu_1 + 3\mu_2 - 5\mu_3$ and $c_2 = \mu_1 - \mu_2$. Note that U_2 and U_3 are orthogonal, so these contrasts form an orthogonal set. The resulting test hypotheses are

1. $H_0 : \dfrac{2\mu_1 + 3\mu_2}{5} = \mu_3$ versus $H_1 : \dfrac{2\mu_1 + 3\mu_2}{5} \neq \mu_3$

2. $H_0 : \mu_1 = \mu_2$ versus $H_1 : \mu_1 \neq \mu_2$

These hypotheses are statistically independent.

A Vital Formula

In writing down the above directions we used the following formula:
The contrast $c = c_1\mu_1 + c_2\mu_2 + \cdots + c_k\mu_k$ is associated with the direction

$$
U_c = \frac{1}{\sqrt{\dfrac{c_1^2}{n_1} + \dfrac{c_2^2}{n_2} + \cdots + \dfrac{c_k^2}{n_k}}}
\begin{bmatrix}
c_1/n_1 \\
\vdots \\
c_2/n_2 \\
\vdots \\
c_k/n_k \\
\vdots
\end{bmatrix}
$$

where $n_1, n_2, \cdots, n_k$ are the sample sizes. This is a more general formula than the one given in §7.3. As in that formula, the contrast coefficients, c_i, must sum to zero.

Fitting the Model

The model is fitted by projecting the observation vector y onto the natural coordinate axes $U_1 = [1, 1, 0, 0, 0, 0, 0]^T/\sqrt{2}$, $U_2 = [0, 0, 1, 1, 1, 0, 0]^T/\sqrt{3}$ and $U_3 = [0, 0, 0, 0, 0, 1, 1]^T/\sqrt{2}$. This yields the usual vector of treatment means, $\bar{y}_{i.}$. The resulting fitted model, in simplified form, is as usual

$$
y - \bar{y}_{..} = (\bar{y}_{i.} - \bar{y}_{..}) + (y - \bar{y}_{i.})
$$

$$
\begin{bmatrix}
-.64 \\
-.04 \\
-.24 \\
.16 \\
-.34 \\
.76 \\
.36
\end{bmatrix}
=
\begin{bmatrix}
-.34 \\
-.34 \\
-.14 \\
-.14 \\
-.14 \\
.56 \\
.56
\end{bmatrix}
+
\begin{bmatrix}
-.30 \\
.30 \\
-.10 \\
.30 \\
-.20 \\
.20 \\
-.20
\end{bmatrix}
$$

The corresponding Pythagorean breakup is

$$
\|y - \bar{y}_{..}\|^2 = \|\bar{y}_{i.} - \bar{y}_{..}\|^2 + \|y - \bar{y}_{i.}\|^2
$$

$$
1.317 = .917 + .400
$$

Testing the Hypotheses

With both approaches to hypothesis formulation, we test each hypothesis using the test statistic $F = (y.U_c)^2/s^2$, where s^2 is the pooled variance estimate.

The ANOVA table for the first approach is shown in Table B.2. With this approach the dependence between the hypothesis tests means that the contrast sums of squares do not add up to the treatment sum of squares. However, the hypothesis tests are easy to interpret.

Source of Variation	df	SS	MS	F
Treatments	2	.917		
$H_0 : \mu_1 + \mu_2 - 2\mu_3 = 0$	1	.904	.904	9.04 (*)
$H_0 : \mu_1 - \mu_2 = 0$	1	.048	.048	0.48 (ns)
Error	4	.400	.100	
Total	6	1.317		

Table B.2: ANOVA table for meaningful, but dependent hypothesis tests.

The ANOVA table for the second approach is shown in Table B.3. Here the contrast sums of squares do add up to the treatment sum of squares. However, the first contrast is quite artificial, and not something one would wish to explain to a group of farmers at a field day!

Source of Variation	df	SS	MS	F
Treatments	2	.917		
$H_0 : 2\mu_1 + 3\mu_2 - 5\mu_3 = 0$	1	.869	.869	8.69 (*)
$H_0 : \mu_1 - \mu_2 = 0$	1	.048	.048	0.48 (ns)
Error	4	.400	.100	
Total	6	1.317		

Table B.3: ANOVA table for less meaningful, but independent hypothesis tests.

Minitab Analysis

With the first approach, it is necessary to use the Oneway command of Minitab to fit the full model and obtain s^2, then use two separate Regress commands to obtain $(y.U_2)^2$ and $(y.U_3)^2$. With the second approach, it suffices to Regress y onto the orthogonal unit vectors U_2 and U_3.

Comparison of Approaches

In our example the two approaches yield quite similar results. This is because the dependence of the tests in the first approach is not too severe, with the angle between U_2 and U_3 being 84^0 instead of 90^0.

In general, statisticians are divided as to which approach is superior. We prefer the first approach because of the easier interpretation of the results.

Summary

In the unequal replications, k populations case the fitted model vector is $\bar{y}_{i\cdot}$, as in the equal replications case. This leads to the familiar simplified decomposition

$$y - \bar{y}_{\cdot\cdot} = (\bar{y}_{i\cdot} - \bar{y}_{\cdot\cdot}) + (y - \bar{y}_{i\cdot})$$

as illustrated in Figure B.1.

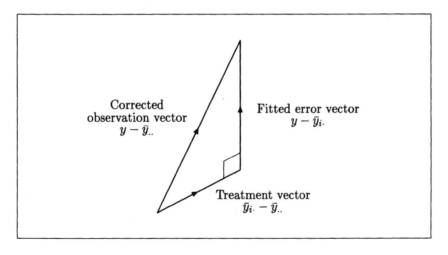

Figure B.1: Orthogonal decomposition of corrected observation vector in unequally replicated, k populations case.

The associated Pythagorean breakup is

$$\|y - \bar{y}_{\cdot\cdot}\|^2 = \|\bar{y}_{i\cdot} - \bar{y}_{\cdot\cdot}\|^2 + \|y - \bar{y}_{i\cdot}\|^2$$

which can also be written algebraically as

$$\sum_{i=1}^{k}\sum_{j=1}^{n_i}(y_{ij} - \bar{y}_{\cdot\cdot})^2 = \sum_{i=1}^{k}n_i(\bar{y}_{i\cdot} - \bar{y}_{\cdot\cdot})^2 + \sum_{i=1}^{k}\sum_{j=1}^{n_i}(y_{ij} - \bar{y}_{i\cdot})^2$$

For the easiest type of contrast, those of the *class comparison* type, there are two approaches to hypothesis testing, as illustrated in the preceding example. The first approach is to specify meaningful contrasts, $c = c_1\mu_1 + \cdots + c_k\mu_k$, between the population means without paying too much attention to the orthogonality of the corresponding unit vectors

$$U_c = \frac{1}{\sqrt{\frac{c_1^2}{n_1} + \cdots + \frac{c_k^2}{n_k}}} \begin{bmatrix} c_1/n_1 \\ \vdots \\ c_k/n_k \\ \vdots \end{bmatrix}$$

Dependent hypotheses of the form $H_0 : c = 0$ are then tested using test statistics of the form $F = (y.U_c)^2/s^2$, where s^2 is obtained by fitting the full model as described above. Here F comes from an $F_{1,N-k}$ distribution if H_0 is true, where $N = n_1 + \cdots + n_k$, assuming a completely randomized layout.

The second approach is to specify less meaningful contrasts which are derived from an orthogonal set of unit vectors U_c. These unit vectors are suggested by the natural classes and subclasses of treatments. For each comparison of one class with another class, the unit vector U_c is written down by keeping the components equal within each of the two classes, and ensuring that the components add to zero. Thus in our example we arrived at the unit vectors

$$U_2 = \frac{1}{\sqrt{\frac{1}{5} + \frac{1}{2}}} \begin{bmatrix} 1/5 \\ 1/5 \\ 1/5 \\ 1/5 \\ 1/5 \\ -1/2 \\ -1/2 \end{bmatrix} \quad \text{and} \quad U_3 = \frac{1}{\sqrt{\frac{1}{2} + \frac{1}{3}}} \begin{bmatrix} 1/2 \\ 1/2 \\ -1/3 \\ -1/3 \\ -1/3 \\ 0 \\ 0 \end{bmatrix}$$

corresponding to comparisons of the class of legumes with the class of non-legumes, and of the class of peas with the class of lupins. The corresponding contrasts, $2\mu_1 + 3\mu_2 - 5\mu_3$ and $\mu_1 - \mu_2$, were derived from these unit vectors using the general formula given earlier in this section. Independent hypotheses of the form $H_0 : c = 0$ are again tested using test statistics of the form $F = (y.U_c)^2/s^2$.

The 95% confidence interval for a contrast $c = c_1\mu_1 + \cdots + c_k\mu_k$ is

$$(c_1\bar{y}_{1.} + \cdots + c_k\bar{y}_{k.}) \pm t_{N-k}(.975)\sqrt{s^2\sum_{i=1}^{k}(c_i^2/n_i)}$$

where $N = n_1 + \cdots + n_k$, assuming a completely randomized layout.

B.2   Factorial Contrasts

In the last section we dealt with contrasts of the class comparison type. There we had a choice between a set of nonorthogonal but easy to interpret contrasts, and a set of orthogonal but less natural contrasts. For contrasts of the *factorial* type the orthogonal set is even less natural, for reasons which we shall not delve into here. In this section we shall therefore discuss only the first approach to hypothesis formulation. This means we use the same contrasts as in the equal replications case, without paying too much attention to their lack of orthogonality. The essential features can be adequately illustrated by looking at a 2 x 2 factorial design, so we now consider a sample data set for such a design.

Study

On a farm with a history of below average lamb growth rates, it was decided to investigate two factors which could be implicated. The factors were disease and mineral deficiency. An experiment was set up on the farm by running 20 lambs separately from the main flock. Five of these lambs, chosen at random, were left untreated as "control" animals, five lambs were injected with antibiotics to treat the suspected disease, five lambs were injected with a mineral mix to correct any mineral deficiency, and five lambs were injected both with antibiotics and with the mineral mix. In short, the experiment had four treatments with a 2 x 2 factorial structure, factor A being antibiotics and factor B being mineral mix.

Owing to a misunderstanding, six lambs from the experimental flock were sent to the meatworks for slaughter when the experiment was only half completed. Fortunately, the lambs sent to the meatworks appeared to have been chosen completely at random, rather than being the heaviest lambs, so that a statistical analysis of the remaining data values was still valid.

Data

The data are weightgains of individual animals over the 3-month period, as in Table B.4. The resulting observation vector is $y = [11.5, \cdots, 14.0]^T$.

Antibiotics (A)	Mineral mix (B)				
	None			Some	
None	11.5	12.0	5.0	7.5	10.5
	6.5	8.5			
Some	10.5	12.5	13.0	15.5	10.5
				12.0	14.0

Table B.4: Weightgains, in kg, for the three month period of the experiment.

Model

The model is

$$
\begin{bmatrix} 11.5 \\ 12.0 \\ 5.0 \\ 6.5 \\ 8.5 \\ 7.5 \\ 10.5 \\ 10.5 \\ 12.5 \\ 13.0 \\ 15.5 \\ 10.5 \\ 12.0 \\ 14.0 \end{bmatrix}
= \mu_1
\begin{bmatrix} 1 \\ 1 \\ 1 \\ 1 \\ 1 \\ 0 \\ 0 \\ 0 \\ 0 \\ 0 \\ 0 \\ 0 \\ 0 \\ 0 \end{bmatrix}
+ \mu_2
\begin{bmatrix} 0 \\ 0 \\ 0 \\ 0 \\ 0 \\ 1 \\ 1 \\ 0 \\ 0 \\ 0 \\ 0 \\ 0 \\ 0 \\ 0 \end{bmatrix}
+ \mu_3
\begin{bmatrix} 0 \\ 0 \\ 0 \\ 0 \\ 0 \\ 0 \\ 0 \\ 1 \\ 1 \\ 1 \\ 0 \\ 0 \\ 0 \\ 0 \end{bmatrix}
+ \mu_4
\begin{bmatrix} 0 \\ 0 \\ 0 \\ 0 \\ 0 \\ 0 \\ 0 \\ 0 \\ 0 \\ 0 \\ 1 \\ 1 \\ 1 \\ 1 \end{bmatrix}
+
\begin{bmatrix} e_{11} \\ e_{12} \\ e_{13} \\ e_{14} \\ e_{15} \\ e_{21} \\ e_{22} \\ e_{31} \\ e_{32} \\ e_{33} \\ e_{41} \\ e_{42} \\ e_{43} \\ e_{44} \end{bmatrix}
$$

where the $e_{ij} = y_{ij} - \mu_i$ are assumed to be independent values from an $N[0, \sigma^2]$ distribution. The model space M is a four dimensional subspace of 14-space, with natural coordinate axes U_1, U_2, U_3 and U_4, where for example $U_1 = [1, 1, 1, 1, 1, 0, 0, 0, 0, 0, 0, 0, 0, 0]^T/\sqrt{5}$.

Test Hypotheses

The questions of interest are:

1. Did the lambs injected with antibiotics gain more or less weight than those not injected with antibiotics?

2. Did the lambs injected with mineral mix gain more or less weight than those not injected with mineral mix?

3. Did the response to antibiotics differ between the lambs injected with mineral mix and those not injected with mineral mix?

Using just the first approach to hypothesis formulation, we set up the corresponding null hypotheses as follows:

$$
\begin{aligned}
H_0 &: c_1 = -\mu_1 - \mu_2 + \mu_3 + \mu_4 = 0 && \text{(Antibiotic effect} = 0) \\
H_0 &: c_2 = -\mu_1 + \mu_2 - \mu_3 + \mu_4 = 0 && \text{(Mineral effect} = 0) \\
H_0 &: c_3 = \mu_1 - \mu_2 - \mu_3 + \mu_4 = 0 && \text{(Interaction} = 0)
\end{aligned}
$$

The directions associated with the three contrasts c_1, c_2 and c_3 are

$$U_2 = \frac{1}{k}\begin{bmatrix} -1/5 \\ \vdots \\ -1/2 \\ \vdots \\ 1/3 \\ \vdots \\ 1/4 \\ \vdots \end{bmatrix}, \quad U_3 = \frac{1}{k}\begin{bmatrix} -1/5 \\ \vdots \\ 1/2 \\ \vdots \\ -1/3 \\ \vdots \\ 1/4 \\ \vdots \end{bmatrix} \quad \text{and} \quad U_4 = \frac{1}{k}\begin{bmatrix} 1/5 \\ \vdots \\ -1/2 \\ \vdots \\ -1/3 \\ \vdots \\ 1/4 \\ \vdots \end{bmatrix}$$

where $k = \sqrt{\frac{1}{5} + \frac{1}{2} + \frac{1}{3} + \frac{1}{4}}$.

Here, for example, $y.U_2 = (-\bar{y}_{1.} - \bar{y}_{2.} + \bar{y}_{3.} + \bar{y}_{4.})/\sqrt{\frac{1}{5} + \frac{1}{2} + \frac{1}{3} + \frac{1}{4}}$, which has an expected value of the form $k(-\mu_1 - \mu_2 + \mu_3 + \mu_4)$, as required. Note that U_2, U_3 and U_4 are not mutually orthogonal, although they do span the treatment space.

Fitting the Model

The model is fitted by projecting y onto the natural coordinate axes $U_1 = [1,1,1,1,1,0,0,0,0,0,0,0,0,0]^T/\sqrt{5}$ and so on. The fitted model vector is again $\bar{y}_{i.}$, and the simplified orthogonal decomposition is again

$$y - \bar{y}_{..} = (\bar{y}_{i.} - \bar{y}_{..}) + (y - \bar{y}_{i.})$$

The corresponding Pythagorean breakup is

$$\|y - \bar{y}_{..}\|^2 = \|\bar{y}_{i.} - \bar{y}_{..}\|^2 + \|y - \bar{y}_{i.}\|^2$$
$$111.80 = 52.00 + 59.80$$

The resulting estimate of the variance σ^2 is

$$s^2 = \|y - \bar{y}_{i.}\|^2/10 = 59.8/10 = 5.98$$

Testing the Hypotheses

Each hypothesis is tested using the test statistic $F = (y.U_c)^2/s^2$. Results are summarized in the usual ANOVA table, shown in Table B.5.

Since our hypothesis tests are dependent the contrast sums of squares do not add up to the treatment sum of squares. The degree of nonorthogonality is reflected by the angles between the unit vectors: U_2 and U_3 are at an angle of 107^0 to one another, U_2 and U_4 are at 80^0, and U_3 and U_4 are at 95^0.

Source of Variation	df	SS	MS	F
Treatments	3	52.00		
Factor A	1	41.53	41.53	6.94 (*)
Factor B	1	1.32	1.32	0.22 (ns)
Interaction	1	0.38	0.38	0.06 (ns)
Error	10	59.80	5.98	
Total	13	111.80		

Table B.5: ANOVA table for meaningful, but dependent hypothesis tests in a 2×2 factorial experiment.

Minitab Analysis

For analysis using Minitab, we firstly fit the full model using the Oneway command, and obtain the variance estimate s^2. We then use three separate Regress commands to obtain the three contrast sums of squares. This enables construction of the above ANOVA table.

Report on Study

Antibiotics have caused a 5% significant increase in lamb growth rate, so disease was presumably involved in the poor lamb growth rates observed on the farm. The experimental results are as summarized in Table B.6. Note that the SED varies with each comparison; however an approximate SED, based on the average replication of 3.5, is $\sqrt{2s^2/3.5} = 1.85$.

Treatment	(No. reps)	Weightgain (kg)
Control	(5)	8.7
Antibiotics	(3)	12.0
Mineral mix	(2)	9.0
Anti. + min.	(4)	13.0
Approx. SED		1.85
Contrasts		
Antibiotics		*
Minerals		ns
Interaction		ns

Table B.6: Report on study for an unequally replicated, 2×2 factorial experiment.

Summary

As in the previous section, the fitted model vector is $\bar{y}_{i\cdot}$, and the simplified decomposition is

$$y - \bar{y}_{\cdot\cdot} = (\bar{y}_{i\cdot} - \bar{y}_{\cdot\cdot}) + (y - \bar{y}_{i\cdot})$$

The common variance, σ^2, is as usual estimated by $s^2 = \|y - \bar{y}_{i\cdot}\|^2/(N-k)$, where $N = n_1 + \cdots + n_k$.

For *factorial* contrasts, the approach suggested in this text is to specify meaningful contrasts $c = c_1\mu_1 + \cdots + c_k\mu_k$ in an identical fashion to that described in Chapter 9. In general these will not be mutually orthogonal. Corresponding unit vectors will be of the form

$$U_c = \frac{1}{\sqrt{\frac{c_1^2}{n_1} + \cdots + \frac{c_k^2}{n_k}}} \begin{bmatrix} c_1/n_1 \\ \vdots \\ c_k/n_k \\ \vdots \end{bmatrix}$$

Dependent hypotheses of the form $H_0 : c = 0$ can then be tested using test statistics of the form $F = (y \cdot U_c)^2/s^2$.

The confidence interval formula is given in the last section.

B.3   Other Cases

We now briefly discuss how inequality of replication affects other cases covered in this textbook.

Polynomial Contrasts

In the case of polynomial contrasts, it is the orthogonal unit vectors, such as $U_2 = (x - \bar{x})/\|x - \bar{x}\|$, which are the fundamental quantities, not the contrasts themselves. The Gram-Schmidt method of orthogonalization, used in Chapter 10 to derive these unit vectors, works equally well in the unequal replications case. If the orthogonal contrasts $c = c_1\mu_1 + \cdots + c_k\mu_k$ are required, they can be obtained from the unit vectors using the formula given in the last two sections. For example, in the case of a boron fertilizer experiment with four 25 kg B/ha plots, one 75 kg B/ha plot, and one 125 kg B/ha plot, the linear unit vector is

$$U_c = \frac{1}{\sqrt{8750}} \begin{bmatrix} 25 - 50 \\ 25 - 50 \\ 25 - 50 \\ 25 - 50 \\ 75 - 50 \\ 125 - 50 \end{bmatrix} = \frac{1}{\sqrt{14}} \begin{bmatrix} -1 \\ -1 \\ -1 \\ -1 \\ 1 \\ 3 \end{bmatrix} = \frac{1}{\sqrt{14}} \begin{bmatrix} -4/4 \\ -4/4 \\ -4/4 \\ -4/4 \\ 1/1 \\ 3/1 \end{bmatrix}$$

so the linear contrast is $c = -4\mu_1 + \mu_2 + 3\mu_3$.

Pairwise Comparisons

In the case of pairwise contrasts, it is only the degree of dependence which is altered by inequality of replication. An example has already been given on the last page of §11.1.

Block Designs

The block designs discussed in Chapters 12, 13 and 14 are dependent upon equality of replication. However, it is allowable for a particular treatment, or treatments, to be replicated several times within each block.

Regression and Covariance

The regression and ANCOVA methods of Chapters 15, 16 and 17 are not dependent on equality of replication.

B.4   Summary

For a completely randomized experiment with unequal replications, the fitted model is $y = \bar{y}_{i\cdot} + (y - \bar{y}_{i\cdot})$, as in the equal replications case. The resulting pooled variance estimate is $s^2 = \|y - \bar{y}_{i\cdot}\|^2/(N-k)$, where $N = n_1 + \cdots + n_k$ is the total sample size.

Class comparisons, factorial contrasts and pairwise comparisons can be written down in an identical manner as for the equal replications case. The unit vector corresponding to a contrast $c = c_1\mu_1 + \cdots + c_k\mu_k$ is

$$U_c = \frac{1}{\sqrt{\frac{c_1^2}{n_1} + \cdots + \frac{c_k^2}{n_k}}} \begin{bmatrix} c_1/n_1 \\ \vdots \\ c_k/n_k \\ \vdots \end{bmatrix}$$

These unit vectors will not in general be orthogonal. Hypotheses of the form $H_0 : c = 0$ can still be tested using the test statistic $F = (y.U_c)^2/s^2$. However, care must be taken with computing; for example, in Minitab separate Regress commands must be used to calculate the $(y.U_c)^2$ terms.

Polynomial contrasts are derived from the corresponding orthogonal unit vectors, using the formula given in the last paragraph, so remain orthogonal in the unequal replications case. These contrasts differ from those obtained in the equal replications case.

For block design experiments, particular treatments may be replicated several times within each block. No other inequality of replication is allowed.

Regression and ANCOVA analyses are unaffected by inequality of replication.

Exercises

(B.1) An experiment was set up to investigate the effect of applications of lime on the productivity of alfalfa. Experimental treatments and numbers of replications were as follows:

1. Lime at 3 tonnes/ha	4 replicates
2. Lime at 4 tonnes/ha	1 replicate
3. Lime at 8 tonnes/ha	3 replicates

The layout followed a completely randomized design.

(a) Work out the linear component unit vector, U_2.

(b) Write down the linear contrast in terms of μ_1, μ_2 and μ_3.

(B.2) An experiment was carried out to determine how well blackberry was controlled by stocking with animals of various types. For the study, ten fields with similar initial blackberry cover were allocated to ten mobs of animals. The data given in the table is the reduction in % blackberry cover for each field, derived from aerial photography assessments carried out before and after one year of treatment.

Treatment	Reduction in % blackberry cover		
Romney sheep	44	40	45
Merino sheep	42	50	46
Angora goats	51	55	
Feral goats	58	52	

The questions of interest in this experiment are:

1. Do sheep differ from goats in their effectiveness as blackberry control agents?

2. Do Romney sheep differ from Merino sheep?

3. Do Angora goats differ from feral goats?

(a) Using the second approach to hypothesis testing as described in §1 of this Appendix, write down the contrasts corresponding to the above three questions.

(b) Write down the corresponding unit vectors U_2, U_3 and U_4.

(c) Are the contrasts mutually orthogonal?

(d) Calculate $(y.U_2)^2$, $(y.U_3)^2$ and $(y.U_4)^2$; also $s^2 = \|y - \bar{y}_{i.}\|^2/(N - k)$.

(e) Set up the corresponding ANOVA table, and test whether the three contrasts are zero. Describe your conclusions.

(f) Check your results using Minitab.

(g) Calculate the 95% confidence interval for the contrast of sheep with goats.

(h) Suppose that there had been three replicate mobs of Romney sheep and Angora goats, and two replicate mobs of Merino sheep and feral goats. Answer (b) and (c) above now.

(B.3) An experiment was conducted last summer to assess the effect of the weedkiller oxadiazon on the early development of Golden Queen peach seedlings. Six control seedlings received no dose, six a half dose, five a single dose and three a triple dose. A single dose was equivalent to .75 kg of oxadiazon per hectare. The four treatments were completely randomized amongst the twenty seedlings. The resulting heights of the seedlings are shown in the table.

Treatment	Height of seedlings, in cm					
Control	79	76	57	105	81	71
Half dose	71	34	35	78	79	59
Single dose	63	60	61	68	44	
Triple dose	11	23	16			

(a) Find $\bar{y}_{..}$, $\bar{y}_{1.}$, $\bar{y}_{2.}$, $\bar{y}_{3.}$ and $\bar{y}_{4.}$. Hence express the observation vector y as a sum of the three orthogonal vectors $\bar{y}_{..}$, $(\bar{y}_{i.} - \bar{y}_{..})$ and $(y - \bar{y}_{i.})$.

(b) Check that $\bar{y}_{..}$ and $(\bar{y}_{i.} - \bar{y}_{..})$ are orthogonal, to the accuracy your figures allow. If time allows, check that $(\bar{y}_{i.} - \bar{y}_{..})$ and $(y - \bar{y}_{i.})$, and $\bar{y}_{..}$ and $(y - \bar{y}_{i.})$, are also orthogonal pairs.

(c) Sketch the relationship of y, $\bar{y}_{..}$, $(\bar{y}_{i.} - \bar{y}_{..})$ and $(y - \bar{y}_{i.})$. Indicate the dimensions of the spaces in which these vectors lie.

(d) Summarize the overall Pythagorean breakup in an ANOVA table.

(e) Write down the contrast between the control and the weedkiller treatments. Calculate the corresponding unit vector U_2, and calculate $(y.U_2)^2$. Hence test the hypothesis that weedkiller had no effect on the seedlings.

(f) Calculate the unit vector, U_3, corresponding to the linear trend in the nonzero rates of weedkiller, 0.5, 1 and 3. Use this to calculate $(y.U_3)^2$, and test the hypothesis of no trend.

Appendix C

Alternative Factorial Notation

Here we introduce the traditional notation for factorial experiments. In the text we decided not to use this more specialized notation since it could cause confusion. We include this appendix, however, to make the link between our textbook and other textbooks, and to clarify the theory for more advanced readers.

We shall introduce the new notation by redisplaying Table 9.21 using the new labels, shown in Table C.1. The 3×4 population means, namely the μ_{ij}, are now labelled by the levels of factors A and B. For example, μ_{23} is the mean for the treatment with the 2nd level of factor A, irrigation at 25% a.s.m., and the 3rd level of factor B, mustard. The observations are similarly labelled y_{ijk}, but with an additional subscript to identify the replication number. If there are three replicates, for example, the observations for this same treatment are labelled y_{231}, y_{232} and y_{233}.

		Factor B (crop)				Factor A
		Legumes		Nonlegumes		means
		Lupins	Peas	Mustard	Barley	
Factor	Dry	μ_{11}	μ_{12}	μ_{13}	μ_{14}	$\mu_{1.}$
A	25% asm	μ_{21}	μ_{22}	μ_{23}	μ_{24}	$\mu_{2.}$
(moisture)	50% asm	μ_{31}	μ_{32}	μ_{33}	μ_{34}	$\mu_{3.}$
Factor B means		$\mu_{.1}$	$\mu_{.2}$	$\mu_{.3}$	$\mu_{.4}$	

Table C.1: Traditional labels for population means and main effect means.

The main advantage of this alternative notation is that we now have convenient labels for the main effect means. For example, $\mu_{.1}$ indicates the average over all populations with a second subscript of one. That is, $\mu_{.1} = (\mu_{11} + \mu_{21} + \mu_{31})/3$, the average of the lupin population means.

Also the corresponding estimated main effect mean has the label $\bar{y}_{.1.}$, the average of the nine observations with a second subscript of one. That is, $\bar{y}_{.1.}$ is the average of the observations from all lupin plots, of which there are nine, corresponding to the three levels of irrigation by the three replications.

Using this notation we can rewrite the main effect contrasts, $c_1, c_2, \ldots, c_5$, from Table C.1 succinctly. These forms of the contrasts and their estimates, $\hat{c}_1, \hat{c}_2, \ldots, \hat{c}_5$, are given in Table C.2. Unfortunately the interaction contrasts, $c_6, \ldots, c_{11}$, are not shortened by our new notation.

Contrasts	**Estimates**
Factor A	
$c_1 = -2\mu_{1.} + \mu_{2.} + \mu_{3.}$	$\hat{c}_1 = -2\bar{y}_{1..} + \bar{y}_{2..} + \bar{y}_{3..}$
$c_2 = \phantom{-2\mu_{1.}} - \mu_{2.} + \mu_{3.}$	$\hat{c}_2 = \phantom{-2\bar{y}_{1..}} - \bar{y}_{2..} + \bar{y}_{3..}$
Factor B	
$c_3 = -\mu_{.1} - \mu_{.2} + \mu_{.3} + \mu_{.4}$	$\hat{c}_3 = -\bar{y}_{.1.} - \bar{y}_{.2.} + \bar{y}_{.3.} + \bar{y}_{.4.}$
$c_4 = -\mu_{.1} + \mu_{.2}$	$\hat{c}_4 = -\bar{y}_{.1.} + \bar{y}_{.2.}$
$c_5 = \phantom{-\mu_{.1} + \mu_{.2}} - \mu_{.3} + \mu_{.4}$	$\hat{c}_5 = \phantom{-\bar{y}_{.1.} + \bar{y}_{.2.}} - \bar{y}_{.3.} + \bar{y}_{.4.}$

Table C.2: Shortened expressions for the main effect contrasts which were given in Table 9.22.

Using the new notation our orthogonal decomposition can be written as

$$\underset{\text{Corrected observation vector}}{y - \bar{y}_{...}} = \underset{\text{Treatment vector}}{(\bar{y}_{ij.} - \bar{y}_{...})} + \underset{\text{Error vector}}{(y - \bar{y}_{ij.})}$$

where i and j vary from 1 to 3 and 1 to 4 respectively. Here the treatment vector lies in the 11 dimensional treatment space, so can be rewritten as

$$\bar{y}_{ij.} - \bar{y}_{...} = (y.U_2)U_2 + \cdots + (y.U_{12})U_{12}$$

where the unit vectors $U_2, \ldots, U_{12}$ correspond to the A and B main effect contrasts and the $A \times B$ interaction contrasts.

The following equalities can be easily proven:

1. The sum of the A main effect projections is
$$(y.U_2)U_2 + (y.U_3)U_3 = \bar{y}_{i..} - \bar{y}_{...}$$

2. The sum of the B main effect projections is
$$(y.U_4)U_4 + (y.U_5)U_5 + (y.U_6)U_6 = \bar{y}_{.j.} - \bar{y}_{...}$$

3. The sum of the $A \times B$ interaction projections is
$$(y.U_7)U_7 + \cdots + (y.U_{12})U_{12} = \bar{y}_{ij.} - \bar{y}_{i..} - \bar{y}_{.j.} + \bar{y}_{...}$$

Hence the treatment vector can be decomposed into an A main effect vector, a B main effect vector, and an $A \times B$ interaction vector as follows:

$$\underset{\substack{\text{Treatment} \\ \text{vector}}}{\bar{y}_{ij\cdot} - \bar{y}_{\cdots}} = \underset{\substack{A \text{ and } B \text{ main effect} \\ \text{vectors}}}{(\bar{y}_{i\cdot\cdot} - \bar{y}_{\cdots}) + (\bar{y}_{\cdot j\cdot} - \bar{y}_{\cdots})} + \underset{\substack{A \times B \text{ interaction} \\ \text{vector}}}{(\bar{y}_{ij\cdot} - \bar{y}_{i\cdot\cdot} - \bar{y}_{\cdot j\cdot} + \bar{y}_{\cdots})}$$

When substituted into the orthogonal decomposition given above, we arrive at the decomposition given in more traditional statistics texts:

$$y - \bar{y}_{\cdots} = (\bar{y}_{i\cdot\cdot} - \bar{y}_{\cdots}) + (\bar{y}_{\cdot j\cdot} - \bar{y}_{\cdots}) + (\bar{y}_{ij\cdot} - \bar{y}_{i\cdot\cdot} - \bar{y}_{\cdot j\cdot} + \bar{y}_{\cdots}) + (y - \bar{y}_{ij\cdot})$$

This is illustrated in Figure C.1.

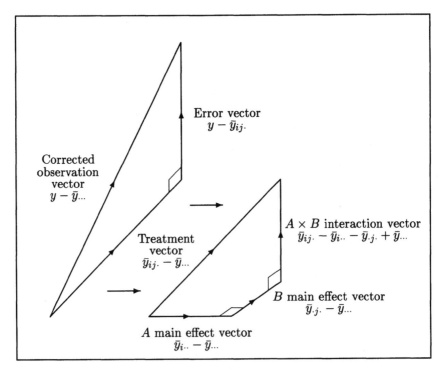

Figure C.1: The fitted model expressed in an alternative notation.

The corresponding Pythagorean breakup is then

$$\|y - \bar{y}_{\cdots}\|^2 = \|\bar{y}_{i\cdot\cdot} - \bar{y}_{\cdots}\|^2 + \|\bar{y}_{\cdot j\cdot} - \bar{y}_{\cdots}\|^2 + \|\bar{y}_{ij\cdot} - \bar{y}_{i\cdot\cdot} - \bar{y}_{\cdot j\cdot} + \bar{y}_{\cdots}\|^2 + \|y - \bar{y}_{ij\cdot}\|^2$$

These are the sums of squares produced by the Twoway command of Minitab.

By the equalities given above it follows that:

1. $\|\bar{y}_{i\cdot\cdot} - \bar{y}_{\cdots}\|^2 = (y.U_2)^2 + (y.U_3)^2$

2. $\|\bar{y}_{\cdot j\cdot} - \bar{y}_{\cdots}\|^2 = (y.U_4)^2 + (y.U_5)^2 + (y.U_6)^2$

3. $\|\bar{y}_{ij\cdot} - \bar{y}_{i\cdot\cdot} - \bar{y}_{\cdot j\cdot} + \bar{y}_{\cdots}\|^2 = (y.U_7)^2 + \cdots + (y.U_{12})^2$

These sums of squares are occasionally used to test overall hypotheses such as "there are no differences in yield induced by varying moisture levels", which can be more formally written as $H_0 : \mu_{1\cdot} = \mu_{2\cdot} = \mu_{3\cdot}$. The test statistic here is

$$F = \frac{\|\bar{y}_{i\cdot\cdot} - \bar{y}_{\cdots}\|^2/2}{s^2}$$

which comes from the $F_{2,24}$ distribution if H_0 is true. This test suffers from a dilution effect caused by pooling tests of large and small contrasts. In practice the researcher usually has more specific ideas in mind, so the overall test is of little interest. For these reasons, we pay scant attention to tests of such hypotheses.

Exercise for the reader

(C.1) A completely randomized design experiment was carried out to determine the effects of phosphate and sulphur fertilizer on the yield of alfalfa. The four treatments consisted of nil or some phosphate factorially combined with nil or some sulphur, in a 2×2 factorial structure.

Four replicates were laid down, so the trial had sixteen plots. Treatment means are given in Table C.3.

	No Sulphur	Sulphur
No Phosphate	6.7	6.9
Phosphate	6.7	10.5

Table C.3: Yield of alfalfa, in tonnes/ha, for individual treatments.

(a) Calculate the phosphate main effect means $\bar{y}_{1\cdot\cdot}$, $\bar{y}_{2\cdot\cdot}$, the sulphur main effect means $\bar{y}_{\cdot 1\cdot}$, $\bar{y}_{\cdot 2\cdot}$ and the overall mean $\bar{y}_{\cdots}$.

(b) Calculate the average response to phosphate, $\bar{y}_{2\cdot\cdot} - \bar{y}_{1\cdot\cdot}$, the average response to sulphur, $\bar{y}_{\cdot 2\cdot} - \bar{y}_{\cdot 1\cdot}$, and the interaction, $\bar{y}_{11\cdot} - \bar{y}_{21\cdot} - \bar{y}_{12\cdot} + \bar{y}_{22\cdot}$.

(c) Write down the sixteen dimensional vectors $\bar{y}_{i\cdot\cdot} - \bar{y}_{\cdots}$, $\bar{y}_{\cdot j\cdot} - \bar{y}_{\cdots}$ and $\bar{y}_{ij\cdot} - \bar{y}_{i\cdot\cdot} - \bar{y}_{\cdot j\cdot} + \bar{y}_{\cdots}$.

(d) Calculate the squared lengths of these vectors.

(e) Given $s^2 = \|y - y_{ij\cdot}\|^2/12 = 1.21$, calculate the F tests of the hypotheses
  (i) $H_0 : \mu_{1\cdot} = \mu_{2\cdot}$.   (average phosphate response $= 0$)
  (ii) $H_0 : \mu_{\cdot 1} = \mu_{\cdot 2}$ (average sulphur response $= 0$)
  (iii) $H_0 : \mu_{22} - \mu_{12} = \mu_{21} - \mu_{11}$ (interaction $= 0$)

(f) How would you present your results?

Solution to the Reader Exercise

(C.1) (a) $\bar{y}_{1\cdot\cdot} = (6.7 + 6.9)/2 = 6.8$, $\bar{y}_{2\cdot\cdot} = 8.6$, $\bar{y}_{\cdot 1\cdot} = 6.7$, $\bar{y}_{\cdot 2\cdot} = 8.7$,
$\bar{y}_{\cdots} = 7.7$.

(b) Average response to P is $8.6 - 6.8 = 1.8$ while the average response to S
is $8.7 - 6.7 = 2.0$. The interaction is $6.7 - 6.7 - 6.9 + 10.5 = 3.6$.

(c) The vectors are

$$
\begin{bmatrix} -.9 \\ \cdot \\ -.9 \\ \cdot \\ .9 \\ \cdot \\ .9 \\ \cdot \end{bmatrix}
\quad
\begin{bmatrix} -1.0 \\ \cdot \\ 1.0 \\ \cdot \\ -1.0 \\ \cdot \\ 1.0 \\ \cdot \end{bmatrix}
\quad
\begin{bmatrix} .9 \\ \cdot \\ -.9 \\ \cdot \\ -.9 \\ \cdot \\ .9 \\ \cdot \end{bmatrix}
$$
$$(\bar{y}_{i\cdot\cdot} - \bar{y}_{\cdots}) \quad (\bar{y}_{\cdot j\cdot} - \bar{y}_{\cdots}) \quad (\bar{y}_{ij\cdot} - \bar{y}_{i\cdot\cdot} - \bar{y}_{\cdot j\cdot} + \bar{y}_{\cdots})$$

(d) $\|\bar{y}_{i\cdot\cdot} - \bar{y}_{\cdots}\|^2 = 16 \times .9^2 = 12.96$, $\|\bar{y}_{\cdot j\cdot} - \bar{y}_{\cdots}\|^2 = 16 \times 1^2 = 16$
$\|\bar{y}_{ij\cdot} - \bar{y}_{i\cdot\cdot} - \bar{y}_{\cdot j\cdot} + \bar{y}_{\cdots}\|^2 = 16 \times .9^2 = 12.96$

(e) (i) $F_{1,12} = 12.96/1.21 = 10.71(**)$
  (ii) $F_{1,12} = 16/1.21 = 13.22(**)$
  (iii) $F_{1,12} = 12.96/1.21 = 10.71(**)$
We therefore reject all three hypotheses at the 1% level of significance.

(f) Since the two factors do not operate independently it makes little sense to
talk of the average response to P or to S. We therefore present our results as
shown in Table C.3, with the addition of the standard error of the difference
between any two means, given by SED $= \sqrt{2s^2/n} = \sqrt{2(1.21)/4} = 0.78$.

Appendix D

Regression Through the Origin

In this appendix we shall outline the method for fitting the model $y = \beta x$, a straight line through the origin. We do this to illustrate the power of the geometric method to handle variants of the standard model. In practice, this model is seldom used in biological work. This may seem odd at first, since the underlying curves often pass through the origin; for example, we know that a zero seeding rate of barley means a zero grain yield, as illustrated in Figure D.1.

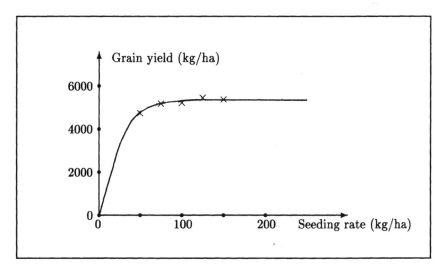

Figure D.1: Relationship between seeding rate and grain yield for malting barley, Example D. In the region of interest, from 50 to 150 kg/ha, the underlying curve can be approximated by a straight line or a quadratic.

The key to the conundrum is that the underlying curve is seldom linear over the full range of x values. For this reason it is ridiculous to fit a straight line through the origin for the dataset of Figure D.1. Instead, we use the straight line $y = \beta_0 + \beta_1(x - \bar{x})$ as a local approximation in the region of interest.

Method of Analysis

We now give a quick summary of the general method.

1. The regression through the origin model is

$$y = \beta x$$

Here the x values are treated as fixed values without error, and the corresponding y values are treated as realized values of independent normal random variables with mean βx, and a common variance σ^2.

2. In vector form the model is

$$\begin{bmatrix} Y_1 \\ \vdots \\ Y_N \end{bmatrix} = \beta \begin{bmatrix} x_1 \\ \vdots \\ x_N \end{bmatrix} + \text{error vector}$$

Hence the model space is of dimension one.

3. The single coordinate axis direction for the model space is $U_1 = [x_1, \ldots, x_N]^T / \|x\|$.

4. The fitted model is

$$\begin{aligned} y &= (y.U_1)U_1 &+& \text{ error vector} \\ &= bx &+& \text{ error vector} \end{aligned}$$

as illustrated in Figure D.2.

5. Here $b = \dfrac{y.U_1}{\|x\|} = \dfrac{\sum_{i=1}^{N} x_i y_i}{\sum_{i=1}^{N} x_i^2}$ is the least squares estimate of β.

6. The variance estimate is $s^2 = \dfrac{\|\text{error vector}\|^2}{N-1}$.

7. The hypothesis $H_0 : \beta = 0$ is tested against $H_0 : \beta \neq 0$ using the test statistic $(y.U_1)^2 / s^2$. This comes from the $F_{1,N-1}$ distribution if H_0 is true.

8. The usual ANOVA table is based on the Pythagorean breakup

$$\|y\|^2 = b^2 \|x\|^2 + \|\text{error vector}\|^2$$

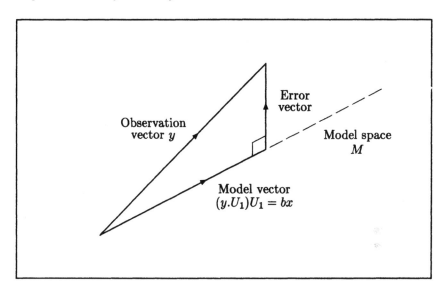

Figure D.2: Orthogonal decomposition of observation vector for the case of regression through the origin.

9. Assumption checking follows the methods given in §15.1.

10. The estimated standard error of the slope is $s/\|x\|$. The resulting 95% confidence interval for the true slope, β, is $b \pm \text{se(slope)} \times t_{N-1}(.975)$.

11. The position of the fitted line at x_0 is bx_0, so the standard error of this fitted value is $sx_0/\|x\|$. The resulting 95% confidence interval is $bx_0 \pm \text{se(fitted value)} \times t_{N-1}(.975)$.

12. To fit the model using Minitab, Regress y on x, using the Noconstant subcommand to suppress the fitting of the constant term.

For an illustrative example, refer to Joan Fisher Box (1978). In her fascinating book she uses this simple case to illustrate how her father, R. A. Fisher, the founder of many modern day statistical methods, derived his results using geometry.

Appendix E

Confidence Intervals

In the main body of the text we have tended to emphasize hypothesis testing rather than the production of confidence intervals. In common with many statisticians, however, we feel that confidence intervals provide an excellent way of expressing the uncertainty associated with the estimate of a parameter. In an effort to redress the imbalance, we shall now discuss confidence intervals in a unified manner, and tabulate the common cases for ease of reference.

E.1 General Theory

A *confidence interval statement* is an assertion that a true parameter value lies within a certain range, with an accompanying degree of confidence. A 95% confidence interval statement will in the long run be true on 95% of occasions; a 99% confidence interval statement will in the long run be true on 99% of occasions, and so on.

Confidence interval statements are based on intelligent gambles. The usual 95% level gamble, based on the t distribution, is that the realized value of t falls between the 2.5 and 97.5 percentiles of the appropriate t distribution. The realized value of t is then rewritten in terms of the unknown parameter and estimated values, enabling calculation of a confidence interval for the parameter.

In this book our t statistic is consistently of the form

$$\frac{Y.U - E(Y.U)}{\sqrt{S^2}}$$

where U is the direction associated with the parameter of interest. Here the random variable $Y.U - E(Y.U)$ is distributed as $N(0, \sigma^2)$, while $S^2 = \left[(Y.U_{p+1})^2 + \cdots + (Y.U_N)^2\right]/(N - p)$ is an average of the squares of $N - p$ random variables $Y.U_i$, each of which is distributed as $N(0, \sigma^2)$. Hence t is a t_{N-p} statistic, as defined in Chapter 3.

For the main ANOVA parameters, Table E.1 gives the parameter of interest, the associated unit vector U, the realized value $y.U$, and the expected value of the estimator $Y.U$. This provides us with the raw materials with which to construct confidence intervals.

Unit vector U	Projection coefficient $y.U$	Expected value $E(Y.U)$
Single population (μ)		
$\frac{1}{\sqrt{n}} \begin{bmatrix} 1 \\ \vdots \\ 1 \end{bmatrix}$	$\sqrt{n}\bar{y}$	$\sqrt{n}\mu$
Two populations, equal reps $(\mu_1 - \mu_2)$		
$\frac{1}{\sqrt{2n}} \begin{bmatrix} 1 \\ \vdots \\ -1 \\ \vdots \end{bmatrix}$	$\dfrac{\sqrt{n}(\bar{y}_{1.} - \bar{y}_{2.})}{\sqrt{2}}$	$\dfrac{\sqrt{n}(\mu_1 - \mu_2)}{\sqrt{2}}$
Two populations, unequal reps $(\mu_1 - \mu_2)$		
$\dfrac{1}{\sqrt{\frac{1}{n_1} + \frac{1}{n_2}}} \begin{bmatrix} 1/n_1 \\ \vdots \\ -1/n_2 \\ \vdots \end{bmatrix}$	$\dfrac{(\bar{y}_{1.} - \bar{y}_{2.})}{\sqrt{\frac{1}{n_1} + \frac{1}{n_2}}}$	$\dfrac{(\mu_1 - \mu_2)}{\sqrt{\frac{1}{n_1} + \frac{1}{n_2}}}$
k populations, equal reps $(c_1\mu_1 + \cdots + c_k\mu_k)$		
$\dfrac{1}{\sqrt{n\sum_{i=1}^{k} c_i^2}} \begin{bmatrix} c_1 \\ \vdots \\ c_k \\ \vdots \end{bmatrix}$	$\dfrac{\sqrt{n}(c_1\bar{y}_{1.} + \cdots + c_k\bar{y}_{k.})}{\sqrt{\sum_{i=1}^{k} c_i^2}}$	$\dfrac{\sqrt{n}(c_1\mu_1 + \cdots + c_k\mu_k)}{\sqrt{\sum_{i=1}^{k} c_i^2}}$
k populations, unequal reps $(c_1\mu_1 + \cdots + c_k\mu_k)$		
$\dfrac{1}{\sqrt{\sum_{i=1}^{k} c_i^2/n_i}} \begin{bmatrix} \frac{c_1}{n_1} \\ \vdots \\ \frac{c_k}{n_k} \\ \vdots \end{bmatrix}$	$\dfrac{(c_1\bar{y}_{1.} + \cdots + c_k\bar{y}_{k.})}{\sqrt{\sum_{i=1}^{k}(c_i^2/n_i)}}$	$\dfrac{(c_1\mu_1 + \cdots + c_k\mu_k)}{\sqrt{\sum_{i=1}^{k}(c_i^2/n_i)}}$

Table E.1: The value $[y.U - E(Y.U)]/s$ comes from a t_{N-p} distribution. This table sets out the terms in the numerator for the listed ANOVA parameters.

In all cases, we can construct a 95% confidence interval from the appropriate raw materials by gambling that

$$t_{N-p}(.025) \leq \frac{y.U - E(Y.U)}{s} \leq t_{N-p}(.975)$$

For example, in the simplest case of a single population, we gamble that

$$t_{n-1}(.025) \leq \frac{y.U_1 - E(Y.U_1)}{s} \leq t_{n-1}(.975)$$

where U_1 is the direction associated with the parameter μ. That is,

$$t_{n-1}(.025) \leq \frac{\sqrt{n}\bar{y} - \sqrt{n}\mu}{s} \leq t_{n-1}(.975)$$

which can be rearranged to read

$$\bar{y} - \frac{s}{\sqrt{n}}t_{n-1}(.975) \leq \mu \leq \bar{y} + \frac{s}{\sqrt{n}}t_{n-1}(.975)$$

as derived in §5.2.

In Table E.2 we tabulate the corresponding raw materials for the simple and polynomial regression parameters. These lead to the appropriate confidence intervals in the same manner.

Unit vector U	Projection coefficient $y.U$	Expected value $E(Y.U)$
Simple regression (β_1)		
$\frac{1}{\|x - \bar{x}\|}\begin{bmatrix} x_1 - \bar{x} \\ \vdots \\ x_N - \bar{x} \end{bmatrix}$	$\frac{\sum_{i=1}^{N} y_i(x_i - \bar{x})}{\|x - \bar{x}\|} = b_1\|x - \bar{x}\|$	$\beta_1\|x - \bar{x}\|$
Quadratic regression (β_2)		
$\frac{1}{\|T_3\|}\begin{bmatrix} p_2(x_1) \\ \vdots \\ p_2(x_N) \end{bmatrix}$	$\frac{\sum_{i=1}^{N} y_i p_2(x_i)}{\|T_3\|} = b_2\|T_3\|$	$\beta_2\|T_3\|$
Polynomial regression (β_l)		
$\frac{1}{\|T_{l+1}\|}\begin{bmatrix} p_l(x_1) \\ \vdots \\ p_l(x_N) \end{bmatrix}$	$\frac{\sum_{i=1}^{N} y_i p_l(x_i)}{\|T_{l+1}\|} = b_l\|T_{l+1}\|$	$\beta_l\|T_{l+1}\|$

Table E.2: Summary of the terms in the numerator for simple, quadratic and general polynomial parameters.

As a second example of the usage of these raw materials, we calculate a 95% confidence interval for the slope, β_1, in a simple regression. As before, we gamble that

$$t_{N-2}(.025) \leq \frac{y.U_2 - E(Y.U_2)}{s} \leq t_{N-2}(.975)$$

where U_2 is the direction associated with the parameter β_1. That is,

$$t_{N-2}(.025) \leq \frac{b_1\|x - \bar{x}\| - \beta_1\|x - \bar{x}\|}{s} \leq t_{N-2}(.975)$$

which can be rearranged to read

$$b_1 - \frac{s}{\|x - \bar{x}\|}t_{N-2}(.975) \leq \beta_1 \leq b_1 + \frac{s}{\|x - \bar{x}\|}t_{N-2}(.975)$$

as derived in Chapter 15.

We conclude this appendix by tabulating the raw materials for the slope parameters for regression through the origin and analysis of covariance (Table E.3).

Unit vector U	Projection coefficient $y.U$	Expected value $E(Y.U)$
Regression through the origin (β)		
$\frac{1}{\|x\|}\begin{bmatrix} x_1 \\ \vdots \\ x_N \end{bmatrix}$	$\frac{\sum_{i=1}^{N} y_i x_i}{\|x\|} = b\|x\|$	$\beta\|x\|$
Analysis of covariance (β_1)		
$\frac{1}{\|x - \bar{x}_{i.}\|}\begin{bmatrix} x_{11} - \bar{x}_{1.} \\ \vdots \\ x_{k1} - \bar{x}_{k.} \\ \vdots \end{bmatrix}$	$\frac{\sum_{i=1}^{k}\sum_{j=1}^{n_i} y_{ij}(x_{ij} - \bar{x}_{i.})}{\|x - \bar{x}_{i.}\|}$ $= b_1\|x - \bar{x}_{i.}\|$	$\beta_1\|x - \bar{x}_{i.}\|$

Table E.3: Summary of the terms in the numerator for regression through the origin and ANCOVA parameters.

Appendix T

Statistical Tables

In this appendix we provide tables of random numbers (Table T.1), percentiles of the $F_{p,q}$ and t_q distributions (Table T.2), and percentiles of the distribution of the correlation coefficient (r) under the null hypothesis $H_0 : \rho = 0$ (Table T.3).

The random numbers in Table T.1 were produced using the computing package *Gauss*. The percentiles in Tables T.2 and T.3 were obtained from Tables 12, 13 and 18 in "Biometrika Tables for Statisticians, Vol. I, Third Edition (1966)", edited by E.S. Pearson and H.O. Hartley, and are reproduced with the kind permission of the Biometrika Trustees.

2	9	1	1	3	6	2	1	7	3	6	5	3	3	0	3	9	3	3
1	3	1	7	6	9	2	9	1	8	7	2	1	4	5	3	6	6	9
5	1	5	2	5	3	1	8	3	0	1	0	4	9	9	8	7	5	9
5	0	1	1	1	6	5	9	2	0	6	8	4	5	2	3	0	4	7
3	4	0	0	4	8	8	3	2	4	3	4	4	1	8	1	4	0	8
6	7	6	5	0	5	6	5	0	1	0	3	0	0	3	2	7	7	6
6	6	4	3	3	3	3	3	9	6	3	5	8	5	8	6	3	1	5
9	8	2	7	7	0	8	1	9	4	1	7	0	2	9	1	2	2	9
3	0	2	1	9	1	4	8	9	2	8	7	0	5	6	6	6	1	2
4	3	2	6	1	9	7	7	8	7	3	8	8	4	7	7	9	5	5
5	1	5	0	1	1	6	1	2	2	1	0	8	2	2	2	0	7	1
0	1	8	6	2	8	0	5	1	1	6	6	5	3	0	5	8	1	1
9	9	3	9	8	2	0	2	3	6	7	3	6	4	1	4	6	8	1
2	8	1	7	9	8	9	0	5	9	1	5	9	3	3	7	2	0	9
2	2	6	5	1	2	3	1	9	6	2	0	7	9	4	0	8	6	5
2	8	1	6	4	4	8	3	4	8	0	3	6	1	3	4	9	2	5
7	7	4	6	2	5	4	7	7	6	3	3	3	9	0	1	3	8	9
3	6	3	3	2	1	3	4	1	5	2	1	5	2	1	7	4	0	5
4	3	5	1	3	6	1	0	8	7	7	2	6	9	6	7	2	5	4
1	2	4	7	5	2	7	2	3	2	5	7	2	6	8	5	8	3	9
2	6	2	0	5	1	2	8	4	2	4	1	8	0	2	1	2	2	4
8	8	0	6	6	1	3	7	2	0	4	7	8	2	2	1	1	2	4
4	8	9	2	3	4	1	5	3	7	1	5	6	2	8	4	6	3	7
9	8	2	5	0	2	7	9	4	6	4	5	9	9	3	9	6	9	5
0	8	8	6	8	0	4	5	3	2	0	7	6	4	7	4	3	8	1
7	3	9	0	8	0	9	6	7	2	1	8	6	7	0	8	6	0	4
6	1	4	9	2	5	7	2	9	7	3	1	6	3	5	5	7	4	4
6	8	4	7	4	4	0	3	0	7	8	0	4	7	8	5	5	9	1
9	0	7	8	6	0	5	9	7	7	0	6	1	1	1	9	4	4	1
8	8	1	5	4	0	5	7	6	4	6	8	8	1	5	0	8	0	3
6	2	9	4	2	2	6	2	5	0	1	1	5	6	3	6	3	8	5
5	2	5	4	9	9	9	0	2	9	6	1	9	8	6	4	0	8	7
6	1	5	5	1	1	8	4	4	9	5	0	4	2	4	8	7	0	4
7	9	5	5	5	3	6	3	0	5	6	7	2	2	2	7	6	4	9
0	1	2	8	5	9	2	0	5	2	1	3	7	8	4	7	0	4	2
5	1	4	5	3	9	2	9	7	4	9	5	6	5	1	4	7	6	3

Table T.1: Table of random numbers. To use the table, begin at a random point and read down individual columns for single-digit numbers, pairs of columns for two-digit numbers, and so on. Ignore numbers which are outside the desired range. When the bottom of the table is reached, go to the top of the next column or group of columns.

Table T.2 gives the 90, 95 and 99 percentiles of selected $F_{p,q}$ distributions. The 95 percentile of the $F_{p,q}$ distribution is illustrated in Figure T.1; it is the number x such that 95% of realized values of $F_{p,q}$ are less than x.

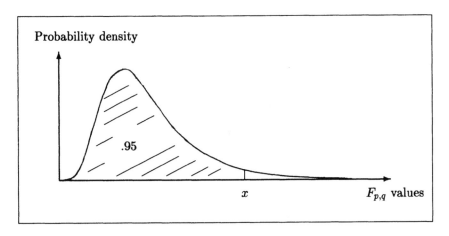

Figure T.1: Illustration of the 95 percentile of the $F_{p,q}$ distribution.

The last two columns of Table T.2 give the 95, 97.5 and 99.5 percentiles of selected t_q distributions. The 97.5 percentile of the t_q distribution is illustrated in Figure T.2; it is the number x such that 97.5% of realized values of t_q are less than x. Note that the t_q percentile is just the square root of the corresponding $F_{1,q}$ percentile, since $t_q^2 = F_{1,q}$.

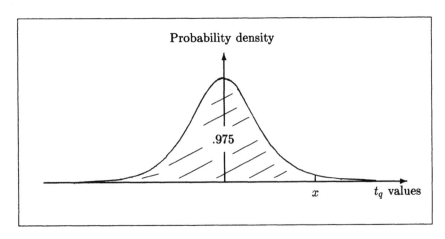

Figure T.2: Illustration of the 97.5 percentile of the t_q distribution.

Denom df (q)		Numerator degrees of freedom (p)						Percentiles of t_q	
		1	2	3	6	10	∞		
	90	40	50	54	58	60	63	95	6.31
1	95	161	200	216	234	242	254	97.5	12.71
	99	4052	5000	5403	5859	6056	6366	99.5	63.66
	90	8.5	9.0	9.2	9.3	9.4	9.5	95	2.920
2	95	18.5	19.0	19.2	19.3	19.4	19.5	97.5	4.303
	99	98.5	99.0	99.2	99.3	99.4	99.5	99.5	9.925
	90	5.54	5.46	5.39	5.28	5.23	5.13	95	2.353
3	95	10.13	9.55	9.28	8.94	8.79	8.53	97.5	3.182
	99	34.12	30.82	29.46	27.91	27.23	26.13	99.5	5.841
	90	4.54	4.32	4.19	4.01	3.92	3.76	95	2.132
4	95	7.71	6.94	6.59	6.16	5.96	5.63	97.5	2.776
	99	21.20	18.00	16.69	15.21	14.55	13.46	99.5	4.604
	90	4.06	3.78	3.62	3.40	3.30	3.10	95	2.015
5	95	6.61	5.79	5.41	4.95	4.74	4.36	97.5	2.571
	99	16.26	13.27	12.06	10.67	10.05	9.02	99.5	4.032
	90	3.78	3.46	3.29	3.05	2.94	2.72	95	1.943
6	95	5.99	5.14	4.76	4.28	4.06	3.67	97.5	2.447
	99	13.75	10.92	9.78	8.47	7.87	6.88	99.5	3.707
	90	3.59	3.26	3.07	2.83	2.70	2.47	95	1.895
7	95	5.59	4.74	4.35	3.87	3.64	3.23	97.5	2.365
	99	12.25	9.55	8.45	7.19	6.62	5.65	99.5	3.499
	90	3.46	3.11	2.92	2.67	2.54	2.29	95	1.860
8	95	5.32	4.46	4.07	3.58	3.35	2.93	97.5	2.306
	99	11.26	8.65	7.59	6.37	5.81	4.86	99.5	3.355
	90	3.36	3.01	2.81	2.55	2.42	2.16	95	1.833
9	95	5.12	4.26	3.86	3.37	3.14	2.71	97.5	2.262
	99	10.56	8.02	6.99	5.80	5.26	4.31	99.5	3.250
	90	3.29	2.92	2.73	2.46	2.32	2.06	95	1.812
10	95	4.96	4.10	3.71	3.22	2.98	2.54	97.5	2.228
	99	10.04	7.56	6.55	5.39	4.85	3.91	99.5	3.169

Table T.2: Table of the 90, 95 and 99 percentiles of the $F_{p,q}$ distributions. For each distribution, the 90 percentile is given first, then the 95 and 99 percentiles. The corresponding percentiles of the t_q distributions are also given in the right-hand column of the table.

Denom		Numerator degrees of freedom (p)						Percentiles	
df (q)		1	2	3	6	10	∞	of t_q	
	90	3.23	2.86	2.66	2.39	2.25	1.97	95	1.796
11	95	4.84	3.98	3.59	3.09	2.85	2.40	97.5	2.201
	99	9.65	7.21	6.22	5.07	4.54	3.60	99.5	3.106
	90	3.18	2.81	2.61	2.33	2.19	1.90	95	1.782
12	95	4.75	3.89	3.49	3.00	2.75	2.30	97.5	2.179
	99	9.33	6.93	5.95	4.82	4.30	3.36	99.5	3.055
	90	3.14	2.76	2.56	2.28	2.14	1.85	95	1.771
13	95	4.67	3.81	3.41	2.92	2.67	2.21	97.5	2.160
	99	9.07	6.70	5.74	4.62	4.10	3.17	99.5	3.012
	90	3.10	2.73	2.52	2.24	2.10	1.80	95	1.761
14	95	4.60	3.74	3.34	2.85	2.60	2.13	97.5	2.145
	99	8.86	6.51	5.56	4.46	3.94	3.00	99.5	2.977
	90	3.07	2.70	2.49	2.21	2.06	1.76	95	1.753
15	95	4.54	3.68	3.29	2.79	2.54	2.07	97.5	2.131
	99	8.68	6.36	5.42	4.32	3.80	2.87	99.5	2.947
	90	3.05	2.67	2.46	2.18	2.03	1.72	95	1.746
16	95	4.49	3.63	3.24	2.74	2.49	2.01	97.5	2.120
	99	8.53	6.23	5.29	4.20	3.69	2.75	99.5	2.921
	90	3.03	2.64	2.44	2.15	2.00	1.69	95	1.740
17	95	4.45	3.59	3.20	2.70	2.45	1.96	97.5	2.110
	99	8.40	6.11	5.18	4.10	3.59	2.65	99.5	2.898
	90	3.01	2.62	2.42	2.13	1.98	1.66	95	1.734
18	95	4.41	3.55	3.16	2.66	2.41	1.92	97.5	2.101
	99	8.29	6.01	5.09	4.01	3.51	2.57	99.5	2.878
	90	2.99	2.61	2.40	2.11	1.96	1.63	95	1.729
19	95	4.38	3.52	3.13	2.63	2.38	1.88	97.5	2.093
	99	8.18	5.93	5.01	3.94	3.43	2.49	99.5	2.861
	90	2.97	2.59	2.38	2.09	1.94	1.61	95	1.725
20	95	4.35	3.49	3.10	2.60	2.35	1.84	97.5	2.086
	99	8.10	5.85	4.94	3.87	3.37	2.42	99.5	2.845

Denom		Numerator degrees of freedom (p)						Percentiles	
df (q)		1	2	3	6	10	∞	of t_q	
	90	2.96	2.57	2.36	2.08	1.92	1.59	95	1.721
21	95	4.32	3.47	3.07	2.57	2.32	1.81	97.5	2.080
	99	8.02	5.78	4.87	3.81	3.31	2.36	99.5	2.831
	90	2.95	2.56	2.35	2.06	1.90	1.57	95	1.717
22	95	4.30	3.44	3.05	2.55	2.30	1.78	97.5	2.074
	99	7.95	5.72	4.82	3.76	3.26	2.31	99.5	2.819
	90	2.94	2.55	2.34	2.05	1.89	1.55	95	1.714
23	95	4.28	3.42	3.03	2.53	2.27	1.76	97.5	2.069
	99	7.88	5.66	4.76	3.71	3.21	2.26	99.5	2.807
	90	2.93	2.54	2.33	2.04	1.88	1.53	95	1.711
24	95	4.26	3.40	3.01	2.51	2.25	1.73	97.5	2.064
	99	7.82	5.61	4.72	3.67	3.17	2.21	99.5	2.797
	90	2.92	2.53	2.32	2.02	1.87	1.52	95	1.708
25	95	4.24	3.39	2.99	2.49	2.24	1.71	97.5	2.060
	99	7.77	5.57	4.68	3.63	3.13	2.17	99.5	2.787
	90	2.91	2.52	2.31	2.01	1.86	1.50	95	1.706
26	95	4.23	3.37	2.98	2.47	2.22	1.69	97.5	2.056
	99	7.72	5.53	4.64	3.59	3.09	2.13	99.5	2.779
	90	2.90	2.51	2.30	2.00	1.85	1.49	95	1.703
27	95	4.21	3.35	2.96	2.46	2.20	1.67	97.5	2.052
	99	7.68	5.49	4.60	3.56	3.06	2.10	99.5	2.771
	90	2.89	2.50	2.29	2.00	1.84	1.48	95	1.701
28	95	4.20	3.34	2.95	2.45	2.19	1.65	97.5	2.048
	99	7.64	5.45	4.57	3.53	3.03	2.06	99.5	2.763
	90	2.89	2.50	2.28	1.99	1.83	1.47	95	1.699
29	95	4.18	3.33	2.93	2.43	2.18	1.64	97.5	2.045
	99	7.60	5.42	4.54	3.50	3.00	2.03	99.5	2.756
	90	2.88	2.49	2.28	1.98	1.82	1.46	95	1.697
30	95	4.17	3.32	2.92	2.42	2.16	1.62	97.5	2.042
	99	7.56	5.39	4.51	3.47	2.98	2.01	99.5	2.750

Denom		Numerator degrees of freedom (p)						Percentiles	
df (q)		1	2	3	6	10	∞	of t_q	
	90	2.84	2.44	2.23	1.93	1.76	1.38	95	1.684
40	95	4.08	3.23	2.84	2.34	2.08	1.51	97.5	2.021
	99	7.31	5.18	4.31	3.29	2.80	1.80	99.5	2.704
	90	2.79	2.39	2.18	1.87	1.71	1.29	95	1.671
60	95	4.00	3.15	2.76	2.25	1.99	1.39	97.5	2.000
	99	7.08	4.98	4.13	3.12	2.63	1.60	99.5	2.660
	90	2.75	2.35	2.13	1.82	1.65	1.19	95	1.658
120	95	3.92	3.07	2.68	2.17	1.91	1.25	97.5	1.980
	99	6.85	4.79	3.95	2.96	2.47	1.38	99.5	2.617
	90	2.71	2.30	2.08	1.77	1.60	1.00	95	1.645
∞	95	3.84	3.00	2.60	2.10	1.83	1.00	97.5	1.960
	99	6.63	4.61	3.78	2.80	2.32	1.00	99.5	2.576

Degrees	Percentile		
of Freedom	95	97.5	99.5
1	.988	.997	1.000
2	.900	.950	.990
3	.805	.878	.959
4	.729	.811	.917
5	.669	.754	.875
6	.621	.707	.834
7	.582	.666	.798
8	.549	.632	.765
9	.521	.602	.735
10	.497	.576	.708
11	.476	.553	.684
12	.457	.532	.661
13	.441	.514	.641
14	.426	.497	.623
15	.412	.482	.606
16	.400	.468	.590
17	.389	.456	.575
18	.378	.444	.561
19	.369	.433	.549
20	.360	.423	.537
25	.323	.381	.487
30	.296	.349	.449
35	.275	.325	.418
40	.257	.304	.393
45	.243	.288	.372
50	.231	.273	.354
60	.211	.250	.325
70	.195	.232	.302
80	.183	.217	.283
90	.173	.205	.267
100	.164	.195	.254

Table T.3: Table of the 95, 97.5 and 99.5 percentiles of the distribution of the correlation coefficient, r, for varying degrees of freedom, under the null hypothesis $H_0 : \rho = 0$.

References

Alvey, N., Galwey, N. and Lane, P. (1982). *An introduction to GENSTAT.* London: Academic Press.

Bates, D.M. and Watts, D.G. (1988). *Nonlinear regression analysis and its applications.* New York: Wiley.

Bourdôt, G.W., Saville, D.J. and Field, R.J. (1984). The response of Achillea millefolium L. (yarrow) to shading. *New Phytologist* 97: 653-663.

Box, G.E.P., Hunter, W.G. and Hunter, J.S. (1978). *Statistics for experimenters: An introduction to design, data analysis and model building.* New York: Wiley.

Box, Joan Fisher (1978). *R.A. Fisher, The life of a scientist.* New York: Wiley.

Bryant, Peter (1984). Geometry, statistics, probability: Variations on a common theme. *The American Statistician* 38: 38-48.

Burdick, D.S., Herr, D.G., O'Fallon, W.M. and O'Neill, B.V. (1974). Exact methods in the unbalanced, two-way analysis of variance - a geometric view. *Communications in Statistics* 3: 581-595.

Cochran, W.G. and Cox, G.M. (1957). *Experimental designs, 2nd edition.* New York: Wiley.

Corsten, L.C.A. (1958). Vectors, a tool in statistical regression theory. *Mededelingen van de Landbouwhogeschool te Wageningen, Nederland* 58: 1-92.

Corsten, L.C.A. (1987). *Lineaire methoden en analyse van multi-stochasten.* Mimeographed notes, Division of Mathematics and Statistics, Agricultural University, Wageningen, The Netherlands.

Corsten, L.C.A. and Van der Laan, P. (1983). *College voortzetting statistiek.* Mimeographed notes, Division of Mathematics and Statistics, Agricultural University, Wageningen, The Netherlands.

Durbin, J. and Kendall, M.G. (1951). The geometry of estimation. *Biometrika* 38: 150-158.

Fisher, R.A. and Yates, F. (1982). *Statistical tables for biological, agricultural and medical research, 6th edition, revised and enlarged*. Harlow, Essex: Longman.

Herr, David G. (1980). On the history of the use of geometry in the general linear model. *The American Statistician* 34: 43-47.

Kruskal, William H. (1975). The geometry of generalized inverses. *Journal of the Royal Statistical Society, Series B* 37: 272-283.

Little, T.M. and Hills, F.J. (1978). *Agricultural experimentation: design and analysis*. New York: Wiley.

Margolis, Marvin S. (1979). Perpendicular projections and elementary statistics. *The American Statistician* 33: 131-135.

McLeod, C.C. (1982). Effects of rates of seeding on barley sown for grain. *New Zealand Journal of Experimental Agriculture* 10: 133-136.

Minitab Inc. (1989). *MINITAB reference manual, release 7*. State College: Minitab.

O'Brien, P.C. (1983). The appropriateness of analysis of variance and multiple comparison procedures. *Biometrics* 39: 787-794.

Pearson, E.S. and Hartley, H.O. (1966). *Biometrika tables for statisticians*. Cambridge: Cambridge University Press, for Biometrika Trustees.

Saville, D.J. (1983). Statistical analysis of the economics of experimental treatments. *New Zealand Journal of Experimental Agriculture* 11: 1-3.

Saville, D.J. (1984). Outside row effects in cereal seeding rate trials. *New Zealand Journal of Experimental Agriculture* 12: 197-202.

Saville, D.J. (1990). Multiple comparison procedures - The practical solution. *The American Statistician* 44: 174-180.

Saville, D.J. and Wood, G.R. (1986). A method for teaching statistics using N-dimensional geometry. *The American Statistician* 40: 205-214.

Scheffé, Henry (1959). *The analysis of variance*. New York: Wiley.

Snedecor, G.W. and Cochran, W.G. (1967). *Statistical methods, 6th edition*. Ames: Iowa State University Press.

Steel, R.G.D. and Torrie, J.H. (1980). *Principles and procedures of statistics, 2nd edition*. New York: McGraw-Hill.

Index

Additivity assumption, 307, 319, 322-323, 343, 349, 366-367
Adjusted means, 470
Alfalfa, 63, 90, 129, 292, 293, 331, 421, 483, 528
Alternative factorial notation, 216, 356
Analysis of covariance, 461, 497
Analysis of variance, 65, 495, 498
    completely randomized, 100, 136
    contrasts, 133, 144
    latin square, 340
    randomized block, 299
    several populations, 133
    single population, 67
    split plot, 354
    two populations, 97
    unequal replication, 504, 515
ANCOVA; see Analysis of covariance
Angle, 20, 409, 446
ANOVA table
    analysis of covariance, 467, 478
    class comparisons, 160
    factorial contrasts, 192
    latin square, 348, 350
    multiple comparisons, 286

pairwise comparisons, 275
polynomial contrasts, 238
polynomial regression, 437, 442, 451
randomized block, 306, 316
regression through origin, 536
simple regression, 392, 397, 401
single population, 75, 81
split plot, 364, 370
two populations, 106, 108, 116
unequal replication, 507
Anscombe's quartet, 418
Antarctic, 416
Antibiotics, 522
Archery, 84
Arthritis, 263
Association between variables, 408, 414
Assumption checking
    additivity, 307, 319, 322-323, 343, 349, 366-367
    analysis of covariance, 463, 471, 474
    class comparisons, 161
    constant variance, 109, 117, 163, 318
    factorial contrasts, 194, 208

general comments, 248
growth studies, 276
independence, 109, 317
latin square, 349
normality of errors, 109, 161, 318
pairwise comparisons, 276
plot of errors against fits, 277, 395, 398, 443
polynomial contrasts, 240
polynomial models, 433, 442, 455
randomized block, 307, 317
several populations, 161
simple regression, 386, 394-396, 397
single population, 76
split plot, 365
two populations, 108, 117

Back transformation, 280, 404
Barley, 138, 139, 141, 294, 374
Bees, 185, 427
Beet, 92, 137, 185, 292, 336
Best linear unbiased estimates, 49, 84
Bias, 472
Blackberry weed example, 179, 528
Block, 297-300, 310, 312-313, 321
orthogonal, 340
space, 302, 310, 315, 324, 347, 496
vector, 303, 311, 315, 324-325
why and how to, 321-324
Boron, 222, 421
Break even point, 242
Bulkometer example, 98, 504
Buttercups, 491

Calculator example, 352
Cancer, 414
Carbon dioxide, 416
Carcase weight, 461
Carnations, 294

Cats, 218, 428, 458
Causality, 414
Cholesterol, 425
Class comparison; *see* Contrast
Class exercises
class comparisons, 176
pairwise comparisons, 291
randomized block, 325
simple regression, 421
single population, 87
two populations, 124
Clover, 383, 490
Coefficient of determination
r^2, 410
R^2, 446
Collinearity, 434, 443, 450
Comparison of models, 498
Competition study, 383
Completely randomized design, 100, 136
Computing, 119
analysis of covariance, 465-466, 477
assumption checking, 119, 281, 309
class comparisons, 165-167, 171
confidence band, 408
factorial contrasts, 197-199, 213
latin square, 348
pairwise comparisons, 281
polynomial contrasts, 243-245
polynomial regression, 443
randomized block, 308-309
simple regression, 393-394
split plot, 364
transformations, 281
two populations, 119-122
unequal replication, 509
Confidence band, 404-405, 406, 442
Confidence intervals
analysis of covariance, 472, 541
class comparisons, 164, 173-174, 539
contrasts, 164, 539

factorial contrasts, 195-196, 210, 216, 539

gamble, 82, 540

general theory, 538-541

orthogonal polynomial coefficients, 257, 540

pairwise comparisons, 279, 539

polynomial contrasts, 241, 539

polynomial regression, 257, 443-445, 540

regression through origin, 537, 541

simple regression, 402-408, 540

single population, 76-77, 82, 539

split plot designs, 368

two populations, 110, 118, 307, 509, 539

unequal replication, 509, 521, 539

Constancy of variance; *see* Variance

Contrast, 133, 144

associated direction, 135, 144, 517-518

class comparisons, 137, 155-186, 169, 172, 515-521

coefficients, 147, 170

factorial, 138, 139, 187-223, 190, 211-215, 312, 522-526

interaction, 190, 205, 531

linear, 252-253, 481

main effect, 205

mixing, 255

non-orthogonal, 273, 517

orthogonal, 135, 153

pairwise, 142, 143, 271-295, 527

polynomial (or rates), 141, 225-270, 526

quadratic, 253-255, 481

strongly prespecified, 173

types, 136

unequal replication, 505

unit vector, 144

weakly prespecified, 173

Controlled experiment, 68, 98, 415

Cook's distance, 417

Coordinate axes, 11, 23

analysis of covariance, 464

block space, 301, 315

change of, 103, 135

column space, 344

error space, 72, 79, 104, 114, 135, 304

mean space, 69, 78

model space, 72, 78, 101, 104, 112, 135, 144-146, 301, 344, 387, 464

orthogonal, 23, 499

row space, 344

simple regression, 387, 400

treatment space, 144-146, 190, 231, 302, 315, 344, 495-496

Correlation coefficient (r), 408-413

range of values, 410

test, 412

true, 409

Cosine, 20-22, 409

Count data, 365

Covariance, 45, 461-492

Covariate, 461

Critical values, 51, 53, 545-549

Crossing contrasts, 213, 314

Curvature, 232, 321

Degrees of freedom (df), 51-53, 75

Design

barley fertilizer study, 138

barley seed rate study, 141

bulkometer study, 98-99

calculator study, 352

competition study, 384

completely randomized, 100, 136

diet study, 300

fungicide study, 68

garlic study, 312

green manure study, 515

greenfeed study, 137

gummy ewes study, 426
lamb fatness study, 462
lamb growth study, 522
latin square, 340-341, 349-353
matrix, 500
optimal, 282, 512
pasture species study, 340
pollution study, 134
randomized block, 299-300, 310-
    313, 321-325, 340-341, 527
soft turnips study, 139
split plot, 354-355, 369-373
stripe rust chemicals study, 143
subtractive, 142, 275
sunflower study, 142
yarrow shading study, 448
df; *see* degrees of freedom
Diets, 300
Dimension, 25
Direction; *see* Hypothesis direction
Distributions, 41
    correlation coefficient, 412, 549
    F, 51, 544-548
    multivariate normal, 84
    normal, 41
    of projection coefficient; *see* Pro-
        jection coefficient
    t, 52, 544-548
Dot product, 21
Drench, 176, 332, 481, 488

Economic analysis, 242
Edge effects, 247
Elevation
    of curve, 232
    of lines, 468
Equally spaced data, 251
Error
    see also Residuals
    coordinate system, 72, 79, 104,
        114, 135, 304
    mean square; *see* Estimation
        of σ^2

space, 49, 62, 72, 79, 104, 114,
    135, 304, 493, 496
Estimate, 48
    best linear unbiased, 84
    maximum likelihood, 84
Estimation of σ^2, 49-50, 76, 82,
    108, 117, 161, 193, 208,
    239, 276, 306, 392, 493-
    494
Estimator, 48
Ewes, 480
Example
    A - winter greenfeeds 137, 155-
        168
    B - fertilizers for barley 138,
        188-202
    C - soft turnips, 139, 203-213,
        356-368
    D - seed rates of barley, 141,
        224-248
    E - sunflowers, 142, 271-282
    F - stripe rust chemicals, 143,
        284-287
Experimental units, 299, 355
Exponential curve, 398, 484
Extrapolation, 415-416

F distribution, 51, 544-548
F test, 63
    critical values, 545-548
    link with r^2, 412
    link with R^2, 447
    link with t-test, 83, 118
    overall, 286
    several populations, 144
    single population, 73, 81
    two populations, 106, 115
Factor, 187
    level, 187
Factorial contrast; *see* Contrast
Factorial experiment, 187-223
    alternative notation, 216, 530-
        534

efficiency of, 201-202
need for caution, 202
Factorial notation, 530-534
Fatness, 425, 461
Fertility gradient, 317, 321, 351
Fertilizer, 128, 182, 220, 337
Field technique, 137, 138, 246-248,
    312-313, 340-342, 351, 384
Fisher, Sir Ronald, 4
Fitted equation, 238, 388, 439, 451-
    452, 465
Fitted values on line, 403
    confidence intervals, 403, 444
Fitting the model, 61
    analysis of covariance, 465, 476
    latin square, 345
    polynomial regression, 436
    randomized block, 303, 310
    several populations, 144, 158
    simple regression, 388
    single population, 71, 78
    split plot, 361
    two populations, 103, 113
Fixed effects, 168
Fodder beet; *see* Beet
Fungicide seed treatment, 67, 221,
    258

Galton, Sir Francis, 383
Garlic, 312, 321
Gaussian distribution, 41
General summary, 493-503
Generation of contrasts
    class comparisons, 169
    factorial, 213
    mixing, 255
    polynomial, 234
GENSTAT, 216
Geometric method, 59
Geometry, 3
    toolkit, 10-31
Giant buttercups, 491
Goats, 481, 488

Golden Queen example, 258-262,
    266
Goodness of fit, 413
Gram-Schmidt method, 228-230, 434
Graphs, the value of, 418
Grass grub, 327
Grazing intensity, 335
Green manure example, 515
Greenfeed example, 137
Growth curve, 449
Gummy ewes, 426, 432

Heartbeat rate class exercises, 87-
    89, 124-127
Herbicide examples, 182, 491
Hierarchical design, 354
Hormone level, 63
Hypothesis
    alternative, 69
    confirmation, 55-57, 289
    formulation, 55-57, 516-518
    generation, 289
    null, 69
    overall null, 286
Hypothesis direction, 58-60, 69, 78,
    102, 113, 135, 144, 518,
    539-541
Hypothesis testing, 58-63
    analysis of covariance, 464, 468,
    475, 478
    contrasts, 135, 144
    regression coefficients, 387, 391
    several populations, 135, 144
    single population, 69, 72
    the key idea, 58
    two populations, 102, 104, 106,
    115
    unequal replication, 505, 516

Ill-thrift in lambs, 221
Inconsistency of tests, 288
Independence, 45, 47-48
    checking assumption, 109, 161,
    317, 395, 455

of projection coefficients, 48, 80
of regression errors, 387
Independent
    lines, 473-485
    samples, 97
    hypothesis tests, 145, 275
Interaction, 189, 196
    contrasts, 205, 212-215
    interpretation of, 200-201
    positive and negative, 201
Interpretation, 403
Irrigation, 369, 374

Kitten, 428, 458
Kiwifruit, 407, 414, 427, 429, 458

Lack of fit, 448, 451, 454
Lambs, 221, 338, 462, 523
Latin square design; *see* Design
Least significant difference, 283, 371
Least squares fitting, 71
Legume, 181, 214, 499, 515, 530
Level of a factor, 187
Lime, 325, 528
Limiting factors, 201
Line fitting, 383
Linear contrasts; *see* Contrast
Linearity assumption, 386, 394, 397
Log transformation; *see* Transforming data
Lurking variable, 415

Main effect, 196, 211
    contrasts, 205, 213-215
    means, 197, 530
Mainplot, 354
    space, 357-358
Mangels, 137, 155, 292
Matrix methods, 499-503
Maximum likelihood, 84, 372
Mean, 42
Mean square, 75
Mean vector, 71
Mineral mix, 522

Minitab; *see also* Computing
    Histogram, 119
    Mplot, 488
    Oneway, 121, 165
    Plot, 281
    Regress, 119
Missing value, 162
Model, 56, 69
    space, 49, 59-60, 69, 78, 101, 112, 134, 144, 156, 189, 226, 272, 301-302, 343, 387, 464, 493-498
    vectors, 59, 69, 101, 134, 156, 189, 226, 272, 301, 343, 360, 387, 464
Model fitting, 61
    latin square, 345
    parallel lines, 465
    polynomials, 248-251, 436, 450-452
    randomized block, 303
    several populations, 144, 158
    single population, 71
    split plot, 361
    straight line, 388
    two populations, 103, 105, 113
Modifying ideas, 55-57, 173
Multiple comparison procedures
    Fisher's restricted LSD, 288-289
    LSD, 284-287
Multiple comparisons, 271, 284-289
Multiple correlation coefficient (R), 446, 455
Multiple regression, 499
Multiplicative model, 278, 398, 449, 484

Nitrogen, 204, 374
Nonorthogonal system, 227-228, 393, 499, 502
Normal distribution, 41
    checking assumption, 109, 161, 318, 394, 455

Notation, 40, 43, 100, 530

Observation vector, 58-61, 69
    corrected for the mean, 107,
        114, 159
Oneway command, 121, 165
Operational definition of measure-
    ment, 98-99, 462
Optimal design, 282, 512
Order of polynomial, 437-439, 451
Origin, 11
    regression through, 535
Orthogonal
    contrast, 133, 153
    coordinate systems, 23, 72, 499
    decomposition, 29, 72, 105, 144,
        149
    model, 228, 435
    polynomials, 231-234, 433-434,
        436
    vectors, 22
Orthogonalization, 228, 434
Outliers, 417
Overall F test, 286
Overall mean space, 105, 495

Pairwise contrasts; *see* Contrasts
Pakihi soil, 263
Parallel straight lines, 461, 463
Parameter, 42, 493
    estimates, 61, 493, 500
Pasture yield, 325, 340
Paw Paw, 428, 458
Peach seedlings, 127, 258, 513, 529
Peaches, 457
Percentage differences, 278, 398, 449,
    484
Percentiles, 51-53, 412, 544-549
Phosphate example, 138, 184, 189,
    514
Pollution example, 134
Polynomial, 226, 431
Polynomial components, 256, 433
Polynomial contrasts; *see* Contrasts

Polynomial regression; *see* Regres-
    sion
Pooled variance, 117
Populations, 39
    derived, 44
    parent, 44
    several, 133
    single, 67
    two, 97
Power of a test, 89, 292, 424
Predicted values on line, 405
    confidence intervals, 405, 445
Prediction example, 407
Probability distribution, 41
Projection
    onto a subspace, 28
    onto a unit vector, 26
    to fit the model, 61
Projection coefficient, 26
    distribution of, 47, 70, 104, 157
Pure error, 431, 448
    no pure error, 431
    sum of squares, 448, 451, 454
Pythagoras' theorem, 30, 62, 72

Quadrat, 384
Quadratic, 226, 433
    contrast, 253
    equation, 239, 439
Quantitative and qualitative treat-
    ments, 224, 381

R, 446, 455
r; *see* Correlation coefficient
R^2, 446, 455
R^2(adj), 448, 455
r^2, 410
Rams, 480
Random
    effects, 168
    numbers, 543
    sample, 40
Random variables, 40
    combinations of, 45
    independence of, 45

Randomization, 100, 351, 371

Randomized block design; *see* Design

Rates; *see* Contrast, polynomial

Ratio of mean squares, 75

Rats, 219

Reference distributions, 51
  F distributions, 51
  t distributions, 52

Refractometer readings, 42

Regress command, 119

Regression, 381, 498
  line, 389
  linear, 383-430, 496
  multiple, 499
  polynomial, 431-460, 497
  simple, 383-430, 496
  through the origin, 535-537

Relationships between variables, 381

Replication, 97, 448
  unequal, 504-529

Report format
  analysis of covariance, 471
  class comparisons, 165
  factorial experiments, 196, 210-211, 525
  pairwise comparisons, 280
  polynomial contrasts, 242, 320
  polynomial regression, 442-443, 455
  several populations, 165
  simple regression, 399
  single population, 77
  splitplots, 367-368
  two populations, 111, 510
  with unequal replication, 510

Research cycle, 56

Residuals (or errors), 69
  checking normality, 109, 397
  plotted against fits, 277, 395, 398, 443
  standardized, 393

Responses, 68

Rice, 266, 330, 377

Right angles, 22

Rule of thumb for linear contrasts, 252

Sample, 40

Sample size, 57

Sample variance, 83, 117

Scattergram, 225-226, 385, 396, 399, 413

Scientific method, 55

SED, 111, 165, 196-197

Seeding rate, 141, 224, 334

Several populations, 133

Simplified decomposition, 107, 159, 390

Single population, 67
  general coordinate system, 79

Slope, 232, 387, 464

Soft turnips example, 139

Source of variation, 75

SP; *see* Sum of products

Span, 18, 59

Split plot design; *see* Design

Split split plot design, 372

Spurious correlation, 414-415

Square root transformation, 209, 365

SS; *see* Sum of squares

Standard deviation, 42

Standard error
  difference of means, 111
  slope, 402

Standard latin squares, 350

Standard polynomial contrasts, 251

Standardized residuals, 393

Statistical significance, 51

Statistical tables
  correlation coefficient, 549
  F and t distributions, 545-548

Statistical toolkit, 39

Stripe rust control example, 143

Subplot, 354
  space, 357, 359-360

Subspace, 17
    dimension, 25
Substituting factors, 201
Subtractive design, 142, 275
Sum of products (SP), 401, 473
Sum of squares (SS), 75, 401, 473
    inflated, 58, 438
Sunflowers, 142, 271
Superphosphate, 325

t distribution, 52, 545-548
t statistic, 52, 538
Tables; *see* Statistical tables
Tagging example, 90, 129
Tea drinking example, 4
Test statistic, 63
    contrast, 135
    several populations, 135
    single population, 73, 75
    slope, 391
    two populations, 62, 106
Tetrahedron, 273
Tilt; *see* Slope
Toolkit
    geometric, 10
    statistical, 39
Transforming data, 209
    using logs, 278-281, 293, 396-
        399, 449, 484
    using square root, 209, 365
Treatment, 68, 100, 137, 224
    space, 105, 114, 136, 145, 159,
        190, 495
    vector, 105, 145, 159, 191
Turnips, 139, 356

Two populations, 97

Unbiased estimate, 48
    best linear, 49, 84
Unequal replication, 504-529
Units, 197
Urea, 328

Value of graphs, 418
Variance, 42
    components, 364
    estimation, 50, 76
    pairwise comparisons of, 163
    pooled estimate, 117
    precision, 472, 482
    testing for equality, 109, 117,
        163, 395
Vector, 11
    addition, 15
    angle between two, 20, 409
    dot product, 21
    length, 13
    orthogonal, 22
    right angles, 22
    scalar multiplication, 16
    transpose, 12
    unit vector, 14
Virtues of estimates, 83, 119

Weevils, 483, 489
Wether, 479, 488
Wheat, 264, 331, 334, 337, 369
    stripe rust disease, 265
Wild oats, 415
Wool, 97

Yarrow, 220, 383, 448, 459, 490

Springer Texts in Statistics *(continued from page ii)*

Santner and Duffy: The Statistical Analysis of Discrete Data
Saville and Wood: Statistical Methods: The Geometric Approach
Sen and Srivastava: Regression Analysis: Theory, Methods, and Applications
Whittle: Probability via Expectation, Third Edition
Zacks: Introduction to Reliability Analysis: Probability Models and Statistical
    Methods